AF389477

# Biological Control of the Plant Pathogens

# Biological Control of the Plant Pathogens

Editors

**Rainer Borriss**
**Chetan Keswani**

Basel • Beijing • Wuhan • Barcelona • Belgrade • Novi Sad • Cluj • Manchester

*Editors*

Rainer Borriss
Institute of Biology
Humboldt University
of Berlin
Berlin, Germany

Chetan Keswani
Academy of Biology and
Biotechnology
Southern Federal University
Rostov-on-Don, Russia

*Editorial Office*
MDPI
St. Alban-Anlage 66
4052 Basel, Switzerland

This is a reprint of articles from the Special Issue published online in the open access journal *Microorganisms* (ISSN 2076-2607) (available at: https://www.mdpi.com/journal/microorganisms/special_issues/BRQXYU022F).

For citation purposes, cite each article independently as indicated on the article page online and as indicated below:

Lastname, A.A.; Lastname, B.B. Article Title. *Journal Name* **Year**, *Volume Number*, Page Range.

**ISBN 978-3-0365-9929-8 (Hbk)**
**ISBN 978-3-0365-9930-4 (PDF)**
**doi.org/10.3390/books978-3-0365-9930-4**

© 2024 by the authors. Articles in this book are Open Access and distributed under the Creative Commons Attribution (CC BY) license. The book as a whole is distributed by MDPI under the terms and conditions of the Creative Commons Attribution-NonCommercial-NoDerivs (CC BY-NC-ND) license.

# Contents

# About the Editors

**Rainer Borriss**

Rainer Borriss is a professor emeritus at Humboldt University, Berlin. He received his Ph.D. degree from Martin Luther University, Halle/Saale, in former East Germany. After working for several years in the industry, he joined the Institute for Plant Breeding (IPK) in Gatersleben and was appointed chair for Bacterial Genetics at Humboldt University in 1992. His research was focused on the sequencing and genetic engineering of microbial glucanases and *Bacillus* genetics. He was involved in the Bacillus subtilis genome project, one of the first bacteria to be wholly sequenced. Within the last two decades, his research has been mainly directed at plant growth-promoting bacteria. After sequencing the genome in 2007, *Bacillus velezensis* FZB42 was established by many of his research contributions as a prototype for plant-associated Gram-positive bacteria capable of suppressing plant pathogens. His scientific work led to 310 items, including 174 articles, and 18 book chapters with 21690 citations, corresponding to an h-index of 67 (Google Scholar). He has presided over many scientific projects sponsored by the Deutsche Forschungsgemeinschaft (DFG) and the German Ministry of Education and Research (BMBF), including international collaboration projects with China and Vietnam. He was a visiting professor at the China Agricultural University in Beijing and the China Northeast Forestry University in Harbin and has received several awards, such as the national award for science and technology (Nationalpreis).

**Chetan Keswani**

Chetan Keswani is the Head of the Agro-Biotechnology Laboratory at the Academy of Biology and Biotechnology, Southern Federal University, Rostov-on-Don, Russia. He has a keen interest in the regulatory, commercialization, and intellectual property aspects of agriculturally important microorganisms in Asia. He is an elected Fellow of the Linnaean Society of London, UK, and received the Best PhD Thesis Award from the Uttar Pradesh Academy of Agricultural Sciences, India, in 2015. He is an editorial board member of several reputable agricultural microbiology journals and has edited ten books on the subject. For his editorial endeavors, he was awarded the Publons Top Peer Review Award in Agricultural Sciences (2018); the Outstanding Editor Award (2019) by Archives of Phytopathology and Plant Protection, Taylor and Francis; and the Springer Society Award (2020) for excellent contribution to the journal *Environmental Sustainability*, Springer Nature. In addition, he has been listed among the World's Top 2

 *microorganisms*

*Editorial*

# Biological Control of the Plant Pathogens

**Rainer Borriss** [1,2,*] **and Chetan Keswani** [3,*]

[1]  Institute of Biology, Humboldt University, 10115 Berlin, Germany
[2]  Institute of Marine Biotechnology e.V., 17489 Greifswald, Germany
[3]  Academy of Biology and Biotechnology, Southern Federal University, 44090 Rostov-on-Don, Russia
*   Correspondence: rainer.borriss@rz.hu-berlin.de (R.B.); kesvani@sfedu.ru (C.K.)

**Citation:** Borriss, R.; Keswani, C. Biological Control of the Plant Pathogens. *Microorganisms* **2023**, *11*, 2930. https://doi.org/10.3390/microorganisms11122930

Received: 4 December 2023
Accepted: 5 December 2023
Published: 6 December 2023

**Copyright:** © 2023 by the authors. Licensee MDPI, Basel, Switzerland. This article is an open access article distributed under the terms and conditions of the Creative Commons Attribution (CC BY) license (https://creativecommons.org/licenses/by/4.0/).

(This article belongs to the Special Issue **Biological Control of the Plant Pathogens.**) The ultimate effects of crop losses manifest in the form of insufficient food production and chronic hunger. The gravity of this issue is rapidly being amplified by the rises in urbanization, climate change, and emerging pests and pathogens, and the deterioration of soil health. As an environmentally friendly alternative to chemical pesticides, microbial biocontrol agents (BCAs) have attracted global attention due to their ability to ensure food security by directly halting pre-harvest crop losses and, thereby, improving crop productivity [1]. Despite the substantial progress achieved in our understanding of plant microbe interactions in recent years, developing more efficient BCAs remains a constant task. The current advances in biocontrol science and technology and the avenues for future research are reflected in the contributions presented in this Special Issue. These can be categorized as follows:

(i)   *Management strategies for major soilborne pathogens in crops*: Ma et al. (contribution 1) review tomato production systems and their challenges. He et al. (contribution 2) have optimized *Metarhizium robertsii* fermentation broth for the efficient management of wolfberry root rot. Fuller et al. (contribution 3) author an informative review on the beneficial and pathogenic microbiota associated with the common alder (*Alnus glutinosa*).

(ii)  *Biocontrol potential of novel volatile compounds and defense inducers*: Recent research has revealed that, besides antimicrobial peptides (AMPs), other natural compounds are involved in biocontrol action. Two examples are given in this Special Issue. Hernández et al. (contribution 4) characterize 65 potential VOCs of the *Kosakonia cowanii* Cp1 strain and demonstrate their role in the effective biocontrol of various economically important phytopathogens. Tripathi et al. (contribution 5) explore defense inducers, including salicylic acid, isonicotinic acid, benzothiadiazole, and lysozymes, as prophylactic and curative sprays for inducing resistance in tomato plants against *Clavibacter michiganensis* subsp. *michiganensis*.

(iii) *Exploring the biocontrol potential of novel microbial isolates and genetically engineered strains*: *Bacillus velezensis* FZB42 is known as a prototype for Gram-positive plant-beneficial bacteria, able to produce durable endospores and suppress plant pathogens [2]. The potential of a novel isolate of *B. velezensis*, strain BV01, as a broad-spectrum BCA against various fungal phytopathogens was assessed (contribution 6). In addition to *B. velezensis*, other representatives of the Bacillaceae family, isolated from Vietnamese crop plants, contain a rich spectrum of compounds efficient against plant pathogens. Numerous members of the *B. cereus* species complex have been comprehensively investigated for the occurrence of the biosynthetic gene clusters (BGCs) involved in the synthesis of secondary metabolites. Antimicrobial peptides, efficient against pathogenic fungi and nematodes, and entomocidal crystal proteins have been detected and partially characterized using mass spectrometry (contribution 7). Moreover, genome mining in plant-associated *Brevibacilli* and *Lysinibacillus* spp. has revealed

36 novel BGCs, not present in the MIBiG data bank, which could be developed as next-generation biocontrol agents (contribution 8). Novel fungal isolates, efficient against grapevine phytopathogens, are described (contribution 9). Interestingly, fungal strains isolated from overwintered tar spot stromata could serve as potential BCAs against tar spot disease in corn (contribution 10). Cui et al. (contribution 11) have engineered an *Escherichia coli* strain to express jasmonic acid carboxyl methyl transferase that catalyzes the conversion from jasmonic acid to methyl jasmonate and demonstrate its biocontrol potential in the management of fungal smut disease.

(iv) *Formulation technology*: Lin et al. (contribution 12) comment on the recent progress in developing "green" adjuvants used for formulating long-living BCAs compatible with chemical pesticides.

Together, the articles published in this Special Issue give an insight into recent and future trends in the development of more powerful and reliable BCAs, and will contribute to sustainable agriculture.

**Author Contributions:** Conceptualization, R.B. and C.K.; writing—original draft preparation, R.B. and C.K.; writing—review and editing, R.B.; funding acquisition, R.B. And C.K. All authors have read and agreed to the published version of the manuscript.

**Funding:** The research of R.B. on this topic was supported through the project ENDOBICA by the Bundesministerium für Bildung und Forschung (BMBF) (grant no. 031B0582A/031B0582B). C.K. gratefully acknowledges support from the Ministry of Science and Higher Education of the Russian Federation project on the development of the Young Scientist Laboratory (no. LabNOTs--21--01AB, FENW-2021-0014) and by the Strategic Academic Leadership Program of the Southern Federal University (Priority 2030).

**Acknowledgments:** The editors express sincere gratitude to all the authors and reviewers for their excellent contributions to this Special Issue. Additional thanks to the *Microorganisms* Editorial Office for their professional assistance and continuous support.

**Conflicts of Interest:** The authors declare no conflict of interests.

**List of Contributions:**

1. Ma, M.; Taylor, P.W.J.; Chen, D.; Vaghefi, N.; He, J.-Z. Major Soilborne Pathogens of Field Processing Tomatoes and Management Strategies. *Microorganisms* **2023**, *11*, 263. https://doi.org/10.3390/microorganisms11020263.
2. He, J.; Zhang, X.; Wang, Q.; Li, N.; Ding, D.; Wang, B. Optimization of the Fermentation Conditions of *Metarhizium robertsii* and its Biological Control of Wolfberry Root Rot Disease. *Microorganisms* **2023**, *11*, 2380. https://doi.org/10.3390/microorganisms11102380.
3. Fuller, E.; Germaine, K.J.; Rathore, D.S. The Good, the Bad, and the Useable Microbes within the Common Alder (*Alnus glutinosa*) Microbiome—Potential Bio-Agents to Combat Alder Dieback. *Microorganisms* **2023**, *11*, 2187. https://doi.org/10.3390/microorganisms11092187.
4. Hernández Flores, J.L.; Martínez, Y.J.; Ramos López, M.Á.; Saldaña Gutierrez, C.; Reyes, A.A.; Armendariz Rosales, M.M.; Cortés Pérez, M.J.; Mendoza, M.F.; RamírezRamírez, J.; Zavala, G.R.; et al. Volatile Organic Compounds Produced by *Kosakonia cowanii* Cp1 Isolated from the Seeds of *Capsicum pubescens* R&P Possess Antifungal Activity. *Microorganisms* **2023**, *11*, 2491. https://doi.org/10.3390/microorganisms11102491.
5. Tripathi, R.; Vishunavat, K.; Tewari, R.; Kumar, S.; Minkina, T.; De Corato, U.; Keswani, C. Defense Inducers Mediated Mitigation of Bacterial Canker in Tomato through Alteration in Oxidative Stress Markers. *Microorganisms* **2022**, *10*, 2160. https://doi.org/10.3390/microorganisms10112160.
6. Huang, T.; Zhang, Y.; Yu, Z.; Zhuang, W.; Zeng, Z. *Bacillus velezensis* BV01 has Broad-Spectrum Biocontrol Potential and the Ability to Promote Plant Growth. *Microorganisms* **2023**, *11*, 2627. https://doi.org/10.3390/microorganisms11112627.
7. Vater, J.; Tam, L.T.T.; Jähne, J.; Herfort, S.; Blumenscheit, C.; Schneider, A.; Luong, P.T.; Thao, L.T.P.; Blom, J.; Klee, S.R.; et al. Plant-Associated Representatives of the *Bacillus cereus* group are a Rich Source of Antimicrobial Compounds. *Microorganisms* **2023**, *11*, 2677. https://doi.org/10.3390/microorganisms11112677.

8.    Jähne, J.; Le Thi, T.T.; Blumenscheit, C.; Schneider, A.; Pham, T.L.; Le Thi, P.T.; Blom, J.; Vater, J.; Schweder, T.; Lasch, P.; et al. Novel Plant-Associated *Brevibacillus* and *Lysinibacillus* Genomo-species Harbor a Rich Biosynthetic Potential of Antimicrobial Compounds. *Microorganisms* **2023**, *11*, 168. https://doi.org/10.3390/microorganisms11010168.
9.    Mannerucci, F.; D'Ambrosio, G.; Regina, N.; Schiavone, D.; Bruno, G.L. New Potential Biological Limiters of the Main Esca-Associated Fungi in Grapevine. *Microorganisms* **2023**, *11*, 2099. https://doi.org/10.3390/microorganisms11082099.
10.   Johnson, E.T.; Dowd, P.F.; Ramirez, J.L.; Behle, R.W. Potential Biocontrol Agents of Corn Tar Spot Disease Isolated from Overwintered *Phyllachora maydis* Stromata. *Microorganisms* **2023**, *11*, 1550. https://doi.org/10.3390/microorganisms11061550.
11.   Cui, G.; Bi, X.; Lu, S.; Jiang, Z.; Deng, Y. A Genetically Engineered *Escherichia coli* for Potential Utilization in Fungal Smut Disease Control. *Microorganisms* **2023**, *11*, 1564. https://doi.org/10.3390/microorganisms11061564.
12.   Lin, F.; Mao, Y.; Zhao, F.; Idris, A.L.; Liu, Q.; Zou, S.; Guan, X.; Huang, T. Towards Sustainable Green Adjuvants for Microbial Pesticides: Recent Progress, Upcoming Challenges, and Future Perspectives. *Microorganisms* **2023**, *11*, 364. https://doi.org/10.3390/microorganisms11020364.

## References

1.    Lugtenberg, B.; Rozen, D.E.; Kamilova, F. Wars between microbes on roots and fruits. *F1000Research* **2017**, *6*, 343. [CrossRef] [PubMed]
2.    Tzipilevich, E.; Russ, D.; Dangl, J.L.; Benfey, P.N. Plant immune system activation is necessary for efficient root colonization by auxin-secreting beneficial bacteria. *Cell Host Microbe* **2021**, *29*, 1507–1520.e4. [CrossRef] [PubMed]

**Disclaimer/Publisher's Note:** The statements, opinions and data contained in all publications are solely those of the individual author(s) and contributor(s) and not of MDPI and/or the editor(s). MDPI and/or the editor(s) disclaim responsibility for any injury to people or property resulting from any ideas, methods, instructions or products referred to in the content.

*microorganisms*

Review

# Major Soilborne Pathogens of Field Processing Tomatoes and Management Strategies

Minxiao Ma, Paul W. J. Taylor, Deli Chen, Niloofar Vaghefi and Ji-Zheng He *

School of Agriculture and Food, Faculty of Science, University of Melbourne, Parkville, VIC 3010, Australia
* Correspondence: jizheng.he@unimelb.edu.au; Tel.: +61-3-90358890

**Abstract:** Globally, tomato is the second most cultivated vegetable crop next to potato, preferentially grown in temperate climates. Processing tomatoes are generally produced in field conditions, in which soilborne pathogens have serious impacts on tomato yield and quality by causing diseases of the tomato root system. Major processing tomato-producing countries have documented soilborne diseases caused by a variety of pathogens including bacteria, fungi, nematodes, and oomycetes, which are of economic importance and may threaten food security. Recent field surveys in the Australian processing tomato industry showed that plant growth and yield were significantly affected by soilborne pathogens, especially *Fusarium oxysporum* and *Pythium* species. Globally, different management methods have been used to control diseases such as the use of resistant tomato cultivars, the application of fungicides, and biological control. Among these methods, biocontrol has received increasing attention due to its high efficiency, target-specificity, sustainability and public acceptance. The application of biocontrol is a mix of different strategies, such as applying antagonistic microorganisms to the field, and using the beneficial metabolites synthesized by these microorganisms. This review provides a broad review of the major soilborne fungal/oomycete pathogens of the field processing tomato industry affecting major global producers, the traditional and biological management practices for the control of the pathogens, and the various strategies of the biological control for tomato soilborne diseases. The advantages and disadvantages of the management strategies are discussed, and highlighted is the importance of biological control in managing the diseases in field processing tomatoes under the pressure of global climate change.

**Keywords:** biocontrol; fungus; oomycete; soilborne pathogen; tomato; tomato disease

**Citation:** Ma, M.; Taylor, P.W.J.; Chen, D.; Vaghefi, N.; He, J.-Z. Major Soilborne Pathogens of Field Processing Tomatoes and Management Strategies. *Microorganisms* **2023**, *11*, 263. https://doi.org/10.3390/microorganisms11020263

Academic Editors: Rainer Borriss and Chetan Keswani

Received: 30 November 2022
Revised: 28 December 2022
Accepted: 28 December 2022
Published: 19 January 2023

**Copyright:** © 2023 by the authors. Licensee MDPI, Basel, Switzerland. This article is an open access article distributed under the terms and conditions of the Creative Commons Attribution (CC BY) license (https://creativecommons.org/licenses/by/4.0/).

## 1. Introduction

Tomato (*Solanum lycopersicum*) has been under global cultivation for several centuries having emerged from western South America [1]. Due to its wide cultivation and unique nutritive value, tomato has become the world's second most cultivated vegetable after potato and the most popular canned vegetable [2,3].

Tomatoes can either be marketed for fresh consumption or as processed products. Around 40 Mt tomatoes are processed annually, making tomatoes the most processed vegetable by weight [4]. The major product of the processing tomato industry is aseptic paste [5], and there are other products like diced and whole peeled tomatoes made into canned food, as well as tomato juice [6]. Processing tomatoes are generally grown in field conditions [7].

Tomatoes are susceptible to over 200 diseases [8] with soilborne diseases being a major component. *Fusarium oxysporum* has been regarded as one of the most important threats to both field and greenhouse-grown tomatoes worldwide with 10–80% of yield loss [8,9]. Another fungal pathogen, *Pyrenochaeta lycopersici*, recently renamed *Pseudopyrenochaeta lycopersici* [10,11], that causes corky root rot, in cooler climates such as northern Europe, can cause up to 75% yield loss of affected tomatoes [12]. Soilborne pathogens are believed to be hard to control for their persistence in soil and wide host range [13].

Traditionally, management methods for tomato diseases use resistant cultivars, chemicals like fungicides and pesticides, physical methods such as soil solarization and soil heating, and cultural methods such as crop rotation and hygiene [9,14,15]. Chemical treatments sometimes have adverse effects on the environment [16,17] and human health and physical control are often laborious and costly.

Breeding for disease resistance is a major objective for most public and private tomato breeders [18]. Breeders usually rely on the genetic diversity within wild tomato species to incorporate desirable traits, especially the resistance to different diseases, by crossing wild tomatoes with cultivated tomatoes [19]. Though genetic resistances usually have high efficiency, the screening for the traits and breeding process can be laborious and time-consuming [20,21], which may hamper the improvement progress of commercial tomato production.

Biological control, also known as biocontrol, is a method of controlling pests and diseases using other organisms (biocontrol agents) through the importation, augmentation, or conservation of agents in the environment [22,23], and is gaining increasing acceptance. Compared with conventional disease/pest control methods, its major advantages are the minimization of the public concerns of impacts on health and environment [24,25], self-sustainability, host-specificity, and cost-efficiency [26,27]. The commercialization of biocontrol for plant disease is relatively a new concept, with the majority of biocontrol agents registered in the United States Environmental Protection Agency after 2000 [28].

This review provides a comprehensive description of the tomato production systems and the major soilborne fungal/oomycete diseases of the global leading processing tomato-producing countries, analysis of the different control strategies for the management of the respective diseases with an emphasis on biocontrol, and proposes a future perspective for disease management within processing tomato industries which are under the threat of emerging pathogens due to climate change.

## 2. The Major Countries Producing Processing Tomatoes and Their Major Soilborne Fungal/Oomycete Diseases

### 2.1. Northern Hemisphere

World tomato production is more concentrated in the northern hemisphere, with an estimated amount of over 34 million metric tonnes (mT) for processing in 2022, with the southern hemisphere producing less than 2.7 million metric tonnes [4]. This is likely to be the result of the combined effects of climate, human and economic factors. The majority of the total processing tomato production (over 65%) is shared by the United States (California), China, and Italy [29].

### 2.1.1. The United States Tomato Industry and Major Soilborne/Fungal Diseases

Though only third in total tomato production size, the United States is the largest producer of processed tomatoes globally. California is the dominant state of US processed tomato production, which accounts for 95% of the nation's tomatoes [30], with the production of 9.8 million mT in 2021 [29]. In California, the production of processing tomatoes is carried out on large farms, with 225 growers owning about 277,000 acres of land [31]. The production starts with the wet soil condition in spring and the harvest and processing in late July until October [32]. Processing tomatoes are mostly irrigated with sprinkler systems, with the increasing application of drip irrigation in the areas threatened by rising saline underground water [33]. All the harvests are performed mechanically, with the fruits graded by inspection before processing [33]. The other USA States with large processing tomato production are Florida and Indiana, but their production scale is very small compared with that of California [34].

Fresh tomato accounts for 8% of total U.S. tomato production [35], which is cultivated nationwide in the USA, with California and Florida being the major production sites, contributing to around 80% of the national production [36]. In California, tomatoes are produced year-round except in winter with the wide application of greenhouse production [36], while in Florida, the growing season is from October to June [35]. In other states, tomato production runs through the summer months when Florida tomatoes are off the market [35].

One important soilborne disease, buckeye rot, is reported to be caused by several oomycete pathogens, including *Phytophthora nicotianae* var. *parasitica*, *P. capsici* and *P. cryptogea* [37–39]. Buckeye rot was first reported in Florida in 1915, causing about 15% yield reduction [40], and was reported in California about two decades later [41]. Buckeye rot was found to cause significant tomato yield reduction in the furrow-irrigated zones of California in 1984, while the major pathogen was *P. parasitica* [39]. However, *P. capsici* was later found to be more virulent to tomatoes, which can also infect *P. parasitica*-resistant tomato varieties [42,43]. In recent years, buckeye rot is most common in the southeast and central states of the U.S. [44].

*P. parasitica* and *P. capsici* can also induce tomato *Phytophthora* root rot by invading plant roots. Reported in California in 1955, the disease had repeated outbreaks in the major processing tomato production zones of the state, leading to the almost complete destruction of the crop [45]. It was also reported that *P. parasitica* was more frequently isolated from Californian processing tomato production sites while *P. capsici* was more abundant in regions growing fresh tomatoes [43,45].

Corky root rot caused by the soilborne fungus *P. lycopersici* is considered to be another major disease limiting Californian tomato production [46]. Campbell [47] tested the pathogen on fresh market tomatoes and harvested 73% fewer large fruits in infected plots without any treatments, which was coherent with the estimation of processing tomato growers consulted in the study. Also, in another study covering nine organic farms and 18 conventional farms in the Central Valley of California, corky root rot was the disease found on most plants in most sampling locations [48]. Recently, *P. lycopersici* was also found to form a disease complex with other pathogens such as *Colletotrichum coccodes* in Ohio, causing severe wilting of tomatoes both in fields and glasshouses [49].

*Fusarium* wilt caused by *F. oxysporum* f. sp. *lycopersici* (*Fol*) also affects tomatoes throughout the United States. All three known physiological races of *Fol* have been reported in the United States. Race 1 was initially reported in 1886, which severely threatened commercial tomato production in Arkansas [50]. Later, *Fol* race 2 which overcame the tomato variety resistant to race 1 was reported in Ohio in 1945 [51]. *Fol* race 3 was reported across the United States [52], causing up to 80% disease incidence in tomatoes resistant to *Fol* race 1 and 2 when first reported in Florida in 1982 [53], while also significantly affecting the processing tomatoes in California [54]. Since the 2000s, *Fusarium* wilt has become a destructive disease in the south-eastern parts of the United States [44].

*Fusarium* crown and root rot (FCRR) is also a major soilborne disease of both U.S. field and greenhouse tomato production. The pathogen, *Fusarium oxysporum* f. sp. *radicis–lycopersici* (*Forl*), was first reported in Japan in 1974 and was later reported across the United States from California to Florida [55,56]. In Florida, the disease was reported to cause a tomato yield reduction of 15–65% [56].

The disease *Verticillium* wilt caused by *Verticillium dahliae* was first reported in California in 1926 [57]. With the resistance *Ve* gene incorporated into commercial tomato varieties, race 2 of *V. dahliae* pathogenic overcoming resistant tomatoes in Ohio and then California was reported [58,59]. Due to the fact that no resistant gene is available in tomato germplasm against *V. dahliae* race 2, this pathogen is now considered one of the diseases limiting the productivity of the Californian tomato industry [46]. In California, the yield loss caused by *V. dahliae* was up to 67% in a susceptible cultivar, with race 2 reducing *V. dahliae* race 1-resistant tomato yield by up to 25% at 100% disease incidence [59,60]. Later, *Verticillium* wilt also became an increasing problem for the Pacific Northwest (Washington, Oregon, and Idaho) states of the U.S., influencing not only tomatoes but also watermelons [61].

### 2.1.2. The Chinese Tomato Industry and Major Soilborne Fungal/Oomycete Diseases

Globally, China ranks first in total tomato production [32] and comes second in terms of estimated 2022 processing tomato production [4]. The major production site of the Chinese tomato industry is Xinjiang province, which accounts for over 80% of the total production size [4]. The major production site of processed tomatoes is the north-western provinces, including Xinjiang, Inner Mongolia and Gansu [32]. The cultivation in Xinjiang is from spring to summer (April to July), while the production is processed between August and October [62].

Tomato for fresh consumption is produced across China, with four provinces, Hebei, Henan, Shandong, and Xinjiang, producing the vast majority [63]. Due to the continuous huge market demand, the fresh tomato industry relies on the utilization of solar-type greenhouses with a year-round production cycle, with Shandong producing over 60% of fresh tomatoes in winter [32].

At least 10 diseases are reported to cause significant tomato yield loss in China [64], with one of them being *Fusarium* wilt. Among the three major physiological races of *Fol*, race 1 is the main pathogen of tomato *Fusarium* wilt in China according to field surveys conducted in several northern provinces including Heilongjiang, Shaanxi and Shanxi [65], while race 2 was only reported in Zhejiang, causing 30–60% tomato yield loss [66].

Recently, symptoms of putative wilt and root rot diseases were reported in several regions growing tomato varieties resistant to *Fol* race 1 and 2 from 2016 to 2018 [67]. Ye et al. [67] conducted field sampling in high-incidence provinces in Zhejiang, Hainan, Shanxi, and Shandong, and found that among the 64 collected *F. oxysporum* isolates, 35 were *Fol* race 3 isolates and 13 were *Forl* isolates. Therefore, *Fol* race 3 and *Forl* seem to have emerged as the major Fusaria pathogens of the Chinese tomato industry, thus more effective management strategies may be required.

### 2.1.3. The Italian Tomato Industry and Major Soilborne Fungal/Oomycete Diseases

In the 2022 industry estimation [4], Italy had the third largest processing tomato industry globally. Over 95% of Italian tomato production is processed [32]. Tomato production is throughout the country, with differences in production methods in different geographical regions. In the northern part of the country, the production is completely mechanized in large farms, with manual harvesting abolished by the early 1990s [68]. This area is the major production site for processing tomatoes with 2.885 million tonnes of estimated yield in 2022 [4]. In southern Italy, farms are usually owned by families, with smaller sizes of 4–10 ha and around 50% of them still practice manual harvesting [32,68]. The center-south region is Italy's second-largest processing tomato production zone, and its estimated yield in 2022 was 2.59 million tonnes [4]. Tomatoes are planted in early May, and harvested in mid-July and the season ends in mid-September.

Tomato production in southern Italy is significantly affected by corky root rot. This disease has been reported since the 1960s in European glasshouse production, which became more severe in later years. In glasshouses in a Mediterranean environment, this disease can cause a 30–40% yield reduction [69].

In northern Italy, *P. capsici* was found to be the causal agent of *Phytophthora* root rot repeatedly observed on grafted tomatoes [70]. This disease was found to cause the sudden collapse of around 25% of the plants within 60 days of transplant [70].

*Fusarium* wilt is also an important tomato disease in Italy. *Fol* race 1 has been long reported in Italy, with race 2 first reported in 1999 [71]. Nowadays, *Fusarium* wilt caused by this pathogen is considered a major cause of economic loss in tomato production in Italy [72–74].

FCRR caused by *Forl* is another important soilborne disease for Italian tomato production. FCRR was reported in northern Italy as early as 1984 and then occurred as an outbreak in Sicily in 1986 [75]. This disease was found to cause substantial yield loss in both greenhouse and soilless fresh market production in several Italian tomato-producing regions such as Calabria, Emilia Romagna, Liguria, Sardinia, and Sicily [76]. FCRR was present in

24% of the greenhouses of Sicily and 66% of greenhouses in Sardinia in surveys conducted during the 1990s [77,78], with a up to 60% mortality rate during severe outbreaks.

### 2.2. Southern Hemisphere

Though tomatoes originated from the Latin American region of the southern hemisphere, the tomato production size of southern hemisphere countries has a comparatively small proportion in the global production. However, when compared with the major producers located in the northern hemisphere, the tomato industry in the southern hemisphere, especially field tomatoes are unique due to several factors. One major factor is the timing of summer in the southern hemisphere, which coincides with winter in the northern hemisphere, thus the most favorable season for tomato production between the two hemispheres is chronologically different. Also, the southern hemisphere generally has a milder climate compared with that of the northern hemisphere due to the higher ocean coverage [79], which may also affect the growth and development of cultivated tomatoes. Therefore, studies on soilborne pathogens threatening tomatoes under the unique climate and production process of the southern hemisphere may also be important for the understanding of the impact of a changing global climate on the disease cycle of these pathogens.

### 2.2.1. The Brazilian Tomato Industry and Major Soilborne Fungal/Oomycete Diseases

In the Southern hemisphere, Brazil is the largest processing tomato producer, with around 1.5 million tonnes in 2021 [29]. Most of Brazil's tomato production is located in the southern provinces close to the coast, such as Goiás, São Paulo, Minas Gerais, Paraná, and Bahia [80]. The processing tomato production is generally based in the central provinces, with Goiás and the Cerrado region contributing to 99% of the production size due to the favorable soil and climatic conditions, and mostly irrigated with conventional methods [81]. To manage the problem caused by whitefly, the Ministry of Agriculture in Brazil set up a rule of a two-month tomato-free period for the processing industry, so the transplanting can only be carried out between 1 February and 30 June, with harvest completed by November [82].

*Fusarium* wilt caused by *Fol* is widely present in Brazil. Race 1 of *Fol* was first reported in São Paulo State in 1941, with only races 1 and 2 detected in all surveys in the 20th century [83]. In 2005, race 3 of *Fol* was also isolated from field samples from wilted hybrid tomato plants with resistance to *Fol* race 1 and 2 from Espirito Santo province [83], which then shortly after spread into other provinces [84]. The severe outbreaks of *Fol* race 3 have led to the widespread replacement of susceptible tomato hybrids with new cultivars carrying the resistant gene *i-3* in Brazil [85].

*Fol* was believed to be the only pathogenic *Fusarium* related to tomato in Brazil until the 2010s. More recently, there were frequent and simultaneous outbreaks of *Fusarium oxysporum* disease capable of infecting *Fol*-resistant tomato cultivars [85]. Subsequently, *Forl* was reported in surveys on field tomato plants showing crown-rot and vascular discoloration symptoms with disease indices ranging from 10 to 50% from three Southeast Brazilian states and two northern states [86].

*Verticillium* wilt is another important soilborne fungal disease for the Brazilian tomato industry. Despite the fact that *V. dahliae* and *V. albo-atrum* were found to cause substantial economic losses in Brazilian vegetable production, only *V. dahliae* was reported on tomatoes [87]. Both races 1 and 2 of *V. dahliae* were reported in the 1980s and late 1990s [88].

### 2.2.2. The Chilean Tomato Industry and Major Soilborne Fungal/Oomycete Diseases

Chile has the second largest tomato production among the southern hemisphere countries, with around 1 million tonnes for processing annually [4]. The majority of Chilean tomato production is for processing, with around 5000 ha for the fresh market and 8000 ha for the processing industry [89,90]. The production area is concentrated in the middle part of the country between 30° and 34° latitude south, benefiting from the local mid-terranean climate optimizing tomato growth [89]. Chilean processing tomato industry is fully contracted, which primarily uses transplanted seedlings sown in July, then transplanted from mid-September to mid-November [91]. Though production was heavily reliant on human labor in the last century, the growing adaptation of mechanized production has led to increased yield and size of individual contracts [91].

Similar to Brazil, *Fusarium oxysporum* is an emerging pathogen for the Chilean tomato industry. Though cultivars resistant to *Fol* race 1 and 2 are commonly used, *Fusarium* wilt still occurred in important production zones such as Azapa valley in northern Chile, and the pathogens were later identified as *Fol* race 3 and *Forl* [92].

*Forl* is also becoming an increasing threat to the Chilean tomato industry, with fresh tomato production affected most severely. FCRR was frequently present in northern Chile and central Chile, with the former practicing both net-house and open-field crops, and the latter growing monocultural tomato crops in polyhouses [93].

### 2.2.3. The Australian Tomato Industry and Major Soilborne Fungal/Oomycete Diseases

Though Australia has a relatively small total tomato production size, it still has the largest processing tomato production outside the Latin America region in the southern hemisphere [4]. Tomatoes in Australia are almost always grown from imported seeds [94]. The fresh and processed tomatoes are produced by two distinctive industries in Australia. The fresh tomatoes are grown in either field or hydroponic environments and are harvested by hand all year round, with Queensland and Victoria as the major production sites [95], while 97% of processing tomatoes are produced in Victoria [96].

The Australian processing tomato industry is sited primarily in the Goulburn/Murray River areas of northern Victoria, with a minority in the Riverina region of New South Wales, which is a seasonal industry (February to April) and harvested by machinery [95]. According to the Australian Processing Tomato Research Council [97], 46% of the total tomato production was sent to the processing industry in 2020.

*Fusarium* wilt caused by *Fol* is considered an economically important disease for the Australian fresh tomato industry, causing huge losses [98]. Also, among the three major races of *Fol* species threatening the global tomato industry, the third race was first reported in Australia in 1979 [99] in Queensland.

Recent disease surveys carried out in the Australian processing tomato industry in Victoria found that processing tomatoes were experiencing an estimated 10% yield loss from soil-borne pathogens, with the two most abundant pathogens being *F. oxysporum* and *Pythium* spp. [100,101]. The *F. oxysporum* strains collected in the survey by Callaghan [100] produced symptoms of rot on tomato roots and crowns such as those caused by *Forl*, a species not reported in Australia, with slight differences in growth temperature, a wider host range and variable pathogenicity [100]. Thus, the disease caused by this isolate was named chocolate streak disease (CSD) to differentiate it from FCRR caused by *Forl*.

Among the eight pathogenic *Pythium* species causing root rot and seedling damping-off, Callaghan et al. [101] found that *P. irrgulare* was one of the most aggressive pathogens, as confirmed in the pre-emergence and early seedling phases of tomato plant growth.

### 3. Control Strategies of the Major Soilborne Fungal/Oomycete Diseases

As discussed above, the major soilborne fungal/oomycete pathogens of the major processing tomato producers are generally similar, with *Fusarium* wilt being the most common disease, followed by FCRR, with *Verticillium* wilt, corky root, and *Phytophthora* root rot also common in the northern hemisphere (Figure 1). In this section, a description of the different diseases and their different management methods is provided.

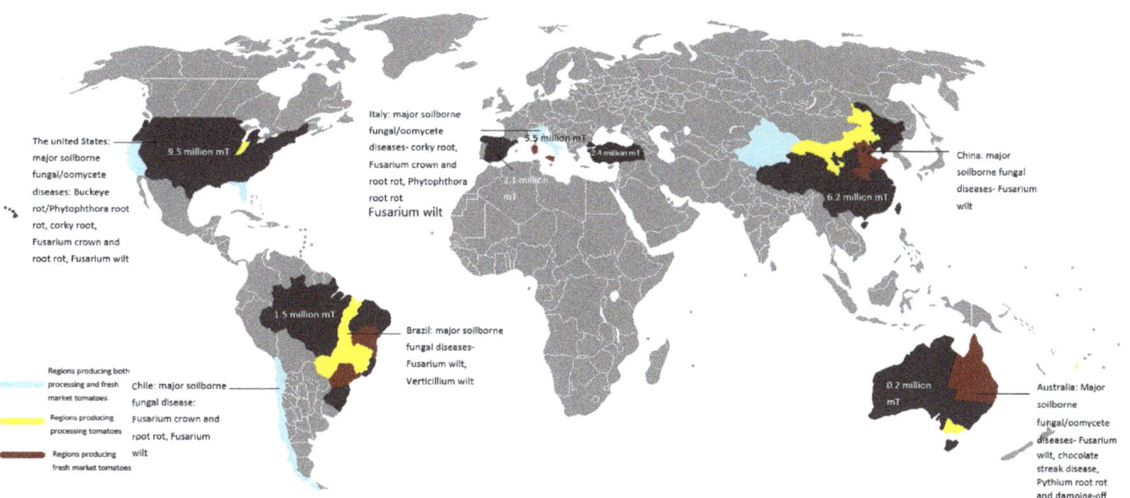


**Figure 1.** Global production of processing tomatoes, major soilborne fungal/oomycete diseases, and the estimated processing tomato production size in 2022.

### 3.1. Tomato Corky Root Rot

Tomato corky root rot is generally caused by the fungus *P. lycopersici*. The roots of infected plants show dark brown, banded lesions [49]. With the development of the symptoms, even larger roots become infected, with extensive swollen and cracked brown lesions, giving them the distinctive corky look appearance [102]. The growth of the plant may be stunted and slow, but the disease usually does not kill plants, resulting in reduced yield [103].

*P. lycopersici* is an ascomycete, which overwinters as microconidia and hyphae, or produces microsclerotia which can withstand harsh conditions, survive in host root and soil, and maintain pathogenicity for up to 15 years [103]. When the environment becomes favorable, the microsclerotia will germinate and produce hyphae [104]. After reaching the host roots, the hyphae penetrate the host epidermis and gradually colonize the whole root [104]. This pathogen is soil-transmitted, which favors monoculture soil without proper disease management [104,105].

#### 3.1.1. Conventional Control Methods
#### Cultural Control

The disease development of corky rot is at optimum at 15.5–20 °C [106]. Thus, it is better to plant tomatoes in spring when the soils start to become warm.

Though effective against many other pathogens, crop rotation alone may not be effective in controlling corky root rot, for *P. lycopersici* has a wide host range including cucumber, eggplant, lettuce, melons, and pepper [103].

#### Physical Control

Soil solarization by covering the field with plastic film for a long period is a practical method for the control of corky root rot. In Italy, Vitale et al. [69] found that solarization performed with ethylene-vinyl-acetate film has an identical level of control effect on corky rot symptoms as compared with fumigation with methyl bromide, which was better than that of metham sodium and metham potassium fumigation. However, the level of success of solarization depends on the combination of high ambient temperatures, maximum solar radiation, and optimum soil moisture as well as the existing inoculum and disease levels [107,108]. Therefore, solarization usually has varying effectiveness, and is generally less effective in climates where high summer temperatures coincide with the rainy season due to the cooling effect of rainfall and extensive clouds blocking the solar radiation [107].

Chemical Control

In fields previously reported to have corky root rot, a preplant treatment with soil fumigation was shown to reduce disease in the subsequent tomato crop [69]. Methyl bromide (MBr) used to be a preferred chemical, but it was proved to be an ozone-depletion agent which is more destructive to stratospheric ozone than chlorine [109], thus its use has been phased out in developed countries by 2005 and in 2015 by the less developed countries as required by Montreal Protocol [16]. Potential alternative chemicals such as chloropicrin, metam sodium, metam potassium, and dazomet [69,110,111] can only provide a lower control level of corky root rot compared with MBr treatment. For example, Vitale et al. [69] found that metham sodium fumigation (MS, 353 litres a.i. ha$^{-1}$) and metham potassium fumigation (MK, 350 litres a.i. ha$^{-1}$) did not reduce the disease incidence of corky root rot in their trial. Therefore, with reduced efficiency of chemical controls, the management of corky root rot may require the addition of more effective methods such as the use of resistant cultivars and biocontrol.

Resistance Breeding

Though breeding for resistant cultivars is a common strategy for the control of crop disease, commercial variants of both processing and fresh consumption tomatoes are susceptible to corky root disease [112]. So far, only one single recessive gene (*pyl*) was shown to confer resistance to corky root rot and was introgressed into *Lycopersicon esculentum* from *L. peruvianum* [113]. The *pyl* gene is later found to possibly be a recessive allele of a susceptibility gene [114] and it has not been cloned yet.

3.1.2. Biological Control

Some fungivorous nematodes have been recorded as potential biocontrol agents for corky root rot. Hasna et al. [115] tested two fungivorous nematodes, *Aphelenchus avenae* and *Aphelenchoides* spp. against *P. lycopersici*, and concluded only *A. avenae* was able to significantly reduce the severity of tomato root rot in greenhouse trials with a population of 3 or 23 nematodes mL$^{-1}$ soil. However, in a later on-farm trial covering two tomato seasons in Sweden, Hasna et al. [105] found even at a higher inoculation rate of 50 nematodes mL$^{-1}$ soil, the application of *A. avanae* into infested soil did not reduce corky root disease severity. Thus, the potential of nematodes to control corky root rot may not be dismissed, but the application method may still need improvements.

In greenhouse trials, bacterial antagonists such as *Streptomyces* spp. have been found to effectively suppress corky root disease of tomatoes and enhance plant growth, resulting in higher yields. Bubici et al. [116] evaluated the antagonism of twenty-six *Streptomyces* spp. against corky root rot on tomatoes in both glasshouse and field conditions and found the most effective strain can reduce disease severity up to 64% in the glasshouse and 48% in the field.

Antagonistic fungi may also be used in the biocontrol against corky root rot. Fiume and Fiume [112] conducted glasshouse trials against corky root rot using *Trichoderma viride*, *Bacillus subtilis*, and *Streptomyces* spp., and concluded that the application of all three microorganisms significantly reduced the corky root symptoms in terms of disease index, with *T. viride* having the best results, followed by *Streptomyces* spp. Besoain et al. [117] performed UV on native *T. harzianum* to obtain mutants and found the mutants ThF1-2 and ThF4-4 inhibited the growth of *P. lycopersici* in vitro by 1.3 and 5 fold, respectively. Sánchez-Téllez et al. [118] further tested the mutant ThF1-2 in greenhouse tomato trials and found applying solid formulation ThF1-2 resulted in a significantly lower root damage caused by *P. lycopersici* compared with a previous trial using MBr. The control of *T. harzianum* against *P. lycopersici* seems to be correlated to the differential expression of extracellular fungal cell wall hydrolytic enzymes between isolates [119].

Organic amendments may also help in the control of corky root rot. Workneh et al. [48] found that the application of green manure and compost reduced the corky root rot severity in organic farm tomatoes by stimulating microbial activities in a field survey. However, *P. lycopersici* responds differently to different amendments. Hasna et al. [120] tested composts consisting of green manure, garden waste, and horse manure against corky root rot in greenhouse tomatoes and found that garden waste compost significantly reduced the disease, whereas horse manure compost significantly stimulated disease, while the green manure compost had no effect on the disease despite the increased microbial activity. It was concluded that the disease severity of corky root rot can be suppressed by composts with a low concentration of ammonium nitrogen and a high concentration of calcium, but further studies may be necessary to further prove this perspective.

### 3.2. Fusarium Crown and Root Rot (FCRR) of Tomato

FCRR is caused by the pathogenic fungus *Fusarium oxysporum* f. sp. *radicis-lycopersici* (*Forl*). Though both being classified as *Fusarium oxysporum*, *Forl* and *Fol*, the agent of tomato *Fusarium* wilt, are two different *formae speciales*, which are informal taxonomic groupings based on the differences in the host range [121]. The pathogen invades the host via wounds and natural openings created by newly emerging roots [122]. Infected seedlings usually show symptoms of stunting and yellowing with premature abscission of lower leaves. With the development of the disease further via the xylem, necrotic lesions may gridle the crown, with the roots becoming rotted and discolored, and the seedlings then develop wilting symptoms [123]. On older plants, the first symptom may be the yellowing and collapse of the oldest leaves. The symptoms then develop upwards and infect young leaves. Older plants may be stunted and wilted, but still alive at harvest [56].

*F. oxysporum* is an ascomycete without an observable sexual stage [124]. This fungus can aggressively colonize both host roots and organic matter [125]. Though being able to persist in the soil as mycelial fragments, the highly durable chlamydospore is the major form of survival of *F. oxysporum* in the absence of a host [124,126]. The germination of the spore is usually triggered by the exudates from growing plant roots [124,126].

### 3.2.1. Conventional Control Methods
#### Cultural Control

Hygiene and sanitation of the seeds and transplant seedlings are important for *Forl* management. For example, Muslim et al. [127] found that plants not challenged with the pathogen still become infected by FCRR, which is probably due to incomplete soil sterilization. It is also strongly recommended that all equipment coming in direct contact with soil is cleaned and disinfected [56]. The pathogen may also use colonized and infected plants as carrying vectors, thus the infected plants and their roots should be removed immediately.

Crop rotation with a non-host crop may also prevent FCRR. Crops susceptible to *Forl* such as eggplants and peppers should be avoided in the rotation [128], while non-hosts such as lettuce may be useful to reduce inoculum levels in the soil [129]. However, the efficiency of crop rotation may be limited for FCRR control, because the pathogen can survive as chlamydospores in the soil for a long time [130].

#### Physical Control

FCRR is favored by cooler temperatures, thus planting in warm periods and using warm water in irrigation is recommended to restrict the development of disease [131]. Soil solarization has also been demonstrated to control FCRR. In studies testing several solarization methods, soil solarization generally reduced populations of *Forl* down to a depth of 5 cm [56].

#### Chemical Control

Before the 2010s, the most effective method for FCRR control was soil disinfection using methyl bromide (MBr) [132,133]. However, MBr has been phased out globally since

2015. The ban on MBr prompted the study of alternative chemicals for the control of soilborne pests including *Forl*. So far, the tested alternatives include 1,3-dichloropropene, chloropicrin, dozamet, fosthiazate, and metam sodium, with similar effects on *Forl* compared with MBr [56,134,135]. For example, McGovern et al. [134] tested the application of metam sodium in field tomatoes and found that rotovation of metam sodium at 935 L/ha into preformed beds consistently reduced FCRR incidence equal to those achieved by methyl bromide-chloropicrin. Also, 1,3-dichloropropene+chloropicrin (60.5% and 33.3%, *w/w*) was tested on Italian field tomatoes [135] and was able to achieve a good tomato yield using drip application in sandy loam soils with slight *Forl* infections and severe infections of *Fol* and galling nematodes, which was similar to those of the plots treated with MBr.

However, there are still several factors that may reduce the efficiency of *Forl* chemical control. For example, *Forl* chlamydospores were found to survive in the soil at a depth beyond 50 cm, which is unreachable by soil fumigation [131]. Also, *Forl* can efficiently colonize sterilized soil [55]. Therefore, soil fumigation may instead create favorable soil conditions for *Forl* colonization by reducing microbial competition.

Resistance Breeding

Resistant tomato varieties can also be used to control FCRR. The resistance of tomatoes to FCRR is found to be controlled by a single dominant locus (*Frl*) on chromosome 9 [122,136]. This gene has been successfully crossed into commercial tomato lines, with many *Forl*-resistant cultivars currently available. However, no additional resistant genes have been identified.

3.2.2. Biocontrol

*Forl* is believed to have poor competitive fitness against other microorganisms [131], thus biocontrol via organic amendments or biocontrol agents may be effective for the management of *Forl*.

Several antagonistic microorganisms have been tested for their properties to control FCRR. Sivan et al. [137] applied *Trichoderma harzianum* as seed coating or wheat-bran/peat in tomatoes grown in FCRR-infested field and recorded a 26.2% increase in yield of treated plots compared with the control, with the control of *Forl* at the highest effect on root tips. Datnoff et al. [138] also applied *T. harzianum* and *Glomus intraradices* into tomato fields with FCRR history and recorded a significant reduction in disease severity and disease incidence of FCRR by applying the fungi both combined and separately. Several hypervirulent binucleate *Rhizoctonia* strains were also found to reduce the vascular discoloration caused by FCRR on tomatoes up to 100% in greenhouse conditions and up to 70% in the field [127]. Moreover, a non-pathogenic endophytic *F. solani* strain was reported to reduce disease incidence of *Forl* when applied alone in glasshouse tomato by 47%, the effects of which improved when combined with certain fungicides [139]. *Pythium oligandurm* was also found to trigger the host defence of greenhouse tomatoes when challenged by *Forl* in the form of deposition of newly formed barriers beyond the infection sites [140].

Several bacteria species may also control FCRR. *Pseudomonas fluorescens* was found to synthesize the antibiotic 2,4-diacetylphloroglucinol, which suppressed the growth of *Forl* in vitro [141]. A further study found that *P. fluorescens* WCS365 used chemotaxis towards *Forl* hyphae, enabling it to efficiently colonize *Forl* and achieve control effects [142]. In a later screening by Kamilova et al. [143], strong competitive biocontrol strains *P. fluorescens* PCL1751 and *P. putida* PCL1760 were found to successfully suppress FCRR under the soil and hydroponic conditions. In addition, Baysal et al. [123] assessed in a greenhouse trial the effect of two *Bacillus subtilis* bacteria strains QST713 and EU07, and concluded that EU07 had a better disease inhibitory effect (disease incidence reduced by 75%) compared with QST713 (disease incidence reduced by 52%), and the inhibition may be achieved by YrvN protein coded in the genome of EU07 as a subunit of protease enzyme. Lytic enzymes, cellulases, proteases, 1,4-b-glucanase, and hydrolases from the secreted proteins from *B. subtilis* EU07 and FZB24 and concluded these essential proteins of *Bacillus* bacteria play an important role in the control of *Forl* [144].

Organic amendments promoting microbial activity may also be used in FCRR management, but they do not have consistent effects in field conditions. Straw was incorporated into the soil to manage FCRR by Jarvis [131], but the *Forl* soil population increased around and inside the straw, which only started to fall when the straw decomposed. However, Kavroulakis et al. [145] concluded that a compost mix made from grape marc wastes and extracted olive press cake can enhance tomato defensive capacity under *Forl* stress by making the pathogen unable to penetrate and colonize the host root, resulting in a 40% reduction in the disease incidence compared to the control. However, the plants in this trial were grown completely in the compost, making large-size commercial applications likely unrealistic.

### 3.3. Fusarium wilt Disease of Tomato

*Fol*, the causal agent of tomato *Fusarium* wilt, is able to penetrate plant cell epidermis, thus infecting tomato plants through the roots and colonizing the xylem for further colonization of the root system [9]. The symptoms are initially characterized as yellowing of the older leaves [50], followed by browning and wilting. Browning will also be visible in the vascular tissue, and this discoloration will extend to the apex of the plant [50]. The infected plant will experience stunted growth and drastically reduced yield, which will often die before maturity [9].

### 3.3.1. Conventional Control Methods
#### Cultural Control

Crop rotation can be used to manage *Fusarium* wilt, and it is recommended not to plant the same or related type of crop for at least four years if one crop is severely infected by *Fusarium* wilt [146]. The recommended crops for rotation are grasses and cereals [147].

Hygiene should also be practiced for *Fol* control. Disease-affected plants should be removed immediately. Used farming tools should be disinfected and cleaned before reuse. The use of sanitized footwear and clothes on the farm may help prevent the transportation of infected soils between paddocks [146]. Fallowing is another strategy for *Fol* control. Briefly, the land is left uncultivated for a period, and for *Fol*, it is recommended to practice fallowing during the summer months to let the high temperature and excessive drying reduce soil levels of *Fol* [9].

#### Physical Control

Soil solarization can also be used to control *Fol* residing in soil, preferably performed in the summertimes. However, since the development of *Fusarium* wilt favors warm temperatures (27–28 °C) [148], this strategy may not work in zones with cool climates.

#### Chemical Control

Soil fumigation with MBr was an effective method for *Fol* management however, with the phase-out of MBr the value of chemical control on *Fol* has drastically reduced. Though alternative chemicals such as chloropicrin, dimethyl disulfide, metam sodium, and 1,3-dichloropropene are available, they all lack the broad-spectrum volatile characteristics of MBr, which made it highly effective [149]. Systemic fungicides such as benomyl, thiabendazole, and thiophanate have also been used to control tomato *Fusarium* wilt [9], but it was believed that there are no fungicides especially effective for the control of this disease [146].

#### Resistance Breeding

The application of tomato cultivars resistant to *Fusarium* wilt is currently the most feasible management method.

The resistance to *Fol* was first identified by Bohn and Tucker in 1939 [150], who identified one single, dominant resistance locus later named *I* gene from one wild tomato accession of *S. pimpinellifolium*, Missouri accession 160 [151]. This gene was crossed into the first commercial *Fol*-resistant tomato cultivar and was located at tomato chromosome 11 [152].

Later, the second race of *Fol*, named *Fol2* was reported to spread widely in Florida in the 1960s [153], which led to another screening for the corresponding resistant gene. The resistant gene was again found in wild tomato relatives- a natural hybrid PI126915, which was name *I-2* and mapped to chromosome 11 [154].

In 1979, the third race of *Fol–Fol3* was reported in Australia in fresh tomato production [99]. McGrath et al. [155] were the first to identify resistance to *Fol3* in the *S. pennellii* accession PI414773 in 1987, and Scott and Jones [156] later identified a dominant *Fol3* resistance locus in the *S. pennellii* accession LA716. This newly discovered gene was later named *I-3* and used as the primary source of *Fol3* resistance in commercial varieties. Gene *I-3* was mapped to chromosome 7 [157], and McGrath et al. located another gene *I-7* gene in chromosome 8 [158].

Three additional genes with partial resistance to *Fol2* were also found by Sela-Buurlage et al. [152]. These researchers studied 53 introgression lines with chromosomes from LA716 and identified alleles *I-4* and locus *I-5* on chromosome 2, with locus *I-6* on chromosome 10 of *S. pennellii*. However, none of these genes have their effects validated nor used for commercial purposes so far.

### 3.3.2. Biological Control

Potential biocontrol agents against *Fol* on tomatoes have been actively tested in a large number of studies. The most commonly used biocontrol agents belonged to various microbial genera including fungi (*Aspergillus* spp., *Chaetomium* spp., *Glomus* spp., non-pathogenic *Fusarium* spp., *Trichoderma* spp. and *Penicillium* spp.) and bacteria (*Bacillus* spp., *Pseudomonas* spp., *Streptomyces* spp., and *Serratia* spp.) [159].

Among the different genera of biocontrol microorganisms, non-pathogenic *Fusarium* strains are of high interest. In 1993, Alabouvette et al. [160] concluded that among the many groups of microorganisms tested for biocontrol activity, only non-pathogenic *Fusarium* species and fluorescent *Pseudomonads* showed consistent responses. In a later review by Ajilogba et al., these strains were found to be involved in most research conducted on plant biological enhancement using fungal endophytes [146]. One representative strain, *F. oxysporum* Fo47, was successfully tested against *Fol* [161–163], with the major mode of function being the induction of systemic resistance and priming of the plant defence reaction.

Another review by Raza et al. [159] analyzed biocontrol trials conducted between 2000 and 2014 and concluded that non-pathogenic *Fusarium* species and *Pseudomonas* species were supported by most research to be more effective in controlling *Fusarium* wilt in natural soil, while *Penicillium*, *Streptomyces*, and *Aspergillus* strains were more effective in growth media. However, the authors also found that 79% of the tests on tomatoes were conducted in greenhouse conditions, with 12% conducted in the field condition. Thus, for processing tomatoes grown predominately in field conditions, further field tests on the efficiency of different biocontrol agents are necessary.

Organic amendments are another group of biocontrol agents. For example, Borrego-Benjumea et al. [164] tested poultry manure, olive residue compost, and pelletised poultry manure for tomatoes grown in natural sandy soil and concluded that the combination of pelletized poultry manure with heating or solarization achieved the greatest reduction in *Fusarium* wilt severity. In a later study by Zhao et al. [165] testing chicken manure, rice straw, and vermicompost in a long-term tomato monocultural soil, vermicompost addition significantly increased soil pH, ammonium nitrogen, soil organic matter, and dissolved organic carbon, which promoted beneficial bacteria suppressing *Fol*. Organic amendments are often applied in combination with biocontrol microorganisms for better effects in different studies [159,166,167]. It was also suggested that the combined application of biocontrol organisms and amendments can increase the biocontrol efficiency of various genera of fungi and bacteria, with the exceptions of *Pseudomonas* and *Penicillium* [159].

### 3.4. Phytophthora Root Rot of Tomato

The oomycete *P. capsici* is the major pathogen causing *Phytophthora* root rot on tomatoes. This soilborne pathogen can invade tomatoes via root, inducing root and crown rot which can be visualized by the brown lesions on the plant's lower part [168]. With sufficient rainfall, the whole plant can be infected. The root infection may cause damping-off of seedlings, stunting, wilting and eventual death in older plants [168]. The pathogen can also infect fruit in contact with the ground or via irrigation splash, and the disease, with symptoms of light-to-dark, water-soaked brown concentric rings on the fruit, is called buckeye rot instead [37,41,169].

As an oomycete, *P. capsci* can reproduce both sexual oospores with antheridium oogonium and asexual zoospores with sporangia [168,170,171]. Chlamydospores are also occasionally produced [168]. Most stages of *P. capsci* require the presence of a host, thus *P. capsica* seems only survive in the soil for long terms as thick-walled oospores [168]. The germination of oospores is facilitated by both chemical and mechanical stimulations, by either growing a germ tube or forming sporangium [168,171]. The sporangium then germinates directly or releases motile zoospores if immersed in water [170]. The zoospores are swimmable, which can be transmitted by rainfall or irrigation, and form germ tubes after contacting hosts [170].

### 3.4.1. Conventional Control Methods
#### Cultural Control

Crop rotation is often used to manage *P. capsici* along with many other soilborne pathogens, but its effectiveness is limited by the long survival of oomycetes in the soil and the wide host range of *P. capsici*. The host range of *P. capsici* was reported to cover at least 45 species of cultivated plants and weeds from 14 families of flowering plants [170], thus the selection of rotation crops for *P. capsici* is very narrow. Also, Lamour and Hausbeck [172] found *P. capsici* can survive as oospores for a 30-month nonhost period during crop rotation. Therefore, long rotations are required even if non-host crops are available, which may make crop rotations economically unfeasible.

It is very difficult to control *P. capsici* once the pathogen becomes established in the field. Thus, most control strategies are aimed at limiting free water to minimize inoculum spread and crop loss, which includes planting at well-drained sites or on a raised bed with controlled irrigation [168].

#### Physical Control

Soil solarization was found to be effective against *Phytophthora* root rot on tomatoes. From a trial in Florida a soil solarization treatment that heated the soil to a maximum of 47 °C at 10-cm depth had similar effects to MBr treatment at the same site in reducing the *P. capsici* population [107].

#### Chemical Control

The application of chemicals has been another approach to managing *P. capsici*. However, the phasing out of MBr has reduced the cost-efficiency of chemical control [173]. Other chemicals frequently applied include cyazofamid, dimethomorph, fluopicolide, fosetyl-Al, mandipropamid and mefenoxam (metalaxyl) [174–177]. Despite the various choices of chemicals, extensive use of fungicide has led to the emergence of resistant *P. capsici* strains, which makes it very hard to protect crops from *P. capsici*. For example, Lamour and Hausbeck [172] collected 141 isolates of *P. capsici* in Michigan and found around 60% to be intermediately sensitive or insensitive to mefenoxam. Even more recent groups of chemicals such as fluopicolide and cyazofamid have resulted in the fast emergence of pathogen resistance. Jackson et al. [175] concluded that among the 40 *P. capsici* isolates tested, all were either intermediately sensitive or resistant to cyazofamid at 100 μg/mL application rate. More recently, Siegenthaler and Hansen [177] found that out of 184 *P. capsici* isolates collected in Tennessee, 84 were resistant to fluopicolide.

Resistance Breeding

Until the 2010s, only several tomato strains moderately resistant to *P. capsici* were commercially available. Quesada-Ocampo and Hausbeck [173] screened 42 tomato cultivars and wild relatives for their resistance against *P. capsici*, and found *Solanum habrochaites* accession LA407, was resistant to all *P. capsici* isolates tested, with four additional cultivars having moderate resistance. However, the authors analyzed the genes of these cultivars and found a lack of correlation between genetic clusters and susceptibility to *P. capsici*, indicating that resistance was distributed in several tomato lineages. In a subsequent study, Quesada-Ocampo et al. [178] generated 62 backcross lines using LA407, and tested their resistance against different *P. capsici* strains and used annotated markers to locate genes related to the resistance. Though the researchers found that the resistance had a good inheritability among the population, they failed to find any annotated markers strongly associated with *P. capsici* resistance, with genes with annotation linked to disease resistance responses mapped to all chromosomes segregated among the population with the exceptions for 8, 9, 11, and 12. Therefore, the resistance of tomatos to *P. capsici* has not been related to specific gene/loci so far, and further studies are required.

### 3.4.2. Biocontrol

With insufficient levels of conventional control measures against *Phytophthora* root rot of tomatoes, antagonistic microbes and organic amendments have been tested to find feasible biocontrol approaches. Bacteria species are frequently studied for their biocontrol properties against *Phytophthora* root rot. Moataza [179] tested five *Pseudomonas fluorescences* strains against *Rhizoctonia solani* and *P. capsici* in tomato pot trials, and concluded that two strains, NRC1 and NRC3 had strong lytic activities leading to the destruction of the pathogens, but the method used in this research was seed coating, which may not be commercially feasible. In another study, Sharma et al. [180] tested 20 *Bacillus* strains against *P. capsici* on tomatoes grown in net house, and found one species, *B. subtilis* showed the best efficiency in terms of decreased disease severity. Furthermore, Syed-Ab-Rahman et al. [181] tested three bacteria- *B. amyloliquefaciens*, *B. velezensis* and *Acinetobacter* sp. on tomato, and concluded all three bacteria promoted tomato growth while significantly reducing the *P. capsici* load in their roots. An oomycete, *Pythium oligandrum* was also tested, and was believed to synthesize two Necrosis- and ethylene-inducing peptide 1 (Nep1)-like proteins PyoINLP5 and PyoINLP7, which induced the expression of antimicrobial tomato defensin genes against *P. capsici* [182].

The application of organic amendments is another approach to biocontrol. For *P. capsici* management, Nicol and Burlakoti [183] aerated compost and water and produced four aerobic compost teas. When tested in the glasshouse, the researchers concluded that if these products were drenched in potting mix before and after *P. capsici* inoculation, the disease progression was reduced by over 70%, with improved plant growth. Other efforts of using composts against *P. capsici* have generally been attempted on pepper [184–186], so the effects of these composts on tomatoes are unknown.

### 3.5. Pythium Root Rot and Damping-Off

The oomycete *Pythium* species tend to infect and cause rot of seeds, rootlets, root tips, and root hairs, with a preference for younger tissue at the root elongation zone and lateral roots [187]. The infection may cause small, brown, water-soaked lesions and can affect the entire root system [188]. With the focus on younger tissues, *Pythium* species infection often causes seedling damping-off at both pre- and post-germination stages [189,190], while infected older plants may also show stunted growth.

*Pythium* species overwinter in the soil as oospores or in plant debris as mycelium [191,192]. The germination of the oospore is facilitated by the exudates of germinating seeds and roots [192]. Similar to *P. capsici*, the oospore can produce a germ tube or form a sporangium, which germinates on its own or releases motile zoospores to contact and invade plant roots [191]. During the invasion, the oomycete hyphae release enzymes to destroy and feed on the host tissues [191]. After the invasion, *Pythium* species can either repeat the infection cycle in a new host with sporangia or survive as mycelia with sexual and asexual structures or dormant oospores until the next growing season [191,192].

### 3.5.1. Conventional Control Methods

#### Cultural Control

The application of pathogen-free seedlings and the control of irrigation are found to be effective forms for tomato *Pythium* disease management [193,194].

For *Pythium* species, crop rotation is generally not considered to be effective in the control of tomato infections because most *Pythium* species have a wide host range [195]. However, one study on wheat found that 3–4-year rotation cycles using wheat, canola and legume resulted in a significantly smaller disease incidence compared with less diverse rotations such as two-year wheat-canola [196]. The reason behind this finding may be that different crops have significantly different susceptibilities to *Pythium* infection, which may restrict the soilborne pathogen inoculum build-up after each crop, and eventually reducing the disease incidence in the next crop.

#### Physical Control

Soil solarization is an effective method for *Pythium* control with a long-period (six weeks to 60 days) of solarization during the summertime having been shown to significantly reduce the soilborne population of *P. aphanidermatum* in tropic zones [197,198]. In a field trial on tomatoes infected by *Pythium* spp., solarized soil showed a significantly lower mean damping-off incidence compared with un-solarized soil (2.15% compared with 68%) [199].

#### Chemical Control

Several chemicals have been used to manage *Pythium* species, including hymexazol, mefenoxam (metalaxyl), phosphonate, thiram and 8-Hydroxyquinoline [200–204]. The chemicals can be applied as seed treatment [205,206] or soil drenching [207] for seedlings of tomato.

In addition to the common economic and environmental concerns of chemical control, several major *Pythium* species collected from the production of various crops have developed resistance against several chemicals, especially mefenoxam. For example, Porter et al. [208] reported over 50% of the *Pythium* soil population consisted of mefenoxam-resistant isolates in ten of 64 potato fields from Oregon and Washington. Del Castillo Munera and Hausbeck [209] tested a total of 202 *Pythium* spp. isolates collected from Michigan, and found 39% of these, mostly *P. ultimum* and *P. cylindrosporum* isolates were intermediate to highly resistant to mefenoxam. For another major species *P. irregulare*, Aegerter et al. [210] tested four *P. irregulare* isolates from a greenhouse extensively applying mefenoxam and found no inhibition of growth of any isolate occurred at mefenoxam concentrations of 10 μg/mL or less. For other *Pythium* species such as *P. aphanidermatum*, resistance to mefenoxam was also reported [211,212]. In a rare case, Garzón et al. [203] even reported that the disease severity of a mefenoxam-resistant *P. aphanidermatum* on geranium can be stimulated by sublethal doses of mefenoxam.

#### Resistance Breeding

Though the deployment of resistant cultivars is a common and effective strategy for crop disease management, currently there is no *Pythium*-resistant tomato. The only potentially useful genetic resource against *Pythium* is the genes encoding pathogenesis-related (PR) proteins, with PR-1 protein showing antifungal activity against oomycetes [213]. Tomato has two related genes, *PR1b1* and *PR1a2*, each encoding a basic and an acidic PR-1 protein [214], but the resistance of PR proteins is not pathogen-specific, with only limited effects against *Pythium* species.

### 3.5.2. Biocontrol

For biocontrol of *Pythium* disease on tomatoes, several bacteria strains have been studied. Postma et al. [215] tested four bacteria strains against *P. aphanidermatum* and found three strains, *Pseudomonas chlororaphis*, *Peanibacillus polymyxa* and *Streptomyces pseudovenezuelae*, significantly controlled *P. aphanidermatum* in under greenhouse conditions. The effect of *Streptomyces* bacteria was also supported by the study of Hassanisaadi et al. [195], who found two root-symbiont *Streptomyces* species significantly decreased disease incidence and improved performance of greenhouse tomato under *P. aphanidermatum* in stress out of the 116 tested species. For *Bacillus* bacteria, Martinez et al. [216] tested one *B. subtilis* strain MBI600 in a peat-based potting mix and concluded the addition of this strain significantly reduce tomato and sweet pepper damping-off and root rot while promoting root growth. Samaras et al. [204] also tested MBI600 on greenhouse tomatoes and concluded that the application of this strain achieved satisfactory control efficacy compared to chemical treatment with 8-Hydroxyquinoline.

For the application of fungal antagonists, the current focus seems to be on the *Trichoderma* species. Caron et al. [217] tested one local *T. harzianum* strain MAUL-20 on greenhouse tomatoes and found that it significantly reduced *P. ultimum* disease incidence, with a better effect compared with Rootshield™, a biofungicide based on *T. harzianum* KRL-AG2. Cuevas et al. [202] also tested *T. parceramosum*, *T. pseudokoningii* and *T. harzianum* respectively, and found the application of the *Trichoderma* pellets into the field before seeding can minimize the activity of *Pythium* spp., with a higher seed germination rate compared with the treatment using chemical fungicide mancozeb. Elshahawy and El-Mohamedy [188] tested the effects of five *Trichoderma* strains on *P. aphanidermatum* damping-off of tomatoes and concluded that under field conditions the combined application of the five isolates reduced by half the root rot severity while almost doubling the survival of tomato. This was thought to be through activating tomato defence enzymes and increasing leaf chlorophyll content, with an increased yield.

Interestingly, even arbuscular mycorrhizal fungi suppressing plant growth may also be used to control *Pythium* species. Larsen et al. [218] pre-treated greenhouse tomato seedlings with *Glomus intraradices*, *G. mosseae*, *G. claroideum*, and then challenged the seedlings with *P. aphanidermatum*, with the hypothesis that the application of growth-suppressive fungi may trigger plant defence response in terms of *PR-1* expression to prepare the plants for *Pythium* infection. However, the application of arbuscular mycorrhizal fungi did not affect *PR-1* gene expression, with only *G. intraradices* reducing the pathogen root infection level of *P. aphanidermatum*, thus the hypothesis was not confirmed.

Several organic amendments have also been tested against *Pythium*, such as canola residues and composts (animal bone charcoal, compost tea, solid green wastes, or green waste +manure) [215,219–221]. Also, Jayaraj et al. [222] found that formulating amendments such as lignite with biocontrol agents such as *B. subtilis* can greatly increase their shelf life, with good effects on *Pythium* suppression and plant growth promotion.

### 3.6. Tomato Verticillium Wilt

*Verticillium. dahliae* and *V. albo-atrum* are soilborne fungi that can induce vascular wilt diseases in over 200 dicotyledonous species, including those economically important such as tomato [223]. These species invade the host through roots [224,225] and then attack the vascular system via xylem vessels. This leads to wilting, vascular discoloration, early senescence, and the eventual death of the infected plant [224].

Though the two *Verticillium* species have similar lifecycles, *V. dahliae* causes monocyclic disease with only one disease cycle in a growing season [224]. In contrast, *V. albo-atrum* can produce conidia on infected plant tissues, which can be airborne and contribute to polycyclic diseases during one growing season [224]. The lifecycles of the two species both have a dormant, a parasitic, and a saprophytic stage [224]. During the dormant stage, the resting structures such as microsclerotia and mycelium in soil or plant debris are under microbiostasis or mycostasis and unable to germinate [224]. The germination of the pathogens is stimulated by root exudates from both host and non-host plants [226]. The pathogens then enter the parasitic stage by invading hosts through the root tip or elongation region to invade the xylem and vascular systems [226]. After the necrosis of the infected tissue, the saprophytic stage begins, in which the pathogens extend their colonization to shoots and roots and produce conidia or microsclerotia to repeat the cycle [224]. The pathogens can be spread by the transport of infected planting stock or by soil cultivation and soil movement by wind or water [226].

### 3.6.1. Conventional Control Methods
#### Cultural Control

Crop rotation with non-host crops is an effective strategy for *Verticillium* wilt management. The known non-host crops include small grain crops such as wheat and corn [227], and long rotations lasting over four years are recommended [44].

Hygiene is also important for *Verticillium* wilt control. pathogen-free seed and disease-free transplants should be used [44], with infected crop debris removed and destroyed away from the field. Equipment and foot ware should be washed to prevent the movement of infested soil between fields. *Verticillium* also prefers humid soil, thus maintaining well-drained soil, and eliminating excessive soil moisture may also limit the development of the pathogen [228].

#### Physical Control

*Verticillium* prefers cool temperatures for survival and developing symptoms, thus heating the soil through solarization could be an effective control method. Currently, solarization against *Verticillium* wilt is practiced generally in Mediterranean, desert, and tropical climates, because these climates allow the accumulation of adequate heat to neutralize the pathogen [229]. However, the data on solarization alone showed poorer performance compared with the MBr application, which can be improved when combined with the fumigation using MBr alternatives [230].

#### Chemical Control

Soil fumigation is also used to control *Verticillium* wilt. MBr alternatives such as chloropicrin (CP) (trichloronitromethane) are traditionally used as in formulations together with MBr to achieve a broader spectrum of activity [230]. In a trial by Gullino et al. [72], CP applied by shank injection at rates $\geq 30$ g/m$^2$ induced a satisfactory and consistent control of tomato *Verticillium* wilt, with no phytotoxicity, but the efficiency was slightly lower than standard MBr application and may have been influenced by soil type and organic matter content. Metam-sodium and 1,3-dichloropropene are other alternative soil fumigants, which have been applied in combination or with metam-sodium alone in the United States to reduce soil populations of *V. dahliae* [231]. Several other chemicals such as fungicides including azoxystrobin, benomyl, captan, thiram, and trifloxystrobin, and a plant defense activator, acibenzolar-S-methyl were also recommended [116,230,232].

#### Resistance Breeding

By far, the most feasible and economic control for *Verticillium* wilt is the application of resistant cultivars. The resistance gene in tomato to *V. dahliae* was first identified as a single dominant factor in the reciprocal crosses between the wilt-resistant variety W6 (Peru Wild × Century) and Moscow, a susceptible variety, and named as *Ve* in 1951 [233]. *Ve* was found to be a locus, which contains two genes, *Ve1* and *Ve2*, with only *Ve1* found to mediate resistance in tomato [223]. The strains of *V. dahliae* resistant to *Ve1* and *V. albo-atrum* were assigned to race 2 [223]. The *Ve1* gene has been incorporated into many commercial cultivars. However, all the current verticillium-resistant gene resources are against *V. dahliae* race 1, thus all race 2 strains of *V. dahliae* and *V. albo-atrum* can still infect the resistant cultivars.

### 3.6.2. Biocontrol

Biological control may be a promising method to control *Verticillium* wilt, given that most current management methods have limited efficiency. Various microorganisms have been tested against *V. dahliae*, such as bacteria *Bacillus subtilis* and *B. velezensis* [234], and fungi including *Burkholderia gladioli* [235], *Gliocladium* spp., *Penicillium* sp. [236,237], *Trichoderma* spp. [238], *Talaromyces flavus* [239], and even *V. klebahnii* and *V. isaacii* with low pathogenicity [240]. Though most of the microorganisms are found to be effective in trials, most of the trials were carried out in greenhouses or with sterilized soil, with only a few verified in field conditions. Larena et al. [237] conducted a field assay using *P. oxalicum* and concluded that seedlings needed to be treated with $10^6$–$10^7$ CFU/g of the biocontrol agent around a week before transplanting to achieve a sufficient level of control, but only in a certain soil type (loam soil, pH = 7.0), and the formulation may not be feasible for tomato mass production due to the high CFU density requirement.

The application of organic amendments is known as another approach for crop disease biocontrol. It has long been known that bloodmeal and fishmeal can eliminate the incidence of *Verticillium* wilt in tomato [230]. Compared to animal-based amendments (manure), plant-based amendments not only support beneficial microbial activities but also have greater efficiency on pathogens due to deleterious chemicals produced by the plants, in addition to supporting beneficial microbial activities [241]. Giotis et al. [12] concluded that fresh Brassica tissue, household waste compost, and composted cow manure significantly reduced soilborne disease severity of tomato *Verticillium* wilt, with enhanced plant growth. Similar results were also achieved by Kadoglidou et al. [242], who applied soil incorporated spearmint and oregano-dried plant material, which caused disease suppression resulting in increased fruit yields of tomatoes inoculated with *V. dahliae*. Moreover, Ait Rahou et al. [243] used compost based on green waste (quackgrass) to greenhouse tomatoes inoculated with *Verticillium* and concluded that growth regulators directly produced by the microorganisms in the compost improved plant growth significantly. However, when Lazarovits et al. [244] applied compost made from sewage sludge to suppress *V. dahliae* in tomato plants, phytotoxicity was detected over one month, which may have been due to the excessive accumulation of plant-toxic heavy metals in soils. To conclude, though organic amendments may be useful for *Verticillium* wilt management, they may also carry toxic compounds which may lead to undesired effects.

## 4. Conclusions and Future Perspective

Among the field processing tomato producing countries covered in this review, the major soilborne fungal/oomycete pathogens affecting their tomato production are *Fol*, *Forl*, *P. lycopersici* and *P. capsici*, with the pathogenic *Verticillium* spp., and *Pythium* spp. being also important in certain countries. The various management methods (Table 1) generally have variable levels of effectiveness on the diseases (Table 2). For cultural controls, hygiene is fundamental to disease control, while the effectiveness of crop rotation is affected by the host range and longevity of the corresponding pathogen in soil and crop debris. Physical control represented by soil solarization is generally effective except for *Fol* which can withstand high temperatures, but the level of success is affected by several environmental and biological factors, thus it may work better as a part of integrated disease management. The situation of chemical control is more problematic. The industry used to rely heavily on the broad-spectrum, cost-efficient MBr soil fumigation effective against all soilborne pathogens, but it turned out to be a lose-lose. MBr heavily damaged the ozone layer, which resulted in it being banned globally, while the ban in turn caused enormous losses to the industry. For example, Cao et al. [132] estimated that in Florida, the phase-out of MBr and the introduction of replacements may have caused a 20% tomato yield reduction and a $1656 decrease in profit per acre, which substantially harmed the competitiveness and sustainability of the tomato industry of Florida. As discussed above, alternative chemicals are available for the control of all diseases, but these generally have reduced spectrum and cost-efficiency, and in some cases, the pathogens have already developed certain levels of resistance. The situation is further exacerbated by the decreasing public acceptance of the chemical application, and the fact that even some alternatives to MBr, such as metam sodium, have been banned or are scheduled for a phase-out in certain areas [245]. Breeding for disease resistance remains highly effective, but it is traditionally laborious and time-consuming, and genes for complete resistance to *P. lycopersici*, *Pythium* sp., *V. dahilae* race 2, and *V. albo-atrum* have not been identified.

**Table 1.** Different management methods and their examples are mentioned in this review.

| Management Methods | Examples |
| --- | --- |
| Cultural control | Crop rotation, farrowing, hygiene |
| Physical control | Soil solarization, soil warming |
| Chemical control | Soil chemical fumigation, application of fungicide |
| Resistance breeding | Crossing the desired traits from wild relatives into cultivated tomato varieties |
| Biological control | Biocontrol agents, organic soil amendments |

**Table 2.** Advantages and disadvantages of the management methods concluded from this review.

| Management Methods | Advantages | Disadvantages |
| --- | --- | --- |
| Cultural control | Basic, easy to be carried out | Limited controlling effects |
| | Can be integrated into other management methods | Laborious |
| Physical control | Effective against pathogens residing in soil | May not be economically feasible |
| | Material highly accessible | Effectiveness depends on the local environment and the biology of the pathogen |
| | | Less effective in deep soil |
| Chemical control | Highly effective-at least in the initial stages | High cost |
| | Broad-spectrum effect of fumigation | Requiring registration |
| | Target-specificity of fungicides | Negative effects on the environment and human health |
| | Industrialized process | Decreasing public acceptance |
| Resistance breeding | Target-specific resistance | Laborious |
| | Sustainability | Time-consuming |
| | Environmentally friendly | Resistance traits against certain pathogens do not exist |
| Biological control | Various mechanisms against specific pathogens | New, largely in in vitro trial stage |
| | Sustainability | Impact on the indigenous microbial community |
| | Environmentally friendly | Requiring registration |
| | Highly levels of disease control effects | |
| | High public acceptance | |
| | Cost-efficiency | |
| | Turns waste into use | |

Though relatively new for crop disease management, biological control seems to show promise. Although most biocontrol agents are still in the greenhouse trial stage, assessment has shown satisfying levels of disease control, many of which directly attack pathogens or initiate the plant's own defence mechanisms, and some directly improve the growth and development of crops. Also, the use of organic amendments may not only improve the plant growth, but also utilize unwanted organic products, which may go to waste otherwise. Therefore, the advantages of biocontrol may minimize the adverse effects on the environment while being highly appealing to the public.

Disease management has long been challenging for the tomato industry. Cultivated tomato is considered to have low genetic diversity due to the population bottleneck effect of domestication and artificial selection [246], making them vulnerable to destructive pathogens. Also, temperate climates and adequate humidity is preferred by cultivated tomatoes [32], which may also facilitate the flourishing of soilborne microorganisms, including pathogens. Moreover, the fast development of international trade in recent years has brought new threats by allowing the emergence of pathogens into new geographical locations through transporting infected products [247], which can adapt to the new environment via mutation or merging with their local native relatives. Also, climate change may also facilitate the natural movement of pathogens. As demonstrated by Bebber et al. [248], global warming has made crop pathogens move poleward, affecting the zones that used to be too cold for them to survive. More importantly, climate change is particularly challenging for the processing tomato industry. Cammarano et al. [249] made predictions of global tomato yield based on five potential future global warming scenarios and concluded that around 6% decline in processing tomato production may take place in the three major producing areas (California, China, and Italy) by 2050 due to the increased temperature, with little differences between the scenarios.

With conventional disease management strategies becoming inadequate for the challenges brought about by pathogen resistance, global trade, and climate change, innovative control strategies are needed. Biocontrol, with its good potential in disease control efficiency, public acceptance, and crop yield improvement, should be incorporated into the integrated disease management of field processing tomato. This will complement cultural practices, physical disease management, and modern breeding techniques such as gene editing and CRISPR/Cas, to develop management practices emphasizing sustainability and the security of both food and the environment, and hence greatly reduced reliance on synthesized chemical application.

**Author Contributions:** Conceptualization, M.M., P.W.J.T. and J.-Z.H.; resources, P.W.J.T., D.C. and J.-Z.H.; writing—original draft preparation, M.M.; writing—review and editing, all authors; supervision, P.W.J.T., D.C., N.V. and J.-Z.H.; project administration, M.M., P.W.J.T. and J.-Z.H.; funding acquisition, D.C. and J.-Z.H. All authors have read and agreed to the published version of the manuscript.

**Funding:** ARC Research Hub for Smart Fertilizers (IH200100023).

**Data Availability Statement:** Not applicable.

**Conflicts of Interest:** The authors declare no conflict of interest.

## References

1. Rick, C.M. The tomato. *Sci. Am.* **1978**, *239*, 76–89. [CrossRef]
2. FAOSTAT. Available online: http://www.fao.org/faostat/en/#home (accessed on 30 November 2022).
3. Barringer, S. Canned tomatoes: Production and storage. In *Handbook of Vegetable Preservation and Processing*, 1st ed.; Hui, Y.H., Ghazala, S., Graham, D.M., Murrell, K.D., Nip, W., Eds.; CRC Press: New York, NY, USA, 2004; pp. 109–120.
4. WPTC. WPTC Crop Update as of 25 October 2022. Available online: https://www.tomatonews.com/force_doc.php?file=c50d465 45467b815c44e583ddcdd7731c1c4ae2f.pdf (accessed on 30 November 2022).
5. Luster, C., III. A rapid and sensitive sterility monitoring technique for aseptically processed bulk tomato paste. *J. Food Sci.* **1978**, *43*, 1046–1048. [CrossRef]
6. Barrett, D.M. Future innovations in tomato processing. In Proceedings of the XIII International Symposium on Processing Tomato, Sirmione, Italy, 8 June 2014; Volume 1081, pp. 49–55.
7. Johnstone, P.R.; Hartz, T.K.; LeStrange, M.; Nunez, J.J.; Miyao, E.M. Managing fruit soluble solids with late-season deficit irrigation in drip-irrigated processing tomato production. *HortScience* **2005**, *40*, 1857–1861. [CrossRef]
8. Singh, V.K.; Singh, A.K.; Kumar, A. Disease management of tomato through PGPB: Current trends and future perspective. *3 Biotech* **2017**, *7*, 1–10. [CrossRef]
9. Bawa, I. Management strategies of *Fusarium* wilt disease of tomato incited by *Fusarium oxysporum* f. sp. *lycopersici*(Sacc.) A Review. *Int. J. Adv. Acad. Res.* **2016**, *2*, 32–42.
10. Valenzuela-Lopez, N.; Cano-Lira, J.F.; Guarro, J.; Sutton, D.A.; Wiederhold, N.; Crous, P.W.; Stchigel, A.M. Coelomycetous *Dothideomycetes* with emphasis on the families *Cucurbitariaceae* and *Didymellaceae*. *Stud. Mycol.* **2018**, *90*, 1–69. [CrossRef]
11. d'Errico, G.; Marra, R.; Crescenzi, A.; Davino, S.W.; Fanigliulo, A.; Woo, S.L.; Lorito, M. Integrated management strategies of *Meloidogyne incognita* and *Pseudopyrenochaeta lycopersici* on tomato using a *Bacillus* firmus-based product and two synthetic nematicides in two consecutive crop cycles in greenhouse. *Crop. Prot.* **2019**, *122*, 159–164. [CrossRef]

12. Giotis, C.; Markelou, E.; Theodoropoulou, A.; Toufexi, E.; Hodson, R.; Shotton, P.; Leifert, C. Effect of soil amendments and biological control agents (BCAs) on soil-borne root diseases caused by *Pyrenochaeta lycopersici* and *Verticillium albo-atrum* in organic greenhouse tomato production systems. *Eur. J. Plant Pathol.* **2009**, *123*, 387–400. [CrossRef]
13. Cheng, H.; Zhang, D.; Ren, L.; Song, Z.; Li, Q.; Wu, J.; Cao, A. Bio-activation of soil with beneficial microbes after soil fumigation reduces soil-borne pathogens and increases tomato yield. *Environ. Pollut.* **2021**, *283*, 117160. [CrossRef]
14. Baysal-Gurel, F.; Gardener, B.M.; Miller, S.A. Soil Borne Disease Management in Organic Vegetable Production. *Org. Agric.* Available online: https://eorganic.org/node/7581. (accessed on 30 November 2022).
15. Spadaro, D.; Gullino, M.L. Improving the efficacy of biocontrol agents against soilborne pathogens. *Crop. Prot.* **2005**, *24*, 601–613. [CrossRef]
16. Gareau, B.J. Lessons from the Montreal Protocol delay in phasing out methyl bromide. *J. Environ. Stud. Sci.* **2015**, *5*, 163–168. [CrossRef]
17. Piel, C.; Pouchieu, C.; Carles, C.; Beziat, B.; Boulanger, M.; Bureau, M.; Baldi, I. Agricultural exposures to carbamate herbicides and fungicides and central nervous system tumour incidence in the cohort AGRICAN. *Environ. Int.* **2019**, *130*, 104876. [CrossRef]
18. Foolad, M.R.; Merk, H.L.; Ashrafi, H. Genetics, genomics and breeding of late blight and early blight resistance in tomato. *Crit. Rev. Plant Sci.* **2008**, *27*, 75–107. [CrossRef]
19. Menda, N.; Strickler, S.R.; Edwards, J.D.; Bombarely, A.; Dunham, D.M.; Martin, G.B.; Mueller, L.A. Analysis of wild-species introgressions in tomato inbreds uncovers ancestral origins. *BMC Plant Biol.* **2014**, *14*, 1–16. [CrossRef] [PubMed]
20. de Toledo Thomazella, D.P.; Brail, Q.; Dahlbeck, D.; Staskawicz, B. CRISPR-Cas9 mediated mutagenesis of a *DMR6* ortholog in tomato confers broad-spectrum disease resistance. *BioRxiv*. Available online: https://www.biorxiv.org/content/10.1101/064824 v3.full. (accessed on 30 November 2022).
21. Panthee, D.R.; Brown, A.F.; Yousef, G.G.; Ibrahem, R.; Anderson, C. Novel molecular marker associated with $Tm2^a$ gene conferring resistance to tomato mosaic virus in tomato. *Plant Breed.* **2013**, *132*, 413–416. [CrossRef]
22. Flint, M.L.; Dreistadt, S.H. *Natural Enemies Handbook: The Illustrated Guide to Biological Pest Control*; Univ of California Press: Berkely, CA, USA, 1998; Volume 3386, pp. 3–6.
23. Landis, D.A.; Orr, D.B. Biological Control: Approaches and Applications. Electronic IPM Textbook. Available online: https://ipmworld.umn.edu/landis (accessed on 30 November 2022).
24. Rechcigl, J.E.; Rechcigl, N.A. (Eds.) *Biological and Biotechnological Control of Insect Pests*; CRC Press: New York, NY, USA, 1999; pp. 3–5.
25. van Lenteren, J.C.; Bolckmans, K.; Köhl, J.; Ravensberg, W.J.; Urbaneja, A. Biological control using invertebrates and microorganisms: Plenty of new opportunities. *BioControl* **2018**, *63*, 39–59. [CrossRef]
26. Singh, S.; Bhatnagar, S.; Choudhary, S.; Nirwan, B.; Sharma, K. Fungi as biocontrol agent: An alternate to chemicals. In *Fungi and Their Role in Sustainable Development: Current Perspectives*, 1st ed.; Gehlot, P., Singh, J., Eds.; Springer: Singapore, 2018; pp. 23–33.
27. Sood, M.; Kapoor, D.; Kumar, V.; Sheteiwy, M.S.; Ramakrishnan, M.; Landi, M.; Sharma, A. *Trichoderma*: The "secrets" of a multitalented biocontrol agent. *Plants* **2020**, *9*, 762. [CrossRef]
28. Junaid, J.M.; Dar, N.A.; Bhat, T.A.; Bhat, A.H.; Bhat, M.A. Commercial biocontrol agents and their mechanism of action in the management of plant pathogens. *Int. J. Mod. Plant Anim. Sci.* **2013**, *1*, 39–57.
29. Hort Innovation. Tomato Topics September 2022. Available online: https://www.aptrc.asn.au/wp-content/uploads/2022/10/Tomato-Topics-September-2022-1.pdf (accessed on 30 November 2022).
30. Pathak, T.B.; Stoddard, C.S. Climate change effects on the processing tomato growing season in California using growing degree day model. *Model. Earth Syst. Environ.* **2018**, *4*, 765–775. [CrossRef]
31. Jones, J.B., Jr. *Tomato Plant Culture: In the Field, Greenhouse, and Home Garden*; CRC Press: New York, NY, USA, 2007; pp. 55–80.
32. Costa, J.M.; Heuvelink, E.P. The global tomato industry. In *Tomatoes*, 2nd ed.; Heuvelink, E., Ed.; CABI: Cambridge, MA, USA, 2018; pp. 1–26.
33. Hartz, T.; Miyao, G.; Mickler, J.; Lestrange, M.; Stoddard, S.; Nuñez, J.; Aegerter, B. Processing Tomato Production in California. Available online: https://escholarship.org/content/qt4hc350c9/qt4hc350c9.pdf (accessed on 30 November 2022).
34. The Top 10 Tomato Producing States in the United States. Available online: https://www.worldatlas.com/articles/the-top-10-tomato-producing-states-in-the-united-states.html (accessed on 30 November 2022).
35. Guan, Z.; Biswas, T.; Wu, F. The US Tomato Industry: An Overview of Production and Trade. Available online: https://journals.flvc.org/edis/article/download/105009/118652/ (accessed on 30 November 2022).
36. Baskins, S.; Bond, J.; Minor, T. Unpacking the Growth in per Capita Availability of Fresh Market Tomatoes. Available online: https://www.ers.usda.gov/webdocs/outlooks/92442/vgs-19c-01.pdf?v=5296.1 (accessed on 30 November 2022).
37. Fernández-Pavía, S.P.; Rodríguez-Alvarado, G.; Sánchez-Yáñez, J.M. Buckeye rot of tomato caused by *Phytophthora capsici* in Michoacan, Mexico. *Plant Dis.* **2003**, *87*, 872. [CrossRef]
38. Gupta, S.K.; Sachin, U.; Sharma, R.C. Biology, epidemiology and management of buckeye rot of tomato. In *Challenging Problems in Horticultural and Forest Pathology*, 1st ed.; Sharma, R.C., Sarma, G.N., Eds.; Indus Publishing Company: New Delhi, India, 2006; pp. 183–199.
39. Hoy, M.W.; Ogawa, J.M.; Duniway, J.M. Effects of irrigation on buckeye rot of tomato fruit caused by *Phytophthora parasitica*. *Phytopathology* **1984**, *74*, 474–478. [CrossRef]
40. Kendrick, J.B. *Phytophthora* rot of tomato, eggplant, and pepper. *Proc. Indiana Acad. Sci.* **1922**, *32*, 299–306.

41. Tompkins, C.M.; Tucker, C.M. Buckeye rot of tomato in California. *J. Agric. Res.* **1941**, *62*, 467–474.
42. Bolkan, H. A technique to evaluate tomatoes for resistance to Phytophthora root rot in the greenhouse. *Plant Dis.* **1985**, *69*, 708–709. [CrossRef]
43. Filho, A.C.; Duniway, J.M. Dispersal of *Phytophthora capsici* and *P. parasitica* in furrow-irrigated rows of bell pepper, tomato and squash. *Plant Pathol.* **1995**, *44*, 1025–1032. [CrossRef]
44. Babadoost, M. Important fungal diseases of tomato in the United States of America. In Proceedings of the III International Symposium on Tomato Diseases, Ischia, Italy, 25–30 July 2010; Volume 914, pp. 85–92.
45. Ioannou, N.; Grogan, R.G. Control of Phytophthora root rot of processing tomato with ethazol and metalaxyl. *Plant Dis.* **1984**, *68*, 429–435. [CrossRef]
46. Stamova, L. Resistance to Two Important Tomato Diseases in California. In Proceedings of the XV Meeting of the EUCARPIA Tomato Working Group, Bari, Italy, 20–23 September 2005; Volume 789, pp. 87–94.
47. Campbell, R.N. Corky Root of Tomato in California Caused by *Pyrenochaeta lycopersici* and Control by Soil Fumigation. *Plant Dis.* **1981**, *66*, 657–661. [CrossRef]
48. Workneh, F.; Van Bruggen AH, C.; Drinkwater, L.E.; Shennan, C. Variables associated with corky root and *Phytophthora* root rot of tomatoes in organic and conventional farms. *Phytopathology* **1993**, *83*, 581–589. [CrossRef]
49. Vrisman, C.M.; Testen, A.L.; Elahi, F.; Miller, S.A. First report of tomato brown root rot complex caused by *Colletotrichum coccodes* and *Pyrenochaeta lycopersici* in Ohio. *Plant Dis.* **2017**, *101*, 247. [CrossRef]
50. Srinivas, C.; Devi, D.N.; Murthy, K.N.; Mohan, C.D.; Lakshmeesha, T.R.; Singh, B.; Srivastava, R.K. *Fusarium oxysporum* f. sp. *lycopersici* causal agent of vascular wilt disease of tomato: Biology to diversity–A review. *Saudi J. Biol. Sci.* **2019**, *26*, 1315–1324. [CrossRef]
51. Alexander, L. Physiologic specialization in the tomato wilt. *J. Agric. Res.* **1945**, *70*, 303.
52. Cai, G.H.; Gale, L.R.; Schneider, R.W.; Kistler, H.C.; Davis, R.M.; Elias, K.S.; Miyao, E.M. Origin of race 3 of *Fusarium oxysporum* f. sp. *lycopersici* at a single site in California. *Phytopathology* **2003**, *93*, 1014–1022. [CrossRef] [PubMed]
53. Volin, R.B.; Jones, J.P. A new race of *Fusarium* wilt of tomato in Florida and sources of resistance. *Proc. Fla. State Hortic. Soc.* **1982**, *95*, 268–269.
54. Davis, R.M.; Kimble, K.A.; Farrar, J.J. A third race of *Fusarium oxysporum* f. sp. lycopersici identified in California. *Plant Dis.* **1988**, *72*, 453. [CrossRef]
55. Benhamou, N.; Charest, P.M.; Jarvis, W.R. Biology and Host-Parasite Relations of *Fusarium oxysporum* f. sp. *radicis-lycopersici*. In *Vascular Wilt Diseases of Plants*, 1st ed.; Tjamos, E.C., Beckman, C.H., Eds.; Springer: Berlin, Germany, 1986; pp. 95–105.
56. Ozbay, N.; Newman, S.E. Fusarium crown and root rot of tomato and control methods. *Plant Pathol. J.* **2004**, *3*, 9–18. [CrossRef]
57. Rudolph, B.A. *Verticillium* wilt of tomatoes in California. sclerotia of *Verticillium albo-atrum* and its influence on infection. *Phytopathology* **1926**, *16*, 234.
58. Alexander, L.J. Susceptibility of certain Verticillium-resistant Tomato varieties to an Ohio isolate of the pathogen. *Phytopathology* **1962**, *52*, 998–1000.
59. Grogan, R.G.; Ioannou, N.; Schneider, R.W.; Sall, M.A.; Kimble, K.A. *Verticillium* wilt on resistant tomato cultivars in California: Virulence of isolates from plants and soil and relationship of inoculum density to disease incidence. *Phytopathology* **1979**, *69*, 1176–1180. [CrossRef]
60. Ashworth, L.J.; Huisman, O.C.; Harper, D.M.; Stromberg, L.K. *Verticillium* wilt disease of tomato: Influence of inoculum density and root extension upon disease severity. *Phytopathology* **1979**, *69*, 490–492. [CrossRef]
61. Buller, S.; Inglis, D.; Miles, C. Plant growth, fruit yield and quality, and tolerance to *verticillium* wilt of grafted watermelon and tomato in field production in the Pacific Northwest. *HortScience* **2013**, *48*, 1003–1009. [CrossRef]
62. Costa, J.M.; Heuvelink, E. Introduction: The tomato crop and industry. In *Tomatoes*, 1st ed.; Heuvelink, E., Ed.; CABI: Cambridge, MA, USA, 2005; pp. 1–20.
63. Zhang, X.X.; Qiu, H.; Huang, Z. Farm structure of the apple and tomato production in the EU and China. In *Apple and Tomato Chains in China and the EU. LEI 26*, 1st ed.; Zhang, X.X., Qiu, H., Huang, Z., Eds.; LEI: The Hague, The Netherlands, 2010; pp. 21–27.
64. Wang, Y.; Zhang, Y.; Gao, Z.; Yang, W. Breeding for resistance to tomato bacterial diseases in China: Challenges and prospects. *Hortic. Plant J.* **2018**, *4*, 193–207. [CrossRef]
65. Chang, Y.D.; Du, B.; Wang, L.; Ji, P.; Xie, Y.J.; Li, X.F.; Wang, J.M. A study on the pathogen species and physiological races of tomato Fusarium wilt in Shanxi, China. *J. Integr. Agric.* **2018**, *17*, 1380–1390. [CrossRef]
66. Xu, Z.; Huang, K.; Shou, W.; Zhou, S.; Li, G.; Chen, L.; Jin, B. The determination of physiological race of *Fusarium oxysporum* f. sp. *lycopersici* of tomato in Zhejiang, China. *Acta Physiol. Plant.* **2000**, *22*, 356–358. [CrossRef]
67. Ye, Q.; Wang, R.; Ruan, M.; Yao, Z.; Cheng, Y.; Wan, H.; Zhou, G. Genetic diversity and identification of wilt and root rot pathogens of tomato in China. *Plant Dis.* **2020**, *104*, 1715–1724. [CrossRef]
68. Perrotta, D. Processing tomatoes in the era of the retailing revolution: Mechanization and migrant labour in northern and southern Italy. In *Migration and Agriculture*, 1st ed.; Corrado, A., de Castro, C., Perrotta, D., Eds.; Routledge: Oxfordshire, UK, 2016; pp. 82–100.
69. Vitale, A.; Castello, I.; Cascone, G.; D'Emilio, A.; Mazzarella, R.; Polizzi, G. Reduction of corky root infections on greenhouse tomato crops by soil solarization in South Italy. *Plant Dis.* **2011**, *95*, 195–201. [CrossRef]

70. Garibaldi, A.; Gilardi, G.; Baudino, M.; Ortu, G.; Gullino, M.L. *Phytophthora capsici*: A soilborne pathogen dangerous on grafted tomato (*Solanum lycopersicum* × *S. hirsutum*) in Italy. *Plant Dis.* **2012**, *96*, 1830. [CrossRef]
71. Stravato, V.M.; Buonaurio, R.; Cappelli, C. First Report of *Fusarium oxysporum* f. sp. *lycopersici* Race 2 on Tomato in Italy. *Plant Dis.* **1999**, *83*, 967. [CrossRef]
72. Gullino, M.L.; Minuto, A.; Gilardi, G.; Garibaldi, A.; Ajwa, H.; Duafala, T. Efficacy of preplant soil fumigation with chloropicrin for tomato production in Italy. *Crop Prot.* **2002**, *21*, 741–749. [CrossRef]
73. Rongai, D.; Pulcini, P.; Pesce, B.; Milano, F. Antifungal activity of pomegranate peel extract against fusarium wilt of tomato. *Eur. J. Plant Pathol.* **2017**, *147*, 229–238. [CrossRef]
74. Srinivasan, K.; Gilardi, G.; Garibaldi, A.; Gullino, M.L. Bacterial antagonists from used rockwool soilless substrates suppress Fusarium wilt of tomato. *J. Plant Pathol.* **2009**, *91*, 147–154.
75. Di Primo, P.; Cartia, G.; Katan, T. Vegetative compatibility and heterokaryon stability in *Fusarium oxysporum* f. sp. *radicis-lycopersici* from Italy. *Plant Pathol.* **2001**, *50*, 371–382. [CrossRef]
76. Garibaldi, A.; Minuto, A.; Minuto, G. Influence of Recycled Substrates on the Severity of Crown Rot on Soilless Tomato. In Proceedings of the XV Meeting of the EUCARPIA Tomato Working Group, Bari, Italy, 20–23 September 2005; Volume 789, pp. 167–170.
77. Cartia, G.; Asero, C. Investigation on spread and pathogenicity of *Fusarium oxysporum* f. sp. *radicis lycopersici* in Sicilian greenhouses. *Colt. Protette* **1994**, *23*, 75–78.
78. Franceschini, A.; Maddau, L.; Corda, P.; Ionta, G. Observations on Fusarium crown and root rot of tomato and its control in Sardinia (Italy). *Dif. Delle Piante* **1996**, *19*, 125–138.
79. Frenzel, B. Atlas of Paleoclimates and Paleoenvironments of the Northern Hemisphere. Available online: https://epic.awi.de/id/eprint/29922/1/Fre1992a.pdf (accessed on 30 November 2022).
80. Elias, M.A.; Borges, F.J.; Bergamini, L.L.; Franceschinelli, E.V.; Sujii, E.R. Climate change threatens pollination services in tomato crops in Brazil. *Agric. Ecosyst. Environ.* **2017**, *239*, 257–264. [CrossRef]
81. Marouelli, W.A.; Silva, W.L. Water tension thresholds for processing tomatoes under drip irrigation in Central Brazil. *Irrig. Sci.* **2007**, *25*, 411–418. [CrossRef]
82. Inoue-Nagata, A.K.; Lima, M.F.; Gilbertson, R.L. A review of geminivirus diseases in vegetables and other crops in Brazil: Current status and approaches for management. *Hortic. Bras.* **2016**, *34*, 8–18. [CrossRef]
83. Reis, A.; Costa, H.; Boiteux, L.S.; Lopes, C.A. First report of *Fusarium oxysporum* f. sp. *lycopersici* race 3 on tomato in Brazil. *Fitopatol. Bras.* **2005**, *30*, 426–428. [CrossRef]
84. Reis, A.; Boiteux, L.S. Outbreak of *Fusarium oxysporum* f. sp. *lycopersici* race 3 in commercial fresh-market tomato fields in Rio de Janeiro State, Brazil. *Hortic. Bras.* **2007**, *25*, 451–454. [CrossRef]
85. Gonçalves, A.M.; Costa, H.; Fonseca, M.E.N.; Boiteux, L.S.; Lopes, C.A.; Reis, A. Variability and geographical distribution of *Fusarium oxysporum* f. sp. *lycopersici* physiological races and field performance of resistant sources in Brazil. In Proceedings of the V International Symposium on Tomato Diseases: Perspectives and Future Directions in Tomato Protection, Málaga, Spain, 13–16 June 2016; Volume 1207, pp. 45–50.
86. Cabral, C.S.; Gonçalves, A.M.; Fonseca ME, N.; Urben, A.F.; Costa, H.; Lourenço, V.; Reis, A. First detection of *Fusarium oxysporum* f. sp. *radicis–lycopersici* across major tomato–producing regions in Brazil. *Phytoparasitica* **2020**, *48*, 545–553. [CrossRef]
87. Suaste-Dzul, A.P.; Costa, H.; Fonseca ME, N.; Boiteux, L.S.; Reis, A. Mating types and physiological races of *Verticillium dahliae* in Solanaceae crops in Brazil. *Eur. J. Plant Pathol.* **2022**, *164*, 1–14. [CrossRef]
88. Suaste-Dzul, A.P.; Veloso, J.S.; Costa, H.; Boiteux, L.S.; Lourenço Jr, V.; Lopes, C.A.; Reis, A. *Verticillium* diseases of vegetable crops in Brazil: Host range, microsclerotia production, molecular haplotype network, and pathogen species determination. *Plant Pathol.* **2022**, *71*, 1417–1430. [CrossRef]
89. Valdés, V. The tomato industry in Chile. In Proceedings of the IV International Symposium on Processing Tomatoes, Mendoza, Argentina, 18–21 February 1991; Volume 301, pp. 59–62.
90. Valenzuela, M.; Fuentes, B.; Alfaro, J.F.; Galvez, E.; Salinas, A.; Besoain, X.A.; Seeger, M. First Report of bacterial speck caused by *Pseudomonas syringae* pv. *tomato* Race 1 affecting tomato in different Regions of Chile. *Plant Dis.* **2021**, *106*, 1979. [CrossRef] [PubMed]
91. Saavedra Del, R.G.; Escaff, G.M.; Cortacáns, P.D.; Ruiz-Tagle, C. Recent developments in processing tomato production in Chile. In Proceedings of the IX International Symposium on the Processing Tomato, Melbourne, Australia, 15–18 November 2004; Volume 724, pp. 335–338.
92. Sepúlveda-Chavera, G.; Huanca, W.; Salvatierra-Martínez, R.; Latorre, B.A. First report of *Fusarium oxysporum* f. sp. *lycopersici* race 3 and *F. oxysporum* f. sp. *radicis-lycopersici* in tomatoes in the Azapa Valley of Chile. *Plant Dis.* **2014**, *98*, 1432. [CrossRef] [PubMed]
93. Elizondo-Pasten, E.; Boix-Ruiz, A.; Gómez-Tenorio, M.A.; Ruiz-Olmos, C.; Marín-Guirao, J.I.; Tello-Marquina, J.C.; Camacho-Ferre, F. Two complementary techniques allow detection of *Fusarium oxysporum* f. sp. *radicis-lycopersici* in soils from two different tomato-cultivated areas of Chile. In Proceedings of the V International Symposium on Tomato Diseases: Perspectives and Future Directions in Tomato Protection, Málaga, Spain, 13–16 June 2016; Volume 1207, pp. 315–318.
94. Constable, F.; Chambers, G.; Penrose, L.; Daly, A.; Mackie, J.; Davis, K.; Gibbs, M. Viroid-infected tomato and capsicum seed shipments to Australia. *Viruses* **2019**, *11*, 98. [CrossRef]

95.  Australian Processing Tomato Grower. Tomatoes. Available online: https://www.accc.gov.au/system/files/public-registers/documents/D11%2B2315841.pdf (accessed on 30 November 2022).
96.  Plant Health Australia. Processing Tomatoes. Available online: https://www.planthealthaustralia.com.au/industries/processing-tomatoes/ (accessed on 30 November 2022).
97.  Australian Processing Tomato Research Council. Annual Industry Survey 2020. Available online: https://www.aptrc.asn.au/wp-content/uploads/2020/12/Industry-Survey-2020-ed.pdf (accessed on 30 November 2022).
98.  Lim, G.T.T.; Wang, G.P.; Hemming, M.N.; Basuki, S.; McGrath, D.J.; Carroll, B.J.; Jones, D.A. Mapping the *I-3* gene for resistance to Fusarium wilt in tomato: Application of an *I-3* marker in tomato improvement and progress towards the cloning of *I-3*. *Australas. Plant Pathol.* **2006**, *35*, 671–680. [CrossRef]
99.  Grattidge, R.; O'Brien, R.G. Occurrence of a third race of Fusarium wilt of tomatoes in Queensland. *Plant Dis.* **1982**, *66*, 165–166. [CrossRef]
100. Callaghan, S.E. Root and Collar Rot Pathogens Associated with Yield Decline of Processing Tomatoes in Victoria, Australia. Doctoral Dissertation, The University of Melbourne, Melbourne, Australia, 2020.
101. Callaghan, S.E.; Burgess, L.W.; Ades, P.K.; Tesoriero, L.A.; Taylor, P.W.J. Diversity and pathogenicity of *Pythium* species associated with reduced yields of processing tomatoes (*Solanum lycopersicum* ) in Victoria, Australia. *Plant Dis.* **2022**, *106*, 1632–1638. [CrossRef]
102. Watterson, J.C. Diseases. In *The Tomato Crop*, 1st ed.; Atherton, J.G., Rudich, J., Eds.; Springer: Dordrecht, Germany, 1986; pp. 443–484.
103. Ekengren, S.K. Cutting the Gordian knot: Taking a stab at corky root rot of tomato. *Plant Biotechnol.* **2008**, *25*, 265–269. [CrossRef]
104. Aragona, M.; Minio, A.; Ferrarini, A.; Valente, M.T.; Bagnaresi, P.; Orrù, L.; Delledonne, M. De novo genome assembly of the soil-borne fungus and tomato pathogen *Pyrenochaeta Lycopersici*. *BMC Genom.* **2014**, *15*, 1–12. [CrossRef]
105. Hasna, M.K.; Ögren, E.; Persson, P.; Mårtensson, A.; Rämert, B. Management of corky root disease of tomato in participation with organic tomato growers. *Crop Prot.* **2009**, *28*, 155–161. [CrossRef]
106. Shankar, R.; Harsha, S.; Bhandary, R. A Practical Guide to Identification and Control of Tomato Diseases. Available online: https://www.researchgate.net/file.PostFileLoader.html?id=589eb904615e2793034a4db2&assetKey=AS%3A460474400677895%401486797060834 (accessed on 30 November 2022).
107. Coelho, L.; Chellemi, D.O.; Mitchell, D.J. Efficacy of solarization and cabbage amendment for the control of *Phytophthora* spp. in North Florida. *Plant Dis.* **1999**, *83*, 293–299. [CrossRef] [PubMed]
108. Yücel, S.; Özarslandan, A.; Colak, A.; Ay, T.; Can, C. Effect of solarization and fumigant applications on soilborne pathogens and root-knot nematodes in greenhouse-grown tomato in Turkey. *Phytoparasitica* **2007**, *35*, 450–456. [CrossRef]
109. Manö, S.; Andreae, M.O. Emission of methyl bromide from biomass burning. *Science* **1994**, *263*, 1255–1257. [CrossRef] [PubMed]
110. Rosskopf, E.N.; Chellemi, D.O.; Kokalis-Burelle, N.; Church, G.T. Alternatives to methyl bromide: A Florida perspective. *Plant Health Prog.* **2005**, *6*, 19. [CrossRef]
111. Locascio, S.J.; Gilreath, J.P.; Dickson, D.W.; Kucharek, T.A.; Jones, J.P.; Noling, J.W. Fumigant alternatives to methyl bromide for polyethylene-mulched. *HortScience* **1997**, *32*, 1208–1211. [CrossRef]
112. Fiume, G.; Fiume, F. Biological control of corky root in tomato. *Commun. Agric. Appl. Biol. Sci.* **2008**, *73*, 233–248.
113. Doganlar, S.; Dodson, J.; Gabor, B.; Beck-Bunn, T.; Crossman, C.; Tanksley, S.D. Molecular mapping of the *py-1* gene for resistance to corky root rot (*Pyrenochaeta lycopersici*) in tomato. *Theor. Appl. Genet.* **1998**, *97*, 784–788. [CrossRef]
114. Milc, J.; Bagnaresi, P.; Aragona, M.; Valente, M.T.; Biselli, C.; Infantino, A.; Pecchioni, N. Comparative transcriptome profiling of the response to *Pyrenochaeta lycopersici* in resistant tomato cultivar Mogeor and its background genotype—Susceptible Moneymaker. *Funct. Integr. Genom.* **2019**, *19*, 811–826. [CrossRef]
115. Hasna, M.K.; Lagerlöf, J.; Rämert, B. Effects of fungivorous nematodes on corky root disease of tomato grown in compost-amended soil. *Acta Agric. Scand. Sect. B-Soil Plant Sci.* **2008**, *58*, 145–153. [CrossRef]
116. Bubici, G.; Marsico, A.D.; D'Amico, M.; Amenduni, M.; Cirulli, M. Evaluation of *Streptomyces* spp. for the biological control of corky root of tomato and *Verticillium* wilt of eggplant. *Appl. Soil Ecol.* **2013**, *72*, 128–134. [CrossRef]
117. Besoain, X.A.; Pérez, L.M.; Araya, A.; Lefever, L.; Montealegre, J.R. New strains obtained after UV treatment and protoplast fusion of native *Trichoderma harzianum*: Their biocontrol activity on *Pyrenochaeta lycopersici*. *Electron. J. Biotechnol.* **2007**, *10*, 604–617. [CrossRef]
118. Sánchez-Téllez, S.; Herrera-Cid, R.A.; Besoain-Canales, X.A.; Pérez-Roepke, L.M.; Montealegre-Andrade, J.R. In vitro and in vivo inhibitory effect of solid and liquid *Trichoderma harzianum* formulations on biocontrol of *Pyrenochaeta lycopersici*. *Interciencia* **2013**, *38*, 425–429.
119. Pérez, L.; Besoaín, X.; Reyes, M.; Pardo, G.; Montealegre, J. The expression of extracellular fungal cell wall hydrolytic enzymes in different *Trichoderma harzianum* isolates correlates with their ability to control *Pyrenochaeta Lycopersici*. *Biol. Res.* **2002**, *35*, 401–410. [CrossRef] [PubMed]
120. Hasna, M.K.; Mårtensson, A.; Persson, P.; Rämert, B. Use of composts to manage corky root disease in organic tomato production. *Ann. Appl. Biol.* **2007**, *151*, 381–390. [CrossRef]
121. Edel-Hermann, V.; Lecomte, C. Current status of Fusarium oxysporum formae speciales and races. *Phytopathology* **2019**, *109*, 512–530. [CrossRef] [PubMed]

122. Mutlu, N.; Demirelli, A.; Ilbi, H.; Ikten, C. Development of co-dominant SCAR markers linked to resistant gene against the *Fusarium oxysporum* f. sp. *radicis-lycopersici*. *Theor. Appl. Genet.* **2015**, *128*, 1791–1798. [CrossRef]
123. Baysal, Ö.; Çalışkan, M.; Yeşilova, Ö. An inhibitory effect of a new *Bacillus subtilis* strain (EU07) against *Fusarium oxysporum* f. sp. *radicis-lycopersici*. *Physiol. Mol. Plant Pathol.* **2008**, *73*, 25–32. [CrossRef]
124. Gordon, T.R. *Fusarium oxysporum* and the *Fusarium* wilt syndrome. *Annu. Rev. Phytopathol.* **2017**, *55*, 23–39. [CrossRef]
125. Gordon, T.R.; Okamoto, D. Population structure and the relationship between pathogenic and nonpathogenic strains of Fusarium oxysporum. *Phytopathology* **1992**, *82*, 73–77. [CrossRef]
126. Olivain, C.; Humbert, C.; Nahalkova, J.; Fatehi, J.; l'Haridon, F.; Alabouvette, C. Colonization of tomato root by pathogenic and nonpathogenic *Fusarium oxysporum* strains inoculated together and separately into the soil. *Appl. Environ. Microbiol.* **2006**, *72*, 1523–1531. [CrossRef]
127. Muslim, A.; Horinouchi, H.; Hyakumachi, M. Control of Fusarium crown and root rot of tomato with hypovirulent binucleate *Rhizoctonia* in soil and rock wool systems. *Plant Dis.* **2003**, *87*, 739–747. [CrossRef] [PubMed]
128. Altinok, H.H.; Yüksel, G.; Altinok, M.A. Pathogenicity and phylogenetic analysis of *Fusarium oxysporum* f. sp. *capsici* isolates from pepper in Turkey. *Can. J. Plant Pathol.* **2020**, *42*, 279–291. [CrossRef]
129. Zhang, S.; Roberts, P.D.; McGovern, R.J.; Datnoff, L.E. Fusarium Crown and Root Rot of Tomato in Florida. Institute of Food and Agricultural Sciences (IFAS), Publication PP52. Available online: https://edis.ifas.ufl.edu/publication/PG082 (accessed on 30 November 2022).
130. McGovern, R.J. Management of tomato diseases caused by Fusarium oxysporum. *Crop. Prot.* **2015**, *73*, 78–92. [CrossRef]
131. Jarvis, W.R. Epidemiology of *Fusarium oxysporum* f. sp. *radicis-lycopersici*. In *Vascular Wilt Diseases of Plants*, 1st ed.; Tjamos, E.C., Beckman, C.H., Eds.; Springer: Berlin, Germany, 1989; pp. 397–411.
132. Cao, X.; Guan, Z.; Vallad, G.E.; Wu, F. Economics of fumigation in tomato production: The impact of methyl bromide phase-out on the Florida tomato industry. *Int. Food Agribus. Manag. Rev.* **2019**, *22*, 589–600. [CrossRef]
133. Horinouchi, H.; Katsuyama, N.; Taguchi, Y.; Hyakumachi, M. Control of Fusarium crown and root rot of tomato in a soil system by combination of a plant growth-promoting fungus, *Fusarium equiseti*, and biodegradable pots. *Crop. Prot.* **2008**, *27*, 859–864. [CrossRef]
134. McGovern, R.J.; Vavrina, C.S.; Noling, J.W.; Datnoff, L.A.; Yonce, H.D. Evaluation of application methods of metam sodium for management of Fusarium crown and root rot in tomato in southwest Florida. *Plant Dis.* **1998**, *82*, 919–923. [CrossRef]
135. Minuto, A.; Gullino, M.L.; Lamberti, F.; D'addabbo, T.; Tescari, E.; Ajwa, H.; Garibaldi, A. Application of an emulsifiable mixture of 1, 3-dichloropropene and chloropicrin against root knot nematodes and soilborne fungi for greenhouse tomatoes in Italy. *Crop Prot.* **2006**, *25*, 1244–1252. [CrossRef]
136. Fazio, G.; Stevens, M.R.; Scott, J.W. Identification of RAPD markers linked to fusarium crown and root rot resistance (*Frl*) in tomato. *Euphytica* **1999**, *105*, 205–210. [CrossRef]
137. Sivan, A.; Ucko, O.; Chet, I. Biological control of Fusarium crown rot of tomato by *Trichoderma harzianum* under field conditions. *Plant Dis.* **1987**, *71*, 587–592. [CrossRef]
138. Datnoff, L.E.; Nemec, S.; Pernezny, K. Biological control of Fusarium crown and root rot of tomato in Florida using *Trichoderma harzianum* and *Glomus intraradices*. *Biol. Control* **1995**, *5*, 427–431. [CrossRef]
139. Malandrakis, A.; Daskalaki, E.R.; Skiada, V.; Papadopoulou, K.K.; Kavroulakis, N. A *Fusarium solani* endophyte vs. fungicides: Compatibility in a *Fusarium oxysporum* f. sp. *radicis-lycopersici*–tomato pathosystem. *Fungal Biol.* **2018**, *122*, 1215–1221. [CrossRef] [PubMed]
140. Benhamou, N.; Rey, P.; Chérif, M.; Hockenhull, J.; Tirilly, Y. Treatment with the mycoparasite *Pythium oligandrum* triggers induction of defense-related reactions in tomato roots when challenged with *Fusarium oxysporum* f. sp. *radicis-lycopersici*. *Phytopathology* **1997**, *87*, 108–122. [CrossRef] [PubMed]
141. Duffy, B.K.; Défago, G. Zinc improves biocontrol of Fusarium crown and root rot of tomato by *Pseudomonas fluorescens* and represses the production of pathogen metabolites inhibitory to bacterial antibiotic biosynthesis. *Phytopathology* **1997**, *87*, 1250–1257. [CrossRef]
142. de Weert, S.; Kuiper, I.; Lagendijk, E.L.; Lamers, G.E.; Lugtenberg, B.J. Role of chemotaxis toward fusaric acid in colonization of hyphae of *Fusarium oxysporum* f. sp. *radicis-lycopersici* by Pseudomonas fluorescens WCS365. *Mol. Plant-Microbe Interact.* **2004**, *17*, 1185–1191. [CrossRef] [PubMed]
143. Kamilova, F.; Validov, S.; Lugtenberg, B. Biological control of tomato foot and root rot caused by *Fusarium oxysporum* f. sp. *radicis-lycopersici* by *Pseudomonas* bacteria. In Proceedings of the II International Symposium on Tomato Diseases, Kusadasi, Turkey, 8–12 October 2007; Volume 808, pp. 317–320.
144. Baysal, Ö.; Lai, D.; Xu, H.H.; Siragusa, M.; Çalışkan, M.; Carimi, F.; Tör, M. A proteomic approach provides new insights into the control of soil-borne plant pathogens by *Bacillus* species. *PLoS ONE* **2013**, *8*, e53182. [CrossRef]
145. Kavroulakis, N.; Ehaliotis, C.; Ntougias, S.; Zervakis, G.I.; Papadopoulou, K.K. Local and systemic resistance against fungal pathogens of tomato plants elicited by a compost derived from agricultural residues. *Physiol. Mol. Plant Pathol.* **2005**, *66*, 163–174. [CrossRef]
146. Ajilogba, C.F.; Babalola, O.O. Integrated management strategies for tomato *Fusarium* wilt. *Biocontrol Sci.* **2013**, *18*, 117–127. [CrossRef]

147. Miller, S.A.; Rowe, R.C.; Riedel, R.M. Fusarium and Verticillium Wilts of Tomato, Potato, Pepper, and Eggplant. The Ohio State University Extension. Available online: https://www.cabdirect.org/cabdirect/abstract/20127800677 (accessed on 30 November 2022).
148. Larkin, R.P.; Fravel, D.R. Effects of varying environmental conditions on biological control of Fusarium wilt of tomato by nonpathogenic *Fusarium* spp. *Phytopathology* **2002**, *92*, 1160–1166. [CrossRef]
149. Yu, J.; Land, C.J.; Vallad, G.E.; Boyd, N.S. Tomato tolerance and pest control following fumigation with different ratios of dimethyl disulfide and chloropicrin. *Pest Manag. Sci.* **2019**, *75*, 1416–1424. [CrossRef]
150. Bohn, G.W.; Tucker, C.M. Immunity to *Fusarium* wilt in the tomato. *Science* **1939**, *89*, 603–604. [CrossRef]
151. Takken, F.; Rep, M. The arms race between tomato and *Fusarium oxysporum*. *Mol. Plant Pathol.* **2010**, *11*, 309–314. [CrossRef] [PubMed]
152. Sela-Buurlage, M.; Budai-Hadrian, O.; Pan, Q.; Carmel-Goren, L.; Vunsch, R.; Zamir, D.; Fluhr, R. Genome-wide dissection of *Fusarium* resistance in tomato reveals multiple complex loci. *Mol. Genet. Genom.* **2001**, *265*, 1104–1111. [CrossRef] [PubMed]
153. Katan, T.; Shlevin, E.; Katan, J. Sporulation of *Fusarium oxysporum* f. sp. *lycopersici* on stem surfaces of tomato plants and aerial dissemination of inoculum. *Phytopathology* **1997**, *87*, 712–719. [CrossRef] [PubMed]
154. Sarfatti, M.; Katan, J.; Fluhr, R.; Zamir, D. An RFLP marker in tomato linked to the *Fusarium oxysporum* resistance gene *I2*. *Theor. Appl. Genet.* **1989**, *78*, 755–759. [CrossRef] [PubMed]
155. McGrath, D.J.; Gillespie, D.; Vawdrey, L. Inheritance of resistance to *Fusarium oxysporum* f. sp. *lycopersici* races 2 and 3 in *Lycopersicon pennellii*. *Aust. J. Agric. Res.* **1987**, *38*, 729–733. [CrossRef]
156. Scott, J.W.; Jones, J.P. Monogenic resistance in tomato to *Fusarium oxysporum* f. sp. lycopersici race 3. *Euphytica* **1989**, *40*, 49–53. [CrossRef]
157. Catanzariti, A.M.; Lim, G.T.; Jones, D.A. The tomato I-3 gene: A novel gene for resistance to *Fusarium* wilt disease. *New Phytol.* **2015**, *207*, 106–118. [CrossRef]
158. Chitwood-Brown, J.; Vallad, G.E.; Lee, T.G.; Hutton, S.F. Breeding for resistance to *Fusarium* wilt of tomato: A review. *Genes* **2021**, *12*, 1673. [CrossRef]
159. Raza, W.; Ling, N.; Zhang, R.; Huang, Q.; Xu, Y.; Shen, Q. Success evaluation of the biological control of *Fusarium* wilts of cucumber, banana, and tomato since 2000 and future research strategies. *Crit. Rev. Biotechnol.* **2017**, *37*, 202–212. [CrossRef]
160. Alabouvette, C.; Lemanceau, P.; Steinberg, C. Recent advances in the biological control of *Fusarium* wilts. *Pestic. Sci.* **1993**, *37*, 365–373. [CrossRef]
161. Aimé, S.; Cordier, C.; Alabouvette, C.; Olivain, C. Comparative analysis of PR gene expression in tomato inoculated with virulent *Fusarium oxysporum* f. sp. *lycopersici* and the biocontrol strain *F. oxysporum* Fo47. *Physiol. Mol. Plant Pathol.* **2008**, *73*, 9–15. [CrossRef]
162. de Lamo, F.J.; Spijkers, S.B.; Takken, F.L. Protection to tomato wilt disease conferred by the nonpathogen *Fusarium oxysporum* Fo47 is more effective than that conferred by avirulent strains. *Phytopathology* **2021**, *111*, 253–257. [CrossRef] [PubMed]
163. Duijff, B.J.; Pouhair, D.; Olivain, C.; Alabouvette, C.; Lemanceau, P. Implication of systemic induced resistance in the suppression of *fusarium* wilt of tomato by *Pseudomonas fluorescens* WCS417r and by nonpathogenic *Fusarium oxysporum* Fo47. *Eur. J. Plant Pathol.* **1998**, *104*, 903–910. [CrossRef]
164. Borrego-Benjumea, A.; Basallote-Ureba, M.J.; Abbasi, P.A.; Lazarovits, G.; Melero-Vara, J.M. Effects of incubation temperature on the organic amendment-mediated control of Fusarium wilt of tomato. *Ann. Appl. Biol.* **2014**, *164*, 453–463. [CrossRef]
165. Zhao, F.; Zhang, Y.; Dong, W.; Zhang, Y.; Zhang, G.; Sun, Z.; Yang, L. Vermicompost can suppress *Fusarium oxysporum* f. sp. *lycopersici* via generation of beneficial bacteria in a long-term tomato monoculture soil. *Plant Soil* **2019**, *440*, 491–505. [CrossRef]
166. Barakat, R.M.; Al-Masri, M.I. *Trichoderma harzianum* in combination with sheep manure amendment enhances soil suppressiveness of *Fusarium* wilt of tomato. *Phytopathol. Mediterr.* **2009**, *48*, 385–395.
167. Mwangi, M.W.; Muiru, W.M.; Narla, R.D.; Kimenju, J.W.; Kariuki, G.M. Effect of soil sterilisation on biological control of *Fusarium oxysporum* f. sp. *lycopersici* and *Meloidogyne javanica* by antagonistic fungi and organic amendment in tomato crop. *Acta Agric. Scand. Sect. B—Soil Plant Sci.* **2018**, *68*, 656–661. [CrossRef]
168. Lamour, K.H.; Stam, R.; Jupe, J.; Huitema, E. The oomycete broad-host-range pathogen *Phytophthora capsici*. *Mol. Plant Pathol.* **2012**, *13*, 329–337. [CrossRef]
169. Wani, A.H. An overview of the fungal rot of tomato. *Mycopath* **2011**, *9*, 33–38.
170. Hausbeck, M.K.; Lamour, K.H. *Phytophthora capsici* on vegetable crops: Research progress and management challenges. *Plant Dis.* **2004**, *88*, 1292–1303. [CrossRef]
171. Ristaino, J.B.; Johnston, S.A. Ecologically based approaches to management of Phytophthora blight on bell pepper. *Plant Dis.* **1999**, *83*, 1080–1089. [CrossRef] [PubMed]
172. Lamour, K.H.; Hausbeck, M.K. Effect of crop rotation on the survival of *Phytophthora capsici* in Michigan. *Plant Dis.* **2003**, *87*, 841–845. [CrossRef] [PubMed]
173. Quesada-Ocampo, L.M.; Hausbeck, M.K. Resistance in tomato and wild relatives to crown and root rot caused by *Phytophthora capsici*. *Phytopathology* **2010**, *100*, 619–627. [CrossRef] [PubMed]
174. Bower, L.A.; Coffey, M.D. Development of laboratory tolerance to phosphorous acid, fosetyl-Al, and metalaxyl in *Phytophthora capsici*. *Can. J. Plant Pathol.* **1985**, *7*, 1–6. [CrossRef]

175. Jackson, K.L.; Yin, J.; Ji, P. Sensitivity of *Phytophthora capsici* on vegetable crops in Georgia to mandipropamid, dimethomorph, and cyazofamid. *Plant Dis.* **2012**, *96*, 1337–1342. [CrossRef] [PubMed]

176. Kousik, C.S.; Keinath, A.P. First report of insensitivity to cyazofamid among isolates of *Phytophthora capsici* from the southeastern United States. *Plant Dis.* **2008**, *92*, 979. [CrossRef] [PubMed]

177. Siegenthaler, T.B.; Hansen, Z.R. Sensitivity of *Phytophthora capsici* from Tennessee to mefenoxam, fluopicolide, oxathiapiprolin, dimethomorph, mandipropamid, and cyazofamid. *Plant Dis.* **2021**, *105*, 3000–3007. [CrossRef]

178. Quesada-Ocampo, L.M.; Vargas, A.M.; Naegele, R.P.; Francis, D.M.; Hausbeck, M.K. Resistance to crown and root rot caused by *Phytophthora capsici* in a tomato advanced backcross of *Solanum habrochaites* and *Solanum lycopersicum*. *Plant Dis.* **2016**, *100*, 829–835. [CrossRef]

179. Moataza, M.S. Destruction of *Rhizoctonia solani* and *Phytophthora capsici* causing tomato root-rot by *Pseudomonas fluorescences* lytic enzymes. *Res. J. Agric. Biol. Sci.* **2006**, *2*, 274–281.

180. Sharma, R.; Chauhan, A.; Shirkot, C.K. Characterization of plant growth promoting *Bacillus* strains and their potential as crop protectants against *Phytophthora capsici* in tomato. *Biol. Agric. Hortic.* **2015**, *31*, 230–244. [CrossRef]

181. Syed-Ab-Rahman, S.F.; Xiao, Y.; Carvalhais, L.C.; Ferguson, B.J.; Schenk, P.M. Suppression of *Phytophthora apsica* infection and promotion of tomato growth by soil bacteria. *Rhizosphere* **2019**, *9*, 72–75. [CrossRef]

182. Yang, K.; Dong, X.; Li, J.; Wang, Y.; Cheng, Y.; Zhai, Y.; Dou, D. Type 2 Nep1-like proteins from the biocontrol oomycete *Pythium oligandrum* suppress *Phytophthora capsici* infection in solanaceous plants. *J. Fungi* **2021**, *7*, 496. [CrossRef] [PubMed]

183. Nicol, R.W.; Burlakoti, P. Effect of aerobic compost tea inputs and application methods on protecting tomato from *Phytophthora capsici*. In Proceedings of the IV International Symposium on Tomato Diseases, Orlando, FL, USA, 24–27 June 2013; Volume 1069, pp. 229–233.

184. González-Hernández, A.I.; Suárez-Fernández, M.B.; Pérez-Sánchez, R.; Gómez-Sánchez M, Á.; Morales-Corts, M.R. Compost tea induces growth and resistance against *Rhizoctonia solani* and *Phytophthora capsici* in pepper. *Agronomy* **2021**, *11*, 781. [CrossRef]

185. Jiang, Z.Q.; Guo, Y.H.; Li, S.M.; Qi, H.Y.; Guo, J.H. Evaluation of biocontrol efficiency of different *Bacillus preparations* and field application methods against *Phytophthora* blight of bell pepper. *Biol. Control* **2006**, *36*, 216–223. [CrossRef]

186. Kim, K.D.; Nemec, S.; Musson, G. Control of *Phytophthora* root and crown rot of bell pepper with composts and soil amendments in the greenhouse. *Appl. Soil Ecol.* **1997**, *5*, 169–179. [CrossRef]

187. Raaijmakers, J.M.; Paulitz, T.C.; Steinberg, C.; Alabouvette, C.; Moënne-Loccoz, Y. The rhizosphere: A playground and battlefield for soilborne pathogens and beneficial microorganisms. *Plant Soil* **2009**, *321*, 341–361. [CrossRef]

188. Elshahawy, I.E.; El-Mohamedy, R.S. Biological control of Pythium damping-off and root-rot diseases of tomato using *Trichoderma* isolates employed alone or in combination. *J. Plant Pathol.* **2019**, *101*, 597–608. [CrossRef]

189. Manjunath, M.; Prasanna, R.; Nain, L.; Dureja, P.; Singh, R.; Kumar, A.; Kaushik, B.D. Biocontrol potential of cyanobacterial metabolites against damping off disease caused by *Pythium aphanidermatum* in solanaceous vegetables. *Arch. Phytopathol. Plant Prot.* **2010**, *43*, 666–677. [CrossRef]

190. Mdee, L.K.; Masoko, P.; Eloff, J.N. The activity of extracts of seven common invasive plant species on fungal phytopathogens. *S. Afr. J. Bot.* **2009**, *75*, 375–379. [CrossRef]

191. Beckerman, J. Pythium Root Rot of Herbaceous Plants. Available online: https://bch.cbd.int/api/v2013/documents/7efc7c33-9c4d-4589-95b0-959650a561fa/attachments/20505/Pythium%20Root%20Rot%20of%20herbaceous%20plants.pdf (accessed on 30 November 2022).

192. Arora, H.; Sharma, A.; Sharma, S.; Haron, F.F.; Gafur, A.; Sayyed, R.Z.; Datta, R. Pythium damping-off and root rot of capsicum annuum l.: Impacts, diagnosis, and management. *Microorganisms* **2021**, *9*, 823. [CrossRef]

193. Male, M.F.; Vawdrey, L.L. Efficacy of fungicides against damping-off in papaya seedlings caused by *Pythium Aphanidermatum*. *Australas. Plant Dis. Notes* **2010**, *5*, 103–104. [CrossRef]

194. Paulitz, T.C.; Zhou, T.; Rankin, L. Selection of rhizosphere bacteria for biological control of *Pythium aphanidermatum* on hydroponically grown cucumber. *Biol. Control* **1992**, *2*, 226–237. [CrossRef]

195. Hassanisaadi, M.; Shahidi Bonjar, G.H.; Hosseinipour, A.; Abdolshahi, R.; Ait Barka, E.; Saadoun, I. Biological control of *Pythium aphanidermatum*, the causal agent of tomato root rot by two Streptomyces root symbionts. *Agronomy* **2021**, *11*, 846. [CrossRef]

196. Harvey, P.; Lawrence, L. Managing Pythium root disease complexes to improve productivity of crop rotations. *Outlooks Pest Manag.* **2008**, *19*, 127. [CrossRef]

197. Triki, M.A.; Priou, S.; El Mahjoub, M. Effects of soil solarization on soil-borne populations of *Pythium aphanidermatum* and *Fusarium solani* and on the potato crop in Tunisia. *Potato Res.* **2001**, *44*, 271–279. [CrossRef]

198. Reddy, G.S.; Rao, V.K.; Sitaramaiah, K.; Chalam, T.V. Soil Solarization for Control of Soil-borne Pathogen Complex due to *Meloidogyne incognita* and *Pythium aphanidermatum*. *Indian J. Nematol.* **2001**, *31*, 136–138.

199. Jayaraj, J.; Radhakrishnan, N.V. Enhanced activity of introduced biocontrol agents in solarized soils and its implications on the integrated control of tomato damping-off caused by *Pythium* spp. *Plant Soil* **2008**, *304*, 189–197. [CrossRef]

200. Abbasi, P.A.; Lazarovits, G. Seed treatment with phosphonate (AG3) suppresses Pythium damping-off of cucumber seedlings. *Plant Dis.* **2006**, *90*, 459–464. [CrossRef]

201. Al-Balushi, Z.M.; Agrama, H.; Al-Mahmooli, I.H.; Maharachchikumbura, S.S.; Al-Sadi, A.M. Development of resistance to hymexazol among *Pythium* species in cucumber greenhouses in Oman. *Plant Dis.* **2018**, *102*, 202–208. [CrossRef]

202. Cuevas, V.C.; Sinohin, A.M. Performance of selected Philippine species of *Trichoderma* as biocontrol agents of damping off pathogens and as growth enhancer of vegetables in farmers' field. *Philipp. Agric. Sci.* **2005**, *88*, 63–71.

203. Garzón, C.D.; Molineros, J.E.; Yánez, J.M.; Flores, F.J.; del Mar Jiménez-Gasco, M.; Moorman, G.W. Sublethal doses of mefenoxam enhance Pythium damping-off of geranium. *Plant Dis.* **2011**, *95*, 1233–1238. [CrossRef] [PubMed]

204. Samaras, A.; Roumeliotis, E.; Ntasiou, P.; Karaoglanidis, G. *Bacillus subtilis* MBI600 promotes growth of tomato plants and induces systemic resistance contributing to the control of soilborne pathogens. *Plants* **2021**, *10*, 1113. [CrossRef] [PubMed]

205. Rajendraprasad, M.; Vidyasagar, B.; Devi, G.U.; Rao, S.K. Biological control of tomato damping off caused by *Pythium debaryanum*. *Int. J. Chem. Stud.* **2017**, *5*, 447–452.

206. Salman, M.; Abuamsha, R. Potential for integrated biological and chemical control of damping-off disease caused by *Pythium ultimum* in tomato. *BioControl* **2012**, *57*, 711–718. [CrossRef]

207. Dukare, A.S.; Prasanna, R.; Dubey, S.C.; Nain, L.; Chaudhary, V.; Singh, R.; Saxena, A.K. Evaluating novel microbe amended composts as biocontrol agents in tomato. *Crop. Prot.* **2011**, *30*, 436–442. [CrossRef]

208. Porter, L.D.; Hamm, P.B.; David, N.L.; Gieck, S.L.; Miller, J.S.; Gundersen, B.; Inglis, D.A. Metalaxyl-M-resistant *Pythium* species in potato production areas of the Pacific Northwest of the USA. *Am. J. Potato Res.* **2009**, *86*, 315–326. [CrossRef]

209. Del Castillo Múnera, J.; Hausbeck, M.K. Characterization of *Pythium* species associated with greenhouse floriculture crops in Michigan. *Plant Dis.* **2016**, *100*, 569–576. [CrossRef]

210. Aegerter, B.J.; Greathead, A.S.; Pierce, L.E.; Davis, R.M. Mefenoxam-resistant isolates of *Pythium irregulare* in an ornamental greenhouse in California. *Plant Dis.* **2002**, *86*, 692. [CrossRef]

211. Lee, S.; Garzón, C.D.; Moorman, G.W. Genetic structure and distribution of *Pythium aphanidermatum* populations in Pennsylvania greenhouses based on analysis of AFLP and SSR markers. *Mycologia* **2010**, *102*, 774–784. [CrossRef]

212. Lookabaugh, E.C.; Kerns, J.P.; Shew, B.B. Evaluating Fungicide Selections to Manage Pythium Root Rot on Poinsettia Cultivars with Varying Levels of Partial Resistance. *Plant Dis.* **2021**, *105*, 1640–1647. [CrossRef]

213. Niderman, T.; Genetet, I.; Bruyere, T.; Gees, R.; Stintzi, A.; Legrand, M.; Mosinger, E. Pathogenesis-related PR-1 proteins are antifungal (isolation and characterization of three 14-kilodalton proteins of tomato and of a basic PR-1 of tobacco with inhibitory activity against *Phytophthora infestans*). *Plant Physiol.* **1995**, *108*, 17–27. [CrossRef] [PubMed]

214. Tornero, P.; Gadea, J.; Conejero, V.; Vera, P. Two *PR-1* genes from tomato are differentially regulated and reveal a novel mode of expression for a pathogenesis-related gene during the hypersensitive response and development. *Mol. Plant-Microbe Interact.* **1997**, *10*, 624–634. [CrossRef] [PubMed]

215. Postma, J.; Clematis, F.; Nijhuis, E.H.; Someus, E. Efficacy of four phosphate-mobilizing bacteria applied with an animal bone charcoal formulation in controlling *Pythium aphanidermatum* and *Fusarium oxysporum* f. sp. *radicis lycopersici* in tomato. *Biol. Control* **2013**, *67*, 284–291. [CrossRef]

216. Martinez, C.; Bourassa, A.; Roy, G.; Desbiens, M.C.; Bussières, P. Efficacy of PRO-MIX® with Biofungicide against Root Diseases caused by *Pythium* spp. and *Rhizoctonia* spp. In Proceedings of the XXVII International Horticultural Congress-IHC2006: International Symposium on Sustainability through Integrated and Organic, Seoul, Republic of Korea, 13–19 August 2006; Volume 767, pp. 185–192.

217. Caron, J.; Laverdière, L.; Thibodeau, P.O.; Bélanger, R.R. Utilisation d'une souche indigène de *Trichoderma harzianum* contre cinq agents pathogènes chez le concombre et la tomate de serre au Québec. *Phytoprotection* **2002**, *83*, 73–87. [CrossRef]

218. Larsen, J.; Graham, J.H.; Cubero, J.; Ravnskov, S. Biocontrol traits of plant growth suppressive arbuscular mycorrhizal fungi against root rot in tomato caused by *Pythium aphanidermatum*. *Eur. J. Plant Pathol.* **2012**, *133*, 361–369. [CrossRef]

219. St Martin, C.C.G.; Dorinvil, W.; Brathwaite, R.A.I.; Ramsubhag, A. Effects and relationships of compost type, aeration and brewing time on compost tea properties, efficacy against *Pythium ultimum*, phytotoxicity and potential as a nutrient amendment for seedling production. *Biol. Agric. Hortic.* **2012**, *28*, 185–205. [CrossRef]

220. Jenana RK, B.; Haouala, R.; Triki, M.A.; Godon, J.J.; Hibar, K.; Khedher, M.B.; Henchi, B. Composts, compost extracts and bacterial suppressive action on *Pythium aphanidermatum* in tomato. *Pak. J. Bot.* **2009**, *41*, 315–327.

221. Postma, J.; Nijhuis, E.H. *Pseudomonas chlororaphis* and organic amendments controlling Pythium infection in tomato. *Eur. J. Plant Pathol.* **2019**, *154*, 91–107. [CrossRef]

222. Jayaraj, J.; Radhakrishnan, N.V.; Kannan, R.; Sakthivel, K.; Suganya, D.; Venkatesan, S.; Velazhahan, R. Development of new formulations of *Bacillus subtilis* for management of tomato damping-off caused by *Pythium aphanidermatum*. *Biocontrol Sci. Technol.* **2005**, *15*, 55–65. [CrossRef]

223. Fradin, E.F.; Zhang, Z.; Juarez Ayala, J.C.; Castroverde, C.D.; Nazar, R.N.; Robb, J.; Thomma, B.P. Genetic dissection of *Verticillium* wilt resistance mediated by tomato *Ve1*. *Plant Physiol.* **2009**, *150*, 320–332. [CrossRef]

224. Fradin, E.F.; Thomma, B.P. Physiology and molecular aspects of *Verticillium* wilt diseases caused by *V. dahliae* and *V. albo-atrum*. *Mol. Plant Pathol.* **2006**, *7*, 71–86. [CrossRef] [PubMed]

225. Mandelc, S.; Timperman, I.; Radišek, S.; Devreese, B.; Samyn, B.; Javornik, B. Comparative proteomic profiling in compatible and incompatible interactions between hop roots and *Verticillium albo-atrum*. *Plant Physiol. Biochem.* **2013**, *68*, 23–31. [CrossRef] [PubMed]

226. Berlanger, I.; Powelson, M.L. Verticillium Wilt. Available online: https://www.apsnet.org/edcenter/disandpath/fungalasco/pdlessons/Pages/VerticilliumWilt.aspx (accessed on 30 November 2022).

227. Butterfield, E.J.; DeVay, J.E.; Garber, R.H. The influence of several crop sequences on the incidence of Verticillium wilt of cotton and on the populations of *Verticillium dahliae* in field soil. *Phytopathology* **1978**, *68*, 1217–1220. [CrossRef]
228. Iott, M.C. Utility of Grafting and Evaluation of Rootstocks for the Management of *Verticillium* Wilt in Tomato Production in Western North Carolina. Master's Thesis, North Carolina State University, Raleigh, NC, USA, 2013.
229. Stapleton, J.J. Soil solarization in various agricultural production systems. *Crop. Prot.* **2000**, *19*, 837–841. [CrossRef]
230. Goicoechea, N. To what extent are soil amendments useful to control *Verticillium* wilt. *Pest Manag. Sci. Former. Pestic. Sci.* **2009**, *65*, 831–839. [CrossRef]
231. Rowe, R.C.; Powelson, M.L. Potato early dying: Management challenges in a changing production environment. *Plant Dis.* **2002**, *86*, 1184–1193. [CrossRef]
232. Ordentlich, A.; Nachmias, A.; Chet, I. Integrated control of *Verticillium dahliae* in potato by *Trichoderma harzianum* and captan. *Crop. Prot.* **1990**, *9*, 363–366. [CrossRef]
233. Baergen, K.D.; Hewitt, J.D.; Clair, D.S. Resistance of tomato genotypes to four isolates of *Verticillium dahliae* race 2. *HortScience* **1993**, *28*, 833–836. [CrossRef]
234. Dhouib, H.; Zouari, I.; Abdallah, D.B.; Belbahri, L.; Taktak, W.; Triki, M.A.; Tounsi, S. Potential of a novel endophytic *Bacillus velezensis* in tomato growth promotion and protection against Verticillium wilt disease. *Biol. Control* **2019**, *139*, 104092. [CrossRef]
235. Elshafie, H.S.; Sakr, S.; Bufo, S.A.; Camele, I. An attempt of biocontrol the tomato-wilt disease caused by *Verticillium dahliae* using *Burkholderia gladioli* pv. *agaricicola* and its bioactive secondary metabolites. *Int. J. Plant Biol.* **2017**, *8*, 7263. [CrossRef]
236. Jabnoun-Khiareddine, H.; Daami-Remadi, M.; Ayed, F.; El Mahjoub, M. Biocontrol of tomato Verticillium wilt by using indigenous *Gliocladium* spp. and *Penicillium* sp. isolates. *Dyn. Soil Dyn. Plant* **2009**, *3*, 70–79.
237. Larena, I.; Sabuquillo, P.; Melgarejo, P.; De Cal, A. Biocontrol of *Fusarium* and *Verticillium* wilt of tomato by *Penicillium oxalicum* under greenhouse and field conditions. *J. Phytopathol.* **2003**, *151*, 507–512. [CrossRef]
238. Jabnoun-Khiareddine, H.; Daami-Remadi, M.; Ayed, F.; El Mahjoub, M. Biological control of tomato *Verticillium* wilt by using indigenous *Trichoderma* spp. *Afr. J. Plant Sci. Biotechnol.* **2009**, *3*, 26–36.
239. Naraghi, L.; Heydari, A.; Rezaee, S.; Razavi, M.; Jahanifar, H.; Khaledi, E. Biological control of tomato Verticillium wilt disease by *Talaromyces flavus*. *J. Plant Prot. Res.* **2010**, *50*, 360–365. [CrossRef]
240. Puri, K.D.; Hu, X.; Gurung, S.; Short, D.P.; Sandoya, G.V.; Schild, M.; Subbarao, K.V. *Verticillium klebahnii* and *V. isaacii* Isolates Exhibit Host-dependent Biological Control of *Verticillium* Wilt Caused by *V. dahliae*. *PhytoFrontiers* **2021**, *1*, 276–290. [CrossRef]
241. Acharya, B.; Ingram, T.W.; Oh, Y.; Adhikari, T.B.; Dean, R.A.; Louws, F.J. Opportunities and challenges in studies of host-pathogen interactions and management of *Verticillium dahliae* in tomatoes. *Plants* **2020**, *9*, 1622. [CrossRef]
242. Kadoglidou, K.; Chatzopoulou, P.; Maloupa, E.; Kalaitzidis, A.; Ghoghoberidze, S.; Katsantonis, D. Mentha and oregano soil amendment induces enhancement of tomato tolerance against soilborne diseases, yield and quality. *Agronomy* **2020**, *10*, 406. [CrossRef]
243. Ait Rahou, Y.; Ait-El-Mokhtar, M.; Anli, M.; Boutasknit, A.; Ben-Laouane, R.; Douira, A.; Meddich, A. Use of mycorrhizal fungi and compost for improving the growth and yield of tomato and its resistance to *Verticillium dahliae*. *Arch. Phytopathol. Plant Prot.* **2021**, *54*, 665–690. [CrossRef]
244. Lazarovits, G.; Conn, K.; Tenuta, M. Control of *Verticillium dahliae* with soil amendments: Efficacy and mode of action. In *Advances in Verticillium Research and Disease Management*, 1st ed.; Tjamos, E.C., Rowe, R.C., Heale, J.B., Fravel, D.R., Eds.; APS Press: St Paul, MN, USA, 2000; pp. 274–291.
245. Yuce, E.K.; Yigit, S.; Tosun, N. Efficacy of solarization combined with metam sodium and hydrogen peroxide in control of *Fusarium oxysporum* f. sp. *radicis-lycopersici* and *Clavibacter michiganensis* subsp. *michiganensis* in Tomato greenhouse. In Proceedings of the III International Symposium on Tomato Diseases, Ischia, Italy, 25–30 July 2010; Volume 914, pp. 385–391.
246. Kulus, D. Genetic resources and selected conservation methods of tomato. *J. Appl. Bot. Food Qual.* **2018**, *91*, 135–144.
247. Fones, H.N.; Bebber, D.P.; Chaloner, T.M.; Kay, W.T.; Steinberg, G.; Gurr, S.J. Threats to global food security from emerging fungal and oomycete crop pathogens. *Nat. Food* **2020**, *1*, 332–342. [CrossRef]
248. Bebber, D.P.; Field, E.; Gui, H.; Mortimer, P.; Holmes, T.; Gurr, S.J. Many unreported crop pests and pathogens are probably already present. *Glob. Chang. Biol.* **2019**, *25*, 2703–2713. [CrossRef] [PubMed]
249. Cammarano, D.; Jamshidi, S.; Hoogenboom, G.; Ruane, A.C.; Niyogi, D.; Ronga, D. Processing tomato production is expected to decrease by 2050 due to the projected increase in temperature. *Nat. Food* **2022**, *3*, 347–444. [CrossRef]

**Disclaimer/Publisher's Note:** The statements, opinions and data contained in all publications are solely those of the individual author(s) and contributor(s) and not of MDPI and/or the editor(s). MDPI and/or the editor(s) disclaim responsibility for any injury to people or property resulting from any ideas, methods, instructions or products referred to in the content.

*microorganisms*

[MDPI]

*Article*

# Optimization of the Fermentation Conditions of *Metarhizium robertsii* and Its Biological Control of Wolfberry Root Rot Disease

Jing He *, Xiaoyan Zhang, Qinghua Wang, Nan Li, Dedong Ding and Bin Wang

College of Forestry, Gansu Agricultural University, Lanzhou 730070, China; zxy1039507106@163.com (X.Z.); wqh1231212@163.com (Q.W.); lilinannan0709@163.com (N.L.); ddd3202022@163.com (D.D.); wangbin@gsau.edu.cn (B.W.)
* Correspondence: hejing268@aliyun.com

**Citation:** He, J.; Zhang, X.; Wang, Q.; Li, N.; Ding, D.; Wang, B. Optimization of the Fermentation Conditions of *Metarhizium robertsii* and Its Biological Control of Wolfberry Root Rot Disease. *Microorganisms* **2023**, *11*, 2380. https://doi.org/10.3390/microorganisms11102380

Academic Editor: Michael J. Bidochka

Received: 5 September 2023
Revised: 21 September 2023
Accepted: 22 September 2023
Published: 23 September 2023

**Copyright:** © 2023 by the authors. Licensee MDPI, Basel, Switzerland. This article is an open access article distributed under the terms and conditions of the Creative Commons Attribution (CC BY) license (https://creativecommons.org/licenses/by/4.0/).

**Abstract:** *Fusarium solani* is the main pathogenic fungus causing the root rot of wolfberry (*Lycium barbarum*). The endophytic fungus *Metarhizium robertsii* has been widely used for the biocontrol of plant pathogenic fungi, but the biocontrol effects of this fungus on wolfberry root rot and its antifungal mechanism against *F. solani* have not been reported. In this study, the antagonism of endophytic fungus *M. robertsii* against *F. solani* was verified. Further, we optimized the fermentation conditions of *M. robertsii* fermentation broth based on the inhibition rate of *F. solani*. In addition, the effects of *M. robertsii* fermentation broth on the root rot of wolfberry and its partial inhibition mechanism were investigated. The results showed that *M. robertsii* exhibited good antagonism against *F. solani*. Glucose and beef extracts were the optimal carbon and nitrogen sources for the fermentation of *M. robertsii*. Under the conditions of 29 °C, 190 rpm, and pH 7.0, the fermentation broth of *M. robertsii* had the best inhibition effect on *F. solani*. Furthermore, the fermentation broth treatment decreased the activities of superoxide dismutase, catalase, and peroxidase of *F. solani*; promoted the accumulation of malondialdehyde; and accelerated the leakage of soluble protein and the decrease in soluble sugar. In addition, inoculation with *M. robertsii* significantly reduced the decay incidence and disease index of wolfberry root rot caused by *F. solani*. These results indicate that *M. robertsii* could be used as a biological control agent in wolfberry root rot disease management.

**Keywords:** endophytic fungus; *Lycium barbarum*; disease control; *Fusarium solani*; antifungal mechanism

## 1. Introduction

As a medicinal and functional food, wolfberry (*Lycium barbarum*) has a long history of planting and cultivation, and it is widely planted in the Nei Monggol, Gansu, Ningxia, Shaanxi, and Qinghai provinces in China [1]. This plant has a high nutritional value and contains a variety of bioactive compounds such as polysaccharides, minerals, carotenoids, and polyphenols, for which their various effects include anticancer, antiaging, and hypoglycemic [2]. However, wolfberry plants are susceptible to phytopathogenic fungi, resulting in decreased fruit quality and yield. Among them, root rot caused by *Fusarium solani* is one of the major soil-borne diseases. The disease occurs in a wide range, spreads rapidly, and is highly destructive. It can cause the yellowing of plant leaves and the shrinking of branches, resulting in a decline in the quality of wolfberry fruit and a decrease in yield. In severe cases, it can lead to the death of the entire plant, causing great economic losses to local wolfberry growers [3]. Therefore, the effective prevention and control of root rot disease are of great significance for the healthy development of the wolfberry industry.

Biological control agents have been widely used to control plant root rot due to their advantages of safety, high efficiency, and low cost. *Trichoderma harzianum*, *Rhizobium japonicum*, and *T. atroviridae* treatments significantly reduce the incidence and severity of peanut and soybean root rot caused by *F. solani*, and they also show good plant growth promotion effects [4–6]. *Metarhizium robertsii* is a common entomopathogenic fungus and has been proven to be a plant endophytic fungus [7–9]; it has a significant and persistent pest control effect on its natural insect enemies. Metarhizium biological agents have been commercialized to some extent in the United States, Brazil, and Europe. The application of *M. robertsii* showed good biological control against banana stem weevil *Odoiporus longicollis* Oliver and European grapevine moth *Lobesia botrana* [10,11]. In addition, *Metarhizium robertsii* also showed good effects in plant disease control and significantly decreased the disease index of soybean root rot caused by *F. solani* [7]. Antioxidant systems play an important role in ROS scavenging, and superoxide dismutase (SOD), catalase (CAT), and peroxidase (POD) are important antioxidant enzymes in pathogenic fungi [12]. The inhibition of antioxidant enzyme activity may disrupt the balance of ROS metabolism, thus affecting the growth and pathogenicity of pathogenic fungi. Malondialdehyde (MDA) is one of the indicators used to measure the degree of oxidative stress, which can reflect the degree of fungal membrane lipid peroxidation. Soluble protein and soluble sugar could act as measures for the level of protein damage and cell carbon metabolism. Our previous study found that *M. robertsii* is also an endophytic fungus of wolfberry, but its control effect on the root rot of wolfberry has not been reported. In addition, whether its antifungal mechanism is related to the reduction in antioxidant enzymes and the destruction of cell membrane structure remains unclear.

The production of secondary metabolites with antifungal effects is the key to biological control. The production of these antifungal substances is not only related to the genetic characteristics of the fungus itself but is also influenced by the medium, nutrient composition, and fermentation conditions. The optimization of medium composition and fermentation conditions can significantly increase the production of antifungal secondary metabolites and enhance the inhibitory effect on pathogenic fungi [13]. In one study, corn steep liquor as a nitrogen source promoted the growth of *M. robertsii* blastospore and increased its virulence relative to corn leafhopper *Dalbulus maidis* [14]. The optimal fermentation conditions obtained using response surface methodology optimization significantly increased the inhibition of *Fulvia fulva* and *Botryosphaeria dothidea* by *Streptomyces lavendulae* fermentation broth [15]. Although *M. robertsii* has been reported to inhibit the root rot of soybeans, the relationship between this fungus's fermentation conditions and its biocontrol potential against *F. solani* is still unclear.

The aims of this study were to (1) analyze the effects of different carbon and nitrogen sources as well as fermentation conditions on the antifungal activity of *M. robertsii*; (2) optimize the fermentation conditions of *M. robertsii* using response surface methodology to increase the antifungal activity of the fermentation broth; and (3) investigate the biocontrol effect of *M. robertsii* on the root rot of wolfberry and possible antifungal mechanism against *F. solani*.

## 2. Materials and Methods

### 2.1. Fungal Strain, Culture Medium, and Wolfberry Plant

The isolation, screening, identification, and preservation of *M. robertsii* (HYC-7 strain) and *F. solani* were carried out at the Forest Protection Laboratory, College of Forestry, Gansu Agricultural University.

Basic fermentation medium: $NaNO_3$ (3 g), $KH_2PO_4$ (1 g), $MgSO_4$ (0.5 g), $KCl$ (0.5 g), $FeSO_4$ (0.01 g), sucrose (30 g), potato (200 g), and distilled water (up to 1 L).

One-year-old healthy wolfberry plants were collected from the economic forest teaching and research practice base of Gansu Agricultural University.

### 2.2. Determination of the Antagonistic Effect of HYC-7 Strain on F. solani

*Fusarium solani* was inoculated on one side of the PDA plate, and the HYC-7 strain was inoculated on the other side at a symmetrical position 3 cm from the edge of the plate. After culturing at 25 °C for 5 d, the antagonistic effect was observed, and the inhibition rate of *F. solani* was calculated: inhibition rate = (colony diameter of control group − colony diameter of treatment group)/(colony diameter of control group) × 100%.

### 2.3. Screening of Optimum Carbon and Nitrogen Sources for Fermentation Medium

Mannitol, glucose, maltose, lactose, and soluble starch were selected as carbon sources to replace sucrose; beef extract, yeast extract paste, L-glutamic acid, carbamide and peptone (purity ≥ 99.7%, Tianjin Guangfu Fine Chemical Research Institute, Tianjin, China) were selected as nitrogen sources to replace sodium nitrate in the basic medium.

Different nitrogen and carbon source liquid fermentation media were each placed into 150 mL conical flasks for 30 min. After cooling, 1 mL of the spore suspension ($1 \times 10^7$ mL$^{-1}$) of the HYC-7 strain was added to each fermentation liquid medium; sterile water and Tween-80 were selected as the control. The medium was incubated in a constant-temperature shaker at 28 °C and 160 rpm for 5 d. The fermentation broth was centrifuged at 4 °C and $9900 \times g$ for 20 min to obtain the supernatant and then filtered through a 0.22 μm microporous membrane for later use. The fermentation broth was mixed with the PDA medium at a ratio of 30% [8]. After cooling, the fungus cake of *F. solani* was inoculated in the center of the plate and cultured at 28 °C in the dark. When the colony diameter in the control group reached 3/4 of the diameter of the plate, it was measured using the crossover method and the inhibition rate was calculated according to the following formula:

$$\text{Inhibition rate (\%)} = \frac{(A - B)}{(A)} \times 100;$$ A and B represent the colony diameter of the control and treatment group, respectively.

### 2.4. Single-Factor Test of the Effect of Different Fermentation Conditions on the Inhibition Rate of HYC-7 Fermentation Broth

Under the basic conditions of temperature (28 °C), pH (6.0), inoculation amount (1 mL), loaded liquid (60 mL), and rotational speed (160 rpm), we kept the other conditions constant and only changed one condition to conduct a single-factor test. The following conditions were set: temperature: 20, 22, 25, 28, and 30 °C; pH: 5.0, 6.0, 7.0, 8.0, 9.0, and 10.0; inoculation amount: 0.5, 1, 1.5, 2, 2.5, and 3 mL; loaded liquid: 40, 50, 60, 70, and 80 mL; rotational speed: 120, 140, 160, 180, and 200 rpm. Six replicates of each treatment were created and incubated for 5 d in a constant-temperature shaker; then, the inhibition rate was determined using the fermentation broth according to the method in Section 2.3.

### 2.5. Response Surface Optimization Test

On the basis of the single-factor test, the three pH (A), rotational speed (B), and temperature (°C) parameters were optimized. The response surface test scheme was designed according to the Box–Behnken method in the Design-Expert 8.0.6 software, as shown in Tables S1 and S2.

### 2.6. Determination of Colony Diameter, Sporulation, Spore Germination Rate, and Germ Tube Length

The HYC-7 strains were inoculated in a basic 60 mL fermentation medium and cultured in a constant-temperature shaker at 190 rpm and 29 °C for 5 d. Then, the fermentation broth was centrifuged at 4 °C and $9900 \times g$ for 20 min, and the supernatant was taken and filtered through a 0.22 μm microporous membrane. A sterile fermentation filtrate was obtained for use.

The colony diameter, sporulation, spore germination rate, and germ tube length were determined according to the method of Li et al. (2020) [16]. The sterile fermentation filtrate was mixed with the PDA medium according to volume fractions of 10%, 20%, 30%, 40%, and 50%, and the basic fermentation medium was used as the control. The fungus cake of *F. solani* was inoculated to the center of the plate at 28 °C in the dark for 7 d. The colony diameter was measured using the crossover method. A plate cultured for 7 d was taken, and 10 mL of sterile water was added. Spores were gently scraped off the plate with a sterilizing coater. After the sterile gauze was filtered, sporulation was counted using a hemocytometer.

The sterile fermentation filtrate was mixed with the PDB medium in volume fractions of 10%, 20%, 30%, 40%, and 50%, and the same amount of basic fermentation medium was added as the control. With a pipette gun, the medium was absorbed and suspended on a hollow glass slide and cultured at 28 °C in the dark. After 6 h, a microscope was used to count the spore germination and measure the length of the germ tubes. Germ tubes longer than half of the spore length were considered as spore germination.

### 2.7. Determination of Antioxidant Enzymes Activity

After *F. solani* were cultured in PDB medium for 7 d, they were filtered and collected; washed with sterile water; mixed with fermentation broth at 1:50 ($w/v$); treated for 0, 3, 6, 9, 12, 24, 36, and 48 h; and stored in liquid nitrogen for later use.

The assays of SOD, CAT, and POD activities were based on the methods previously described by Zhang et al. (2022) and Wang et al. (2021) [17,18]. In total, 0.2 g of frozen mycelium was ground in 5 mL of precooled phosphate buffer (50 mM and pH 7.8) and centrifuged at 4 °C and $9900 \times g$ for 20 min, and the supernatant was the crude enzyme of SOD and CAT. The SOD determination reaction system included 1.5 mL of phosphate buffer (50 mM, pH 7.8), 0.3 of mL L-methionine (130 mM), 0.3 mL nitroblue tetrazolium chloride (750 µM), 0.3 mL of EDTA-Na$_2$ (100 µM), 0.5 mL of distilled water, 0.3 mL of riboflavin (20 µM), and 100 µL of crude enzyme solution. Subsequently, the mixture was placed in 25 °C and 3000–4000 lx light conditions for 20 min and then placed in the dark to terminate the reaction; the absorbance value was immediately measured at 560 nm. One activity unit (U) of the SOD enzyme inhibits 50% of the photochemical reduction of NBT and is expressed as $U \cdot g^{-1}$ FW. The CAT determination reaction system included 2.9 mL of H$_2$O$_2$ (0.067%) and 0.1 mL of enzyme solution, with distilled water as the control. The change in OD$_{240}$ within 2 min was recorded. A decrease of 0.01 OD$_{240}$ per minute was defined as one unit (U) and expressed as $U \cdot g^{-1}$ FW. The POD determination reaction system included 2.6 mL of guaiacol (0.3%), 0.3 mL of H$_2$O$_2$ (0.6%), and 0.1 mL of enzyme solution, with distilled water as the control. The change in OD$_{470}$ within 2 min was recorded. An increase of 0.01 OD$_{470}$ per minute was defined as one unit (U) and expressed as $U \cdot g^{-1}$ FW.

### 2.8. Determination of Content of Malondialdehyde, Soluble Protein, and Soluble Sugar

Malondialdehyde content was determined according to the method of Li et al. (2020) [15]. The 0.2 g of frozen mycelium was ground in 5 mL of TCA (10%) in an ice bath and centrifuged at 4 °C and $9900 \times g$ for 20 min, and the supernatant was the crude extract. In total, 1 mL of crude extract was added to 2 mL of 0.67% thiobarbituric acid in a boiling water bath for 30 min and then centrifuged after rapid cooling. The absorbance value of the supernatant was measured at 450 nm, 532 nm, and 600 nm. The MDA content was expressed as $mmol \cdot g^{-1}$ FW.

The soluble protein content was determined according to the method of Bradford. (1976) [19]. Then, 0.2 g of frozen mycelium was ground in 5 mL of distilled water in an ice bath and centrifuged at 4 °C and $9900 \times g$ for 20 min; the supernatant was a protein extraction solution. An amount of 0.5 mL of the extract was added to 0.5 mL of distilled water and 5 mL of Coomas bright blue G-250 reagent. After standing for 2 min, the absorbance value was measured at 595 nm. The soluble protein content was calculated using a standard curve and expressed as $mg \cdot g^{-1}$ FW.

The content of soluble sugar was determined according to the method of Dai et al. (2017) [20]. Then, 0.2 g of frozen mycelium was ground in 5 mL of distilled water and transferred to a test tube, boiled for 30 min, naturally cooled, and filtered into a 25 mL volumetric flask. In total, 0.5 mL of filtrate was added to 1.5 mL of distilled water, 0.5 mL of anthrone solution, and 5 mL of concentrated sulfuric acid successively. After the reaction solution became transparent, the absorbance value was determined at 620 nm. The soluble sugar content was expressed as $mg \cdot g^{-1}$ FW.

*2.9. Determination of Decay Incidence and Disease Index of Wolfberry*

Healthy wolfberry root tissues were selected and washed with running water to remove soil. Then, they were soaked in 75% alcohol for 20 s, 0.1% mercuric chloride solution for 30 s, and finally rinsed with sterile water to remove the disinfectant residue. The roots were cut into 10 mm slices with a sterile blade, and the surface of the roots was pierced evenly with a sterile needle. HYC-7 strain, *F. solani*, and HYC-7 strain + *F. solani* were inoculated as treatment groups, and sterile water was used as the control. The specific experimental steps were as follows: The injured wolfberry root tissues were soaked in the fermentation broth of the HYC-7 strain with a concentration of $1 \times 10^7$ spores/mL for 30 min and then placed in a Petri dish with moist sterile filter paper. *Fusarium solani* with a concentration of $1 \times 10^6$ spores/mL were sprayed uniformly on the surface of the root tissues after 24 h and then cultured at 28 °C in the dark. According to the severity of the root's decay, they were divided into five grades, where 0 = no root rot symptoms; 1 = less than 5% root area rotted; 2 = 5–25% root area rotted; 3 = 26–50% root area rotted; 4 = 51–75% root area rotted; and 5 = more than 75% root area rotted. After 7 d, the decay incidence and disease index were determined according to the method of Sasan and Bidochka. (2013) and calculated as the following formulas [7].

$$\text{Decay incidence (\%)} = \frac{\text{Number of diseased plants}}{\text{Investigation of total number of plants}} \times 100$$

$$\text{Disease index (\%)} = \frac{\Sigma(\text{Number of disease at all levels} \times \text{Disease grade})}{\text{Survey the total number of plants} \times \text{Highest grade}} \times 100$$

*2.10. Data Analysis*

All determinations in this study were repeated at least three times. Data were expressed as means and standard errors, and Origin 2022b was used for mapping. The significance analysis of Duncan's multiple differences was performed using SPSS 22.0 (SPSS, Chicago, IL, USA).

## 3. Results

*3.1. HYC-7 Strain on F. solani*

The results of the PDA plate confrontation showed that the HYC-7 strain had a significant antagonistic effect on *F. solani*, with an inhibition rate of 39. 8% at the 5th d of culture (Figure 1).

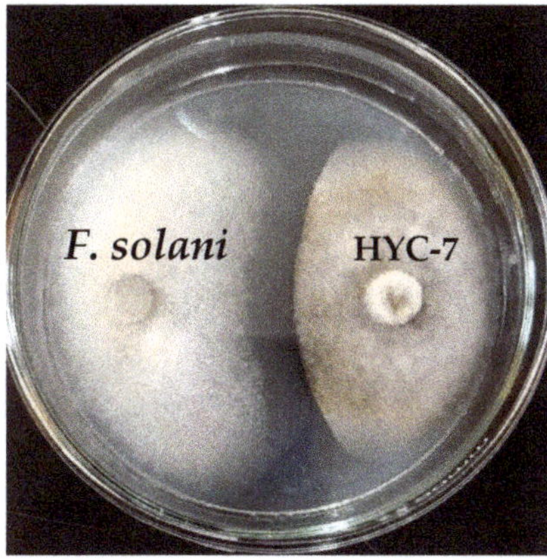


**Figure 1.** HYC-7 strain inhibits the colony growth of *F. solani*.

### 3.2. Screening of Optimum Nitrogen and Carbon Sources for Fermentation Medium

Nitrogen and carbon sources are important nutrients for fungal growth. The results showed that the inhibition rate of the HYC-7 fermentation broth was significantly different in different nitrogen and carbon source media. When beef extract and glucose were used as nitrogen and carbon sources, the inhibition rates of the fermentation broth were 40.35% and 35.58%, respectively, which were significantly higher than those of other nitrogen and carbon sources ($p < 0.05$) (Figure 2). Therefore, the fermentation medium with beef extract as the nitrogen source and glucose as the carbon source was selected for the single-factor test and the collection of HYC-7 strain fermentation broth.

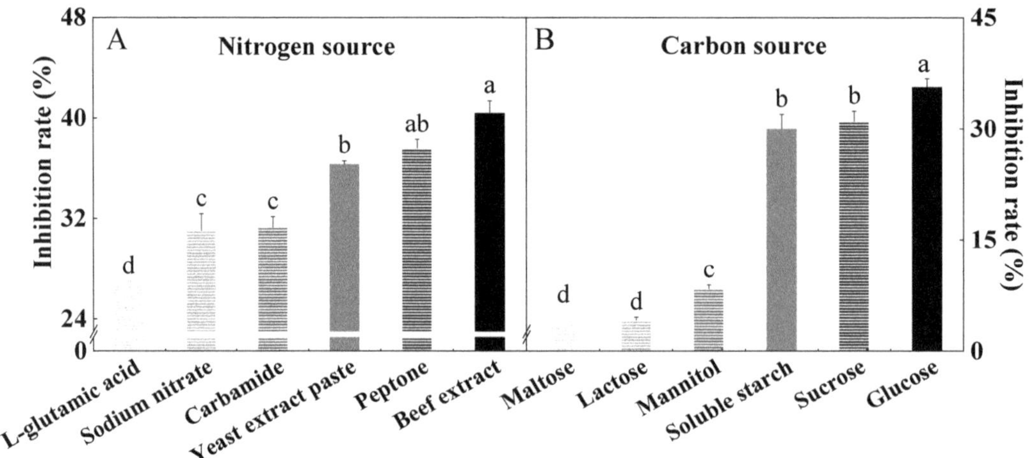


**Figure 2.** Effect of different nitrogen (**A**) and carbon sources (**B**) on the inhibition rate of HYC-7 strain fermentation broth. Vertical bars represent standard error ($\pm$SE), and different letters indicate significant differences between groups ($p < 0.05$).

### 3.3. Single-Factor Test

The inhibition rate of the HYC-7 fermentation broth showed a first increasing and then decreasing trend with an increase in pH. When the pH was 9, the highest inhibition rate was 38.9% (Figure 3A). When the inoculation amount was within the range of 0.5–3.0 mL, the inhibition rate of the HYC-7 fermentation broth showed an obvious first increasing and then decreasing trend. When the inoculation amount was 2 mL, the inhibition rate was 28.11%, which was significantly higher than other groups ($p < 0.05$) (Figure 3B). The loaded liquid had a significant effect on the inhibition rate of the HYC-7 fermentation broth. When the loaded liquid was 70 mL, the maximum inhibition rate was 30.75% (Figure 3C). The inhibition rate of the fermentation broth increased with an increase in rotational speed. The maximum inhibition rate was 38.39% at 200 rpm (Figure 3D). The inhibition rate of the fermentation broth decreased first and then increased with an increase in temperature and reached a maximum value of 38.02% at 28 °C (Figure 3E).

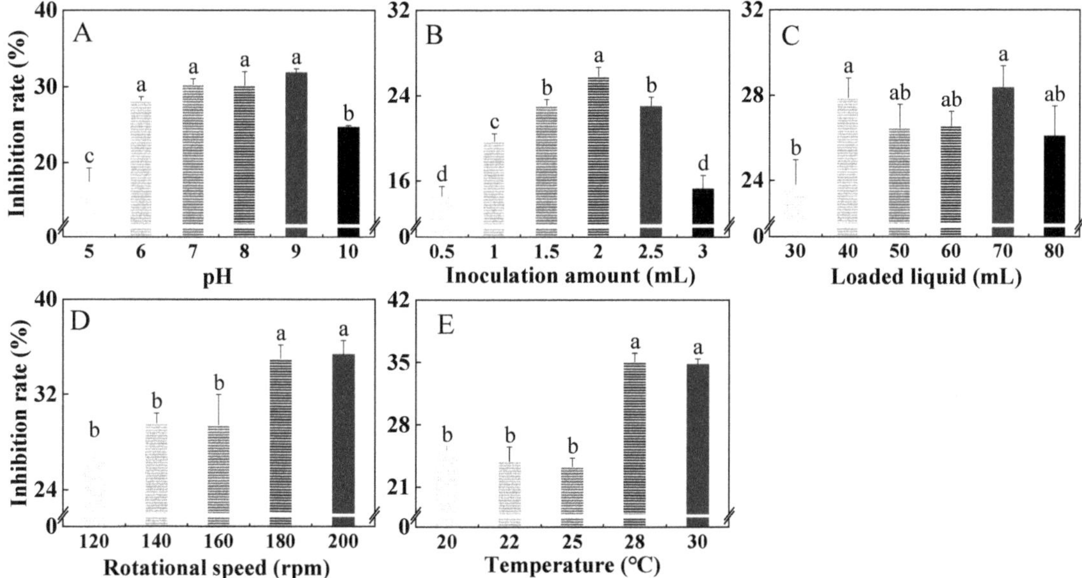

**Figure 3.** Effects of pH (**A**), inoculation amount (**B**), loaded liquid (**C**), rotational speed (**D**), and temperature (**E**) on the inhibition rate of the HYC-7 strain fermentation broth. Vertical bars represent standard error (±SE), and different letters indicate significant differences between groups ($p < 0.05$).

### 3.4. Response Surface Test Optimization Results

The response surface test's design and results are shown in Table S2.

### 3.4.1. Regression Equation Fitting and Analysis of Variance

A mathematical model with the regression equation was established via statistical analyses of the experimental data: the inhibition rate (%) = 51.71 − 3.09 A − 0.10 B + 3.37 C + 2.96 AB + 0.45AC + 0.80 BC − 4.38 A$^2$ − 8.39 B$^2$ − 7.51 C$^2$.

Variance analysis and the significant difference test were conducted for the regression model, and the results are shown in Table S3. The regression of the model was significant ($p < 0.0001$). The loss of quasi-item $p = 0.1179 > 0.05$ was not significant, indicating that the model was suitable with a high degree of fit, indicating that the test results were consistent with the model, while other unknown factors had little interference in the test results; the model was suitable, and the fitting degree was high, so the establishment of the regression model had a certain guiding significance. At the same time, the first A, C, and AB terms and the second A$^2$, B$^2$, and C$^2$ terms all had significant antifungal activity. The correlation of the model was high with a regression coefficient of R$^2$ = 0.9777. The regression coefficient was R$^2$ = 0.9777; this showed that the correlation of the model was high. The F value represents the importance of each influencing factor to the test index. The larger the F value, the stronger the influence of the factor on the inhibition rate. The results showed that the influence of three factors on the inhibition rate was in the order of temperature (C, °C) > pH (A) > rotational speed (B, rpm).

### 3.4.2. Response Surface Analysis of Interaction of Various Factors

The response surface diagram below could more intuitively reflect the interaction of the three main factors and their influence on the inhibition rate. It can be observed in the surface diagram and contour lines in (Figure 4) that the interaction between A (pH) and B (rotational speed) had a significant impact on the antifungal activity of the strain. This is consistent with the results shown in Table S3, such as AB = 0.0102 < 0.05. The optimal conditions were obtained using a quadratic multinomial regression fitting equation: pH—6.96; rotational speed—189.40 rpm; and fermentation temperature—29.21 °C. Under these conditions, the predicted inhibition rate was 52.62%. The optimal fermentation conditions were a pH of 7.0; a rotational speed of 190 rpm; and a culture temperature of 29 °C for simple and feasible operation.

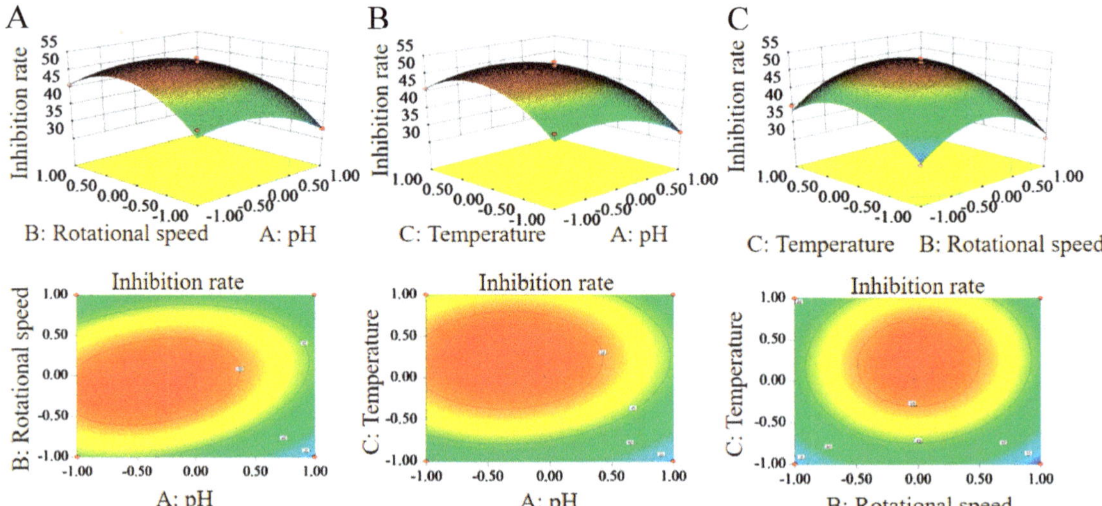

**Figure 4.** The response surface methodology and contour plots of the effects of the interaction between rotational speed and pH (**A**), temperature and pH (**B**), and temperature and rotational speed (**C**) on the inhibition rate of the HYC-7 strain fermentation broth. Note: The incline degree of the surface diagram is directly proportional to the influence degree of factors on the response value. The larger the focal length of the contour, the stronger the interaction between parameters.

The fermentation broth was prepared under the optimal fermentation conditions (pH 7.0, 190 rpm, and 29 °C) and repeated three times to verify the accuracy of the model. The average inhibition rate was 51.80%, which was consistent with the predicted value of 52.62%.

*3.5. Effects of HYC-7 Fermentation Broth on the Growth and Development of F. solani*

The colony diameter could visually reflect the amount of mycelium growth. Compared with the control, the colony diameter of *F. solani* decreased continuously with the increase in fermentation broth concentration. When the concentration of fermentation broth was 50%, the colony diameter was the smallest, which was 65.2% lower than the control ($p < 0.05$) (Figure 5A). Sporulation, spore germination rate, and germ tube length are important indicators for measuring fungal reproductive capacity and spore viability. Compared with the control, the lowest sporulation was observed when the fermentation broth concentration was 30%, which was 82.1% lower than the control ($p < 0.05$) (Figure 5B). When the fermentation broth concentration was 30%, 40%, and 50%, there were no significant differences in sporulation. Similarly, both the spore germination rate and germ tube length of *F. solani* gradually decreased with the increase in fermentation broth concentration. When the concentration of the broth was 50%, the spore germination rate and germ tube length were the lowest, and they were 75.8% and 62.6% lower than the control ($p < 0.05$) (Figure 5C,D).

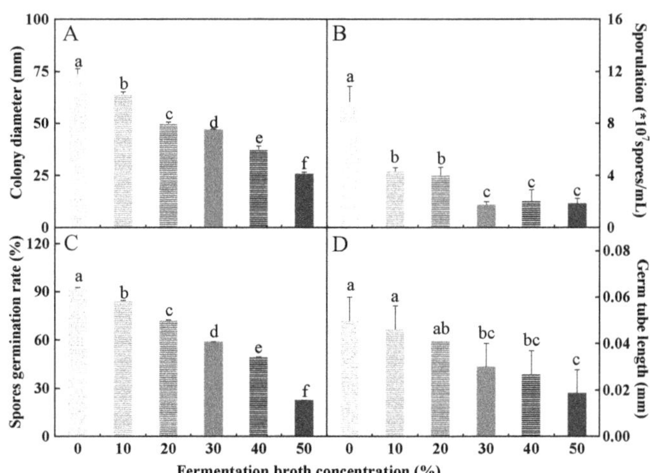

**Figure 5.** Effects of the HYC-7 strain fermentation broth with different concentrations on the colony diameter (**A**), sporulation (**B**), spore germination rate (**C**), and germ tubes length (**D**) of *F. solani*. Vertical bars represent standard error (±SE), and different letters indicate significant differences between groups ($p < 0.05$).

### 3.6. Effects of HYC-7 Strain Fermentation Broth on the Activities of SOD, CAT, and POD and the Contents of MDA, Soluble Protein, and Soluble Sugar

SOD, CAT, and POD are important antioxidant enzymes in phytopathogenic fungi and play an important role in the scavenging of excess reactive oxygen species (ROS). The SOD activity of both the HYC-7 fermentation treatment and the control *F. solani* increased and then decreased during the treatment period, reaching a peak at 24 h. The HYC-7 fermentation broth treatment significantly decreased the SOD activity of *F. solani* and was lower than the control during the treatment period. At 24 h and 48 h, they were 29.89% and 64.01% lower than the control, respectively ($p < 0.05$) (Figure 6A). The activity of CAT in both the HYC-7 fermentation broth treatment and the control *F. solani* also increased and then decreased during the treatment period, reaching a peak at 24 h. The HYC-7 fermentation broth treatment significantly decreased the CAT activity of *F. solani* in the later treatment period, and it was lower than the control by 20.25% and 41.66% at 24 h and 48 h, respectively ($p < 0.05$) (Figure 6B). The POD activity of the control *F. solani* increased continuously during the treatment, but the HYC-7 fermentation broth treatment increased first and then decreased, always being lower than that of the control. At 24 h and 48 h, these two were 25.73% and 78.17% lower than the control ($p < 0.05$) (Figure 6C).

The MDA content of both the HYC-7 fermentation broth treatment and control *F. solani* increased continuously during the treatment period. The HYC-7 fermentation broth treatment significantly increased the MDA content of *F. solani* and was higher than the control during the treatment period. At 24 h and 48 h, it was 13.15% and 32.15% higher than the control, respectively ($p < 0.05$) (Figure 6D). The soluble protein content of the HYC-7 fermentation broth treatment and control *F. solani* increased first and then decreased during the treatment period. The control reached the peak at 36 h, but the broth treatment reached the peak at 24 h and then began to decrease rapidly. At 48 h, it was significantly lower than the control: 30.07% ($p < 0.05$) (Figure 6E). Similarly, the soluble sugar content of the broth treatment and control *F. solani* decreased continuously during the treatment period. The HYC-7 fermentation broth accelerated the decrease in soluble sugar content, which was 28.23% and 40.95% lower than that of the control at 24 h and 48 h, respectively ($p < 0.05$) (Figure 6F).

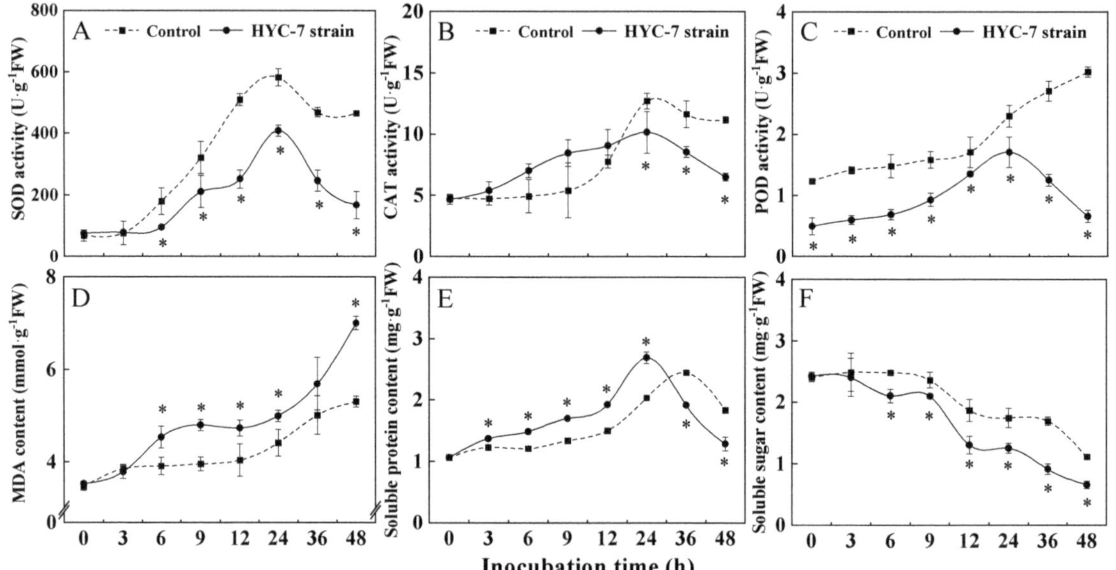

**Figure 6.** Effects of the HYC-7 strain fermentation broth on the activities of SOD (**A**), CAT (**B**), and POD (**C**) and the contents of MDA (**D**), soluble protein (**E**), and soluble sugar (**F**) of *F. solani*. Vertical bars represent standard error (±SE), and asterisks indicate significant differences between the treatment and the control at the same time ($p < 0.05$). SOD: Superoxide dismutase; CAT: catalase; POD: peroxidase; MDA: malondialdehyde.

### 3.7. Effect of HYC-7 Fermentation Broth on the Decay Incidence and Disease Index of Wolfberry Root Rot

As shown in Figure 7, the control group also showed a certain degree of decay incidence. Compared with the control, HYC-7 strain inoculation significantly decreased the decay incidence and disease index, which were 38.5% and 25.5% lower, respectively, than those of the control after 7 d of inoculation ($p < 0.05$). *Fusarum solani* inoculation resulted in a more serious decay incidence of wolfberry root rot. Similarly, compared with the *F. solani* inoculation, HYC-7 strain+*F. solani* inoculation significantly decreased the incidence and disease index by 57.5% and 41.8%, respectively, after 7 d of inoculation ($p < 0.05$).

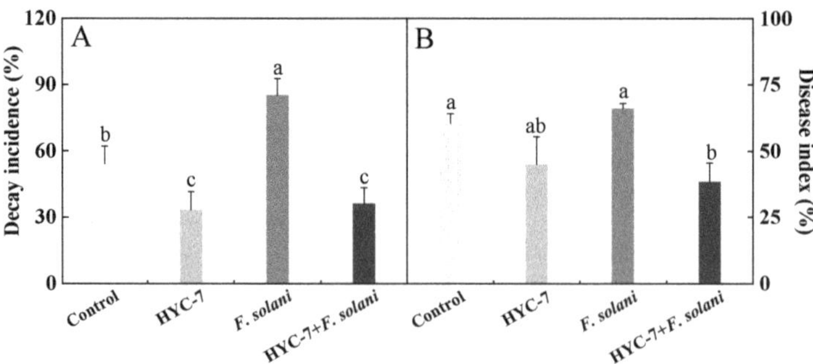

**Figure 7.** Effect of the HYC-7 strain fermentation broth on the decay incidence (**A**) and disease index (**B**) of wolfberry root rot. Vertical bars represent standard error (±SE), and different letters indicate significant differences between groups ($p < 0.05$).

## 4. Discussion

In this study, we found that the antifungal activity of *M. robertsii* fermentation broth was strongest when glucose and beef paste were used as carbon and nitrogen sources. This result is similar to the previous results obtained by optimizing the composition of the *Bacillus pumilus* fermentation medium [21]. It was also shown that the best carbon and nitrogen sources for *M. anisopliae* to produce the secondary metabolite destruxin were maltose and peptone, respectively, which is inconsistent with the results obtained in the present study possibly due to differences among the strains [22]. The response surface results indicated that temperature, pH, and rotational speed had a greater effect on the antifungal activity of *M. robertsii* fermentation broth. The optimal fermentation conditions were a temperature of 29 °C, a pH of 7.0, a rotational speed of 190 rpm, and 60 mL of loaded liquid; the inhibition rate was 51.80% under these conditions. Previous studies have shown that the evaluation of biocontrol effectiveness is based on the content of antimicrobial substances or the inhibition rate of the target pathogenic fungi [23]. Using the antifungal ability of the fermentation broth as a response variable, the response surface methodology optimized the optimal medium volume, initial pH, and fermentation temperature of the DS-R5 strain, which significantly improved the inhibition of its fermentation broth against *F. solani* [24]. Similarly, the optimization of the fermentation conditions of *Xenorhabdus nematophila* using the response surface methodology significantly improved its antibiotic activity [13].

In this study, it was shown that the HYC-7 strain, isolated from healthy wolfberry inter-root soil, had a significant biological control effect against root rot disease caused by *F. solani*. Previous studies have shown that the key to the biocontrol effect of microorganisms is the production of secondary metabolites, and the main metabolites in the fermentation broth of *M. anisopliae* are the nonribosomal cyclic peptides of destruxins. Destruxins exhibit a variety of acute toxic effects against insects. In addition, serinocyclin, subglutinol, and swainsonine were also identified in the secondary metabolites of *Metarhizium* spp. [25]. Serinocyclin A showed entomophagous activity as the exposed mosquito larvae to this compound exhibited abnormal swimming as they were unable to control the position of their heads. Swainsonine, as a mycotoxin, has the effect of inhibiting lysosomal $\alpha$-mannosidase [26,27]. In this study, the results indicated that the HYC-7 fermentation broth treatment with different concentrations had a significant inhibitory effect on the growth and development of *F. solani*, and the inhibition increased with the increasing concentration of fermentation broth volume. This may be due to the increase in the concentration of antifungal secondary metabolites in the broth with respect to the increase in the concentration volume.

Pathogenic fungi are subjected to severe oxidative stress when infesting plants or under unfavorable environmental conditions, which can limit their colonization or normal growth and development [28]. However, such fungi have the ability to scavenge reactive oxygen species (ROS) to neutralize excess ROS from normal physiological processes or environmental stresses. SOD catalyzes the conversion of $O_2^-$ to $H_2O_2$, followed by the further decomposition of $H_2O_2$ to $H_2O$ and $O_2$ by CAT, thereby reducing ROS-induced oxidative stress. POD acts synergistically with CAT and SOD to form an antioxidant enzyme system that is also involved in the detoxification of $H_2O_2$ [29]. Previous studies found that citral and cinnamaldehyde treatments significantly reduced SOD, CAT, and POD activities and inhibited the growth of *F. sambucinum*. Our study indicated that HYC-7 fermentation broth treatment significantly inhibited the SOD, CAT, and POD activities of *F. solani*. The inhibition of antioxidant enzyme activity may lead to the disruption of the balance of ROS metabolism and the massive accumulation of intracellular reactive oxygen species, which exacerbate cellular membrane lipid peroxidation and disrupted membrane integrity, resulting in a significant reduction in pathogenicity [28]. Previous studies have shown that when the cell membrane is destroyed by antibacterial substances, it will cause changes in permeability, resulting in a large accumulation of MDA and the release of macromolecular proteins in cells [30]. In this study, we found that the HYC-7 fermentation broth treatment promoted the accumulation of MDA and accelerated the leakage of soluble protein and the decrease in soluble sugar content. Therefore, we inferred that the HYC-7 fermentation broth caused serious damage to the cell membrane and the leakage of soluble protein and inhibited the carbon metabolism of *F. solani*, limiting its normal life metabolism and thus effectively inhibiting the growth of *F. solani*. This also indirectly supported the results that the HYC-7 fermentation broth treatment significantly inhibited the activities of antioxidant enzymes. These results were consistent with previous reports that *Bacillus pumilus* HR10 fermentation broth treatment promoted MDA accumulation and inhibited soluble protein and soluble sugar content in *Sphaeropsis sapinea* [31]. In summary, we suggest that the phenomenon of HYC-7 fermentation broth reducing the occurrence of wolfberry root rot is related to inhibiting the growth and development of *F. solani*, promoting MDA accumulation and accelerating the leakage of soluble protein and the decrease in soluble sugar content; however, the specific mechanism needs to be further studied.

## 5. Conclusions

In this study, the fermentation broth of *M. robertsii* had the best inhibitory effect on *F. solani* when glucose and beef extract were selected as carbon and nitrogen sources. The fermentation factors affecting the inhibitory effect of *M. robertsii* fermentation broth were temperature > pH > rotational speed, and the optimal fermentation conditions were a temperature of 29 °C, a pH of 7.0, a rotational speed of 190 rpm, and 60 mL of loaded liquid; the inhibition rate was 51.80% under these conditions. *Metarhizium robertsii* fermentation broth treatment inhibited colony growth, sporulation, spore germination, and germ tube elongation of *F. solani*. *Metarhizium robertsii* fermentation broth treatment also decreased the SOD, CAT, and POD activities of *F. solani*; promoted the accumulation of MDA; accelerated the leakage of soluble protein; and reduced the soluble sugar content. In addition, *M. robertsii* inoculation significantly decreased the decay incidence and disease index of wolfberry root rot. In conclusion, we believe that *M. robertsii* has a good control effect on the root rot of wolfberry and could be used for the development of biological control agents.

**Supplementary Materials:** The following supporting information can be downloaded at https://www.mdpi.com/article/10.3390/microorganisms11102380/s1. Table S1: Codes and levels of factors for response surface methodology experiments. Table S2: Experimental design and results of strain culture conditions. Table S3: Regression analysis of experimental results based on the Box–Behnken design.

**Author Contributions:** Conceptualization, J.H., X.Z. and B.W.; methodology, X.Z.; software, N.L. and B.W.; validation, J.H.; formal analysis, J.H. and N.L.; investigation, N.L. and D.D.; resources, N.L.; data curation, N.L. and D.D.; writing—original draft preparation, J.H., N.L., and B.W.; writing—review and editing, J.H., X.Z., N.L. and B.W.; visualization, Q.W.; supervision, J.H.; project administration, J.H.; funding acquisition, J.H. All authors have read and agreed to the published version of the manuscript.

**Funding:** This research was funded by the National Natural Science Foundation of China (32060341), the Natural Science Foundation of Gansu Province (20JR10RA508), and the Supporting Fund for Yongth Mentor of Gansu Agricultural University (GAU-QDFC-2018-07).

**Data Availability Statement:** No new data were created or analyzed in this study. Data sharing is not applicable to this article.

**Conflicts of Interest:** The authors declare no conflict of interest.

## References

1. Zhang, D.; Xia, T.; Dang, S.; Fan, G.; Wang, Z. Investigation of Chinese wolfberry (*Lycium* spp.) germplasm by restriction site-associated DNA sequencing (RAD-seq). *Biochem. Genet.* **2018**, *56*, 575–585. [CrossRef] [PubMed]
2. Gao, Y.; Wei, Y.; Wang, Y.; Gao, F.; Chen, Z. *Lycium barbarum*, a traditional Chinese herb and promising anti-aging agent. *Aging Dis.* **2017**, *8*, 778–791. [CrossRef] [PubMed]
3. Uwaremwe, C.; Yue, L.; Liu, Y.; Tian, Y.; Zhao, X.; Wang, Y.; Xie, Z.K.; Zhang, Y.B.; Cui, Z.T.; Wang, R.Y. Molecular identification and pathogenicity of *Fusarium* and *Alternaria* species associated with root rot disease of wolfberry in Gansu and Ningxia provinces, China. *Plant Pathol.* **2021**, *70*, 397–406. [CrossRef]

4.    Rojo, F.G.; Reynoso, M.M.; Ferez, M.; Sofía, N.; Chulze, S.N.; Torres, A.M. Biological control by *Trichoderma* species of *Fusarium solani* causing peanut brown root rot under field conditions. *Crop Prot.* **2007**, *26*, 549–555. [CrossRef]

5.    Al-Ani, R.A.; Adhab, M.A.; Mahdi, M.H.; Abood, H.M. *Rhizobium japonicum* as a biocontrol agent of soybean root rot disease caused by *Fusarium solani* and *Macrophomina phaseolina*. *Plant Protect Sci.* **2012**, *48*, 149–155. [CrossRef]

6.    Toghueo, R.; Eke, P.; Zabalgogeazcoab, Í.Z.; Aldana, B.D.; Nana, L.W.; Boyom, F.F. Biocontrol and growth enhancement potential of two endophytic *Trichoderma* spp. from *Terminalia catappa* against the causative agent of common bean root rot (*Fusarium solani*). *Biol. Control* **2016**, *96*, 8–20. [CrossRef]

7.    Sasan, R.K.; Bidochka, M.J. Antagonism of the endophytic insect pathogenic fungus *Metarhizium robertsii* against the bean plant pathogen *Fusarium solani* f. sp. phaseoli. *Can. J. Plant Pathol.* **2013**, *35*, 288–293. [CrossRef]

8.    Zhang, X.Y.; He, J.; Hou, C.X.; Zhang, S.H. Screening and identification of antagonistic strains of wolfberry root rot. *Acta Agric. Zhejiangensis* **2020**, *32*, 858–865.

9.    Sasan, R.K.; Bidochka, M.J. The insect-pathogenic fungus *Metarhizium robertsii* (Clavicipitaceae) is also an endophyte that stimulates plant root development. *Am. J. Bot.* **2012**, *99*, 101–107. [CrossRef]

10.   Velavan, V.; Rangeshwaran, R.; Sivakumar, G.; Sasidharan, T.O.; Sundararaj, R.; Kandan, A. Occurrence of *Metarhizium* spp. isolated from forest samples in south india and their potential in biological control of banana stem weevil *Odoiporus longicollis* oliver. *Egypt J. Biol. Pest Control* **2021**, *31*, 131. [CrossRef]

11.   Plantey, R.L.; Papura, D.; Couture, C.; Thiéry, D.; Lucero, G.S. Characterization of entomopathogenic fungi from vineyards in argentina with potential as biological control agents against the European grapevine moth *Lobesia botrana*. *BioControl* **2019**, *64*, 501–511. [CrossRef]

12.   Heller, J.; Tudzynski, P. Reactive oxygen species in phytopathogenic fungi: Signaling, development, and disease. *Annu. Rev. Phytopathol.* **2011**, *49*, 369–390. [CrossRef] [PubMed]

13.   Wang, Y.H.; Feng, J.T.; Zhang, Q.; Zhang, X. Optimization of fermentation condition for antibiotic production by *Xenorhabdus nematophila* with response surface methodology. *J. Appl. Microbiol.* **2008**, *104*, 735–744. [CrossRef]

14.   Wanicki, S.A.; Mascarin, G.M.; Moreno, S.G.; Eilenberg, J.; Júnior, I.D. Growth kinetic and nitrogen source optimization for liquid culture fermentation of *Metarhizium robertsii* blastospores and bioefficacy against the corn leafhopper *Dalbulus maidis*. *World J. Microbiol. Biot.* **2020**, *36*, 71. [CrossRef] [PubMed]

15.   Gao, X.; He, Q.; Yun, J.; Huang, L. Optimization of nutrient and fermentation parameters for antifungal activity by *Streptomyces lavendulae* xjy and its biocontrol efficacies against *Fulvia fulva* and *Botryosphaeria dothidea*. *J. Phytopathol.* **2016**, *164*, 155–165. [CrossRef]

16.   Li, M.L.; Zheng, T.T.; Chen, Y.Q.; Sui, Y.; Ding, R.H.; Hou, L.W.; Zheng, F.L.; Zhu, C.Y. The antagonistic mechanisms of *Streptomyces sioyaensis* on the growth and metabolism of poplar canker pathogen *Valsa sordida*. *Biol. Control* **2020**, *151*, 104392. [CrossRef]

17.   Zhang, M.; Zhang, Y.D.; Li, Y.C.; Bi, Y.; Mao, R.Y.; Yang, Y.Y.; Jiang, Q.Q.; Prusky, D. Cellular responses required for oxidative stress tolerance of the necrotrophic fungus *Alternaria alternata*, causal agent of pear black spot. *Microorganisms* **2022**, *10*, 621. [CrossRef]

18.   Wang, B.; Li, Z.C.; Han, Z.H.; Xue, S.L.; Bi, Y.; Prusky, D. Effects of nitric oxide treatment on lignin biosynthesis and texture properties at wound sites of muskmelons. *Food Chem.* **2021**, *362*, 130193. [CrossRef]

19.   Bradford, M.M. A rapid and sensitive method for the quantitation of microgram quantities of protein utilizing the principle of protein-dye binding. *Anal. Biochem.* **1976**, *72*, 248–254. [CrossRef]

20.   Dai, T.T.; Xu, Z.Y.; Zhou, X.; Li, B.; Mao, S.F. The inhibitory effect of the plant alkaloid camptothecin on the rice sheath blight pathogen Rhizoctonia solani. *Int. J. Agric. Biol.* **2017**, *19*, 558–562. [CrossRef]

21.   Dai, Y.; Wang, Y.H.; Li, M.; Zhu, M.L.; Wen, T.Y.; Wu, X.Q. Medium optimization to analyze the protein composition of *Bacillus pumilus* HR10 antagonizing *Sphaeropsis sapinea*. *AMB Express* **2022**, *12*, 61. [CrossRef] [PubMed]

22.   Liu, B.L.; Chen, J.W.; Tzeng, Y.M. Production of cyclodepsipeptides destruxin A and B from *Metarhizium anisopliae*. *Biotechnol. Progr.* **2000**, *16*, 993–999. [CrossRef] [PubMed]

23.   Zhao, X.; Han, Y.; Tan, X.Q.; Wang, J.; Zhou, Z.J. Optimization of antifungal lipopeptide production from *Bacillus* sp. BH072 by response surface methodology. *J. Microbiol.* **2014**, *52*, 324–332. [CrossRef] [PubMed]

24.   Sa, R.B.; He, S.; Han, D.D.; Liu, M.J.; Yu, Y.X.; Shang, R.G.; Song, M.M. Isolation and identification of a new biocontrol bacteria against *Salvia miltiorrhiza* root rot and optimization of culture conditions for antifungal substance production using response surface methodology. *BMC Microbiol.* **2022**, *22*, 231. [CrossRef] [PubMed]

25.   Donzelli, B.; Krasnoff, S.B. Molecular genetics of secondary chemistry in *Metarhizium* fungi. *Adv. Genet.* **2016**, *94*, 365–436.

26.   Yadav, R.N.; Mahtab Rashid, M.; Zaidi, N.W.; Kumar, R.; Singh, H.B. *Secondary Metabolites of Metarhizium spp. and Verticillium spp. and Their Agricultural Applications. Secondary Metabolites of Plant Growth Promoting Rhizomicroorganisms*; Singh, H., Keswani, C., Reddy, M., Sansinenea, E., García-Estrada, C., Eds.; Springer: Singapore, 2019.

27.   Xu, Y.J.; Luo, F.; Li, B.; Shang, Y.F.; Wang, C.H. Metabolic conservation and diversification of *Metarhizium* species correlate with fungal host-specificity. *Front. Microbiol.* **2016**, *7*, 2020. [CrossRef]

28.   Zhang, Z.Q.; Chen, Y.; Li, B.Q.; Chen, T.; Tian, S.P. Reactive oxygen species: A generalist in regulating development and pathogenicity of phytopathogenic fungi. *Comput. Struct. Biotec.* **2020**, *18*, 3344–3349. [CrossRef]

29.   Marschall, R.; Tudzynski, P. Reactive oxygen species in development and infection processes. *Semin. Cell Dev. Biol.* **2016**, *57*, 138–146. [CrossRef]

30. Fang, S.W.; Li, C.F.; Shih, D.Y.C. Antifungal activity of chitosan and its preservative effect on low-sugar Candied Kumquat. *J. Food Prot.* **1994**, *57*, 136–140. [CrossRef]
31. Dai, Y.; Wu, X.Q.; Wang, Y.H.; Zhu, M.L. Biocontrol potential of *Bacillus pumilus* HR10 against Sphaeropsis shoot blight disease of pine. *Biol. Control* **2021**, *152*, 104458. [CrossRef]

**Disclaimer/Publisher's Note:** The statements, opinions and data contained in all publications are solely those of the individual author(s) and contributor(s) and not of MDPI and/or the editor(s). MDPI and/or the editor(s) disclaim responsibility for any injury to people or property resulting from any ideas, methods, instructions or products referred to in the content.

 *microorganisms*

*Review*

# The Good, the Bad, and the Useable Microbes within the Common Alder (*Alnus glutinosa*) Microbiome—Potential Bio-Agents to Combat Alder Dieback

**Emma Fuller [1,2], Kieran J. Germaine [1] and Dheeraj Singh Rathore [2,*]**

[1] EnviroCore, Dargan Research Centre, Department of Applied Science, South East Technological University, Kilkenny Road, R93 V960 Carlow, Ireland; emma.fuller@teagasc.ie (E.F.); kieran.germaine@setu.ie (K.J.G.)
[2] Teagasc, Forestry Development Department, Oak Park Research Centre, R93 XE12 Carlow, Ireland
* Correspondence: dheeraj.rathore@teagasc.ie

**Abstract:** Common Alder (*Alnus glutinosa* (L.) Gaertn.) is a tree species native to Ireland and Europe with high economic and ecological importance. The presence of Alder has many benefits including the ability to adapt to multiple climate types, as well as aiding in ecosystem restoration due to its colonization capabilities within disturbed soils. However, Alder is susceptible to infection of the root rot pathogen *Phytophthora alni*, amongst other pathogens associated with this tree species. *P. alni* has become an issue within the forestry sector as it continues to spread across Europe, infecting Alder plantations, thus affecting their growth and survival and altering ecosystem dynamics. Beneficial microbiota and biocontrol agents play a crucial role in maintaining the health and resilience of plants. Studies have shown that beneficial microbes promote plant growth as well as aid in the protection against pathogens and abiotic stress. Understanding the interactions between *A. glutinosa* and its microbiota, both beneficial and pathogenic, is essential for developing integrated management strategies to mitigate the impact of *P. alni* and maintain the health of Alder trees. This review is focused on collating the relevant literature associated with Alder, current threats to the species, what is known about its microbial composition, and Common Alder–microbe interactions that have been observed worldwide to date. It also summarizes the beneficial fungi, bacteria, and biocontrol agents, underpinning genetic mechanisms and secondary metabolites identified within the forestry sector in relation to the Alder tree species. In addition, biocontrol mechanisms and microbiome-assisted breeding as well as gaps within research that require further attention are discussed.

**Keywords:** biocontrol agents; *Phytophthora*; microbiome; beneficial microbes; pathogens; PGPR; PGPF

**Citation:** Fuller, E.; Germaine, K.J.; Rathore, D.S. The Good, the Bad, and the Useable Microbes within the Common Alder (*Alnus glutinosa*) Microbiome—Potential Bio-Agents to Combat Alder Dieback. *Microorganisms* **2023**, *11*, 2187. https://doi.org/10.3390/microorganisms11092187

Academic Editors: Chetan Keswani and Rainer Borriss

Received: 17 July 2023
Revised: 24 August 2023
Accepted: 25 August 2023
Published: 30 August 2023

**Copyright:** © 2023 by the authors. Licensee MDPI, Basel, Switzerland. This article is an open access article distributed under the terms and conditions of the Creative Commons Attribution (CC BY) license (https://creativecommons.org/licenses/by/4.0/).

## 1. Introduction

Over the last two decades, there has been an elevated interest regarding the significance of native tree species, such as Common Alder (*Alnus glutinosa* (L.) *Gaertn*) [1], also known as Black Alder, European Alder, or just Alder. The genus *Alnus* belongs to the family *Betulaceae*, the birch clade [2]. Alder is known to be a short-lived, light-demanding, and water-demanding species [1]. In addition, Alder preferentially grows in non-shaded areas with little competition; however, it has the ability to naturally self-prune, as branches tend to die with lack of light [3,4]. This species typically contains short stalked reddish-brown winter buds, small flat waxy winged seeds, and woody cone-shaped fruit (catkins) [4,5]. There are more than thirty *Alnus* species worldwide, including shrubs and trees, with *A. glutinosa* being the primary native species within Europe [6,7]. While some other well-known Alder species include Italian Alder (*A. cordata*), Grey Alder (*A. incana*), Red Alder (*A. rubra*), White Alder (*A. rhombifolia*), Japanese Alder (*A. japonica*), and American Green Alder (*A. viridis*) [1,5]. Alder has demonstrated phenomenal resilience due to its adaptation to a variety of climates around the world [8]. This resilient species tolerates frost and survives in an extensive range of temperatures [9]. *A. glutinosa* has been found on a broad range of sites; however, it favours wet, riverine, and fertile areas with high humidity [3,10]. Preferable growing sites include riverbanks, ponds, surrounding lakes, streams, marshy waterlogged areas, shaded mountainous regions, wet woodlands, wet soils, and highlands with sufficient moisture content [10,11]. This species tends not to grow well on calcareous soils, acidic peat, arid sandy soils, and areas with stagnant water [1,12]. Alder has been found in and/or introduced to regions of Africa, Australia, Canada, India, Japan, Russia, North America, South America, and New Zealand [5,12–14].

Historically, Alder was considered to generate an inferior class of wood with low economic value [15]. A black dye, known as 'poor man's dye', was produced from Alder bark and catkins and generally used for tanning leather [15]. Alder had an association with war and was known as the 'tree of death' since when cut, the light-coloured wood swiftly oxidises to a vivid red, giving a 'bleeding' effect [15,16]. Alder wood was used for items such as war shields, wheels, bowls, tubs, and troughs [15]. Alder has been used within many regions worldwide for various traditional health and healing medicinal practices, due to the presence of chemical constituents and biological components such as flavonoids, terpenes, phenols, saponins, and steroids [17]. For example, Common Alder bark has been shown to have medicinal benefits and has been used for treating burns, diseases, and infections, as well as potential antitumor activity [17]. Acero et al. (2012) conducted an ethno-pharmacological study of Common Alder bark and observed that it potentially has anti-oxidant and anti-inflammatory properties [18]. Fresh Alder catkins have shown antioxidant, antimicrobial, and anti-inflammatory properties due to the presence of polyphenols and certain microorganism strains [19].

Alder tends to have an extensive rooting system and forms large nitrogen-fixing root nodules via symbiosis with nitrogen-fixing filamentous bacteria, such as the genus *Frankia* [20]. Thus, this actinorhizal species has the pioneering ability to improve soil conditions by increasing organic matter and nitrogen content within soils to improve fertility, contributing to biodiversity so other plant populations can grow in the area. Furthermore, Alder can colonise areas that have previously been subject to significant disturbance and/or degradation, playing a vital role in ecosystem and land restoration [1,21]. Other ecological benefits of Alder roots include the reduction of flooding and soil erosion, water filtration/purification, and riverbank stabilisation [4,5]. Alder is a biodiverse habitat providing shelter for numerous plants, animals, and microbes. For example, otters and fish tend to use Alder roots surrounding waterbodies for nesting purposes as well as shelter to reduce predation [11]. Alder provides an important food source for several birds, fungi, lichens, mosses, and approximately 140 insects and mites, resulting in an additional food source for fish if these insects feed on trees planted beside waterbodies [3,22]. For example, Alder leaves provide a food source for plant-eating invertebrates and Alder catkins provide a source of pollen for bees as well as nectar and seeds for many bird species. The nitrogen-fixing capabilities of Alder allow for the production of nitrogen-rich leaf litter due to the presence of invertebrate detritivores and microbial decomposition, which provides a primary food source for invertebrates that are later eaten by bigger organisms, creating multiple food chains [23].

Its porous wood has various uses for carpentry, building, furniture production, biomass production, instrument manufacturing, cabinetry, dyes, and manufacturing charcoal [3,5,16]. Its timber has high durability and decay resistance beneath water, so it is frequently used for bridge piles, small boats, water structure supports, and jetties [4,24]. The climatic adaptation of *A. glutinosa*, coupled with the worldwide distribution of this species has substantially increased its importance and promising utilisation [8]. As a result, Alder breeding programs have been established; therefore, greater amounts of *A. glutinosa* are being cultivated within commercial forests in order to generate trees to produce timber [10,16]. As part of the European Union (EU) forest strategy for 2030, policies on broadleaf and biodiversity have been put in place to improve the quantity and quality of European forestry; thus, since Alder is a broadleaved species, the volumes of Alder plantations throughout Europe has increased [25].

There are numerous existing threats to Alder cultivation, some of which are abiotic such as drought [26,27], whilst others can be biological such as pathogenic infections [28]. Research regarding the relationship between pathogenic microbes and native trees has increased because of the elevated levels of disease spread throughout the world, which has caused damage to and the depletion of numerous tree species. Thus, a greater understanding of *Alnus*-associated beneficial microbes could provide an ecological, environmentally sustainable solution to the biological control of *Phytophthora* infection. Therefore, to gain a better understanding of the beneficial and pathogenic microbiota associated with *A. glutinosa*, this review reports the known microbiome of Alder, pathogenic threats of Alder, as well as any biological solutions available to date for the control of these pathogenic threats.

## 2. The Microbiome of Alder Species

Microorganisms are extremely diverse and plentiful within the environment. The type and abundance of microbial communities that exist within an organism are influenced by the environmental conditions surrounding plant communities as well as soil type, indicating the presence of selective communities associated with different plant species. Forest trees exist in close association with a diverse range of microbial organisms that play a crucial role in maintaining tree health, nutrient conditions, and ecosystem functions. This association can be mutualistic, parasitic, or symbiotic. Combined, these microbial communities associated with the tree are known as its microbiome. The composition of the microbial community can vary due to a range of factors including climate, edaphic conditions, anthropogenic activities, silviculture management practices, and various other events and stressors [29]. Organic matter and associated decomposition products also cause changes within forest soils, causing soil physicochemical alterations, acidification, and supply/leaching of nutrients [30]. The long-lived nature of trees provides a secure food source for beneficial microbiota as well as parasites and pathogens. In comparison to annual crops, underground microbial communities related to trees may be more consistent and have stable interactions due to the deep rooting system, and a constant energy flow system ranging from photosynthesates being pumped into the soil to the abscission of leaf/flower and fruit material, leading to organic matter build up in the soil and the absence of soil disturbance [29].

Tree roots and rhizospheric soil tend to have a particular and plentiful microbial community due to the occurrence of nutrient and mineral exchange between the soil microbiota and tree-component microbiota present in root tissues, ectomycorrhizal fungal roots, arbuscular mycorrhiza roots, and fungal mycelia [31]. These roots have the ability to alter the composition of the microbial community within the soil due to the compounds and root cells that are released into the soil [29]. Throughout tree development, some soil physiochemical properties can change, which causes alterations to the microbial communities present in the rhizosphere that modify tree morphology, promote growth, and enhance nutrient and mineral content [32]. Furthermore, beneficial and/or pathogenic bacterial and fungal communities are also present within the endosphere and phyllosphere microbiota of forest trees, which create ecological interactions such as mutualism, commensalism, and/or antagonism [33]. Beneficial microbiota include plant growth-promoting rhizobacteria (PGPR), plant growth-promoting fungi (PGPF), and biocontrol agents that have the ability to economically and efficiently alter the metabolism of tree substances in order to improve the growth, performance, and resistive properties of the species. Moreover, beneficial microbiota are necessary for the natural degradation of plant residues, since greater amounts of microbial assemblages are present within residues, which additionally degrade them into soil organic matter containing beneficial macromolecules and micromolecules. This process provides increased soil protection and sustains nutrient capacity. Furthermore, when plants develop under stressful environmental conditions, the plant-associated microbiota tends to increase levels of hormones such as abscisic acid and jasmonic acid, which help plants regulate growth and adapt to extreme conditions [33]. Plant growth-promoting microbes tend to be found on the root surface or as endophytes within plant tissues, which have the ability to act as biocontrol agents, bio-fertilisers, and/or bio-stimulants [34]. Due to the complex relationships and interactions within forest ecosystems, there are still many unknowns and potential beneficial microbes associated with the microbiome of trees; therefore, it is important that research on this topic continues.

Plant growth-promoting rhizobacteria (PGPR) play a significant role in the forest ecosystem by releasing numerous regulatory molecules, aiding the growth and development of forest trees [35]. Deciduous forest soils tend to contain a wide array of bacterial phyla, which include Acidobacteria, Actinobacteria, Proteobacteria, Bacteroidetes, and Firmicutes [36,37]. PGPR can be found in the rhizosphere, phyllosphere, and/or endosphere and have the ability to promote plant growth in different ways including nitrogen fixation, the production of the plant hormones such as auxins (e.g., indole acetic acid (IAA)), enzymes (1-aminocyclopropane-1-carboxylic acid (ACC) deaminase, cellulase, chitinases, etc.) and siderophores, as well as the ability to compete with pathogenic microbes to minimise the spread of infection [33]. IAA is a plant hormone that regulates growth and development [38]. ACC deaminase is produced in the presence of ACC, the immediate precursor to the plant stress hormone ethylene, when plants grow within stressful conditions [39]. ACC deaminase aids plant survival by decreasing the amount of ethylene present, as excessive amounts of this plant hormone can have negative effects on plant growth [40]. Siderophores have the ability to uptake iron from the rhizosphere to promote plant growth and contribute to nutrition whilst minimising the amount of iron available for harmful pathogens [41]. Cellulase has been known to promote the breakdown of fungal and plant cell walls, aid soil fertility by speeding up the decomposition of plant residues, reduce spore germination, and reduce fungal growth [42–44]. Richter et al. (2010) analysed cellulase activity to suppress *P. cinnamomi*, a root rot pathogen of avocados and reported that sporangia production was reduced when cellulase enzymes were present, indicating that cellulase enzymes may have the ability to decelerate the spread of *P. cinnamomi* within avocado sites [45]. Pectinase is an essential enzyme for plant growth and development as it breaks down pectin within cell walls, allowing new plant growth to occur as well as providing entry for beneficial microbes to live endophytically [46]. Other mechanisms that PGPR induce in order to promote the health and growth of plants include the production of exopolysaccaride biofilms, organic acids to increase phosphate solubilisation, hydrogen cyanide to help with biotic stressors, and the production of cytokines and gibberellins which aid growth and development [35]. In addition, some PGPR can reduce oxidative stress by producing abscisic acid as well as stimulating growth by altering physiological mechanisms using emitted bacterial volatile organic components [35]. Some PGPR studies in particular have focused on analysing the interaction between the *Betulaceae* family and identified beneficial microbes including Actinobacteria genera (*Frankia, Streptomyces*) [47–50], an endophytic bacterial genus (*Rhizobium*) [51], *Bacillus* isolates (*B. licheniformis, B. pumilus, B. megaterium*, and *B. longisporus*) [52–54], and Gammaproteobacteria (*Pseudomonas*) [50,55]. Other known PGPR particularly in agriculture include *Arthrobacter, Azospirillum, Azorhizobium, Burkholderia, Flavobacterium, Mesorhizob*, and *Methylobacterium* [56]. With the evident advantages of these microbes in agriculture, there is great potential to exploit these PGPR with further research in the forestry sector.

The inoculation of plants with plant growth-promoting fungi (PGPF) has demonstrated many health benefits including enhanced seed germination, seed vigour, root morphogenesis, pathogenic suppression, the reduction of (a)biotic stressors, as well as an improved process of photosynthesis and mineralization [57]. The roots from the *Alnus* species have a tendency to form stable long-term biological interactions with mycorrhizae, including ectomycorrhizal fungi and/or arbuscular mycorrhizal fungi [58]. This symbiosis helps to increase the uptake of water and nutrients through the roots, improve tree health, growth, and reproductive ability, as well as provide a degree of tolerance to (a)biotic stressors [58]. For example, Thiem et al. (2020) showed that the inoculation of Alder seedlings with an ectomycorrhizal fungus, *Paxillus involutus* OW-5, promoted growth and increased tolerance in saline soils [59]. Culturable dark septate endophytes (DSEs) have the potential to promote plant growth particularly in metal-contaminated soils due to their capacity to provide a degree of phytoremediation/phytostabilization of heavy metals, increasing nutrients within soils, and produce the IAA hormone [2,60]. Fungal melanin tends to be found alongside some PGPF, which may help promote plant growth by providing a higher tolerance/resistance to environmental stress and promoting colonisation [58,61]. Fungal melanin exhibits beneficial biological functions including photo-protection, metal binding, mechanical protection, energy harvesting, cell development, antioxidant functions, anti-desiccant functions, chemical protection, as well as thermoregulation [61]. Some examples of identified PGPF that interact with the *Betulaceae* family include arbuscular mycorrhizal fungi (*Gigaspora rosea*), ectomycorrhizal fungi (*Hebeloma* sp., *Helotiales* sp., *Geopora* sp., *Thelephora* sp., *Tomentella* spp., *Paxillus involutus, Tylospora, Leccinum,* and *Rhizopogon*), endophytic fungi (*Cryptosporiopsis* spp., *Rhizoscyphus* spp.), DSE (*Phialocephala*), and ericoid fungi (*Oidiodendron*) [2,29,47,58,62–65]. In addition to all the above-mentioned PGPF, other known growth-promoting microbes used in agriculture belong to the genera *Alternaria, Chaetomium, Penicillum, Phoma, Serendipita,* and *Trichoderma* [66]. There has been a major growth of interest regarding PGPF acting as bio-fertilisers in agriculture due to their ecological benefits and plant improvement. In particular, the genera *Trichoderma, Penicillum,* and *Aspergillus* are the most studied within agriculture as they can promote crop growth and act as eco-friendly biological control agents to promote resistance/tolerance to biotic stresses [67]. These fungal genera may have potential plant growth-promoting properties within forestry, if not already used within this sector. Furthermore, exploring such unknown genera in forest tree species and soils would lead to novel microbes that can benefit both forestry and agriculture for sustainable production.

*What Is Known about the Microbiome of Alder?*

To date, there have been few studies regarding microbes associated with *A. glutinosa.* Much of the information that is available focuses on the microbial communities present within *Alnus* roots and their associated rhizosphere. These studies particularly focus on soil salinity [68], inoculation with nitrogen-fixing bacteria [69], colonisation capabilities [70], plant growth promotion [30], antagonistic effects against pathogenic microbes [71], as well as adaptation to extreme environmental conditions [72]. Alder-associated bacterial microbes that have been identified belong to the phyla Actinobacteria (*Frankia alni*), Acidobacteria (*Granulicella*), Alphaproteobacteria (*Caulobacteraceae, Rhizobiales,* and *Sphingomonas*), Bacteroidetes (*Chitinophagaceae, Flavobacteriaceae,* and *Sphingobacteriaceae,*), Betaproteobacteria, Firmicutes (*Bacillus licheniformis* and *Bacillus pumilus*), Gammaproteobacteria, Pseudomonadota (*Bradyrhizobium, Oxalobacteraceae, Pseudomonas, Rhizobium, Rhodanobacter,* and *Xanthomoadaceae*), and Thermoleophilia [20,29,30,36,47,48,50,52,53,65,73]. Phyla of fungal microbes associated with Alder species include Ascomycota (*Cryptosporiopsis*), Basidiomycota (*Tomentella, Thelephora*), Ectomycorrhizal fungi (*Alnicola, Lactarius,* and *Phialocephala*), Glomeromycota (*Gigaspora rosea*), and Zygomycota (*Rhizocyphus*) [2,20,47,58,60,65,74]. See Figure 1 for a summary of the microbes identified in the different compartments of the *Alnus* species. With the economic and environmental importance of Alder coupled with its pathogenic threat, research regarding the root and rhizospheric microbiomes may potentially grow in the future.

Very little is known about the leaf, bark, and catkin microbiomes of Alder trees. A greater focus has been placed on analysing the chemical and biological components, as well as the secondary metabolites present in the bark, leaf, and fruit due to the medicinal value of these components [39–41]. Sukhikh et al. (2022) examined the antioxidant properties of *A. glutinosa* female catkins and found high levels of methanol extracts, ellagic acid, and ethyl acetate indicating their potential use as natural antioxidants [75]. Thiem et al. (2020) analysed the correlation between salt stress and mycorrhizal fungi from *A. glutinosa* growing in saline soil. Three fungal species were isolated from *A. glutinosa* catkins, which were *Amanita muscaria, Paxillus involutus,* and *Gymnopus.* Seedlings were later inoculated with each fungus to determine their effects against salt stress. It was observed that *P. involutus* aided in promoting the growth of seedlings and showed a degree of salt tolerance [59]. There have been studies on the metabolites and medicinal extracts of Alder bark, but no studies were found for the analysis of Alder tree bark microbiomes. Alder leaves have been studied regarding their medicinal potential. For example, Mushkina (2021) investigated the ability of *A. glutinosa* leaf tinctures to be used as wound-healing gels. It was determined that this gel had the ability to regenerate and heal wounds due to the presence of flavonoids, tannins, and phenolic acid [76]. Leaf extracts have been used to analyse their antagonistic effects against pathogens such as *Staphylococcus aureus*, *Bacillus subtilis*, *Escherichia coli*, *Pseudomonas aeruginosa*, and *Candida albicans* [77]. Concerning the microbial community within Alder leaves, there are few studies regarding this topic. Kayini and Pandey (2010) isolated fungal microbes from Nepalese Alder leaves including *Alternaria alternata*, *A. raphani*, *Aspergillus niger*, *Cladosporium cladosporioides*, *Epicoccum purpurascens*, *Fusarium oxysporum*, *Gliocladium roseum*, *Mucor hiemalis*, and *Pestalotiopsis* sp. [78]. A greater focus is needed regarding the microbial communities present within Alder leaves, bark, and catkins.

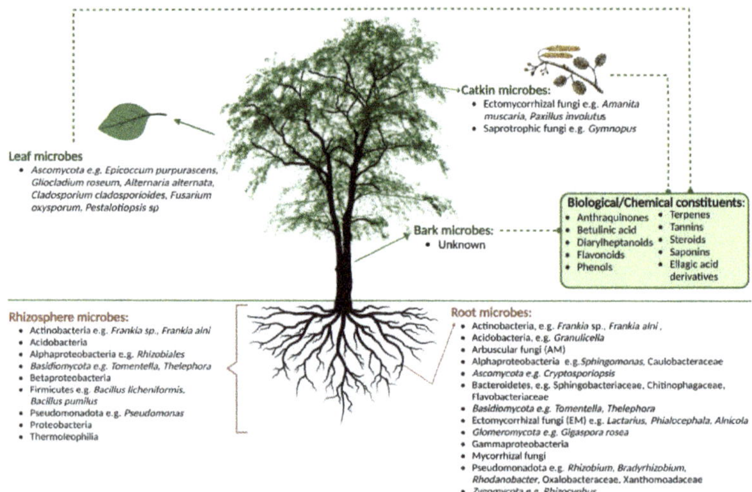


**Figure 1.** Overview of microbial communities present in different compartments of the Alnus species based on published literature (Created with BioRender.com, accessed on 1 February 2023).

## 3. Microbial Pathogens of Alder Species

There are several pests and microbial pathogens associated with Alder. For example, Alder yellows phytoplasma disease (caused by a phytoplasma bacterial parasite) has been identified within numerous European countries, which causes stunted growth, yellowing of leaves, reduction of leave size and amount, as well as dieback and/or death [79–82]. The Ascomycota fungus, *Taphrina alni*, is a causal agent of Alder tongue galls on female catkins and has been identified throughout Europe. These galls are known as Alder tongues, as they are green-red elongated structures (depending on the season) [83]. The fungus *Septoria alnifolia* has been known to form leaf spots, stem cankers, and stem breakage on Alder trees [84]. Furthermore, a rust fungus, *Melampsoridium hiratsukanum*, causing yellow–brown spotting on Alder leaves, followed by early leaf fall, crown thinning, and/or death, has been identified [85]. The fungal pathogen *Mycopappus alni* tends to cause brown blotches on leaves and defoliation of Alder trees [86]. The bacterium *Erwinia alni*, has been found to cause bark cankers and bleeding and eventually kill off branches and/or the tree as a whole if the infection is severe [87]. The bacterium *Pseudomonas syringae* has been reported to cause leaf necrosis and the dieback of Common Alder [88]. With regards to insects, for example, the Alder leaf beetle, *Agelastica alni*, has been identified in Europe causing damage and defoliation of Common Alder [89]. Several species of Alder sawflies have been identified feeding on Common Alder such as *Monsoma pulveratum* [90], *Eriocampa ovata* [91], *Hernichroa crocea*, and *Cimbex conatus* [92]. A leaf miner, *Fenusa dohrnii*, was identified as a pest of Alder that causes damage [93]. There is an extensive list of pests and pathogens associated with Alder; however, not all tend to have damaging effects on this genus. For example, Sims (2014) discusses different types of fungal pathogens and insects that have been identified as associated with red, white, and thin-leaf Alder in western Oregon [94]. McVean (1953) extensively lists the insects and mites that associate with Common Alder as a food source [9], however not all of them are of economic importance and do not pose a severe threat to Alder trees. One of the major pathogens of Alder trees is the plant pathogenic oomycete species *Phytophthora*.

The species *Phytophthora*, meaning 'plant destroyer', has been associated with the dieback and decline of Alder across Europe [95]. There are approximately 200 identified and accepted species of *Phytophthora*, with more species unnamed/unidentified and likely to be discovered [96]. The pathogen typically infects the tree roots, causing them to rot, and spreads throughout the tree to cause damaging effects such as crown rot [97]. Infection can be caused via *Phytophthora* chlamydospores, hyphae, oospores, sporangium, and more frequently zoospores. Flagellated *Phytophthora* zoospores have a similar morphology to filamentous fungi but mainly have diploid hyphae and a cell wall comprising cellulose and/or β-glucans [97]. *Phytophthora* show phylogenetic similarities to algae and diatoms due to the structural composition of zoospores as well as the release of similar sexual antheridia (male) and oogonia (female) [96,98]. *Phytophthora* are soil-borne water moulds that have a strong dependency on water and humidity; therefore, the zoospores typically disperse and spread throughout water systems, eventually infecting tree roots [96,99]. In terms of the pathogen's life cycle, sporangia tend to form when chlamydospores (asexually) and/or oospores (sexually) are germinated within wet environments which results in zoospores being released from the mature sporangium [97,100]. Oospores are formed when oogonia are fertilised by antheridia of the *Phytophthora* species. Zoospores swim through water within wet soils where they have the ability to form cysts on susceptible tree roots and bark near the root collar [100]. These cysts germinate, forming mycelia, which allows the pathogen to grow and spread biotrophically throughout the plant tissues where reproduction occurs and the life cycle begins again by producing chlamydospores, oospores, and sporangium via germination within plant tissues, spreading from the roots and further throughout the tree (Figure 2) [97,100]. Infection via tree roots results in damage which causes a reduction in water and nutrient uptake, impairing their availability for the rest of the tree compartments. The lack of minerals causes harmful effects such as leaf stomata closure and a reduction in photosynthesis [101].

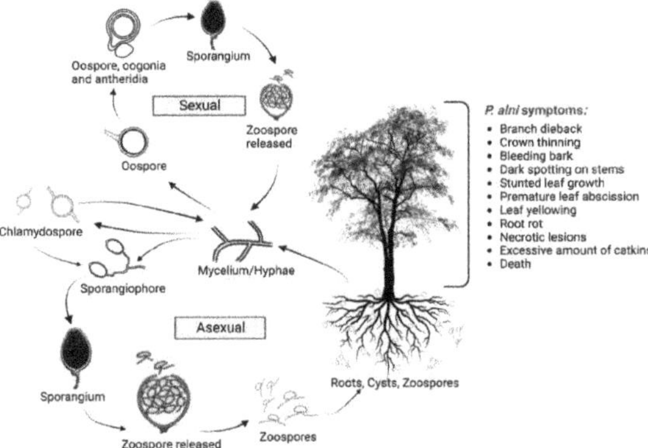


**Figure 2.** Life cycle of *Phytophthora* showing sexual and asexual phases and associated symptoms on an Alder tree. (Adapted from [97,102] and created with BioRender.com accessed on 17 June 2023).

Nave et al. (2021) inoculated Alder roots with *Phytophthora alni*, and after three weeks, oogonia, oospore, sporangium, antheridium, and mycelium were found at different stages of the reproductive cycle throughout the plant tissues [103]. *P. alni* hyphae has the ability to grow from Alder roots into the bark via secondary growth, travelling throughout the tree and infecting the cambium layer and adjacent phloem and xylem by continuously reproducing [101]. *P. alni* infects Alder bark via lenticels and/or bark wounds and spreads via secondary growth throughout the plant tissues by medullary rays into the bark xylem [101]. When the pathogen spreads to the bark, damaging effects include the deterioration of phloem tissue and a reduction in mineral transportation. It is unknown whether *P. alni* directly infects Alder leaves and catkins; however, infection of the roots and bark has damaging effects on all compartments of the tree.

Since the 1990s, *P. alni* has been a major issue within many countries and has continued to spread worldwide today, causing a range of damage to the *Alnus* species, which is believed to have emerged due to hybridisation within Europe [104–107]. The parent species of *P. alni* are *P. uniformis* (diploid) and *P. multiformis* (haploid), with the ploidy level showing that *P. alni* contains half of each parental genome; therefore *P. alni* is a triploid homoploid-type taxon [107]. According to Gibbs (2005), this pathogen was first detected in Southern England and has continued to spread throughout Europe [108,109] (Figure 3). It is hypothesised that the European spread of *P. alni* is due to clonal dispersion, since several mitochondrial haplotype variations have been observed, which may imply that multiple sexual hybridisation incidents generated a hybrid from several clones [107]. *P. alni* ssp. *lat.* was formally named by Brasier et al. (2004) and is now divided into three subspecies: *P. alni* ssp. *alni*, *P. alni* ssp. *uniformis*, and *P. alni* ssp. *multiformis*, considering different variants and hybrids amongst the species, with *P. alni* ssp. *alni* being the most aggressive one [110–112]. The *P. alni* species complex was later renamed by Husson et al. (2015) to *P.* × *alni*, *P.* × *multiformis, and P. uniformis* [112]. Only *P. alni* ssp. *uniformis* has been identified in North America to date [94,109]. The pathogen has also been identified in Ireland [110]. Symptoms of the *P. alni* complex species include dieback of branches, crown thinning, bleeding bark, dark spotting on stems, stunted leaf growth, premature leaf abscission, leaf yellowing, root rot, necrotic lesions on Alder bark, an excessive number of catkins on the species due to stress, and/or death [4,11]. Alder trees situated around riverbanks and flood plains have greater exposure to infection as the soil-borne pathogen spreads through water, infecting Alder bark and the roots, thus spreading the disease to a greater extent [3,4]. The pathogen spreads via streams, irrigation, flooding events, canals, drainage water, and other pathways via water systems [113]. Alternative pathways may include the plantation of Alder that originated from an affected nursery, human recreational activities, as well as animal activity [23].

Jung et al. (2018) conducted a thorough review of forestry diseases causing decline and suggested that other *Phytophthora* species such as *P. acerina, P. cactorum, P. chlamydospora, P. gonapodyides, P. lacustris, P. plurivora, P. polonica, P. pseudocryptogea,* and *P. siskiyouensis* may also be responsible for diseases associated with Alder dieback, although *P. alni* complex species appears to be the main cause of Alder decline [28,94,114–116]. For example, O'Hanlon et al. (2020) first reported the presence of *P. lacustris* causing disease within Alder trees in Ireland, indicating that Alder is at risk of dieback due to other *Phytophthora* species [28]. In Ukraine, Matsiakh et al. (2020) analysed rhizosphere soils of declining Alder where *P. lacustris, P. plurivora,* and *P. polonica* were detected, which further indicates that these pathogens play a role in Alder dieback [117]. Tkaczyk et al. (2023) identified *P. plurivora* and *P. cactorum* within soil, root, and water samples associated with declining Alder in Slovakia [118]. Bregant et al. (2020) reported the species *P. plurivora*, which was isolated from declining Alder trees within areas of Turkey, Italy, Spain, and Poland [111]. Furthermore, Bregant et al. (2020) revealed that *P. chlamydospora, P. gonapodyides,* and *P. siskiyouensis* potentially caused disease within Alder trees in North America. Zamora-Ballesteros et al. (2016) found that *P. plurivora* caused a high mortality rate within inoculated Alder seedlings, indicating that this species is pathogenic to Alder seedlings [119]. Haque and Diez (2015) conducted in vitro pathogenicity tests of Alder leaves, branches, and twigs with *P. × alni*, *P. × multiformis*, and *P. uniformis*, and it was observed that the pathogen caused foliar necrosis, indicating that the roots and root collar may not be the only source of pathogen inoculum [120].

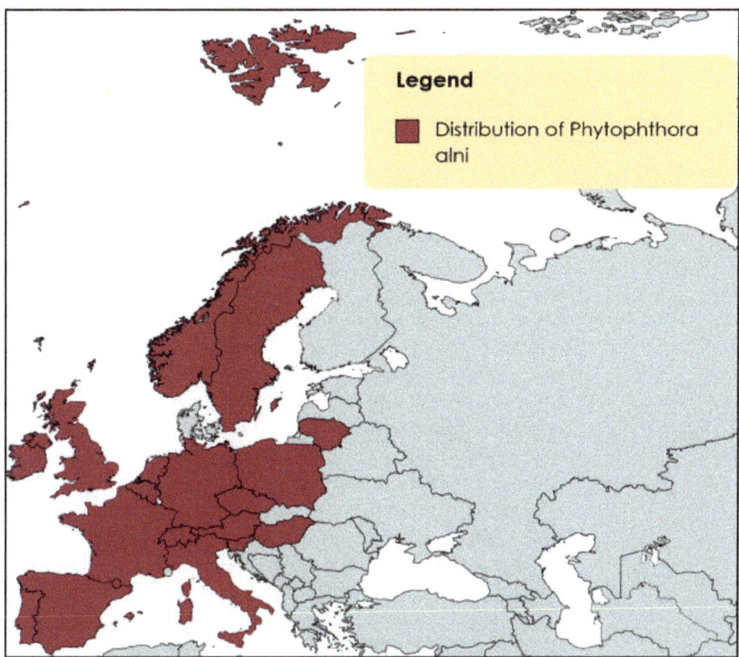


**Figure 3.** Distribution of *Phytophthora alni* across Europe [109,121].

## 4. The Potential of Microbial Biocontrol Agents in the Fight against Alder Diseases

Biocontrol agents tend to be any beneficial organism, including PGPR and PGPF, which have the ability to provide biological control against harmful pathogenic disease, as well as assisting in acquiring nutrients and amplifying resistive traits within a plant species [33]. Furthermore, the use of PGPR/F can act as a biocontrol agent by enhancing the health of a tree species, thus, a stronger tree can fight/withstand pathogenic infections. Moreover, *A. glutinosa* has many antioxidant, anti-inflammatory, antitumour, and inhibition properties due to the presence of anthraquinones, betulinic acid, diarylheptanoids, flavonoids, phenols, terpenes, tannins, steroids, and saponins; thus, these substances have the potential as BCAs within forestry [17,18,122]. Emerging diseases within forestry are a cause for concern, therefore, the advancement of BCAs within this sector is desirable. BCAs and bio-fertilisers have been used in agriculture for many years for sustainable crop production to minimise the use of harmful pesticides and fertilisers and reduce pathogenic infections, as well as to promote the health and growth of crops [123,124]. This approach is used intermittently within forestry with a few successful applications. Through research, several potential BCAs have been identified within forest ecosystems to reduce pathogenic spread and associated symptoms including dieback and root rot. Utkhede et al. (1997) analysed the antagonistic effects of the bacterium *Enterobacter aerogenes* against the pathogen *P. lateralis*, which tends to affect tree roots within soil and water, causing root rot disease of Lawson cypress trees [125]. D'Souza et al. (2005) investigated several potential BCAs including *Acacia extensa*, *A. stenoptera*, *A. alata*, and *A. pulchella* against the root rot disease caused by *P. cinnamomi*, to observe that these genera have the potential to provide protection to susceptible species [126]. With the *Alnus* species being under threat from the harmful effects of *Phytophthora*, such studies could potentially aid in determining the association between *P. alni* and the *Alnus* species and thus determining potential BCAs to prevent infections.

Several bacterial strains could potentially provide BCAs against emerging tree pathogens [127]. For example, for the pathogen *Hymenoscyphus fraxineus*, a study performed on the microbial community associated with ash tree leaves (*Fraxinus excelsior*) discovered elevated amounts of isolates within the genera *Luteimonas*, *Aureimonas*, Pseudomonas, *Bacillus*, and *Paenibacillus* present within tolerant trees, which potentially inhibit the invasion and spread of ash dieback disease (*H. fraxineus*) by direct competition or inducing systemic resistance, as well as the potential production of antagonistic metabolites and/or antifungal compounds [128,129]. Similarly, Becker et al. (2022) explored the microbiome of tolerant ash trees, and it was determined that the bacterium *Aureimonas altamirensis* had the ability to reduce *H. fraxineus* infections by niche exclusion and inducing mechanisms such as exopolysaccharides production and protein secretion [130]. The plant growth-promoting members of the genus *Streptomyces* have proven to be a prime choice as BCAs since these are readily available in nature and have the capacity to control pathogenic infections [9,131–133]. Liu et al. (2009) discovered that the BCA metabolite *daidzein* was produced by the bacterial genus *Streptomyces* and isolated it from a root of *A. nepalensis* alongside the *Streptomyces* bacterium, which may have potential antioxidant, anti-inflammatory, and anticancer effects [134,135]. The genus *Streptomyces* can produce the enzyme chitinase, which has been shown to have antagonistic effects against fungi due to the ability to degrade chitin present in the cell walls of fungi as well as reduce spore germination [136].

Biocontrol modes of action can be direct (interaction between pathogens and BCAs) or indirect (interaction between BCAs and plants to enhance their health and resistance to (a)biotic stressors) [137]. Direct biocontrol factors include antimicrobial metabolites, hydrolytic enzymes, quorum sensing, quenching, and resource competition, as well as siderophores. Indirect biocontrol factors include organisms triggering enhanced immunity for plants, environmental adaption, and hormone modulation. Microorganisms have the ability to produce microbe-associated molecular patterns (MAMPs) or damage-associated molecular patterns (DAMPs) like flagellin, glucan, xylan, or chitin [56]. These patterns are recognised by pattern-recognition receptors (PRRs) or other elicitors such as volatile organic compounds (VOCs) or siderophores, which are detected by specific receptors. Activation of these receptors initiates various signalling mechanisms that serve as precursors for the production of phytohormones, triggering defensive pathways. The kinase pathway can phosphorylate transcription factors that regulate the expression of early and late response genes [56]. There are many genes that have an association with biocontrol agents and defence-related genes [138]. Each BCA has thousands of genes present within its genome (depending on its genome size); therefore, genes with biocontrol traits can differ from one species to another. Nelkner et al. (2019) used transcription and genome mining in order to identify genes that are related to biocontrol traits and biosynthetic genes within a *Pseudomonas* strain [139]. In this study, biocontrol activity included the production of siderophores, secondary metabolites, and antibiotics (2,4-Diacetylphloroglucinol (DAPG), hydrogen cyanide (HCN) synthesis, pathogen inhibition genes (iron acquisition, exoprotease activity, and chitinase activity), resistance genes (PRRs), exopolysaccharides genes, lipopolysaccharides genes, metabolism genes, detoxification genes, and genes related to ISR and growth-promoting compounds (VOCs, acetoin, 2,3-butanediol, growth regulators/plant hormones, and phosphate solubilisation) [139]. Genes also involved in plant defence include the phenylpropanoid pathway gene-expressing phenylalanine ammonia-lyase (PAL), which facilitates the deamination process when phenylalanine is converted into cinnamate and ammonia, as well as the lipoxygenase pathway gene encoding hydroxyperoxide lyase (HPL) [56]. Mitogen-activated protein kinase (MAPK)- and cyclic adenosine monophosphate (cAMP)-associated molecules represent widespread intracellular signalling pathways that integrate extracellular stimuli, alter the expression and functionality of receptors, and regulate processes such as cell survival and neuroplasticity [140]. TGA transcription factors are vital regulators of diverse cellular processes which link to hormonal pathways, interacting proteins, and regulatory elements [141]. In addition, the phytohormones brassinosteroids (BRs) and jasmonates (JAs) also aid in the regulation of plant growth, development, and defensive response [142].

Chitinase has been used in agriculture during crop cultivation for disease management, improved growth, and greater yields [143]. Kumar et al. (2018) highlighted various ways in which chitinase gene expression has been used in agriculture to combat fungal diseases [143]. Auxins such as IAA and phenylacetic acid (PAA) are well-known bio-agents within agriculture due to their plant growth-promoting abilities and provide potential control against *Fusarium* pathogens, which have been found within soils, causing root rot [144,145]. IAA has been abundantly produced from PGPR isolates associated with forest trees; therefore they may have the potential for BCAs within this sector since they are proven to be advantageous in agriculture [38]. Trinh et al. (2022) discovered that the rhizobacteria *Bacillus subtilis*, isolated from the rhizosphere roots of black pepper, has potential as a bio-agent for *Fusarium* antagonism [144]. Fungal endophytes could potentially provide BCAs against emerging tree pathogens [146]. For example, Kosawang et al. (2017) performed in vitro antagonistic assays of *Sclerostagonospora* sp., *Setomelanomma holmii*, *Epicoccum nigrum*, *Boeremia exigua*, and *Fusarium* sp. against the dieback-causing fungus, *H. fraxineus*, and it was observed that these endophytes are potential BCAs [147]. Halecker et al. (2020) analysed the antagonistic effects of the fungal endophyte *Hypoxylon rubiginosum* on *H. fraxineus*, and it was determined that this endophyte has potential as a BCA due to the production of metabolites that are toxic to the pathogen [128,148]. This investigative approach is beneficial to determine potential pathogen-fighting microbes present within tolerant tree species that can aid in the protection of trees that are susceptible to certain diseases. See Table 1 for a summary of BCAs.

The most commonly used biocontrol agents belong to the genera *Pseudomonas*, *Bacillus*, and *Trichoderma* [29]. For example, *Pseudomonas* (*P. pudida* 06909) was used as a biocontrol agent against *Phytophthora* root rot within citrus orchids [149], *P. chlororaphis* has antagonistic effects against the cacao rot pathogen *Ph. Palmivora* [150], and *Bacillus amyloliquefaciens* isolated from ginseng rhizosphere induced systemic resistance to *Ph. cactorum* [151]. In addition, *Trichoderma virens*, *Trichoderma harzianum*, *Trichoderma asperellum*, and *Trichoderma spirale* isolated from cocoa show antagonistic effects against *Ph. Palmivora* [152], and *Trichoderma saturnisporum* also showed antagonistic effects against the *Phytophthora* spp. [29,153]. Abbas et al. (2022) conducted an extensive review regarding the identification of multiple biocontrol encoding genes within the *Trichoderma* spp. against *R. solani* including secondary metabolite genes, siderophores genes, signalling molecular genes, cell wall degradation enzymatic genes, and plant growth regulatory genes [154]. Furthermore, G-protein coupled receptor (GPCR) genes, adenylate cyclase genes, protein kinase-A genes, and transcription factor proteins are important genes found in *Trichoderma* for biocontrol against *R. solani* [154]. Moreover, *Trichoderma* has the ability to generate chitinase, which can minimise the survival of *R. solani* through the activation of the expression of chitinase genes [155]. Hernando José et al. (2021) discuss several isolated endophytic bacteria, fungi, and metabolites that have shown antagonistic and biocontrol activities as well as induced mechanisms against *Phytophthora* pathogens [156]. Some bacterial examples include *Pseudomonas fluorescence* isolated from the flowering vine saw greenbrier, which showed antagonistic effects against *P. parasitica*, *P. cinnamomi* and *P. palmivora*, as well as *Burkholderia* spp. isolated from the herb huperzine showing the ability to inhibit the growth of *P. capsici*. The isolate *Acinebacter calcoaceticus* from soybeans showed antagonistic effects against *P. sojae*. An isolate from a tomato plant (*Bacillus cereus*) helped to reduce the infection severity of *P. capsici* associated with cacao trees. Several strains of bacteria isolated from cucurbits including *Bacillus*, *Cronobacer*, *Enterobacteriaceae*, *Lactococcus*, *Pantoea*, and *Pediococus* showed antagonistic activity against *P. capsici* using various biocontrol mechanisms. Some fungal examples include species of *Trichoderma*, *Pestalotiopsis*, and *Fusarium* which were isolated from cacao trees and illustrated antagonism against *P. palmivora*. The fungus *Muscodor crispans* was isolated from a pineapple plant and showed inhibition properties against *P. cinnamomi* and *P. palmivora*. Hernando José et al. (2021) discuss numerous other bacterial and fungal strains that showed biocontrol activities and mechanisms against *P. capsici*, *P. infestans*, *P. citricola*, *P. cactorum*, and *P. pini* [156]. It is evident that there are various studies associated with the biocontrol of several *Phytophthora* species, but there are minimal studies regarding *P. alni*.

*4.1. Improving Plant Resistance via Microbe Inoculation and Genetic Resistant Breeding*

In recent studies, a greater focus has been placed on ways to detect and diagnose tree diseases, understand the interactions between trees and pathogens, develop disease-resistant trees, and ultimately optimise soil and tree microbiomes to improve plant health and behaviour [157–159]. The microbial communities associated with forest trees provide vital information relative to species health since numerous beneficial microbes have the ability to increase growth, development, productivity, and ecosystem function, as well as improve soil structure, and provide some resistance to pests and pathogenic diseases [30,33]. Alder trees may be inoculated with beneficial microbes in order to improve disease tolerance and environmental stressors, as well as enhance growth and quality. For example, Chandelier et al. (2015) discuss several inoculation methods in order to screen *A. glutinosa* resistance to *P. alni* [160]. These included wound inoculation, stem inoculation, seed and seedling inoculation, inoculation by flooding of root cuttings, and zoospore suspension inoculation. It was determined that the combination of zoospore suspension with flooding root systems was the most reliable since unwounded trees better mimicked natural environmental conditions [70]. Zaspel et al. (2014) investigated a way to increase the resistance of Common Alder against *P. alni* using in vitro and in planta analysis with a cyclolipopeptide (CLP)-producing *Pseudomonas veronii* metabolite (PAZ1) [70]. It was observed that inoculation with PAZ1 showed some inhibition effects as well as growth-promoting effects on the Alder species.

In order to aid the protection and integrity of forestry, natural genetic resistance breeding programmes of several species against biotic stressors have been studied [161]. Examples of resistance programmes include white pine and white pine blister rust resistance, Port Orford cedar and *P. lateralis*, Sitka spruce and white pine weevil resistance, Loblolly pine and fusiform rust resistance, *Pinus radiata* and *Dothistroma pini* resistance, Koa and koa wilt resistance, American beech and beech bark disease resistance, and Dutch elm disease [161–163]. Wei and Jousset (2017) suggest an alternative foundation to reach economically novel phenotypes by altering genetic information coupled with plant-associated microbiota [33,164]. Since pathogenic organisms have caused significant economic losses within forestry, greater research and screening techniques are required in order to expand bio-control procedures against forestry pathogens. Marco et al. (2022) extensively review microbe-assisted breeding programmes to date, particularly those associated with crops, and how this has improved the agricultural sector [165]. The Institute of Forest Genetics in Waldsieversdorf, Germany, was working on a selective breeding programme to improve the resistance of *A. glutinosa* to *P. alni*; however, this project appears to be complete since 2018 [166]. A greater focus on the development of natural genetic resistance breeding programmes specifically focused on the *Alnus* genus coupled with identifying species-specific biocontrol agents to minimise the infection of the pathogen *P. alni* will greatly aid in the protection of the *Alnus* species.

**Table 1.** Bio-control agents (BCAs) and the plant diseases they suppress via associated biocontrol mechanisms.

| Potential BCA | Classification | Pathogens They Potentially Control | Mechanism Expressed Showing Bio-Control Abilities | In Vitro/In Planta/In Vivo/Field/ Commercial | Reference |
|---|---|---|---|---|---|
| *A. extensa, A. stenoptera, A. alata, A. pulchella* | Legume (*Acacia*) | *Phytophthora cinnamomi* | Suppression and containment of *Phytophthora cinnamomi* inoculum in soil-infested areas to protect susceptible species and enhance species diversity | In planta | [125] |
| *Bacillus amyloliquefaciens* | Firmicutes | *Phytophthora cactorum* | Induces systemic resistance | In vitro | [151] |
| *Pseudomonas chlororaphis* | Gammaproteobacteria | *Phytophthora palmivor* | Antibiosis and suppression; reduce symptom severity; production of lytic enzymes, siderophores, and bio-surfactants | In vitro/in planta | [150] |
| *Pseudomonas putida 06909* | Gammaproteobacteria | *Phytophthora* spp. | Hypovirulence and suppression; colonise pathogenic hyphae and reduce pathogenic populations | Field | [149] |
| *Enterobacter aerogenes* | Proteobacteria | *Phytophthora lateralis* | Suppression; reduce disease symptoms and promote plant growth | Field | [124] |
| *Hypoxylon rubiginosum* | Ascomycetes | *Hymenoscyphus fraxineus* | Antibiosis; production of metabolites that are toxic to *H. fraxineus* | In vitro/in planta | [148] |
| *Aureimonas altamirensis* | Alphaproteobacteria | *Hymenoscyphus fraxineus* | Colonisation resistance; exopolysaccharide production; protein secretion; stress adaptation genes | In vitro/in planta | [129] |
| *Bacillus subtilis* | Firmicutes | *Fusarium* sp. | Antibiosis and antagonistic effects | In vitro | [144] |
| *Pseudomonas fluorescens* | Pseudomonas | *Fusarium oxysporum, Rhizoctonia solani, Macrophomira phaseolina, Fusarium* sp. | Antibiosis, hypovirulence, and suppression; inhibit plant disease, minimise fungal infections, protect seeds and roots from infection; production of secondary metabolites | In vitro/field | [51,52,167,168] |
| *daidzein* | Isoflavones | No specific pathogen | Antibiosis; antioxidant, anti-inflammatory, and anticancer effects | In vitro/in vivo | [133,134] |
| *Streptomyces* | Actinobacteria | *Acidovorax* spp., *Fusarium* sp., *Ralstonia solanacearum, Xanthomonas* spp., *Sclerotinia sclerotiorum, Erwinia amylovora* | Antibiosis and induced resistance; source of bioactive compounds like antimetabolites, antibiotics, extracellular enzymes, and antitumor agents | In vitro/in vivo | [48,130–133] |
| Chitinase | Enzyme | *Phytophthora cinnamomi* | Suppression; reduced spread, reduced sporangia production, applied in agriculture during crop cultivation for disease management, improved growth, and greater yields | Commercial/in vitro | [142,143] |
| IAA | Auxin Plant Hormone | *Fusarium* sp. | Systemic resistance; regulate growth and development processes | Commercial/in vitro | [38,53,145] |

*4.2. Microbiome Engineering and Its Potential Phytophtoria Control*

The practice of microbiome engineering has been applied recently to enhance human well-being, agricultural efficiency and combat challenges of bioremediation and contamination within the environment [169,170]. This involves various methods for the modification of host-associated microbiota to improve its health, resilience, and disease tolerance, as well as to benefit surrounding ecosystems [170]. These methods include enrichment, artificial selection, direct evolution, population control, pairwise interactions, microbiome transfer, synthetic microbes, and engineered interactions which are used to manipulate a microbiome to obtain desired characteristics [169,170]. In agriculture, a method used to manage plant diseases involves root microbiome transfer by combining fertile disease-suppressive soils with less fertile disease-conductive soils [171]. For example, Mendes et al. (2011) showed that by mixing these soil types, sugar beet soil was suppressive to the pathogen *R. solani* due to the presence of several species of Proteobacteria, Firmicutes, and Actinobacteria, which play a role in disease suppression. Santhanam et al. (2015) showed that synthetic root-associated microbiota transplants using a mixture of native bacterial isolates (species from *Arthrobacter*, *Bacillus*, and *Pseudomonas*) were effective at reducing the wilt disease of tobacco *Nicotiana attenuate* and that they increased crop resilience [172]. Mukherjee et al. (2022) used a bio-inoculant containing two bacterial strains isolated from chickpeas, *Enterobacter hormaechei* and *Brevundimonas naejangsanensis*, to enhance the productivity of inoculated chickpeas seeds, and it was observed that the consortium increased plant-growth attributes, yields, nutritional content, levels of IAA, siderophore, ammonia, phosphate solubilisation, and potassium solubilisation, as well as antagonistic activity against *Fusarium* sp. due to successful manipulation of the plant microbiome [173]. Wicaksono et al. (2017) inoculated wounds of kiwi fruit plants with *Pseudomonas* strains isolated from the medicinal plant Mānuka, and it was determined that the bacterial strains aided the pathogenic resistance of *P. syringae*, indicating that microbe transfer can be successful in reducing disease severity [174]. There is a great potential for the use of microbiome engineering in forestry as it offers an innovative solution to address various challenges in sustainable forest management. The manipulation of tree-associated microbiota can potentially enhance tree growth and pathogenic resistance as well as optimise nutrient cycles and improve tree tolerance to environmental stressors.

## 5. Conclusions and Future Perspectives

The Common Alder plays a significant ecological role and has many commercial uses; however, the native species is under threat of decline due to the known causal agent *Phytophthora alni*. Other *Phytophthora* species may also contribute to Alder decline, as well as other pests and pathogens to a lesser extent. Understanding the beneficial and pathogenic microbiota associated with Alder is crucial for developing biological solutions to control these threats. Forest trees have many relationships with a wide variety of microbial organisms, which are essential for tree health, nutrient conditions, and ecosystem functionality. Beneficial microbiota, such as PGPR/F, are present in the tree's endosphere, phyllosphere, and rhizosphere, providing benefits such as nutrient fixation, hormone production, pathogen suppression, and improved plant growth. Research associated with the microbiota of *Alnus* species is limited and mainly focuses on the root and rhizosphere soil. Bacterial phyla such as Actinobacteria, Acidobacteria, Proteobacteria, Bacteroidetes, and Firmicutes, as well as fungal phyla including Ascomycota, Basidiomycota, Ectomycorrhizal fungi, Glomeromycota, and Zygomycota, have been identified in association with the roots and rhizosphere of Alder trees.

Although, there is limited information available on the microbial communities in other parts of Alder trees, such as the leaves, bark, and catkins. Further research is needed to explore the microbial communities within different compartments of Alder trees and their potential interactions. Understanding the core microbiome of Alder and its functional roles will help improve our knowledge of Alder tree health, growth, and tolerance to (a)biotic stressors. Additionally, exploring the use of plant growth-promoting microbes in the forestry sector may result in beneficial applications for sustainable tree production and protection against pathogens. A greater global focus is needed regarding the breeding of microbe-optimised Alder trees that have a degree of resistance to infection caused by *P. alni*. Biocontrol agents have the ability to induce mechanisms to suppress pathogenic infections by direct attack on the pathogen, competition for resources, and indirectly due to a systemic stress response resulting in improved defence and resistance. The identification of species-specific biocontrol agents will aid in the protection of Alder trees from this disease threat. Also, the application of PGPR and PGPF to Alder affected by this disease could potentially pave the way to control its spread. The identification of biocontrol agents which can persist in sufficiently large populations within the microbial communities associated with Alder, and are capable of effectively expressing their biocontrol genes, is a very promising approach to minimise the severity and spread of devastating diseases of Alder and other important tree species. Another current knowledge gap that exists within this sector includes a detailed understanding of the systemic Alder microbiome and its seasonal and geographical dynamics. It is important to understand the different varieties and species of Alder as well as the type of microbes present in tolerant versus susceptible genotypes that can be exploited to improve the health of future Alder stands. Furthermore, having the ability to identify effective *Phytophthora* BCAs that can systematically colonise the tree and persist for long periods of time will aid the long-term survival of Alder to (a)biotic stressors. However, greater knowledge is required to identify effective methods of BCA inoculation of new and existing Alder stands, as well as what positive/negative effects these BCAs have on non-target microbial species within Alder stands.

**Author Contributions:** Conceptualization, E.F., K.J.G. and D.S.R.; writing—original draft preparation, E.F., K.J.G. and D.S.R.; writing—review and editing, E.F., K.J.G. and D.S.R.; supervision, K.J.G. and D.S.R. All authors have read and agreed to the published version of the manuscript.

**Funding:** This research was funded by the Irish Research Council Government of Ireland Postgraduate Scholarship GOIPG/2022/2262.

**Data Availability Statement:** Not applicable.

**Acknowledgments:** This review article was developed with the help of Kieran J. Germaine and Dheeraj Singh Rathore.

**Conflicts of Interest:** The authors declare no conflict of interest.

## References

1. Savill, P. The Silviculture of Trees Used in British Forestry. *CABI Int.* **2013**, *2*, 29–37.
2. Lalancette, S.; Lerat, S.; Roy, S.; Beaulieu, C. Fungal Endophytes of *Alnus incana* ssp. *rugosa* and *Alnus alnobetula* ssp. *crispa* and their potential to tolerate heavy metals and to promote plant growth. *Mycobiology* **2019**, *47*, 415–429. [CrossRef] [PubMed]
3. Claessens, H.; Oosterbaan, A.; Savill, P.; Rondeux, J. A review of the characteristics of black alder (*Alnus glutinosa (L.) Gaertn.*) and their implications for silvicultural practices. *For. Int. J. For. Res.* **2010**, *83*, 163–175. [CrossRef]
4. Durrant, T.; de Rigo, D.; Caudullo, G. Alnus glutinosa in Europe: Distribution, habitat, usage and threats. In *European Atlas of Forest Tree Species*; San-Miguel-Ayanz, J., de Rigo, D., Caudullo, G., Houston Durrant, T., Mauri, A., Eds.; Publication Office of the European Union: Luxembourg, 2016; pp. 64–65.
5. Petruzzello, M. *Alder*. Encyclopedia Britannica. 2021. Available online: https://www.britannica.com/plant/alder (accessed on 26 September 2022).
6. Mabberley, D.J. *Mabberley's Plant-Book: A Portable Dictionary of Plants, Their Classifications and Uses*; Cambridge University Press: Cambridge, UK, 2008; Volume 3, pp. 1–1021.
7. Parnell, J.; Curtis, T.G. *Webb's an Irish Flora*; Cork University Press: Cork, Ireland, 2012; pp. 1–556.
8. Roy, S.; Khasa, D.; Greer, C. Combining alders, frankiae, and mycorrhizae for the revegetation and remediation of contaminated ecosystems. *Can. J. Bot.* **2011**, *85*, 237–251. [CrossRef]
9. McVean, D.N. *Alnus glutinosa* (L.) Gaertn. *J. Ecol.* **1953**, *41*, 447–466. [CrossRef]
10. Buckley, D.J. Observations on Two Non-Native Alder Species (Betulaceae) Naturalising in Ireland. *Br. Ir. Bot.* **2021**, *3*, 90–98. [CrossRef]
11. Woodland Trust. *Alder (Alnus glutinosa)*. 2022. Available online: https://www.woodlandtrust.org.uk/trees-woods-and-wildlife/british-trees/a-z-of-british-trees/alder/ (accessed on 27 September 2022).

12. Keet, J.H.; Robertson, M.P.; Richardson, D.M. *Alnus glutinosa (Betulaceae)* in South Africa: Invasive potential and management options. *South Afr. J. Bot.* **2020**, *135*, 280–293. [CrossRef]
13. Funk, D.T. Alnus glutinosa (L.) Gaertn. European Alder. In *Agriculture Handbook*; U.S. Department of Agriculture, Forest Service: Washington, DC, USA, 1990; Volume 654, pp. 239–256.
14. Kajba, D.; Gračan, J. *EUFORGEN Technical Guidelines for Genetic Conservation and Use Black Alder (Alnus glutinosa)*; International Plant Genetic Resources Institute (IPGRI): Rome, Italy, 2003; p. 4.
15. Moore, C. Alder, Why the tree of death? *The Pallasboy Project.* 2015. Available online: https://thepallasboyvessel.wordpress.com/2015/11/18/alder-why-the-tree-of-death/ (accessed on 9 December 2022).
16. Sheridan, O. Alder (Alnus glutinosa) More Than Just a Native Species. 2020. Available online: https://www.teagasc.ie/news{-}{-}events/daily/forestry/alder-alnus-glutinosa-more-than-just-a-native-species.php (accessed on 19 September 2022).
17. Sati, S.C.; Sati, N.; Sati, O.P. Bioactive constituents and medicinal importance of genus *Alnus*. *Pharmacogn. Rev.* **2011**, *5*, 174–183. [CrossRef]
18. Acero, N.; Muñoz-Mingarro, D. Effect on tumor necrosis factor-$\alpha$ production and antioxidant ability of black alder, as factors related to its anti-inflammatory properties. *J. Med. Food* **2012**, *15*, 542–548. [CrossRef]
19. Nawirska-Olszańska, A.; Zaczyńska, E.; Czarny, A.; Kolniak-Ostek, J. Chemical Characteristics of Ethanol and Water Extracts of Black Alder (*Alnus glutinosa* L.) Acorns and Their Antibacterial, Anti-Fungal and Antitumor Properties. *Molecules* **2022**, *27*, 2804. [CrossRef] [PubMed]
20. Thiem, D.; Gołębiewski, M.; Hulisz, P.; Piernik, A.; Hrynkiewicz, K. How Does Salinity Shape Bacterial and Fungal Microbiomes of *Alnus glutinosa* Roots? *Front. Microbiol.* **2018**, *9*, 651. [CrossRef]
21. Fennessy, J. Common alder (*Alnus glutinosa*) as a forest tree in Ireland. *Coford Connect.* **2004**, 80–84. Available online: http://www.coford.ie/media/coford/content/publications/projectreports/cofordconnects/Alder-reprod.pdf (accessed on 2 November 2022).
22. Kennedy, C.E.J.; Southwood, T.R.E. The Number of Species of Insects Associated with British Trees: A Re-Analysis. *J. Anim. Ecol.* **1984**, *53*, 455–478. [CrossRef]
23. Bjelke, U.; Boberg, J.; Oliva, J.; Tattersdill, K.; Mckie, B.G. Dieback of riparian alder caused by the *Phytophthora alni* complex: Projected consequences for stream ecosystems. *Freshw. Biol.* **2016**, *61*, 565–579. [CrossRef]
24. Klaassen, R.K.W.M.; Creemers, J.G.M. Wooden foundation piles and its underestimated relevance for cultural heritage. *J. Cult. Herit.* **2012**, *13*, 123–128. [CrossRef]
25. European Commission. New EU forest Strategy for 2030. 2023. Available online: https://environment.ec.europa.eu/strategy/forest-strategy_en#:~:text=New%20EU%20forest%20strategy%20for%202030&text=The%20strategy%20will%20contribute%20to,and%20climate%20neutrality%20by%202050 (accessed on 18 July 2023).
26. Valor, T.; Camprodon, J.; Buscarini, S.; Casals, P. Drought-induced dieback of riparian black alder as revealed by tree rings and oxygen isotopes. *For. Ecol. Manag.* **2020**, *478*, 1–9. [CrossRef]
27. Teshome, D.T.; Zharare, G.E.; Naidoo, S. The Threat of the Combined Effect of Biotic and Abiotic Stress Factors in Forestry Under a Changing Climate. *Front. Plant Sci.* **2020**, *11*, 1874. [CrossRef]
28. O'Hanlon, R.; Wilson, J.; Cox, D. Investigations into *Phytophthora* dieback of alder along the river Lagan in Belfast, Northern Ireland, ed. B. Tobin. *Ir. For. Soc. Ir. For.* **2020**, *77*, 33–43.
29. Mercado-Blanco, J.; Abrantes, I.; Barra Caracciolo, A.; Bevivino, A.; Ciancio, A.; Grenni, P.; Hrynkiewicz, K.; Kredics, L.; Proença, D.N. Belowground Microbiota and the Health of Tree Crops. *Front. Microbiol.* **2018**, *9*, 1006. [CrossRef] [PubMed]
30. Gałązka, A.; Marzec-Grządziel, A.; Varsadiya, M.; Niedźwiecki, J.; Gawryjołek, K.; Furtak, K.; Przybyś, M.; Grządziel, J. Biodiversity and Metabolic Potential of Bacteria in Bulk Soil from the Peri-Root Zone of Black Alder (*Alnus glutinosa*), Silver Birch (*Betula pendula*) and Scots Pine (*Pinus sylvestris*). *Int. J. Mol. Sci.* **2022**, *23*, 2633. [CrossRef] [PubMed]
31. Baldrian, P. Forest microbiome: Diversity, complexity and dynamics. *FEMS Microbiol. Rev.* **2016**, *41*, 109–130. [CrossRef] [PubMed]
32. Lakshmanan, V.; Selvaraj, G.; Bais, H.P. Functional soil microbiome: Belowground solutions to an aboveground problem. *Plant Physiol* **2014**, *166*, 689–700. [CrossRef] [PubMed]
33. Tian, L.; Lin, X.; Tian, J.; Ji, L.; Chen, Y.; Tran, L.-S.P.; Tian, C. Research Advances of Beneficial Microbiota Associated with Crop Plants. *Int. J. Mol. Sci.* **2020**, *21*, 1792. [CrossRef] [PubMed]
34. Gafur, A. Plant growth promoting microbes (PGPM) for the sustainability of tropical plantation forests in Indonesia. *Acad. Open* **2021**, 1–3. [CrossRef]
35. Vocciante, M.; Grifoni, M.; Fusini, D.; Petruzzelli, G.; Franchi, E. The Role of Plant Growth-Promoting Rhizobacteria (PGPR) in Mitigating Plants Environmental Stresses. *Appl. Sci.* **2022**, *12*, 1231. [CrossRef]
36. Lladó, S.; López-Mondéjar, R.; Baldrian, P. Forest Soil Bacteria: Diversity, Involvement in Ecosystem Processes, and Response to Global Change. *Microbiol. Mol. Biol. Rev.* **2017**, *81*, e00063-16. [CrossRef]
37. Lauber, C.L.; Hamady, M.; Knight, R.; Fierer, N. Pyrosequencing-based assessment of soil pH as a predictor of soil bacterial community structure at the continental scale. *Appl. Environ. Microbiol.* **2009**, *75*, 5111–5120. [CrossRef]
38. Fu, S.F.; Wei, J.Y.; Chen, H.W.; Liu, Y.Y.; Lu, H.Y.; Chou, J.Y. Indole-3-acetic acid: A widespread physiological code in interactions of fungi with other organisms. *Plant Signal. Behav.* **2015**, *10*, e1048052. [CrossRef]
39. Polko, J.K.; Kieber, J.J. 1-Aminocyclopropane 1-Carboxylic Acid and Its Emerging Role as an Ethylene-Independent Growth Regulator. *Front. Plant Sci.* **2019**, *10*, 1602. [CrossRef]

40. Glick, B.R. Bacteria with ACC deaminase can promote plant growth and help to feed the world. *Microbiol. Res.* **2014**, *169*, 30–39. [CrossRef]
41. Sulochana, M.B.; Jayachandra, S.Y.; Kumar, S.A.; Parameshwar, A.B.; Reddy, K.M.; Dayanand, A. Siderophore as a potential plant growth-promoting agent produced by *Pseudomonas aeruginosa* JAS-25. *Appl. Biochem. Biotechnol.* **2014**, *174*, 297–308.
42. Behera, B.C.; Sethi, B.K.; Mishra, R.R.; Dutta, S.K.; Thatoi, H.N. Microbial cellulases—Diversity & biotechnology with reference to mangrove environment: A review. *J. Genet. Eng. Biotechnol.* **2017**, *15*, 197–210. [PubMed]
43. Bhattacharyya, C.; Banerjee, S.; Acharya, U.; Mitra, A.; Mallick, I.; Haldar, A.; Haldar, S.; Ghosh, A.; Ghosh, A. Evaluation of plant growth promotion properties and induction of antioxidative defense mechanism by tea rhizobacteria of Darjeeling, India. *Sci. Rep.* **2020**, *10*, 15536.
44. Phitsuwan, P.; Laohakunjit, N.; Kerdchoechuen, O.; Kyu, K.L.; Ratanakhanokchai, K. Present and potential applications of cellulases in agriculture, biotechnology, and bioenergy. *Folia Microbiol.* **2013**, *58*, 163–176.
45. Richter, B.S.; Ivors, K.; Shi, W.; Benson, D.M. Cellulase activity as a mechanism for suppression of *phytophthora* root rot in mulches. *Phytopathology* **2011**, *101*, 223–230.
46. Confortin, T.C.; Spannemberg, S.S.; Todero, I.; Luft, L.; Brun, T.; Albornoz, E.; Kuhn, R.C.; Mazutti, M. Chapter 21—Microbial Enzymes as Control Agents of Diseases and Pests in Organic Agriculture. In *New and Future Developments in Microbial Biotechnology and Bioengineering*; Gupta, V.K., Pandey, A., Eds.; Elsevier: Amsterdam, The Netherlands, 2019; pp. 321–332.
47. Callender, K.L.; Roy, S.; Khasa, D.P.; Whyte, L.G.; Greer, C.W. Actinorhizal Alder Phytostabilization Alters Microbial Community Dynamics in Gold Mine Waste Rock from Northern Quebec: A Greenhouse Study. *PLoS ONE* **2016**, *11*, e0150181.
48. McEwan, N.R.; Wilkinson, T.J.; Girdwood, S.E.; Snelling, T.J.; Peate, T.; Dougal, K.; Jones, D.L.; Godbold, D.L. Evaluation of the microbiome of decaying alder nodules by next generation sequencing. *Endocytobiosis Cell Res.* **2017**, *28*, 14–19.
49. Zappelini, C.; Alvarez-Lopez, V.; Capelli, N.; Guyeux, C.; Chalot, M. *Streptomyces* Dominate the Soil Under *Betula* Trees That Have Naturally Colonized a Red Gypsum Landfill. *Front. Microbiol.* **2018**, *9*, 1772. [PubMed]
50. Elo, S.; Maunuksela, L.; Salkinoja-Salonen, M.; Smolander, A.; Haahtela, K. Humus bacteria of Norway spruce stands: Plant growth promoting properties and birch, red fescue and alder colonizing capacity. *FEMS Microbiol. Ecol.* **2000**, *31*, 143–152. [PubMed]
51. Dukunde, A.; Schneider, D.; Schmidt, M.; Veldkamp, E.; Daniel, R. Tree Species Shape Soil Bacterial Community Structure and Function in Temperate Deciduous Forests. *Front. Microbiol.* **2019**, *10*, 1519.
52. Probanza, A.; Lucas, J.A.; Acero, N.; Gutierrez Mañero, F.J. The influence of native rhizobacteria on European alder (*Alnus glutinosa* (L.) *Gaertn.*) Growth. *Plant Soil* **1996**, *182*, 59–66.
53. Gutiérrez-Mañero, F.J.; Ramos, B.; Probanza, A.; Mehouachi, J.; Tadeo, F.R.; Talon, M. The plant-growth-promoting rhizobacteria *Bacillus pumilus* and *Bacillus licheniformis* produce high amounts of physiologically active gibberellins. *Physiol. Plant.* **2001**, *111*, 206–211.
54. Egamberdiyeva, D. Effect of plant growth promoting bacteria on growth of Scots pine and silver birch seedlings. *Uzb. J. Agric.* **2005**, *1*, 66–69.
55. Imperato, V.; Kowalkowski, L.; Portillo-Estrada, M.; Gawronski, S.W.; Vangronsveld, J.; Thijs, S. Characterisation of the Carpinus *betulus* L. Phyllomicrobiome in Urban and Forest Areas. *Front. Microbiol.* **2019**, *10*, 1110.
56. Lahlali, R.; Ezrari, S.; Radouane, N.; Kenfaoui, J.; Esmaeel, Q.; El Hamss, H.; Belabess, Z.; Barka, E.A. Biological Control of Plant Pathogens: A Global Perspective. *Microorganisms* **2022**, *10*, 596.
57. Dąbrowska, G.B.; Garstecka, Z.; Trejgell, A.; Dąbrowski, H.P.; Konieczna, W.; Szyp-Borowska, I. The Impact of Forest Fungi on Promoting Growth and Development of *Brassica napus* L. *Agronomy* **2021**, *11*, 2475.
58. Thiem, D.; Piernik, A.; Hrynkiewicz, K. Ectomycorrhizal and endophytic fungi associated with *Alnus glutinosa* growing in a saline area of central Poland. *Symbiosis* **2018**, *75*, 17–28.
59. Thiem, D.; Tyburski, J.; Golebiewski, M.; Hrynkiewicz, K. Halotolerant fungi stimulate growth and mitigate salt stress in *Alnus glutinosa Gaertn.* *Dendrobiology* **2020**, *83*, 30–42.
60. Xu, R.; Li, T.; Cui, H.; Wang, J.; Yu, X.; Ding, Y.; Wang, C.; Yang, Z.; Zhao, Z. Diversity and characterization of Cd-tolerant dark septate endophytes (DSEs) associated with the roots of Nepal alder (*Alnus nepalensis*) in a metal mine tailing of southwest China. *Appl. Soil Ecol.* **2015**, *93*, 11–18.
61. Cordero, R.J.; Casadevall, A. Functions of fungal melanin beyond virulence. *Fungal Biol. Rev.* **2017**, *31*, 99–112.
62. Hrynkiewicz, K.; Szymańska, S.; Piernik, A.; Thiem, D. Ectomycorrhizal Community Structure of Salix and *Betula* spp. at a Saline Site in Central Poland in Relation to the Seasons and Soil Parameters. *Water Air Soil Pollut.* **2015**, *226*, 99. [PubMed]
63. Sousa, N.R.; Franco, A.R.; Ramos, M.A.; Oliveria, R.S.; Castro, P.M.L. The response of *Betula* pubescens to inoculation with an ectomycorrhizal fungus and a plant growth promoting bacterium is substrate-dependent. *Ecol. Eng.* **2015**, *81*, 439–443.
64. Badalamenti, E.; Catania, V.; Sofia, S.; Sardina, M.T.; Sala, G.; La Mantia, T.; Quatrini, P. The Root Mycobiota of *Betula aetnensis* Raf., an Endemic Tree Species Colonizing the Lavas of Mt. Etna (Italy). *Forests* **2021**, *12*, 1624.
65. Orfanoudakis, M.; Wheeler, C.T.; Hooker, J.E. Both the arbuscular mycorrhizal fungus *Gigaspora rosea* and *Frankia* increase root system branching and reduce root hair frequency in *Alnus glutinosa*. *Mycorrhiza* **2010**, *20*, 117–126.
66. Karunarathna, S.C.; Ashwath, N.; Jeewon, R. Editorial: The Potential of Fungi for Enhancing Crops and Forestry Systems. *Front. Microbiol.* **2021**, *12*, 813051. [PubMed]

67. Argumedo-Delira, R.; Gómez-Martínez, M.J.; Mora-Delgado, J. Plant Growth Promoting Filamentous Fungi and Their Application in the Fertilization of Pastures for Animal Consumption. *Agronomy* **2022**, *12*, 3033.
68. Aguiar-Pulido, V.; Huang, W.; Suarez-Ulloa, V.; Cickovski, T.; Mathee, K.; Narasimhan, G. Metagenomics, Metatranscriptomics, and Metabolomics Approaches for Microbiome Analysis. *Evol. Bioinform Online* **2016**, *12* (Suppl. S1), 5–16.
69. Pujic, P.; Alloisio, N.; Miotello, G.; Armengaud, J.; Abrouk, D.; Fournier, P.; Normand, P. The Proteogenome of Symbiotic *Frankia alni* in *Alnus glutinosa* Nodules. *Microorganisms* **2022**, *10*, 651.
70. Gomes Marques, I.; Faria, C.; Conceição, S.I.R.; Jansson, R.; Corcobando, T.; Milanovic, S.; Laurent, Y.; Bernez, I.; Dufour, S.; Mandák, B.; et al. Germination and seed traits in common alder (*Alnus* spp.): The potential contribution of rear-edge populations to ecological restoration success. *Restor. Ecol.* **2022**, *30*, e13517.
71. Zaspel, I.; Naujoks, G.; Krüger, L.; Pham, L.H. Promotion of resistance of black alder clones (*Alnus glutinosa (L.) Gaertn.*) against *Phytophthora alni* ssp. *alni* by cyclolipopeptide producing bacteria. *Silvae Genet.* **2014**, *63*, 222–229.
72. Vacek, Z.; Vacek, S.; Cukor, J.; Bulušek, D.; Slávik, M.; Lukáčik, I.; Štefančík, I.; Sitková, Z.; Eşen, D.; Ripullone, F.; et al. Dendrochronological data from twelve countries proved definite growth response of black alder ([L.] *Gaertn.*) to climate courses across its distribution range. *Cent. Eur. For. J.* **2022**, *68*, 139–153. [CrossRef]
73. Asghari, R.; Rahimian, H.; Babaeizad, V. Isolation of *Rhizobium* strains from Alder root nodules in Mazandaran. In Proceedings of the 22nd Iran Plant Protection Congress, Karaj, Iran, 27–30 August 2016; Volume 2, p. 103.
74. Bogar, L.M.; Dickie, I.A.; Kennedy, P.G. Testing the co-invasion hypothesis: Ectomycorrhizal fungal communities on *Alnus glutinosa* and *Salix fragilis* in New Zealand. *Divers. Distrib.* **2015**, *21*, 268–278. [CrossRef]
75. Sukhikh, S.; Ivanova, S.; Skrypnik, L.; Bakhtiyarova, A.; Larina, V.; Krol, O.; Prosekov, A.; Frolov, A.; Povydysh, M.; Babich, O. Study of the Antioxidant Properties of *Filipendula ulmaria* and *Alnus glutinosa*. *Plants* **2022**, *11*, 2415. [CrossRef]
76. Mushkina, V. Effect of wound healing in gels containing tinctures of *Alnus glutinosa* (L) leaves. *Clin. Phytoscience* **2021**, *7*, 62. [CrossRef]
77. Altınyay, Ç.; Eryılmaz, M.; Yazgan, A.N.; Sever Yılmaz, B.; Altun, M.L. Antimicrobial activity of some *Alnus* species. *Eur. Rev. Med. Pharmacol. Sci.* **2015**, *19*, 4671–4674.
78. Kayini, A.; Pandey, R.R. Phyllosphere Fungi of *Alnus nepalensis*, *Castanopsis hystrix* and *Schima walichii* in a Subtropical Forest of North East India. *J. Am. Sci.* **2010**, *6*, 118–123.
79. Berges, R.; Seemüller, E. Impact of phytoplasma infection of common alder (*Alnus glutinosa*) depends on strain virulence. *For. Pathol.* **2002**, *32*, 357–363. [CrossRef]
80. Atanasova, B.; Spasov, D.; Jakovljević, M.; Jović, J.; Krstić, O.; Mitrović, M.; Cvrković, T. First Report of Alder Yellows Phytoplasma Associated with Common Alder (*Alnus glutinosa*) in the Republic of Macedonia. *Plant Dis.* **2014**, *98*, 1268. [CrossRef] [PubMed]
81. Marcone, C.; Franco-Lara, L.; Toševski, I. Major Phytoplasma Diseases of Forest and Urban Trees. In *Phytoplasmas: Plant Pathogenic Bacteria—I*; Rao, G., Bertaccini, A., Fiore, N., Liefting, L., Eds.; Springer: Singapore, 2018; pp. 287–312.
82. Cvrković, T.; Jović, J.; Mitrović, M.; Petrović, A.; Krnjajić, S.; Malembic-Maher, S.; Toševski, I. First report of alder yellows phytoplasma on common alder (*Alnus glutinosa*) in Serbia. *Plant Pathol.* **2008**, *57*, 773. [CrossRef]
83. Forbes, H. Plant of the Week—16 November—Alder tongue (Taphrina alni) (fungus), in Botany in Scotland. 2020. Available online: https://botsocscot.wordpress.com/2020/11/15/plant-of-the-week-date-alder-tongue-taphrina-alni-fungus/ (accessed on 25 March 2023).
84. Sims, L. Alder (Alnus spp.)-Leaf Spot. 2012. Available online: https://pnwhandbooks.org/plantdisease/host-disease/alder-alnus-spp-leaf-spot (accessed on 25 March 2023).
85. Moricca, S.; Benigno, A.; Oliveira Longa, C.M.; Cacciola, S.O.; Maresi, G. First Documentation of Life Cycle Completion of the Alien Rust Pathogen *Melampsoridium hiratsukanum* in the Eastern Alps Proves Its Successful Establishment in This Mountain Range. *J. Fungi* **2021**, *7*, 617. [CrossRef]
86. Tomoshevich, M.; Kirichenko, N.; Holmes, K.A.; Kenis, M.; Stenlid, J. Foliar fungal pathogens of European woody plants in Siberia: An early warning of potential threats? *For. Pathol.* **2013**, *43*, 345–359. [CrossRef]
87. Surico, G.; Mugnai, L.; Pastorelli, R.; Giovannetti, L.; Stead, D.E. Erwinia alni, a New Species Causing Bark Cankers of Alder (Alnus Miller) Species. *Int. J. Syst. Evol. Microbiol.* **1996**, *46*, 720–726. [CrossRef]
88. Scortichini, M. Leaf necrosis and sucker and twig dieback of *Alnus glutinosa* incited by *Pseudomonas syringae pv. syringae*. *Eur. J. For. Pathol.* **1997**, *27*, 331–336. [CrossRef]
89. Dolch, R.; Tscharntke, T. Defoliation of alders (*Alnus glutinosa*) affects herbivory by leaf beetles on undamaged neighbours. *Oecologia* **2000**, *125*, 504–511. [CrossRef] [PubMed]
90. The Sawflies (*Symphyta*) of Britain and Ireland. *Monsoma pulveratum (Retzius 1783)*. 2021. Available online: https://www.sawflies.org.uk/monsoma-pulveratum/ (accessed on 25 March 2023).
91. The Sawflies (*Symphyta*) of Britain and Ireland. *Eriocampa ovata (Linnaeus 1760)*. 2021. Available online: https://www.sawflies.org.uk/eriocampa-ovata/ (accessed on 25 March 2023).
92. Edmunds, H.A.; Springate, N.D. Cimbex connatus (Schrank) (Hymenoptera: Cimbicidae): A rare species of sawfly in the British Isles. *Br. J. Entomol. Nat. Hist.* **1998**, *11*, 65–68.
93. Érsek, L. Fenusa Dohrnii (Tischbein 1846) European Alder Leafminer. 2019. Available online: https://bladmineerders.nl/parasites/animalia/arthropoda/insecta/hymenoptera/symphyta/tenthredinoidea/tenthredinidae/heterarthrinae/fenusa/fenusa-dohrnii/ (accessed on 26 March 2023).

94. Sims, L.L. *Phytophthora* Species and Riparian Alder Tree Damage in Western Oregon State University. Ph.D. Thesis, Botany and Plant Pathology, Western Oregon State University, Monmouth, OR, USA, 2014; pp. 1–12.
95. Kroon, L.P.; Brouwer, H.; de Cock, A.W.A.M.; Govers, F. The genus *Phytophthora* anno 2012. *Phytopathology* **2012**, *102*, 348–364. [CrossRef]
96. Brasier, C.; Scanu, B.; Cooke, D.; Jung, T. *Phytophthora*: An ancient, historic, biologically and structurally cohesive and evolutionarily successful generic concept in need of preservation. *IMA Fungus* **2022**, *13*, 12. [CrossRef]
97. Haque, M. Identification, characterization and pathogenicity of *Phytophthora* spp. associated with the mortality of *Alnus glutinosa* in Spain. Ph.D. Thesis, University of Valladolid, Valladolid, Spain, 2014; p. 6.
98. Cai, G.; Hillman, B.I. Chapter Twelve—Phytophthora Viruses. In *Advances in Vrius Research*; Ghabrial, S.A., Ed.; Academic Press: Cambridge, MA, USA, 2013; Volume 86, pp. 327–350.
99. Wingfield, M.J. Pathology | Disease Affecting Exotic Plantation Species. In *Encyclopedia of Forest Sciences*; Burley, J., Ed.; Elsevier: Oxford, UK, 2004; pp. 816–822.
100. Abad, Z.G.; Burgess, T.I.; Redford, A.J.; Bienapfl, J.C.; Srivastava, S.; Mathew, R.; Jennings, K. IDphy: An international online resource for molecular and morphological identification of *Phytophthora*. *Plant Dis.* **2023**, *107*, 987–998. [CrossRef]
101. Oßwald, W.; Fleischmann, F.; Rigling, D.; Coelho, A.C.; Cravador, A.; Diez, J.; Dalio, R.J.; Horta Jung, M.; Pfanz, H.; Robin, C.; et al. Strategies of attack and defence in woody plant–*Phytophthora* interactions. *For. Pathol.* **2014**, *44*, 169–190. [CrossRef]
102. Beckerman, J.; Creswell, T. Phytophthora Diseases in Ornamentals. Plant Pathology in the Landscape Series. 20201-6. Available online: https://www.extension.purdue.edu/extmedia/BP/BP-215-W.pdf (accessed on 11 January 2023).
103. Nave, C.; Schwan, J.; Werres, S.; Riebesehl, J. *Alnus glutinosa* Threatened by Alder *Phytophthora*: A Histological Study of Roots. *Pathogens* **2021**, *10*, 977. [CrossRef]
104. Gibbs, J.; van Dijk, C.; Webber, J. Phytophthora disease of alder in Europe. In *Forestry Commission Bulletin 126*; Forestry Commission: Edinburgh, UK, 2003; pp. 1–82.
105. Thoirain, B.; Husson, C.; Marçais, B. Risk factors for the *phytophthora*-induced decline of alder in northeastern France. *Phytopathology* **2007**, *97*, 99–105. [CrossRef]
106. Černý, K.; Strnadová, V. *Phytophthora* Alder Decline: Disease Symptoms, Causal Agent and its Distribution in the Czech Republic. *Plant Prot. Sci.* **2010**, *46*, 12–18. [CrossRef]
107. Aguayo, J.; Halkett, F.; Husson, C.; Nagy, Z.Á.; Szigethy, A.; Bakonyi, J.; Frey, P.; Marçais, B. Genetic Diversity and Origins of the Homoploid-Type Hybrid *Phytophthora alni*. *Appl. Environ. Microbiol.* **2016**, *82*, 7142–7153. [CrossRef]
108. Gibbs, J.N. *Phytophthora* root disease of alder in Britain. *EPPO Bull.* **1995**, *25*, 661–664. [CrossRef]
109. Marçais, B. *Phytophthora Alni Species Complex (Alder Phytophthora)*; CABI International; CABI Compendium: Wallingford, UK, 2022. [CrossRef]
110. Brasier, C.M.; Kirk, S.A.; Delcan, J.; Cooke, D.E.; Jung, T.; Man in't Veld, W.A. *Phytophthora alni* sp. *nov.* and its variants: Designation of emerging heteroploid hybrid pathogens spreading on Alnus trees. *Mycol. Res.* **2004**, *108 Pt 10*, 1172–1184. [CrossRef] [PubMed]
111. Bregant, C.; Sanna, G.P.; Bottos, A.; Maddau, L.; Montecchio, L.; Linaldeddu, B.T. Diversity and Pathogenicity of *Phytophthora* Species Associated with Declining Alder Trees in Italy and Description *of Phytophthora alpina* sp. *nov.* *Forests* **2020**, *11*, 848. [CrossRef]
112. Husson, C.; Aguayo, J.; Revellin, C.; Frey, P.; Ioos, R.; Marçais, B. Evidence for homoploid speciation in *Phytophthora alni* supports taxonomic reclassification in this species complex. *Fungal Genet. Biol.* **2015**, *77*, 12–21. [CrossRef]
113. Trzewik, A.; Maciorowski, R.; Orlikowska, T. Pathogenicity of *Phytophthora & times; alni* isolates obtained from symptomatic trees, soil and water against Alder. *Forests* **2022**, *13*, 20.
114. Jung, T.; Pérez-Sierra, A.; Durán, A.; Horta Jung, M.; Balci, Y.; Scanu, B. Canker and decline diseases caused by soil- and airborne *Phytophthora* species in forests and woodlands. *Persoonia* **2018**, *40*, 182–220. [CrossRef]
115. Seddaiu, S.; Linaldeddu, B.T. First Report of *Phytophthora acerina, plurivora*, and *P. pseudocryptogea* Associated with Declining Common Alder Trees in Italy. *Plant Dis.* **2020**, *104*, 1874. [CrossRef]
116. Hansen, E.; Reeser, P.; Rooney-Latham, S. Forest Phytophthoras of the World: *Phytophthora siskiyouensis*. *For. Phytophthoras* **2012**, *1*. [CrossRef]
117. Matsiakh, I.; Kramarets, V.; Cleary, M. Occurrence and diversity of *Phytophthora* species in declining broadleaf forests in western Ukraine. *For. Pathol.* **2021**, *51*, e12662. [CrossRef]
118. Tkaczyk, M.; Sikora, K.; Galko, J.; Kunka, A. Occurrence of *Phytophthora* species in riparian stands of black alder (*Alnus glutinosa*) in Slovakia. *For. Pathol.* **2023**, *53*, e12800. [CrossRef]
119. Zamora-Ballesteros, C.; Haque, M.M.U.; Diez, J.J.; Martín-García, J. Pathogenicity of *Phytophthora alni* complex and *P. plurivora* in *Alnus glutinosa* seedlings. *For. Pathol.* **2017**, *47*, e12299. [CrossRef]
120. Haque, M.M.U.; Martín-García, J.; Diez, J.J. Variation in pathogenicity among the three subspecies of *Phytophthora alni* on detached leaves, twigs and branches of *Alnus glutinosa*. *For. Pathol.* **2015**, *45*, 484–491. [CrossRef]
121. Downing, M.C.; Jung, T.; Thomas, V.; Blaschke, M.; Tuffly, M.F.; Reich, R. *Estimating the Susceptibility to Phytophthora alni Globally Using Both Statistical Analyses and Expert Knowledge*; General Technical Report; Pacific Northwest Research Station, USDA Forest Service: Washington, DC, USA, 2010; Volume 802, pp. 559–570.

122. Smeriglio, A.; D'Angelo, V.; Cacciola, A.; Ingegneri, M.; Raimondo, F.M.; Trombetta, D.; Germanò, M.P. New Insights on Phytochemical Features and Biological Properties of *Alnus glutinosa* Stem Bark. *Plants* **2022**, *11*, 2499. [CrossRef] [PubMed]
123. Pirttilä, A.M.; Mohammad Parast Tabas, H.; Baruah, N.; Koskimäki, J.J. Biofertilizers and Biocontrol Agents for Agriculture: How to Identify and Develop New Potent Microbial Strains and Traits. *Microorganisms* **2021**, *9*, 817. [CrossRef]
124. Köhl, J.; Kolnaar, R.; Ravensberg, W.J. Mode of Action of Microbial Biological Control Agents against Plant Diseases: Relevance beyond Efficacy. *Front. Plant Sci.* **2019**, *10*, 845. [CrossRef] [PubMed]
125. Utkhede, R.; Stephen, B.; Wong, S. Control of *Phytophthora lateralis* root rot of Lawson cypress with *Enterobacter aerogenes*. *J. Arboric.* **1997**, *23*, 144–146. [CrossRef]
126. D'Souza, N.; Colquhoun, I.J.; Sheared, B.L.; Hardy, G.E. Assessing the potential for biological control of *Phytophthora cinnamomi* by fifteen native Western Australian jarrah-forest legume species. *Australas. Plant Pathol.* **2005**, *34*, 533–540. [CrossRef]
127. Bonaterra, A.; Badosa, E.; Daranas, N.; Francés, J.; Roselló, G.; Montesinos, E. Bacteria as Biological Control Agents of Plant Diseases. *Microorganisms* **2022**, *10*, 1759. [CrossRef]
128. Ulrich, K.; Becker, R.; Behrendt, U.; Kube, M.; Ulrich, A. A Comparative Analysis of Ash Leaf-Colonizing Bacterial Communities Identifies Putative Antagonists of *Hymenoscyphus fraxineus*. *Front. Microbiol.* **2020**, *11*, 966. [CrossRef]
129. Prospero, S.; Botella, L.; Santini, A.; Robin, A. Biological control of emerging forest diseases: How can we move from dreams to reality? *For. Ecol. Manag.* **2021**, *496*, 119377. [CrossRef]
130. Becker, R.; Ulrich, K.; Behrendt, U.; Schneck, V.; Ulrich, A. Genomic Characterization of *Aureimonas altamirensis* C2P003—A Specific Member of the Microbiome of *Fraxinus excelsior* Trees Tolerant to Ash Dieback. *Plants* **2022**, *11*, 3487. [CrossRef] [PubMed]
131. Olanrewaju, O.S.; Babalola, O.O. *Streptomyces*: Implications and interactions in plant growth promotion. *Appl. Microbiol. Biotechnol.* **2019**, *103*, 1179–1188. [CrossRef] [PubMed]
132. Le, K.D.; Yu, N.H.; Park, A.R.; Park, D.J.; Kim, C.J.; Kim, J.C. *Streptomyces* sp. AN090126 as a Biocontrol Agent against Bacterial and Fungal Plant Diseases. *Microorganisms* **2022**, *10*, 791. [CrossRef]
133. Ghanem, G.A.M.; Gebily, D.A.S.; Ragab, M.M.; Ali, A.M.; Soliman, N.E.D.K.; El-Moity, T.H.A. Efficacy of antifungal substances of three *Streptomyces* spp. against different plant pathogenic fungi. *Egypt. J. Biol. Pest Control* **2022**, *32*, 112. [CrossRef]
134. Liu, N.; Wang, H.; Lie, M.; Gu, Q.; Zheng, W.; Huang, Y. *Streptomyces alni* sp. nov., a daidzein-producing endophyte isolated from a root of Alnus nepalensis D. Don. *Int. J. Syst. Evol. Microbiol.* **2009**, *59*, 254–258. [CrossRef]
135. Alshehri, M.M.; Sharifi-Rad, J.; Herrera-Bravo, J.; Jara, E.L.; Salazar, L.A.; Kregiel, D.; Uprety, Y.; Akram, M.; Iqbal, M.; Martorell, M.; et al. Therapeutic Potential of Isoflavones with an Emphasis on Daidzein. *Oxid. Med. Cell Longev.* **2021**, *2021*, 6331630. [CrossRef]
136. Ekundayo, F.O.; Folorunsho, A.E.; Ibisanmi, T.A.; Olabanji, O.B. Antifungal activity of chitinase produced by *Streptomyces* species isolated from grassland soils in Futa Area, Akure. *Bull. Natl. Res. Cent.* **2022**, *46*, 95. [CrossRef]
137. Legein, M.; Smets, W.; Vandenheuvel, D.; Eilers, T.; Muyshondt, B.; Prinsen, E.; Samson, R.; Lebeer, S. Modes of Action of Microbial Biocontrol in the Phyllosphere. *Front. Microbiol.* **2020**, *11*, 1619. [CrossRef]
138. Daguerre, Y.; Siegel, K.; Edel-Hermann, V.; Steinberg, C. Fungal proteins and genes associated with biocontrol mechanisms of soil-borne pathogens: A review. *Fungal Biol. Rev.* **2014**, *28*, 97–125. [CrossRef]
139. Nelkner, J.; Tejerizo, G.T.; Hassa, J.; Lin, T.W.; Witte, J.; Verwaaijen, B.; Winkler, A.; Bunk, B.; Spröer, C.; Overmann, J.; et al. Genetic Potential of the Biocontrol Agent *Pseudomonas brassicacearum* (Formerly P. trivialis) 3Re2-7 Unraveled by Genome Sequencing and Mining, Comparative Genomics and Transcriptomics. *Genes* **2019**, *10*, 601. [CrossRef] [PubMed]
140. Funk, A.J.; McCullumsmith, R.E.; Haroutunian, V.; Meador-Woodruff, J.H. Abnormal activity of the MAPK- and cAMP-associated signaling pathways in frontal cortical areas in postmortem brain in schizophrenia. *Neuropsychopharmacology* **2012**, *37*, 896–905. [CrossRef] [PubMed]
141. Tomaž, Š.; Gruden, K.; Coll, A. TGA transcription factors—Structural characteristics as basis for functional variability. *Front. Plant Sci.* **2022**, *13*, 935819. [CrossRef]
142. Liao, K.; Peng, Y.J.; Yuan, L.B.; Dai, Y.S.; Chen, Q.F.; Yu, L.J.; Bai, M.Y.; Zhang, W.Q.; Xie, L.J.; Xiao, S. Brassinosteroids Antagonize Jasmonate-Activated Plant Defense Responses through BRI1-EMS-SUPPRESSOR1 (BES1). *Plant Physiol.* **2020**, *182*, 1066–1082. [CrossRef]
143. Kumar, M.; Brar, A.; Yadav, M.; Chawade, A.; Vivekanand, V.; Pareek, N. Chitinases—Potential Candidates for Enhanced Plant Resistance towards Fungal Pathogens. *Agriculture* **2018**, *8*, 88. [CrossRef]
144. Trinh, T.H.T.; Nguyen, V.B.; Tran, D.M.; Doan, C.T.; Tran, T.N.; Wang, S.L.; Kosta, K.; Szkladanyi, S.; Le, M.H.; Nguyen, A.D. A Potent *Fusarium* Antagonistic Bacterium *Bacillus subtilis* RB.CJ41 Isolated from the Rhizosphere Roots of Black Pepper (*Piper nigrum* L.). unpublished manuscript.
145. Petti, C.; Reiber, K.; Ali, S.S.; Berney, M.; Doohan, F.M. Auxin as a player in the biocontrol of *Fusarium* head blight disease of barley and its potential as a disease control agent. *BMC Plant Biol.* **2012**, *12*, 224. [CrossRef]
146. Rabiey, M.; Hailey, L.E.; Roy, S.R.; Grenz, K.; Al-Zadjali, M.A.S.; Barrett, G.A.; Jackson, R.W. Endophytes vs tree pathogens and pests: Can they be used as biological control agents to improve tree health? *Eur. J. Plant Pathol.* **2019**, *155*, 711–729. [CrossRef]
147. Kosawang, C.; Amby, D.B.; Bussaban, B.; McKinney, L.V.; Xu, J.; Kjær, E.D.; Collinge, D.B.; Nielsen, L.R. Fungal communities associated with species of *Fraxinus* tolerant to ash dieback, and their potential for biological control. *Fungal Biol.* **2018**, *122*, 110–120. [CrossRef]

148. Halecker, S.; Wennrich, J.P.; Rodrigo, S.; Andrée, N.; Rabsch, L.; Baschien, C.; Steinert, M.; Standler, M.; Surup, F.; Schulz, B.J. Fungal endophytes for biocontrol of ash dieback: The antagonistic potential of *Hypoxylon rubiginosum*. *Fungal Ecol.* **2020**, *45*, 100918. [CrossRef]

149. Steddom, K.; Becker, O.; Menge, J.A. Repetitive Applications of the Biocontrol Agent *Pseudomonas putida* 06909 and Effects on Populations of *Phytophthora parasitica* in Citrus Orchards. *Phytopathology* **2002**, *92*, 850–856. [CrossRef] [PubMed]

150. Acebo-Guerrero, Y.; Hernández-Rodríguez, A.; Vandeputte, O.; Miguélez-Sierra, Y.; Heydrich-Pérez, M.; Ye, L.; Cornelis, P.; Bertin, P.; El Jaziri, M. Characterization of *Pseudomonas chlororaphis* from *Theobroma cacao* L. rhizosphere with antagonistic activity against *Phytophthora palmivora* (Butler). *J. Appl. Microbiol.* **2015**, *119*, 1112–1126. [CrossRef]

151. Lee, B.D.; Dutta, S.; Ryu, H.; Yoo, S.J.; Suh, D.S.; Park, K. Induction of systemic resistance in *Panax ginseng* against *Phytophthora cactorum* by native *Bacillus amyloliquefaciens* HK34. *J. Ginseng. Res.* **2015**, *39*, 213–220. [CrossRef] [PubMed]

152. Mpika, J.; Kébé, I.B.; Issali, A.E.; N'Guessan, F.K.; Druzhinina, S.; Komon-Zélazowska, M.; Kubicek, C.P.; Aké, S. Antagonist potential of *Trichoderma* indigenous isolates for biological control *of Phytophthora palmivora* the causative agent of black pod disease on cocoa (*Theobroma cacao* L.) in Côte d'Ivoire. *Afr. J. Biotechnol.* **2009**, *8*, 5280–5293.

153. Diánez Martínez, F.; Santos, M.; Carretero, F.; Marín, F. Trichoderma saturnisporum, a new biological control agent. *J. Sci. Food Agric.* **2016**, *96*, 1934–1944. [CrossRef]

154. Abbas, A.; Mubeen, M.; Zheng, H.; Sohail, M.A.; Shakeel, Q.; Solanki, M.K.; Iftikhar, Y.; Sharma, S.; Kashyap, B.K.; Hussain, S.; et al. *Trichoderma* spp. Genes Involved in the Biocontrol Activity against *Rhizoctonia solani*. *Front. Microbiol.* **2022**, *13*, 884469. [CrossRef]

155. Palmieri, D.; Ianiri, G.; Del Grosso, C.; Barone, G.; De Curtis, F.; Castoria, R.; Lima, G. Advances and Perspectives in the Use of Biocontrol Agents against Fungal Plant Diseases. *Horticulturae* **2022**, *8*, 577. [CrossRef]

156. Hernando José, B.-A.; González-Rodríguez, V.E.; Almeida, G.R.; Izquierdo-Bueno, I.; Moraga, J.; Carbú, M.; Cantoral, J.M.; Garrido, C. Endophytic Microorganisms as an Alternative for the Biocontrol of *Phytophthora* spp. In *Agro-Economic Risks of Phytophthora and an Effective Biocontrol Approach*; Waleed Mohamed Hussain, A., Ed.; IntechOpen: Rijeka, Croatia, 2021; pp. 1–12.

157. Naidoo, S.; Slippers, B.; Plett, J.M.; Coles, D.; Oates, C.N. The Road to Resistance in Forest Trees. *Front. Plant Sci.* **2019**, *10*, 273. [CrossRef]

158. Moricca, S.; Panzavolta, T. Recent Advances in the Monitoring, Assessment and Management of Forest Pathogens and Pests. *Forests* **2021**, *12*, 1623.

159. Rabiey, M.; Welch, T.; Sanchez-Lucas, R.; Stevens, K.; Raw, M.; Kettles, G.J.; Catoni, M.; McDonald, M.C.; Jackson, R.W.; Luna, E. Scaling-up to understand tree–pathogen interactions: A steep, tough climb or a walk in the park? *Curr. Opin. Plant Biol.* **2022**, *68*, 102229.

160. Chandelier, A.; Husson, C.; Druart, P.; Marçais, B. Assessment of inoculation methods for screening black alder resistance to *Phytophthora* × *alni*. *Plant Pathol.* **2016**, *65*, 441–450.

161. Sniezko, R.A.; Koch, J. Breeding trees resistant to insects and diseases: Putting theory into application. *Biol. Invasions* **2017**, *19*, 3377–3400.

162. Woodcock, P.; Marzano, M.; Quine, C.P. Key lessons from resistant tree breeding programmes in the Northern Hemisphere. *Ann. For. Sci.* **2019**, *76*, 51.

163. Pike, C.C.; Koch, J.; Nelson, C.D. Breeding for Resistance to Tree Pests: Successes, Challenges, and a Guide to the Future. *J. For.* **2020**, *119*, 96–105.

164. Wei, Z.; Jousset, A. Plant Breeding Goes Microbial. *Trends Plant Sci.* **2017**, *22*, 555–558. [PubMed]

165. Marco, S.; Loredana, M.; Riccardo, V.; Raffaella, B.; Walter, C.; Luca, C. Microbe-assisted crop improvement: A sustainable weapon to restore holobiont functionality and resilience. *Hortic. Res.* **2022**, *9*, uhac160.

166. Bubner, B. Resistant alder; Selection of Black alder (*Alnus glutinosa* (L.) resistant against *Phytophthora alni*. Thünen Institute. 2018. Available online: https://www.thuenen.de/en/institutes/forest-genetics/research-groups/pathogen-resistance-and-seed-quality-research/completed-projects/breeding-of-black-alder (accessed on 26 March 2023).

167. Rathore, R.; Vakharia, D.N.; Rathore, D.S. In vitro screening of different *Pseudomonas fluorescens* isolates to study lytic enzyme production and growth inhibition during antagonism of *Fusarium oxysporum f.* sp. *cumini*, wilt causing pathogen of cumin. *Egypt. J. Biol. Pest Control* **2020**, *30*, 57.

168. Ganeshan, G.; Manoj Kumar, A. *Pseudomonas fluorescens*, a potential bacterial antagonist to control plant diseases. *J. Plant Interact.* **2005**, *1*, 123–134.

169. Foo, J.L.; Ling, H.; Lee, Y.S.; Chang, M.W. Microbiome engineering: Current applications and its future. *Biotechnol. J.* **2017**, *12*, 1600099.

170. Hu, H.; Wang, M.; Huang, Y.; Xu, Z.; Xu, P.; Nie, Y.; Tang, H. Guided by the principles of microbiome engineering: Accomplishments and perspectives for environmental use. *mLife* **2022**, *1*, 382–398.

171. Mendes, R.; Kruijt, M.; de Bruijn, I.; Dekkers, E.; van der Voort, M.; Schneider, J.H.; Piceno, Y.M.; DeSantis, T.Z.; Andersen, G.L.; Bakker, P.A.; et al. Deciphering the rhizosphere microbiome for disease-suppressive bacteria. *Science* **2011**, *332*, 1097–1100. [PubMed]

172. Santhanam, R.; Luu, V.T.; Weinhold, A.; Goldberg, J.; Oh, Y.; Baldwin, I.T. Native root-associated bacteria rescue a plant from a sudden-wilt disease that emerged during continuous cropping. *Proc. Natl. Acad Sci. USA* **2015**, *112*, E5013–E5020. [PubMed]

173. Mukherjee, A.; Singh, S.; Gaurav, A.K.; KumarChouhan, G.; Jaiswal, D.K.; Pereira, A.P.A.; KumarPassari, A.; Abdel-Azeem, A.; Verma, J.P. Harnessing of phytomicrobiome for developing potential bio stimulant consortium for enhancing the productivity of chickpea and soil health under sustainable agriculture. *Sci. Total Environ.* **2022**, *836*, 2–11.
174. Wicaksono, W.A.; Jones, E.; Casonato, S.; Monk, J.; Ridgway, H.J. Biological control of *Pseudomonas syringae* pv. *actinidiae* (Psa), the causal agent of bacterial canker of kiwifruit, using endophytic bacteria recovered from a medicinal plant. *Biol. Control* **2018**, *116*, 103–112.

**Disclaimer/Publisher's Note:** The statements, opinions and data contained in all publications are solely those of the individual author(s) and contributor(s) and not of MDPI and/or the editor(s). MDPI and/or the editor(s) disclaim responsibility for any injury to people or property resulting from any ideas, methods, instructions or products referred to in the content.

 *microorganisms*

*Article*

# Volatile Organic Compounds Produced by *Kosakonia cowanii* Cp1 Isolated from the Seeds of *Capsicum pubescens* R & P Possess Antifungal Activity

José Luis Hernández Flores [1], Yomaiko Javier Martínez [2], Miguel Ángel Ramos López [2], Carlos Saldaña Gutierrez [3], Aldo Amaro Reyes [2], Mariem Monserrat Armendariz Rosales [2], Maraly Jazmin Cortés Pérez [2], Mayela Fosado Mendoza [2], Joanna Ramírez Ramírez [2], Grecia Ramírez Zavala [2], Paola Lizeth Tovar Becerra [2], Laila Valdez Santoyo [2], Karen Villasana Rodríguez [2], José Alberto Rodríguez Morales [4,*] and Juan Campos Guillén [2,*]

[1] Centro de Investigación y de Estudios Avanzados del IPN, Irapuato 6824, Mexico; jose.hernandezf@cinvestav.com.mx
[2] Facultad de Química, Universidad Autónoma de Querétaro, Cerro de las Campanas S/N, Querétaro 76010, Mexico; yomaiko_javier@outlook.com (Y.J.M.); agromyke@gmail.com (M.Á.R.L.); aldo.amaro@uaq.edu.mx (A.A.R.); mararmendarizr@gmail.com (M.M.A.R.); mcortes15@alumnos.uaq.mx (M.J.C.P.); mfosado10@alumnos.uaq.mx (M.F.M.); jramirez175@alumnos.uaq.mx (J.R.R.); gramirez52@alumnos.uaq.mx (G.R.Z.); ptovar29@alumnos.uaq.mx (P.L.T.B.); lvaldez24@alumnos.uaq.mx (L.V.S.); kvillasana07@alumos.uaq.mx (K.V.R.)
[3] Facultad de Ciencias Naturales, Universidad Autónoma de Querétaro, Av. de las Ciencias S/N, Querétaro 76220, Mexico; carlos.saldana@uaq.mx
[4] Facultad de Ingeniería, Universidad Autónoma de Querétaro, Cerro de las Campanas S/N, Querétaro 76010, Mexico
[*] Correspondence: josealberto970@hotmail.com (J.A.R.M.); juan.campos@uaq.mx (J.C.G.)

**Citation:** Hernández Flores, J.L.; Martínez, Y.J.; Ramos López, M.Á.; Saldaña Gutierrez, C.; Reyes, A.A.; Armendariz Rosales, M.M.; Cortés Pérez, M.J.; Mendoza, M.F.; Ramírez Ramírez, J.; Zavala, G.R.; et al. Volatile Organic Compounds Produced by *Kosakonia cowanii* Cp1 Isolated from the Seeds of *Capsicum pubescens* R & P Possess Antifungal Activity. *Microorganisms* **2023**, *11*, 2491. https://doi.org/10.3390/microorganisms11102491

Academic Editors: Rainer Borriss and Chetan Keswani

Received: 23 September 2023
Accepted: 1 October 2023
Published: 4 October 2023

**Copyright:** © 2023 by the authors. Licensee MDPI, Basel, Switzerland. This article is an open access article distributed under the terms and conditions of the Creative Commons Attribution (CC BY) license (https://creativecommons.org/licenses/by/4.0/).

**Abstract:** The *Kosakonia cowanii* Cp1 strain was isolated from seeds of *Capsicum pubescens* R. & P. cultivated in Michoacan, Mexico. Genetic and ecological role analyses were conducted for better characterization. The results show that genome has a length of 4.7 Mbp with 56.22% G + C and an IncF plasmid of 128 Kbp with 52.51% G + C. Furthermore, pathogenicity test revealed nonpathogenic traits confirmed by the absence of specific virulence-related genes. Interestingly, when fungal inhibitory essays were carried out, the bacterial synthesis of volatile organic compounds (VOCs) with antifungal activity showed that *Sclerotinia sp.* and *Rhizoctonia solani* were inhibited by 87.45% and 77.24%, respectively. Meanwhile, *Sclerotium rolfsii*, *Alternaria alternata*, and *Colletotrichum gloeosporioides* demonstrated a mean radial growth inhibition of 52.79%, 40.82%, and 55.40%, respectively. The lowest inhibition was by *Fusarium oxysporum*, with 10.64%. The VOCs' characterization by headspace solid–phase microextraction combined with gas chromatography–mass spectrometry (HS-SPME-GC–MS) revealed 65 potential compounds. Some of the compounds identified with high relative abundance were ketones (22.47%), represented by 2-butanone, 3-hydroxy (13.52%), and alcohols (23.5%), represented by ethanol (5.56%) and 1-butanol-3-methyl (4.83%). Our findings revealed, for the first time, that *K. cowanii* Cp1 associated with *C. pubescens* seeds possesses potential traits indicating that it could serve as an effective biocontrol.

**Keywords:** *Enterobacteriaceae*; 2-butanone; 3-hydroxy; *Capsicum pubescens* seeds; fungal inhibition; biocontrol; genome sequencing

## 1. Introduction

Pepper (*Capsicum* spp.) has an exciting history of domestication and has seen more than 6000 years of use in diverse foodstuffs in Mexico, such as in fresh, dried, or processed products. It is believed that this crop originated in South America [1–3]. Therefore, Mexico and Central America are considered genetic diversity hotspots that have generated important domesticated varieties of pepper. These peppers have relevance in the agriculture and food industry, specifically those industries that are linked to the use of pigments, shape, size, appearance, flavor, pungency, and nutritive components [4,5]. In addition, historical records indicate that *Capsicum* was introduced to Spain by Christopher Columbus after discovering America in 1493, and then to the rest of Europe and—eventually—to India, Asia, and Africa [6]. Geographical displacement and domestication processes such as artificial and natural selection in agricultural environments have led to the existence of a great number of species in the *Capsicum* genus. However, there are five important cultivated species of economical relevance that are considered here: *C. annuum* L., *C. frutescens* L., *C. chinense* Jacq., *C. baccatum* L., and *C. pubescens* R. & P. [1–6].

Therefore, all agronomic management approaches for pepper cultivation during historical domestication are important to consider in terms of obtaining fruits and seeds with high quality for diverse purposes [3]. From these significant traits, genetic lines with resistances to diverse pathogens, such as viruses, bacteria, and fungi, are of ecological and economical relevance during plant and seed development stages [7–10]. In addition to modern breeding programs, recently, diverse methodologies, such as metagenomics, transcriptomics, proteomics, or metabolomics approaches, have been relevant to understanding microorganism–plant interactions or plant development during the lifecycle of *Capsicum* [11–15].

Throughout the last decade, the study of microbiomes on the seeds from diverse plant species, including *Capsicum*, has increased due to its ecological importance [16–18]. These studies highlight the metabolic activities of microorganism endophytes or epiphytes on seed germination, seedling establishment, plant growth, and plant development stages and as a biocontrol [16–18]. Moreover, it is of special interest that the variations in the diversity of seed-associated microorganisms (SAMs) could be determined by their plant genotype as well as by environmental and soil management practices [16]. Therefore, the characterization of the SAMs on *Capsicum* with beneficial or detrimental traits is necessary during agricultural practices for an assessment of seed quality and plant development.

Although the origin or transmission route of SAMs is under study, it is interesting to understand the recruitment of diverse communities of microorganisms during plant seed development and to evaluate their metabolic potential through molecule production and characterization, which may indicate ecological significance and agricultural applications [16]. Thus, for instance (and of particular interest for our research work), the diverse members of the family *Enterobacteriaceae* were isolated from seeds of a xerophytic plant, *Lactuca serriola*. Among them is *Kosakonia cowanii*, which possesses the ability to produce a high concentration of exopolysaccharides with effects on plant drought tolerance [19]. In addition, *K. cowanii* has been isolated from diverse crops with plant-promoting activities through auxin (IAA) and siderophore production [20,21]. Also, the great metabolic ability of the *Kosakonia* genus has been characterized by the diverse strains isolated from insects, plants, human, or other sources [22–30]. Recently, our research group reported the presence of *K. cowanii* in chili powder via 16s rRNA library sequencing [31], and the isolation of a *K. cowanii* Ch1 strain with the ability to produce active VOCs against certain fungal pathogens and colonize the seeds of *C. annuum* L. [32] was also achieved. These results allow us to hypothesize that *K. cowanii* could be associated with fruits or seeds of *C. annuum* L.

Although systematic research of SAMs in *Capsicum* is limited, this plant is cultivated around the world, and its production is close to 34.5 million tons annually. Furthermore, in Mexico, it is one of the most important economic and agricultural crops, cultivated on a total of 150 million hectares [33]. Therefore, the high diversity of genetic varieties of *Capsicum* plants that are grown in diverse soils and environmental conditions make it necessary to explore SAMs as a source of beneficial microorganisms. In the present report, we hypothesize that *K. cowanii* could be associated with *Capsicum* seeds. Therefore, our results show, for the first time, that the presence of *K. cowanii*, specifically, the strain associated with *Capsicum pubescens* seeds, has an important ecological role in terms of controlling pathogenic fungi. In addition, this opens the possibility of a further exploration of the SAMs in the *Capsicum* that are cultivated under different environmental conditions in Mexico.

## 2. Materials and Methods

### 2.1. Bacterial Strain Isolation

Serrano pepper (*Capsicum annuum* L.) and manzano pepper (*Capsicum pubescens*) fruits without visual damage or infections were obtained from cultivated plants in Querétaro and Michoacán, México. Then, they were surface disinfected with sodium hypochlorite (10% *v/v*) for 10 min; this was followed by disinfection with ethanol (50% *v/v*) for 10 min. Then, they were air dried at room temperature in a biosafety cabinet. The seeds were placed on tryptic soy agar (TSA) medium (Difco Laboratories, Detroit, MI, USA) with 100 µg/mL of ampicillin and incubated at 37 °C for 24 h. Bacterial colonies were isolated in the same culture medium. For taxonomy identification, the 16s rDNA gene was amplified by PCR, and amplicon was sequenced at Macrogen Inc. (Seoul, Republic of Korea). The sequences obtained were analyzed by neighbor-joining method in MEGA X software and compared with the sequences of *K. cowanii* obtained from NCBI database [34–37].

### 2.2. Pathogenicity Test

A pathogenicity test was conducted using five bacterial strains isolated from *C. pubescens* seeds (which were identified as *K. cowanii*), following the methodology in our previous report [32]. In general, seeds of *Capsicum annuum* L. var. serrano were disinfected in sodium hypochlorite (10%, *v/v*) for 5 min and rinsed with sterile distilled water. Germination was carried out in seedling trays under greenhouse conditions. After 45 to 50 days post-germination, seedlings were transplanted into pots. Plants with 13–14 true leaves were used for pathogenicity test. On the other hand, fruits without damage or infection were disinfected with sodium hypochlorite (10%, *v/v*) for 10 min, ethanol (50%, *v/v*) for 10 min, and, finally, rinsed with sterile distilled water. Bacterial suspension (10 µL) with approximately $1 \times 10^8$ CFU/mL was inoculated on six leaves of each plant in triplicate and on three pepper fruits; both were damaged with a sterile needle and inoculated at that zone. Negative controls were inoculated with sterile distilled water. Inoculated plants were kept in greenhouse conditions while inoculated fruits were placed inside of sealed container at room temperature. The plants were observed for 5–7 days.

### 2.3. Genome Sequencing and Assembly

We randomly selected the *K. cowanii* strain Cp1 based on pathogenicity test and fungal inhibitory essays to isolate its genomic DNA for sequencing. DNA was extracted with the ZymoBIOMICS™ DNA Miniprep Kit (Zymo Research, Irvine, CA, USA) according to the manufacturer's instructions. We used the Genomic Sequencing Service at Zymo Research, Irvine, CA, USA, which used the NovaSeq® (Illumina, San Diego, CA, USA) platform. Sequence reads were processed at the Bacterial and Viral Bioinformatics Resource Center (BV-BRC). First, a Fastq was used to filtrate the sequences [38]. Second, for genome annotation, the RAST tool kit (RASTtk) was used [39]. Lastly, the circular genome map was modeled in CIRCOS software [40]. The genome statistics were as follows: coarse consistency (98.7%), fine consistency (98%), CheckM completeness (99.7%), and CheckM contamination (0.1%). The bacterial phylogenetic tree was constructed by using the pipelines at BV-BRC, where Mash/MinHash [41], PGFams [38], MUSCLE v5 [37], RaxML v8.2.11 [37], and fast bootstrapping [38] were included in order to conduct the phylogenetic analysis. PlasmidSPAdes version v3.13.0, with default settings, was used for the plasmid assembly [42]. DNA sequences were deposited in NCBI as BioProject ID PRJNA1003013. The genome accession number is JAUZWC000000000.

### 2.4. Antibiotic Susceptibility Testing

The antimicrobial phenotype of *K. cowanii* Cp1 and *Escherichia coli* XL1 blue as a control were evaluated using the guidelines of CLSI [43]. In general, bacterial strains were inoculated in tryptic soy broth (TSB) medium and growth at 37 °C and shaken until an optical density (OD) between 0.4 and 0.5 at 600 nm was achieved. Then, 100 μL of bacterial culture was spread on Mueller–Hinton agar (Bioxon). The antimicrobial test discs (Oxoid) were as follows: penicillin (10 U), ampicillin (10 μg), carbenicillin (100 μg), dicloxacillin (1 μg), and amoxicillin/clavulanic acid (20/10 μg); cephalosporins were cephalothin (30 μg) and cefotaxime (30 μg); fluoroquinolones were ciprofloxacin (5 μg) and norfloxacin (10 μg); aminoglycosides were amikacin (30 μg), gentamicin (10 μg), and netilmicin (30 μg); macrolides were clindamycin (30 μg) and erythromycin (15 μg); carbapenems were imipenem (10 μg) and meropenem (10 μg); chloramphenicol (30 μg); nitrofurantoin (300 μg), tetracycline (30 μg); trimethoprim-sulfamethoxazole (25 μg) and the glycopeptide vancomycin (30 μg). Culture plates were incubated overnight at 37 °C and the inhibitory or clear zones around the disc were recorded in mm, according to guidelines.

## 2.5. Inhibitory Effects of VOCs

The following fungal strains were provided by the Laboratory of Plants and Agriculture Biotechnology at Queretaro University, México: *Alternaria alternata*, *Colletotrichum gloeosporioides*, *Fusarium oxysporum*, *Rhizoctonia solani*, *Sclerotium rolfsii*, and *Sclerotinia* sp. The growing conditions were as follows: potato dextrose agar (PDA) medium (Difco Laboratories, Detroit, MI, USA) kept at 28 °C over a 5- to 7-day period, conducted according to the growth rate of each fungal strain.

The inhibitory assays were evaluated by radial mycelial growth according to the methodology described previously in [32]. In general, the two-compartment plastic plate device was sealed with parafilm to avoid VOC loss. On the lower side of the device, which was used with TSA medium (Difco Laboratories, Detroit, MI, USA), a bacterial suspension ($1 \times 10^8$ CFU/mL) was inoculated. On the upper side of the device, which was used with PDA medium (Difco Laboratories, Detroit, MI, USA), a fungal growth disk was inoculated. Each control was tested for fungi and growth on PDA medium (Difco Laboratories, Detroit, MI, USA). Each device was an experimental unit and was incubated at 28 °C to obtain the radial mycelial growth according to the following equation: mycelial growth inhibition (%) = $[(dc-dt)/dc] \times 100$, where dc and dt represent the mycelial growth diameters (in mm) of the control and treatment groups, respectively. The experiments were conducted in triplicate for a statistical analysis of variance and for Duncan's multiple range test ($p = 0.05$), which were performed using DPS V12.01 software. Alternatively, bacterial antagonism toward *S. rolfsii* was tested by dual-culture assay on PDA to prove additional mechanism of inhibition. Bacterial strain (10 μL, $1 \times 10^8$ CFU/mL) was inoculated close to and around fungal disk. The essay was kept at 28 °C over a 2- to 3-day period.

## 2.6. HS-SPME-GC–MS Analysis of VOCs

As the first step in obtaining a bacterial culture with volatile organic compounds (which have inhibitory effects on *Sclerotium rolfsii*), a cell-free filtrate was obtained of *K. cowanii* Cp1 grown on Tryptic soy broth (TSB) medium, with continuous shaking (100 rpm) at 37 °C, which was sampled every 6 h up to a total of 48 h. TSB medium without *K. cowanii* Cp1 strain was used as the control. The cell-free filtrate was obtained via centrifugation at 14,000 rpm for 5 min and after being filtered through a 0.2 μm membrane (Sigma-Aldrich). Then, 500 μL of the sample was placed on PDA medium (Difco Laboratories, Detroit, MI, USA), and a 7 mm mycelial disk of *Sclerotium rolfsii* was inoculated. The radial mycelial that was grown was evaluated according to the equation mentioned above. Furthermore, the sample that was obtained after 18 h of bacterial growth displayed the highest inhibitory effect; thus, it was used for VOC characterization. To recover the VOCs from the bacterial culture that was grown over 18 h on TSB medium, the sample was processed according to the methodology, and using the device, that was previously reported in [32].

Among the bacterial volatiles obtained, and based on the commercial availability (J.T., Baker, Fermont, Meyer, Sigma-Aldrich, Toluca, Mexico), seven standard compounds were evaluated individually on the inhibition of radial mycelial growth. Using the previous two-compartment plastic plate device, different amounts of individual VOCs were added to sterile filter papers on one side and on the upper side of the device, and an *S. rolfsii* growth disk of 8 mm was inoculated on PDA medium. The amounts VOCs assayed were: 50, 100, and 150 µL/plate of 2-Butanone; 50, 100, and 150 µL/plate of 2-Butanone, 3-hydroxy; 10, 20, and 40 µL/plate of 2-Nonanone; 50, 100, and 150 µL/plate of acetone; 5, 10, and 20 µL/plate of acetic acid (unidentified compound as VOC component in *K. cowanii* Cp1); 50, 100, and 150 µL/plate of benzyl alcohol; and 50, 100, and 200 µL/plate of ethanol. Sterile distilled water was used as control. Each device was incubated at 28 °C for a 3-day period to obtain the radial mycelial growth, according to equation previously described. The experiments were conducted in triplicate.

## 3. Results

### 3.1. Isolation of K. cowanii from Capsicum pubescens Seeds

Seeds from *C. annuum* and *C. pubescens* were obtained and processed for bacterial isolation on TSA medium. In a previous work [32], we isolated a *K. cowanii* strain that was resistant to β-lactam antibiotics; as such, in this case, we decided to supplement the TSA medium with 100 µg/mL of ampicillin in order to determine if this could be a similar case. From diverse seeds that were obtained from the fruits of *Capsicum* spp., a bacterial growth was noted on the seeds of manzano peppers (*C. pubescens*) that were obtained from Michoacan, Mexico. In 100% of the seeds of one fruit, there was a similar phenotype to *K. cowanii* (Figure 1). The bacterial isolates from the seeds were streaked on the same culture medium so as to obtain isolated colonies. Five bacterial colonies were selected and identified by using the 16s rDNA gene as a molecular marker. Sequence analysis confirmed them as *K. cowanii*. Then, we conducted a pathogenicity test for the *K. cowanii* strains to evaluate the phytopathogenic traits, as previously reported [32]. The infection assays on the fruits and leaves of *Capsicum annuum* L. showed no infection symptoms (Figure 2).

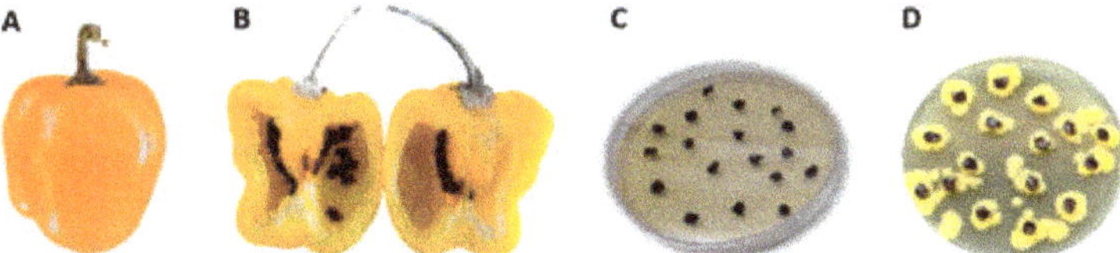

**Figure 1.** Isolation of *Kosakonia cowanii* Cp1. (**A**) *Capsicum pubescens* fruits with no infection symptoms; (**B**) interior of *Capsicum pubescens* fruits displaying seeds with no infection symptoms; (**C**) *Capsicum pubescens* seeds on TSA medium that was supplemented with 100 µg/mL of ampicillin; (**D**) a growth of *Kosakonia cowanii* after 24 h of incubation.

**Figure 2.** Pathogenicity test of *K. cowanii* Cp1 in a serrano pepper (*Capsicum annuum* L.). (**A**) Assay of *K. cowanii* Cp1 that grew on serrano pepper leaves; (**B**) assay of *K. cowanii* Cp1 on the external part of pepper fruits; (**C**) assay of *K. cowanii* Cp1 on the internal part of pepper fruits. No lesions were observed in any case.

### 3.2. General Genome and the Plasmid Features of K. cowanii Strain Cp1

From these bacterial isolates, based on fungal inhibitory essays (explained below), one of them was selected randomly for genomic characterization. The genome of *K. cowanii* Cp1 has a total length of 4,765,484 bp, 56.22% of G + C, and a plasmid (pCp1) of 128,063 bp with 52.51% of G + C(Figure 3). The annotation results indicated 4509 protein coding sequences (CDS) and 89 RNA genes (84 tRNA and 5 rRNA). The genome functional subsystem categories are shown in Figure 3, where the metabolism category is represented at 37.88%; protein processing at 9.57%; the stress response, defense, and virulence category at 8.43%; energy at 13.43%; membrane transport at 6.76%; cellular processes at 9.21%; DNA processing at 4.49%; RNA processing at 3.44%; cell envelope at 4.31%; and regulation and cell signaling at 1.31%. On the pCp1 plasmid, 105 hypothetical proteins and 59 proteins with subsystem functional assignments were detected. The BLASTN analysis of plasmid pCp1 showed a 49% alignment and 94.22% similarity with the plasmid 888-76-1 identified in the *K. cowanii* JCM 10956 strain. The plasmid Wem22 identified in the *K. cowanii* SMBL-WEM22 strain showed a 46% alignment and 94.22% similarity. The plasmid unnamed1 that was identified in the phytopathogenic *K. cowanii* strain Pa82 showed a 45% alignment and 95.0% similarity. When the pCh1 plasmid identified in the *K. cowanii* Ch1 strain that was isolated from chili powder was compared with the pCp1 plasmid, the sequence coverage and similarity was 100%. The pCp1 and pCh1 are IncF plasmids that have VapBC toxin–antitoxin systems; therefore, the presence or association of IncF plasmids with the *K. cowanii* isolates obtained from chili powder and *Capsicum pubescens* could be important. As such, additional research is necessary to understand IncF plasmid's functionality.

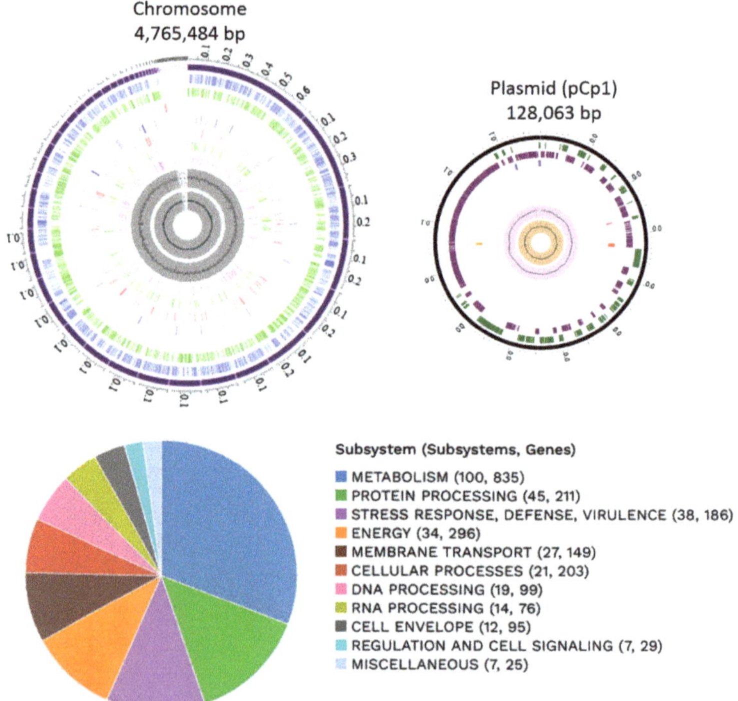



**Figure 3.** Graphical circular map of the chromosome and plasmid of *K. cowanii* Cp1. From the outside to the center rings are the ensembled contigs, CDS on forward, CDS on reverse, non-CDS, AMR genes, VF genes, transporters, drug targets, GC content, and GC skew. The subsystem functional assignments are represented in the figure below.

Additionally, according to the reference database for bacterial virulence factors (VDFB), virulence-related genes are principally classified as flagellar, iron uptake, and siderophore and vir/avr genes were not identified (Table 1).

**Table 1.** Virulence-related genes as predicted by VFDB source.

| Virulence Factor Classification | Gene Name | Putative Function |
| --- | --- | --- |
| Endotoxin | *gtrB* | Bactoprenol glucosyl transferase |
| Iron uptake, siderophore | *entA* | 2,3-dihydro-2,3-dihydroxybenzoate dehydrogenase 2,3-dihydro-2,3-dihydroxybenzoate dehydrogenase of siderophore biosynthesis |
| | *entB* | Isochorismatase of siderophore biosynthesis |
| | *fepB* | Ferric enterobactin-binding periplasmic protein FepB |
| | *fepC* | Ferric enterobactin transport ATP-binding protein FepC |
| | *fepD* | Ferric enterobactin transport system permease protein FepD |
| | *fepG* | Ferric enterobactin transport system permease protein FepG |
| | *entS* | Enterobactin exporter EntS |
| Secretion system, invasion, motility | *flgC* | Flagellar basal-body rod protein FlgC |
| | *motA* | Flagellar motor rotation protein MotA |
| | *flgH* | Flagellar L-ring protein FlgH |
| | *flgB* | Flagellar basal-body rod protein FlgB |
| | *flgG* | Flagellar basal-body rod protein FlgG |
| | *fliG* | Flagellar motor switch protein FliG |
| | *fliP* | Flagellar biosynthesis protein FliP |
| | *cheY* | Chemotaxis regulator; transmits chemoreceptor signals to flagellar motor components CheY |
| | *cheW* | Positive regulator of CheA protein activity (CheW) |
| | *fliM* | Flagellar motor switch protein FliM |
| | *fliC* | Flagellin FliC |
| | *fliA* | RNA polymerase sigma factor for flagellar operon |

On the other hand, the antibiotic resistance (AMR) genes predicted in the genome were classified into eight categories (Figure 4). However, the genotype must be proven with an AMR phenotype. In this regard, as a first approximation, the AMR phenotype analysis showed that *K. cowanii* Cp1 has a resistance to the penicillin antibiotic class, such as ampicillin, penicillin G, and carbenicillin, probably due to the TEM β-lactamase detected. In addition, the bacterial strain showed resistance to erythromycin, clindamycin, and vancomycin. We suspect that this is most likely due to the genes being classified as efflux pumps (Figure 4). Interestingly, the rest of the antibiotics caused growth inhibition in *K. cowanii* Cp1, supported by a clear zone of inhibition (Supplementary Figure S2), which demonstrated the importance of the phenotype versus the genotype in the analysis. Therefore, additional AMR analysis is necessary to understand the complete AMR of *K cowanii* Cp1.

Based on the previous genome sequences of the *Kosakonia* genus isolates from diverse sources and geographical regions, we decided to perform a phylogenomic analysis that was based on 100 core genes for *K. cowanii* Cp1 and on the related strains of the *Kosakonia* genus (Figure 5). The results showed that *K. cowanii* Cp1 is clustered with *K. cowanii strain 888-76, K. cowanii JCM 10956, K. cowanii Ch1, K. cowanii strain PF-104, K. cowanii Pa82*, and *K. cowanii strain Esp Z*. Although it is closely related to *K. cowanii* Ch1—an isolate from Mexican chili powder—the high genetic similarity between both strains means that there is strong evidence to suspect that the *K cowanii* detected in chili powder is due to its spread through *Capsicum* seeds.

**Figure 4.** Antibiotic resistance genes. The AMR mechanism and its related genes are indicated with a color diagram.

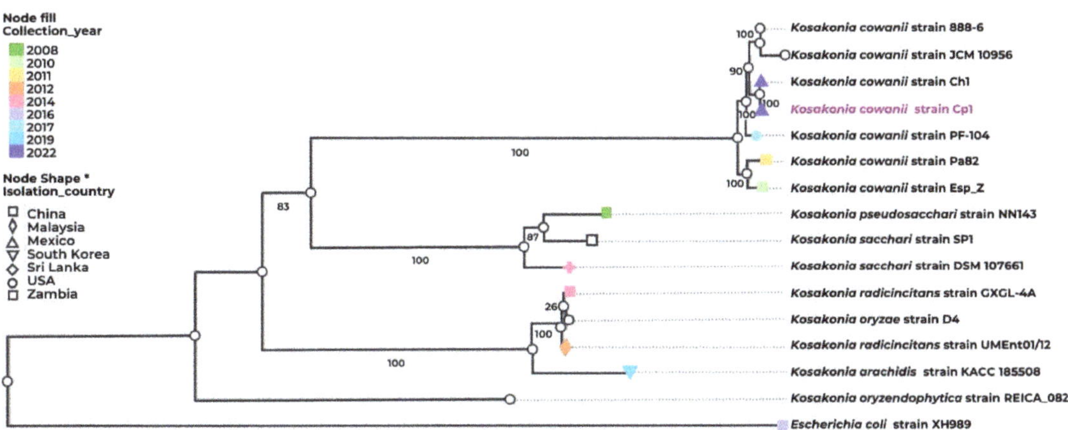

**Figure 5.** Phylogenomic analysis of *K. cowanii* Cp1. Phylogenetic analysis was performed in BV-BRC. The following strains were included: *K. cowanii* strain 888-76, *K. cowanii* JCM 10956, *K. cowanii* Ch1, *K. cowanii* strain PF-104, *K. cowanii* Pa82, *K. cowanii* strain Esp Z, *K. pseudosacchari* strain NN143, *K. sacchari* SP1, *K. sacchari* strain DSM 107661, *K. radicincitans* strain GXGL-4A, *K. oryzae* strain D4, *K. radicincitans* UMEnt01/12, *K. arachidis* strain KACC 18508, *K. oryzendophytica* strain REICA 082, and—as an outgroup—*Escherichia coli* XH989. The label color represents the collection year. The node fill and shape represent the country from which the isolates were gathered, including Mexico, China, Malaysia, Republic of Korea, Sri Lanka, USA, and Zambia.

### 3.3. Effects of VOCs on Mycelial Growth

To investigate the effect of the VOCs on mycelial growth, each one of the fungi was exposed to *K. cowanii* Cp1 in a dual-plate culture. The results in Figure 6 show that the VOCs produced by *K. cowanii* Cp1 during growth on TSA medium had a significant inhibitory effect on the mycelial growth of the tested fungi. The colony diameter of the six-fungus plates that were treated with *K. cowanii* Cp1 VOCs were significantly lower than the diameter of those grown on the control plates (Figure 6). Thus, for instance, the highest inhibition rate caused by the VOCs of all the tested fungi was observed in *Sclerotinia* sp., which exhibited an 87.45% inhibition, followed by *R. solani*, with a 77.24% inhibition. *S. rolfsii*, *A. alternata*, and *C. gloeosporioides* demonstrated a mean radial growth inhibition of 52.79%, 40.82%, and 55.40%, respectively. The lowest inhibition percentage demonstrated was by *F. oxysporum*, with a 10.64% inhibition. Alternatively, to prove additional mechanisms on fungal inhibition, a dual-culture assay on PDA was tested. At first approximation, *S. rolfsii* demonstrated a bacterial strain; however, mycelial growth was not affected (Supplementary Figure S1), suggesting that VOCs produced during fermentation could be the principal mechanism of fungal inhibition. Based on this result, we decided to analyze the VOCs' composition.

### 3.4. Characterization of VOCs

Characterization of the VOCs produced by *K. cowanii* Cp1 can clarify the diverse routes of metabolism that are used by this bacterial strain. With this in mind, we decided to analyze the VOCs by HS-SPME-GC–MS, which was achieved by using the bacterial culture that had 18 h of growth and was based on a 50% inhibition rate in a cell-free filtrate against *Sclerotium rolfsii*. The volatile organic compounds produced by *K. cowanii* Cp1 were compared with the control, and the results showed 65 VOCs (Table 2). They were then classified in the following functional groups (according to their total relative peak area (Figure 7)): alcohols (23.5%), ketones (22.97%), pyrazines (10.43%), esters (10.32%), hydrocarbons (9.5%), aldehydes (2.31%), acids (1.81%), and aromatics (0.11%).

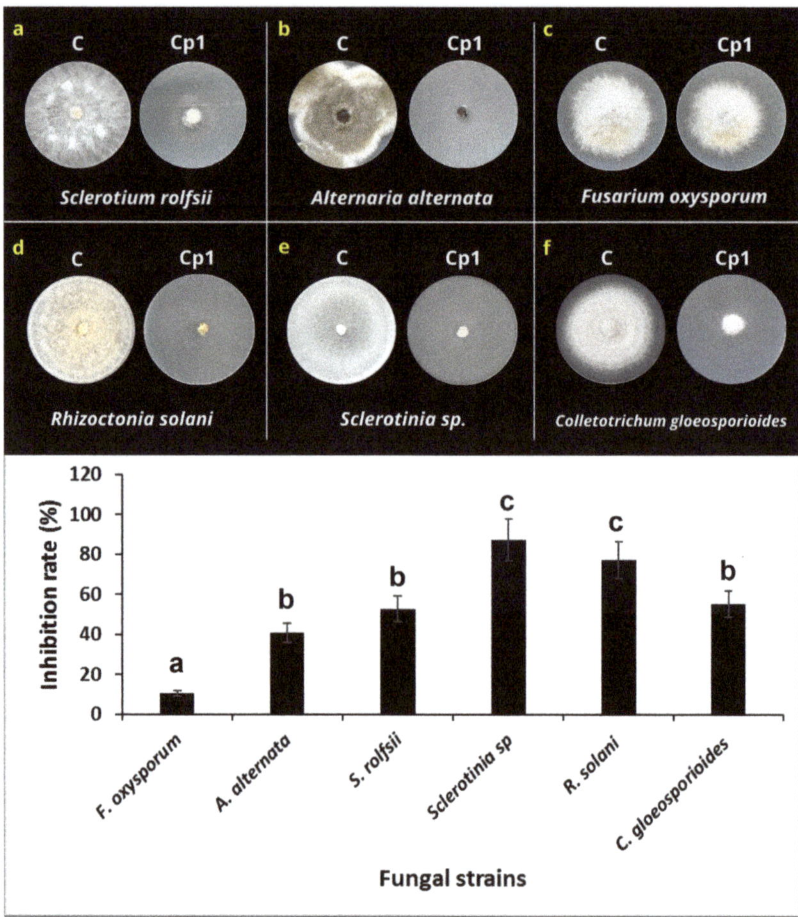



**Figure 6.** Effect of the VOCs produced by *K. cowanii* Cp1 on mycelial growth. (**a–f**) show the radial mycelial growth for the untreated controls (indicated by C) and those that were VOC-treated (indicated by Cp1). The lower graph shows the inhibition mycelial rate (%). Statistical differences are represented by letters.

**Table 2.** The VOCs produced by *K. cowanii* Cp1 and detected by HS-SPME-GC–MS.

| Compounds | Retention Time (min) | Relative Peak Area (%) | Chemical Classes | Compounds | Retention Time (min) | Relative Peak Area (%) | Chemical Classes |
|---|---|---|---|---|---|---|---|
| Carbon dioxide | 4.041 | 1.81 | Hydrocarbons | 1-Hexanol, 2-ethyl- | 39.459 | 2.22 | Alcohols |
| Ethyl ether | 4.466 | 0.70 | Esters | | | | |
| Methanethiol | 4.703 | 0.42 | Alcohols | Benzaldehyde | 40.287 | 0.94 | Aldehydes |
| Acetaldehyde | 4.854 | 0.54 | Aldehydes | 1-Octanol | 42.356 | 1.62 | Alcohols |
| Acetone | 6.176 | 0.61 | Ketones | 2-Undecanone | 44.335 | 1.12 | Ketones |
| Butanal | 7.610 | 0.32 | Aldehydes | 2-Acetylthiazole | 45.502 | 0.54 | Other compounds |
| Ethyl Acetate | 8.037 | 0.29 | Esters | Acetophenone | 45.741 | 0.74 | Ketones |
| 2-Butanone | 8.337 | 0.90 | Ketones | Pyrazine, 2,5-dimethyl-3-(3-methylbutyl)- | 46.713 | 0.55 | Pyrazine |
| Butanal, 3-methyl- | 9.088 | 0.23 | Aldehydes | Acetic acid, decyl ester | 48.558 | 0.77 | Esters |

**Table 2.** *Cont.*

| Compounds | Retention Time (min) | Relative Peak Area (%) | Chemical Classes | Compounds | Retention Time (min) | Relative Peak Area (%) | Chemical Classes |
|---|---|---|---|---|---|---|---|
| Ethanol | 9.566 | 5.56 | Alcohols | 4,4-Dimethyl-2-cyclopenten-1-one | 49.412 | 0.08 | Other compounds |
| 2,3-Butanedione | 12.247 | 2.58 | Ketones | Cyclodecane | 53.105 | 3.10 | Hydrocarbons |
| Trichloroethylene | 12.871 | 0.21 | Other compounds | 2-Tridecanone | 57.172 | 1.25 | Ketones |
| Trichloromethane | 14.529 | 0.33 | Other compounds | 3-Decen-1-ol, acetate, (Z)- | 58.325 | 1.20 | Alcohols |
| Toluene | 15.273 | 0.30 | Hydrocarbons | Benzyl alcohol | 60.204 | 0.68 | Alcohols |
| Disulfide, dimethyl | 16.91 | 1.35 | Other compounds | Butanoic acid, butyl ester | 62.141 | 4.54 | Esters |
| p-Xylene | 21.171 | 0.12 | Other compounds | Propanoic acid, 2-methyl-, 2,2-dimethyl-1-(2-hydroxy-1-methylethyl) propyl ester | 63.486 | 2.12 | Esters |
| 1-Butanol | 21.744 | 1.42 | Alcohols | Phenylethyl Alcohol | 63.944 | 1.99 | Alcohols |
| Ethanol, 2-methoxy- | 23.279 | 0.95 | Alcohols | Ethanol, 2,2'-oxybis- | 68.575 | 0.51 | Alcohols |
| 2-Heptanone | 24.063 | 0.26 | Ketones | Cyclododecane | 71.619 | 4.29 | Hydro-carbons |
| 1,3-Diazine | 24.981 | 0.48 | Other compounds | 2-Propenal, 3-phenyl- | 73.169 | 0.28 | Aldehydes |
| 1-Butanol, 3-methyl- | 25.259 | 4.83 | Alcohols | 2-Nonadecanone | 74.724 | 0.58 | Ketones |
| Pyrazine, methyl- | 27.947 | 1.15 | Pyrazine | Octanoic Acid | 76.216 | 0.68 | Acids |
| 2-Butanone, 3-hydroxy- | 29.244 | 13.52 | Ketones | Triacetin | 76.667 | 0.84 | Other compounds |
| 2-Propanone, 1-hydroxy- | 29.817 | 0.09 | Ketones | 2,4,7,9-Tetramethyl-5-decyn-4,7-diol | 78.388 | 1.90 | Alcohols |
| Pyrazine, 2,5-dimethyl- | 30.931 | 6.47 | Pyrazine | Benzoic acid, 4-tert-butyl-, 3,5-dichloro-4-pyridyl ester | 79.388 | 0.19 | Esters |
| Pyrazine, 2,6-dimethyl- | 31.245 | 0.27 | Pyrazine | Benzoic acid, 2-amino-, methyl ester | 83.037 | 0.12 | Esters |
| 1-Hexanol | 32.875 | 0.20 | Alcohols | | | | |
| Dimethyl trisulfide | 33.789 | 0.20 | Other compounds | Methyl 8-methyl-nonanoate | 84.445 | 0.51 | Esters |
| Pyrazine, 2-ethyl-6-methyl- | 34.218 | 0.12 | Pyrazine | Hexanedioic acid, dibutyl ester | 85.423 | 0.44 | Esters |
| Pyrazine, 2-ethyl-5-methyl- | 34.484 | 0.43 | Pyrazine | Dodecanoic acid, 3-hydroxy- | 85.693 | 1.13 | Acids |
| 2-Nonanone | 34.997 | 1.32 | Ketones | Diethyl phthalate | 88.452 | 0.39 | Other compounds |
| Pyrazine, 2-methyl-5-(1-methylethyl)- | 35.597 | 0.14 | Pyrazine | Indole | 90.108 | 0.11 | Aromatics |

**Table 2.** *Cont.*

| Compounds | Retention Time (min) | Relative Peak Area (%) | Chemical Classes | Compounds | Retention Time (min) | Relative Peak Area (%) | Chemical Classes |
|---|---|---|---|---|---|---|---|
| Pyrazine, 3-ethyl-2,5-dimethyl- | 37.062 | 1.30 | Pyrazine | Phthalic acid, hex-2-yn-4-yl isobutyl ester | 95.053 | 1.10 | Other compounds |
| Octanoic acid, ethyl ester | 37.407 | 0.43 | Esters | | | | |
| Acetic acid, octyl ester | 39.183 | 0.21 | Esters | | | | |

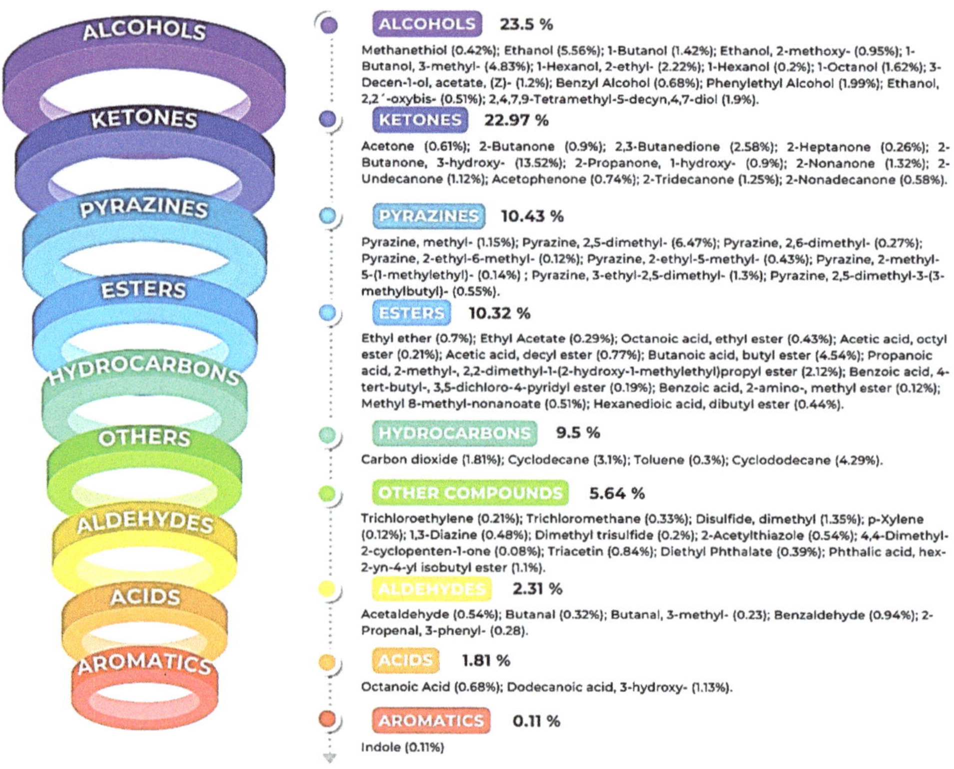


**Figure 7.** The VOCs detected in *K. cowanii* Cp1. Each figure color shows the class family of the compounds and their percentage according to their relative peak-area profile.

As is indicated in Figure 8, the six compounds with the highest relative peak-area percentages were 2-butanone, 3-hydroxy (13.52%); pyrazine-2,5-dimethyl (6.47%); ethanol (5.56%); 1-butanol-3-methyl (4.83%); butanoic acid and butyl ester (4.54%); and cyclododecane (4.29%). All of the results show that, during the growth of *K. cowanii* Cp1 on TSA medium, there was a fermentation process that produced a mixture of VOCs—some of them, most likely, exerted effects on mycelial growth inhibition. However, further analysis is necessary to understand the potential mechanisms leading to the production of each VOC and its effects on cellular processes.

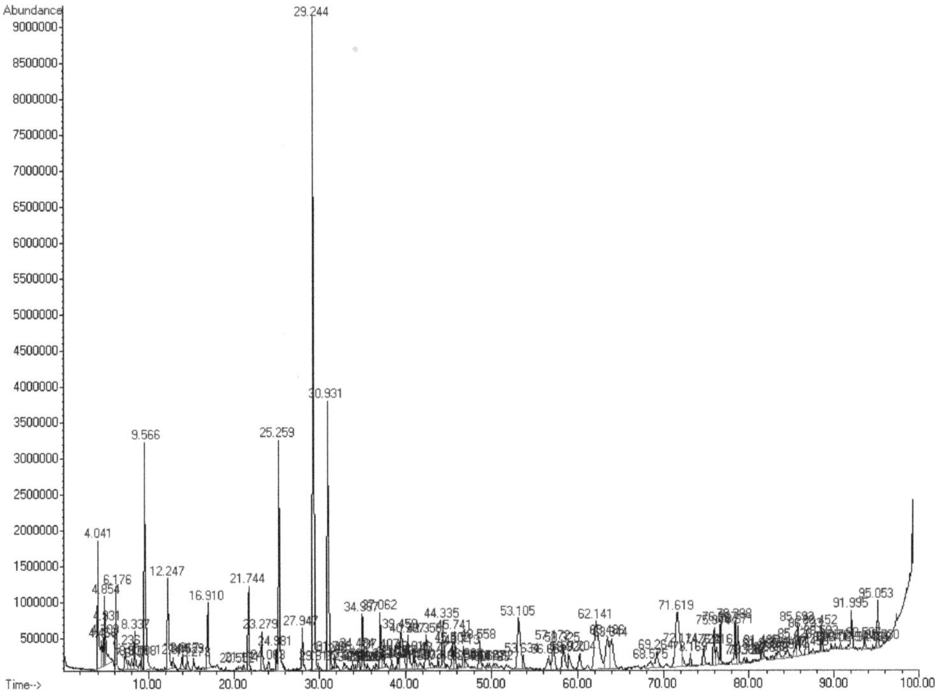

**Figure 8.** The VOC profiles of *K. cowanii* Cp1. The relative peak abundance and retention time are shown.

### 3.5. Evaluation of Standard VOCs

According to the identified VOCs, we decided to analyze some compounds. The mycelial inhibition rate of the seven compounds were as follows: the strongest antifungal effect against *S. rolfsii* was benzyl alcohol, with 82.5% using 50 µL/plate and 100% using 100 and 150 µL/plate. Then, acetic acid (unidentified compound as VOC component in *K. cowanii* Cp1 and used as control positive) showed 85 and 100% inhibition using 10 and 20 µL/plate, respectively, and only 21% inhibition using 5 µL/plate. For 2-nonanone, a low inhibition of 25% was observed at 10 µL/plate, while an inhibition of 72.5 and 90% was observed with 20 and 40 µL/plate, respectively. With 2-butanone, 3-hydroxy, an inhibition of 70 and 82.5% was observed with 100 and 150 µL/plate, respectively, while the inhibition observed with 50 µL/plate was only 37.5%. With 2-butanone, inhibition was 15, 56.2, and 72.5% using 50, 100, and 150 µL/plate, respectively. Acetone only showed 25% inhibition using the highest volume (150 µL/plate). Ethanol was the compound that showed the lowest inhibition rate; only at the highest volume of 200 µL/plate was 21.25% inhibition observed (Figure 9).

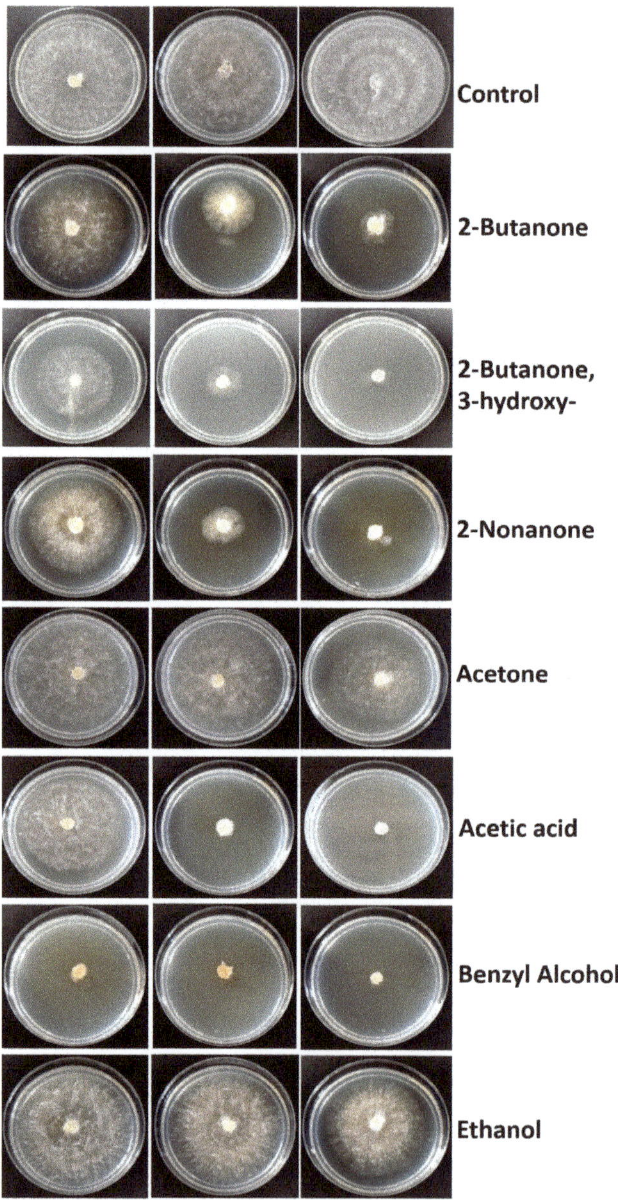



**Figure 9.** Inhibitory effect of standard VOCs on the mycelia growth of *S. rolfsii*. The amount for each compound assayed were, from left to right: 50, 100, and 150 µL/plate of 2-Butanone; 50, 100, and 150 µL/plate of 2-Butanone, 3-hydroxy; 10, 20, and 40 µL/plate of 2-Nonanone; 50, 100, and 150 µL/plate of acetone; 5, 10, and 20 µL/plate of acetic acid; 50, 100, and 150 µL/plate of benzyl alcohol; and 50, 100, and 200 µL/plate of ethanol. Sterile distilled water was used in the control.

## 4. Discussion

According to the hypothesis established in this research work, the findings reported here present, for the first time, the association of the *K. cowanii* strain Cp1 with seeds of *C. pubescens*, a plant that is cultivated in different agroecological zones in Michoacan,

Mexico [44]. Although *C. annuum* L. seeds from Queretaro, Mexico, were analyzed, bacterial strains related to the *Kosakonia* genus were not detected; however, several other isolated bacterial genera are under study in our laboratory. Although we analyzed a relatively limited number of samples of *Capsicum* seeds, it is necessary to continue exploring in order to understand the seed association of *K. cowanii* with the diverse species of the *Capsicum* genus in the diverse geographical regions of Mexico so as to determine if the plant genotype, plant species, environment, and soil management practices have an effect on the presence of this bacterial species. This should be performed because previous results have detected the presence of *K. cowanii* in diverse samples of chili powder [16–18,31].

Throughout the last decade, *K. cowanii* has been isolated and associated with the microbiome of diverse organisms and processed products. Its huge metabolic versatility has shown an extensive resistance to diverse environmental factors, such as pH, salinity, temperature, and, most likely, host biomolecules, which could promote diverse bacterial interactions with diverse organisms such as insects, plants, and humans [20–32]. In this sense, it is widely recognized that the fruits and seeds of *C. pubescens* are a source of capsaicinoids and phenolic compounds [45]. Research works have demonstrated their antimicrobial properties as a consequence of structural or functional disruptions to the bacterial cell membrane [46]. The fact that *K. cowanii* has been isolated from *C. pubescens* seeds demonstrates that it possesses mechanisms of resistance to these compounds. In this sense, the multidrug efflux pumps of MdtABC-TolC that were detected in *K. cowanii* Cp1 have been characterized as important components of *Salmonella* sp. during colonization in the intestine and of virulence in mice [47]. In addition, this multidrug efflux pump in the phytopathogenic *Erwinia amylovora* plays an important role during colonization and resistance to plant metabolites [48] and during insect infection by *Photorhabdus luminescens* [49]. Also, the expression regulation of *mdtABC* by the primary regulator BaeR has been observed in vitro in *E. coli* and *Salmonella* as a response to a wide range of stress conditions, including plant metabolites such as tannins and flavonoids [49–51]. With this in mind, it would be interesting, in the future, to understand these regulatory molecular mechanisms as well as the chemiotaxis pathways that could serve as a route of dispersion or interaction of the *K. cowanii* Cp1 strain with these biological *Capsicum* seed compounds.

The genome sequencing results of the *K. cowanii* Cp1 strain showed a close phylogenetic relationship with *K. cowanii* Ch1, including the IncF plasmid. Both strains presented virulence genes that are principally related with iron uptake, siderophore, invasion, and motility (Table 1). Interestingly, secretion system component genes were not detected, which is especially relevant because the *K. cowanii* Pa82 strain encodes the type VI secretion system as a virulence factor that causes plant disease [25]. Based on the results observed in the pathogenic test of *K. cowanii* Cp1 in *C. annuum* L. plants and fruits (Figure 2), we can conclude that *K. cowanii* Cp1 is not able to act as a phytopathogen despite the virulence related genes detected. Similar results were observed in *K. cowanii* Ch1 [32], which suggests that both bacterial strains most likely do not represent a potential risk of affecting *Capsicum* plants. However, the fact that the spread of the *Kosakonia* genus could be associated with *Capsicum* seeds is important to consider if the phytopathogenic strains of the *Kosakonia* genus associated with other cultivated plants eventually might cause damage in the *Capsicum* genus through different spread routes. Thus, for instance, *K. cowanii* has been reported as the causal agent of bacterial wilt in tomatoes and patchouli plants and, recently, as an emergent phytopathogen affecting soybean (*Glycine max* Willd) [25,26]. In Eucalyptus trees, symptoms of bacterial blight [27] are presented, and in *Mabea fistulifera* Mart. (Euphorbiaceae), necrotic spots can occur on leaves [28]. Therefore, the monitoring of the microorganism-associated seeds of the *Capsicum* genus in either fresh, dried, or processed products could be relevant in terms of avoiding the bacterial spread that comes with a potential risk of causing infection in susceptible *Capsicum* plants.

On the other hand, it is recognized that the antibiotic resistance in bacteria is part of their natural genetic characteristics and evolution processes, which play important roles during environmental competition [52]. However, it is also one of the greatest emergent problems in human or animal health due to the increase in the prevalence of multidrug-resistant (MDR) bacteria, which have arisen through natural selection due to the abusive overuse of antibiotics in various human activities [52,53]. Therefore, the presence of antibiotic-resistant mechanisms in *K. cowanii* Cp1, such as TEM β-lactamase (which allows for phenotypical resistance to penicillin G, ampicillin, and carbenicillin) is a pressing concern. In addition, the antibiotic efflux pump system that was detected and represented by *acrAB-tolC* and *acrAD-tolC*, respectively, is an intrinsic mechanism of multidrug resistance in the diverse members of Gram-negative bacteria. The antibiotic profile of the pump system includes tetracycline, novobiocin, chloramphenicol, fluoroquinolone, fusidic acid, nalidixic acid, lipophilic antibiotics, and β-lactam antibiotics [54]. Another tripartite efflux system identified was *emrAB-tolC*, which has been demonstrated to induce overexpression in *Escherichia coli*, which causes a resistance to nalidixic acid, thiolactomycin, and nitroxoline [55]. Also, the macrolide transporter *macAB-tolC* has been identified as the most likely cause of erythromycin resistance [56]. The transporter *mdfA/cmr* that has also been identified contributes to multidrug resistance in Gram-negative bacteria [57]. Additionally, the other genes that are linked to the resistance mechanism that was identified—such as transcription factors *marAB*, transcriptional repressor *marR*, target modification, and the protein-altering cell wall charge conferring antibiotic resistance—are shown in Figure 4. All of these resistance mechanisms that were detected in *K. cowanii* Cp1 make it necessary to conduct further work so as to evaluate any potential environmental risks. This is required because of its presence in *Capsicum* seeds and chili powder that are used in human consumption.

The colonization ability of microorganisms depends on the metabolic strategies that are ligated to the environmental factors that promote genetic expression regulation so as to synthetize biomolecules with diverse functions during competition in an ecological niche with a community structure [58]. Thus, for instance, the competition for nutrients could promote a change in growth rate and could generate metabolites with certain effects on their competitors. One class of these compounds are the VOCs, which are a huge variety of molecules that are produced by the different metabolic pathways that have been studied in a diverse array of microorganisms [59]. In this sense, the experimental strategy used in *K. cowanii* Cp1 during a dual experiment with six tested phytopatogenic fungi suggested that the volatile organic compounds produced during the growth of this bacterial strain on TSA medium has inhibitory effects on the mycelial rate growth of *Sclerotinia* sp., *R. solani*, *S. rolfsii*, *A. alternata*, and *C. gloeosporioides*, and this occurs by different mechanisms when compared with *F. oxysporum* (which showed a lower degree of effect). Our evidence, obtained by HS-SPME-GC–MS, demonstrates that the diverse chemical classes of VOCs were produced by *K. cowanii* Cp1 during its growth fermentation on TSA medium (Figure 7). The compounds with a high relative abundance were alcohols (23.5%), which were represented by ethanol (at 5.56%) and 1-butanol-3-methyl (at 4.83%) as the most abundant examples, and ketones (22.97%), where the most abundant was 2-butanone, 3-hydroxy, at 13.52%. In addition, those with a lower relative abundance were the diverse compounds detected from the chemical family of pyrazines, esters, aromatics, acids, hydrocarbons, aldehydes, and aromatics. Interestingly, similar results of VOC production were found in the *K. cowanii* Ch1 strain, where 2-butanone, 3-hydroxy was the most abundant compound (at 10.60%), as well as ethanol (at 5.40%), and 1-butanol-3-methyl (4.88%) [32]. According to diverse evidence, some of these identified compounds may affect plants through a diverse array of mechanisms and through the functionality of fungi cells (thereby causing a disruption of membrane fluidity or wall integrity) as well as alterations in metabolism and redox balance [59]. Although 2-butanone, 3-hydroxy could be one of the principal compounds with antifungal properties in *K. cowanii* Cp1, it is too complicated to understand because it is part of a heterogeneous mix of VOCs that have fluctuating concentrations during the fermentation processes that were tested. This interpretation was confirmed using standard VOCs, where 2-butanone, 3-hydroxy showed an inhibition rate of 82.5% when the highest volume was used (150 μL/plate) compared with acetic acid, benzyl alcohol, or 2-nonanone, which demonstrated high inhibition activities when using a lower volume. In addition, relative peak areas identified were 0.68% for benzyl alcohol, 1.32% for 2-nonanone, and 0.90% for 2-butanone compared with 13.52% for 2-butanone, 3-hydroxy or 5.56% for ethanol (Table 2). Therefore, a mix of VOCs has more impact than individual compounds during biocontrol.

According to the genome sequencing analysis of *K. cowanii* Cp1, metabolic pathways were present that produce 2-butanone, 3-hydroxy through the enzymatic activities of aceto-lactate decarboxylase and 2,3-butanediol dehydrogenase via the diverse carbon sources that are used in fermentation [60]. Evidence suggests that this molecule, functionally, has significant effects in plants, e.g., during growth development [61], drought tolerance [62], and defense against pathogens [63]. On the one hand, the VOC results suggest that, during fermentation pathways, aldehyde dehydrogenase and alcohol dehydrogenase enzymes could be necessary for ethanol production as well as for the butanal (butyraldehyde) path-way that is required to obtain butanol (which were detected in relatively high abundance in the VOCs from *K. cowanii* Cp1). These alcohol classes are components of the VOCs that are present in a diverse array of microorganisms with antifungal activity [64]. Another interest-ing compound that was identified as a component of the VOCs was Pyrazine-2,5-dimethyl, with a high relative peak area (6.47%). This compound is produced in diverse bacterial species, and it engages in antifungal activity against *Magnaporthe oryzae*, *Phytophthora capsici*, and *A. solani* [59]. Although other VOCs were produced that had a lower relative-peak abundance (Table 2), these could, nevertheless, be more important, because their antifungal activity has been registered in the mVOC 2.0 Database [65]. Therefore, additional work is necessary to understand the impact of each VOC, on its own or as part of a mixture, on the growth of phytopathogenic fungi strains.

In conclusion, the finding of *K. cowanii* Cp1 as a microorganism that is associated with *Capsicum pubescens* seeds and which has antifungal properties through VOC production opens up the possibility of further exploration of important ecological roles as well as of the plant–bacteria interactions with diverse molecules such as capsaicinoids and phe-nolic compounds. Additionally, the *Capsicum* genus, cultivated throughout all Mexican territory under different environmental conditions, remains to be explored in order to understand the microbiomes associated with the seeds that are of particular interest in terms of finding potential bacterial strains, such as *Kosakonia cowanii*, for the development of biocontrol agents.

**Supplementary Materials:** The following supporting information can be downloaded at: https://www.mdpi.com/article/10.3390/microorganisms11102491/s1, Figure S1. Dual-culture assay on PDA. *K. cowanii* Cp1 and *S. rolfsii* were confronted. Mycelial and bacterial growth was register during 2 days of growth at 28 °C. Figure S2. Antibiotic resistance profile of *K. cowanii* CP1 strain

**Author Contributions:** J.C.G. and J.A.R.M.; methodology and software, Y.J.M., M.M.A.R., M.J.C.P., M.F.M., J.R.R., G.R.Z., P.L.T.B., L.V.S. and K.V.R.; validation and formal analysis, J.C.G., J.L.H.F., M.Á.R.L. and J.A.R.M.; writing—review and editing, J.C.G., C.S.G., J.L.H.F. and A.A.R.; visualization, supervision, and project administration, J.C.G.; funding acquisition, J.C.G. All authors have read and agreed to the published version of the manuscript.

**Funding:** This research was partially funded by the Universidad Autónoma de Querétaro (FONDEC-UAQ-2022; FOPER-2022).

**Data Availability Statement:** The data presented in this work are available from the corresponding authors upon request.

**Conflicts of Interest:** The authors declare no conflict of interest.

## References

1. Kraft, K.H.; Brown, C.H.; Nabhan, G.P.; Luedeling, E.; de Luna Ruiz, J.; Coppens d'Eeckenbrugge, G.; Hijmans, R.J.; Gepts, P. Multiple lines of evidence for the origin of domesticated chili pepper, *Capsicum annuum*, in Mexico. *Proc. Natl. Acad. Sci. USA* **2014**, *111*, 6165–6170. [CrossRef] [PubMed]

2.  Tripodi, P.; Rabanus-Wallace, M.T.; Barchi, L.; Kale, S.; Esposito, S.; Acquadro, A.; Schafleitner, R.; van Zonneveld, M.; Prohens, J.; Diez, M.J.; et al. Global range expansion history of pepper (*Capsicum* spp.) revealed by over 10,000 genebank accessions. *Proc. Natl. Acad. Sci. USA* **2021**, *118*, e2104315118. [CrossRef]
3.  Swamy, K.R.M. Origin, distribution, taxonomy, botanical description, genetic diversity and breeding of capsicum (*Capsicum annuum* L.). *Int. J. Dev. Res.* **2023**, *13*, 61956–61977.
4.  McCoy, J.; Martínez-Ainsworth, N.; Bernau, V.; Scheppler, H.; Hedblom, G.; Adhikari, A.; McCormick, A.; Kantar, M.; McHale, L.; Jardón-Barbolla, L.; et al. Population structure in diverse pepper (*Capsicum* spp.) accessions. *BMC Res. Notes* **2023**, *16*, 20. [CrossRef]
5.  Pereira-Dias, L.; Vilanova, S.; Fita, A.; Prohens, J.; Rodríguez-Burruezo, A. Genetic diversity, population structure, and relationships in a collection of pepper (*Capsicum* spp.) landraces from the Spanish centre of diversity revealed by genotyping-by-sequencing (GBS). *Hortic. Res.* **2019**, *6*, 54. [CrossRef]
6.  Zhang, X.-m.; Zhang, Z.-h.; Gu, X.-z.; Mao, S.-l.; Li, X.; Chadœuf, J.; Palloix, A.; Wang, L.-h.; Zhang, B.-x. Genetic diversity of pepper (*Capsicum* spp.) germplasm resources in China reflects selection for cultivar types and spatial distribution. *J. Integr. Agric.* **2016**, *15*, 1991–2001. [CrossRef]
7.  Sarath Babu, B.; Pandravada, S.R.; Pasada Rao, R.D.V.J.; Anitha, K.; Chakrabarty, S.K.; Varaprasad, K.S. Global sources of pepper genetic resources against arthropods, nematodes and pathogens. *Crop Prot.* **2011**, *30*, 389–400. [CrossRef]
8.  Ros, C.; Lacasa, C.M.; Martínez, V.; Bielza, P.; Lacasa, A. Response of pepper rootstocks to co-infection of Meloidogyne incognita and *Phytophthora* spp. *Eur. J. Hortic. Sci.* **2014**, *79*, 22–28.
9.  Özkaynak, E.; Devran, Z.; Kahveci, E.; Doğanlar, S.; Başköylü, B.; Doğan, F.; Yüksel, M. Pyramiding multiple genes for resistance to PVY, TSWV and PMMoV in pepper using molecular markers. *Eur. J. Hortic. Sci.* **2014**, *79*, 233–239.
10. Parisi, M.; Alioto, D.; Tripodi, P. Overview of Biotic Stresses in Pepper (*Capsicum* spp.): Sources of Genetic Resistance, Molecular Breeding and Genomics. *Int. J. Mol. Sci.* **2020**, *21*, 2587. [CrossRef]
11. Lozada, D.N.; Bosland, P.W.; Barchenger, D.W.; Haghshenas-Jaryani, M.; Sanogo, S.; Walker, S. Chile Pepper (*Capsicum*) Breeding and Improvement in the "Multi-Omics" Era. *Front. Plant Sci.* **2022**, *13*, 879182. [CrossRef]
12. Kim, T.J.; Hyeon, H.; Park, N.I.; Yi, T.G.; Lim, S.H.; Park, S.Y.; Ha, S.H.; Kim, J.K. A high-throughput platform for interpretation of metabolite profile data from pepper (*Capsicum*) fruits of 13 phenotypes associated with different fruit maturity states. *Food Chem.* **2020**, *331*, 127286. [CrossRef] [PubMed]
13. Vijay, A.; Kumar, A.; Islam, K.; Momo, J.; Ramchiary, N. Functional genomics to understand the tolerance mechanism against biotic and abiotic stresses in *Capsicum* species. In *Transcriptome Profiling*; Academic Press: Cambridge, MA, USA, 2023; pp. 305–332.
14. Zhang, L.; Qin, C.; Mei, J.; Chen, X.; Wu, Z.; Luo, X.; Cheng, J.; Tang, X.; Hu, K.; Li, S.C. Identification of microRNA targets of *Capsicum* spp. using miRTrans—A trans-omics approach. *Front. Plant Sci.* **2017**, *8*, 495. [CrossRef]
15. Reimer, J.J.; Shaaban, B.; Drummen, N.; Sanjeev Ambady, S.; Genzel, F.; Poschet, G.; Wiese-Klinkenberg, A.; Usadel, B.; Wormit, A. *Capsicum* leaves under stress: Using multi-omics analysis to detect abiotic stress network of secondary metabolism in two species. *Antioxidants* **2022**, *11*, 671. [CrossRef] [PubMed]
16. Jonkers, W.; Gundel, P.E.; Verma, S.K.; White, J.F. (Eds.) *Seed Microbiome Research*; Frontiers Media SA: Lausanne, Switzerland, 2022.
17. Tkalec, V.; Mahnic, A.; Gselman, P.; Rupnik, M. Analysis of seed-associated bacteria and fungi on staple crops using the cultivation and metagenomic approaches. *Folia Microbiol.* **2022**, *67*, 351–361. [CrossRef] [PubMed]
18. Dowarah, B.; Agarwal, H.; Krishnatreya, D.B.; Sharma, P.L.; Kalita, N.; Agarwala, N. Evaluation of seed associated endophytic bacteria from tolerant chilli cv. Firingi Jolokia for their biocontrol potential against bacterial wilt disease. *Microbiol. Res.* **2021**, *248*, 126751. [CrossRef]
19. Jeong, S.; Kim, T.M.; Choi, B.; Kim, Y.; Kim, E. Invasive *Lactuca serriola* seeds contain endophytic bacteria that contribute to drought tolerance. *Sci. Rep.* **2021**, *11*, 13307. [CrossRef]
20. Jan-Roblero, J.; Cruz-Maya, J.A.; Barajas, C.G. Chapter 12—Kosakonia. In *Beneficial Microbes in Agro-Ecology*; Amaresan, N., Annapurna, K., Sankaranarayanan, A., Kumar, M.S., Kumar, K., Eds.; Academic Press: Cambridge, MA, USA, 2020; pp. 213–231.
21. Romano, I.; Ventorino, V.; Ambrosino, P.; Testa, A.; Chouyia, F.E.; Pepe, O. Development and Application of Low-Cost and Eco-Sustainable Bio-Stimulant Containing a New Plant Growth-Promoting Strain *Kosakonia pseudosacchari* TL13. *Front. Microbiol.* **2020**, *11*, 2044. [CrossRef]
22. Yang, X.-J.; Wang, S.; Cao, J.M.; Hou, J.H. Complete genome sequence of human pathogen *Kosakonia cowanii* type strain 888-76. *Braz. J. Micro-Biol.* **2018**, *49*, 16–17. [CrossRef]
23. Bhatti, M.D.; Kalia, A.; Sahasrabhojane, P.; Kim, J.; Greenberg, D.E.; Shelburne, S.A. Identification and Whole Genome Sequencing of the First Case of *Kosakonia radicincitans* Causing a Human Blood-stream Infection. *Front. Microbiol.* **2017**, *8*, 62. [CrossRef]
24. Berinson, B.; Bellon, E.; Christner, M.; Both, A.; Aepfelbacher, M.; Rohde, H. Identification of *Kosakonia cowanii* as a rare cause of acute cholecystitis: Case report and review of the literature. *BMC Infect. Dis.* **2020**, *20*, 366. [CrossRef]
25. Zhang, Y.; Wang, B.; Li, Q.; Huang, D.; Zhang, Y.; Li, G.; He, H. Isolation and Complete Genome Sequence Analysis of *Kosakonia cowanii* Pa82, a Novel Pathogen Causing Bacterial Wilt on Patchouli. *Front. Microbiol.* **2022**, *12*, 818228. [CrossRef]
26. Krawczyk, K.; Borodynko-Filas, N. Kosakoniacowanii as the New Bacterial Pathogen Affecting Soybean (*Glycine max* Willd.). *Eur. J. Plant Pathol.* **2020**, *157*, 173–183. [CrossRef]

27. Brady, C.L.; Venter, S.N.; Cleenwerck, I.; Engelbeen, K.; de Vos, P.; Wingfield, M.J.; Telechea, N.; Coutinho, T.A. Isolation of *Enterobacter cowanii* from Eucalyptus showing symptoms of bacterial blight and dieback in Uruguay. *Lett. Appl. Microbiol.* **2009**, *49*, 461–465. [CrossRef]

28. Furtado, G.Q.; Guimarães, L.M.S.; Lisboa, D.O.; Cavalcante, G.P.; Arriel, D.A.A.; Alfenas, A.C.; Oliveira, J.R. First Report of *Enterobacter cowanii* Causing Bacterial Spot on *Mabea fistulifera*, a Native Forest Species in Brazil. *Plant Dis.* **2012**, *96*, 1576. [CrossRef] [PubMed]

29. Petrzik, K.; Brázdová, S.; Krawczyk, K. Novel Viruses That Lyse Plant and Human Strains of *Kosakonia cowanii*. *Viruses* **2021**, *13*, 1418. [CrossRef]

30. Dennison, N.J.; Jupatanakul, N.; Dimopoulos, G. The mosquito microbiota influences vector competence for human pathogens. *Curr. Opin. Insect Sci.* **2014**, *3*, 6–13. [CrossRef]

31. Hernández Gómez, Y.F.; González Espinosa, J.; Ramos López, M.Á.; Arvizu Gómez, J.L.; Saldaña, C.; Rodríguez Morales, J.A.; García Gutiérrez, M.C.; Pérez Moreno, V.; Álvarez Hidalgo, E.; Nuñez Ramírez, J.; et al. Insights into the Bacterial Diversity and Detection of Opportunistic Pathogens in Mexican Chili Pow-der. *Microorganisms* **2022**, *10*, 1677. [CrossRef] [PubMed]

32. González Espinosa, J.; Hernández Gómez, Y.F.; Javier Martínez, Y.; Flores Gallardo, F.J.; Pacheco Aguilar, J.R.; Ramos López, M.Á.; Arvizu Gómez, J.L.; Saldaña Gutierrez, C.; Rodríguez Morales, J.A.; García Gutiérrez, M.C.; et al. *Kosakonia cowanii* Ch1 Isolated from Mexican Chili Powder Reveals Growth Inhibition of Phytopathogenic Fungi. *Microorganisms* **2023**, *11*, 1758. [CrossRef] [PubMed]

33. Aguilar-Rincón, V.H.; Corona Torres, P.T.; López López, L.; Latournerie Moreno, M.; Ramírez Meraz, H.; Villalón Mendoza, Y.J.; Aguilar Castillo, A. *Los Chiles de México y su Distribución*; Rev Fitotecnia Mex SINAREFI, Colegio de Postgraduados; INIFAP, ITConkal, UANL, UAN; Sociedad Mexicana de Fitogenética A.C.: Montecillo, Texcoco, México, 2010; p. 114.

34. Saitou, N.; Nei, M. The neighbor-joining method: A new method for reconstructing phylogenetic trees. *Mol. Biol. Evol.* **1987**, *4*, 406–425.

35. Felsenstein, J. Confidence Limits on Phylogenies: An Approach Using the Bootstrap. *Evolution* **1985**, *39*, 783–791. [CrossRef] [PubMed]

36. Jukes, T.H.; Cantor, C.R. Evolution of protein molecules. In *Mammalian Protein Metabolism*; Munro, H.N., Ed.; Academic Press: Cambridge, MA, USA, 1969; pp. 21–132.

37. Kumar, S.; Stecher, G.; Li, M.; Knyaz, C.; Tamura, K. MEGA X: Molecular Evolutionary Genetics Analysis across Computing Platforms. *Mol. Biol. Evol.* **2018**, *35*, 1547–1549. [CrossRef] [PubMed]

38. Olson, R.D.; Assaf, R.; Brettin, T.; Conrad, N.; Cucinell, C.; Davis, J.J.; Dempsey, D.M.; Dickerman, A.; Dietrich, E.M.; Kenyon, R.W.; et al. Introducing the Bacterial Viral Bioinformatics Resource Center (BV-BRC): A resource combining PATRIC, IRD and ViPR. *Nucleic Acids Res.* **2022**, *51*, D678–D689. [CrossRef] [PubMed]

39. Brettin, T.; Davis, J.J.; Disz, T.; Edwards, R.A.; Gerdes, S.; Olsen, G.J.; Olson, R.; Overbeek, R.; Parrello, B.; Pusch, G.D.; et al. RASTtk: A modular and extensible implementation of the RAST algorithm for building custom annotation pipelines and annotating batches of genomes. *Sci. Rep.* **2015**, *5*, 8365. [CrossRef] [PubMed]

40. Naquin, D.; d'Aubenton-Carafa, Y.; Thermes, C.; Silvain, M. CIRCUS: A package for Circos display of structural genome variations from paired-end and mate-pair sequencing data. *BMC Bioinform.* **2014**, *15*, 198. [CrossRef]

41. Ondov, B.D.; Treangen, T.J.; Melsted, P.; Mallonee, A.B.; Bergman, N.H.; Koren, S.; Phillippy, A.M. Mash: Fast genome and metagenome distance estimation using MinHash. *Genome Biol.* **2016**, *17*, 132. [CrossRef]

42. Dmitry, A.; Mikhail, R.; Alla, L.; Pevzner, P.A. Plasmid detection and assembly in genomic and metagenomic data sets. *Genome Res.* **2019**, *29*, 961–968.

43. Clinical and Laboratory Standards Institute. *Performance Standards for Antimicrobial Disk Susceptibility Tests*, approved standard, 12th ed.; Clinical and Laboratory Standards Institute: Wayne, PA, USA, 2018.

44. Escalera-Ordaz, A.K.; Guillén-Andrade, H.; Lara-Chávez, M.B.N.; Lemus-Flores, C.; Rodríguez-Carpena, J.G.; Valdivia-Bernal, R. Characterization of cultivated varieties of *Capsicum pubescens* in Michoacán, Mexico. *Rev. Mex. Cienc. Agríc.* **2019**, *10*, 239–251.

45. Meckelmann, S.W.; Jansen, C.; Riegel, D.W.; van Zonneveld, M.; Ríos, L.; Peña, K.; Mueller-Seitz, E.; Petz, M. Phytochemicals in Native Peruvian *Capsicum pubescens* (Rocoto). *Eur. Food Res. Technol.* **2015**, *241*, 817–825. [CrossRef]

46. Romero-Luna, H.E.; Colina, J.; Guzmán-Rodríguez, L.; Sierra-Carmona, C.G.; Farías-Campomanes, Á.M.; García-Pinilla, S.; González-Tijera, M.M.; Malagón-Alvira, K.O.; Peredo-Lovillo, A. Capsicum fruits as functional ingredients with antimicrobial activity: An emphasis on mechanisms of action. *J. Food Sci. Technol.* **2022**, *60*, 1–11. [CrossRef]

47. Nishino, K.; Latifi, T.; Groisman, E.A. Virulence and drug resistance roles of multidrug efflux systems of *Salmonella enterica* serovar Typhimurium. *Mol. Microbiol.* **2006**, *59*, 126–141. [CrossRef]

48. Pletzer, D.; Weingart, H. Characterization and regulation of the resistance-nodulation-cell division-type multidrug efflux pumps MdtABC and MdtUVW from the fire blight pathogen *Erwinia amylovora*. *BMC Microbiol.* **2014**, *14*, 185. [CrossRef]

49. Abi Khattar, Z.; Lanois, A.; Hadchity, L.; Gaudriault, S.; Givaudan, A. Spatiotemporal expression of the putative MdtABC efflux pump of *Phtotorhabdus luminescens* occurs in a protease-dependent manner during insect infection. *PLoS ONE* **2019**, *14*, e0212077. [CrossRef] [PubMed]

50. Hirakawa, H.; Inazumi, Y.; Masaki, T.; Hirata, T.; Yamaguchi, A. Indole induces the expression of multidrug exporter genes in *Escherichia coli*. *Mol. Microbiol.* **2005**, *55*, 1113–1126. [CrossRef]

51. Nishino, K.; Nikaido, E.; Yamaguchi, A. Regulation of multidrug efflux systems involved in multidrug and metal resistance of *Salmonella enterica* serovar Typhimurium. *J. Bacteriol.* **2007**, *189*, 9066–9075. [CrossRef]
52. Larsson, D.G.J.; Flach, C.F. Antibiotic resistance in the environment. *Nat. Rev. Microbiol.* **2022**, *20*, 257–269. [CrossRef] [PubMed]
53. Koch, N.; Islam, N.F.; Sonowal, S.; Prasad, R.; Sarma, H. Environmental antibiotics and resistance genes as emerging contaminants: Methods of detection and bioremediation. *Curr. Res. Microb. Sci.* **2021**, *2*, 100027. [CrossRef] [PubMed]
54. Chowdhury, N.; Suhani, S.; Purkaystha, A.; Begum, M.K.; Raihan, T.; Alam, M.J.; Islam, K.; Azad, A.K. Identification of AcrAB-TolC Efflux Pump Genes and Detection of Mutation in Efflux Repressor AcrR from Omeprazole Responsive Multidrug-Resistant *Escherichia coli* Isolates Causing Urinary Tract Infections. *Microbiol. Insights* **2019**, *12*, 1178636119889629. [CrossRef]
55. Yousefian, N.; Ornik-Cha, A.; Poussard, S.; Decossas, M.; Berbon, M.; Daury, L.; Taveau, J.C.; Dupuy, J.W.; Đorđević-Marquardt, S.; Lambert, O.; et al. Structural characterization of the EmrAB-TolC efflux complex from *E. coli*. *Biochim. Biophys. Acta Biomembr.* **2021**, *1863*, 183488. [CrossRef]
56. Lu, S.; Zgurskaya, H.I. MacA, a periplasmic membrane fusion protein of the macrolide transporter MacAB-TolC, binds lipopolysaccharide core specifically and with high affinity. *J. Bacteriol.* **2013**, *195*, 4865–4872. [CrossRef]
57. Swick, M.C.; Morgan-Linnell, S.K.; Carlson, K.M.; Zechiedrich, L. Expression of multidrug efflux pump genes acrAB-tolC, mdfA, and norE in *Escherichia coli* clinical isolates as a function of fluoroquinolone and multidrug resistance. *Antimicrob. Agents Chemother.* **2011**, *55*, 921–924. [CrossRef] [PubMed]
58. Moreno, R.; Rojo, F. The importance of understanding the regulation of bacterial metabolism. *Environ. Microbiol.* **2023**, *25*, 54–58. [CrossRef] [PubMed]
59. Almeida, O.A.C.; de Araujo, N.O.; Dias, B.H.S.; de Sant'Anna Freitas, C.; Coerini, L.F.; Ryu, C.-M.; de Castro Oliveira, J.V. The power of the smallest: The inhibitory activity of microbial volatile organic compounds against phytopathogens. *Front. Microbiol.* **2023**, *13*, 951130. [CrossRef]
60. Marquez-Villavicencio, M.d.P.; Weber, B.; Witherell, R.A.; Willis, D.K.; Charkowski, A.O. The 3-Hydroxy-2-Butanone Pathway Is Required for *Pectobacterium carotovorum* Pathogenesis. *PLoS ONE* **2011**, *6*, e22974. [CrossRef]
61. Ryu, C.M.; Farag, M.A.; Hu, C.H.; Reddy, M.S.; Wei, H.X.; Paré, P.W.; Kloepper, J.W. Bacterial volatiles promote growth in Arabidopsis. *Proc. Natl. Acad. Sci. USA* **2003**, *100*, 4927–4932. [CrossRef] [PubMed]
62. Ryu, C.-M.; Farag, M.A.; Hu, C.-H.; Reddy, M.S.; Kloepper, J.W.; Paré, P.W. Bacterial Volatiles Induce Systemic Resistance in Arabidopsis. *Plant Physiol.* **2004**, *134*, 1017–1026. [CrossRef]
63. Cho, S.M.; Kang, B.R.; Han, S.H.; Anderson, A.J.; Park, J.-Y.; Lee, Y.-H.; Cho, B.H.; Yang, K.-Y.; Ryu, C.-M.; Kim, Y.C. 2R,3R-Butanediol, a Bacterial Volatile Produced by Pseudomonas chlororaphis O6, Is Involved in Induction of Systemic Tolerance to Drought in *Arabidopsis thaliana*. *Mol. Plant Microbe Interact.* **2008**, *21*, 1067–1075. [CrossRef]
64. Zhao, X.; Zhou, J.; Tian, R.; Liu, Y. Microbial volatile organic compounds: Antifungal mechanisms, applications, and challenges. *Front. Microbiol.* **2022**, *13*, 922450. [CrossRef]
65. Lemfack, M.C.; Gohlke, B.-O.; Toguem, S.M.T.; Preissner, S.; Piechulla, B.; Preissner, R. mVOC 2.0: A database of microbial volatiles. *Nucleic Acids Res.* **2017**, *46*, D1261–D1265. [CrossRef] [PubMed]


**Disclaimer/Publisher's Note:** The statements, opinions and data contained in all publications are solely those of the individual author(s) and contributor(s) and not of MDPI and/or the editor(s). MDPI and/or the editor(s) disclaim responsibility for any injury to people or property resulting from any ideas, methods, instructions or products referred to in the content.

 *microorganisms*

*Article*

# Defense Inducers Mediated Mitigation of Bacterial Canker in Tomato through Alteration in Oxidative Stress Markers

Ruchi Tripathi [1,*], Karuna Vishunavat [1], Rashmi Tewari [1], Sumit Kumar [2], Tatiana Minkina [3], Ugo De Corato [4] and Chetan Keswani [3,*]

[1] Department of Plant Pathology, College of Agriculture, G B Pant University of Agriculture and Technology, Pantnagar 263145, India
[2] Department of Mycology and Plant Pathology, Institute of Agricultural Sciences, Banaras Hindu University, Varanasi 221005, India
[3] Academy of Biology and Biotechnology, Southern Federal University, Rostov-on-Don 44090, Russia
[4] Department of Bioenergy, Biorefinery and Green Chemistry (TERIN-BBC-BIC)-Italian National Agency for the New Technologies, Energy and Sustainable Economic Development (ENEA)-Territorial Office of Bari, Via Giulio Petroni 15/F, 70124 Bari, Italy
* Correspondence: assidious.rt35685@gmail.com (R.T.); kesvani@sfedu.ru (C.K.); Tel.: +7-98-8991-9786 (C.K.)

**Abstract:** The bacterial canker disease of tomato caused by *Clavibacter michiganensis* subsp. *michiganensis* (*Cmm*) has been reported to adversely affect the tomato cultivation in the NE hilly regions of India. Defense inducers such as salicylic acid (SA), isonicotinic acid (INA), benzothiadiazole (BTH) and lysozyme were used as prophylactic and curative sprays at different concentrations to test their efficacy in inducing resistance in tomato plants against *Cmm* under protected conditions. The induced resistance was studied through the alteration in the activities of oxidative stress marker enzymes (PAL, PO, PPO, TPC and PR-2 protein), hydrogen peroxide formation in leaf tissues and lignin accumulation in stem tissues, as well as through the reduction in disease severity under glasshouse conditions. The results of the present study revealed that the enzymatic activity, hydrogen peroxide formation and lignin production were significantly higher in the BTH (500 ppm)-treated leaves than in those observed in the control. The lowest disease incidence was recorded when BTH was applied as a prophylactic spray (27.88%) in comparison to being applied as a curative spray (53.62%), thereby suggesting that a defense inducer, BTH, shows antibacterial activity against *Cmm*, reduces disease incidence severity and induces defense responses in the tomato plant.

**Keywords:** *Cmm*; *Solanum lycopersicum*; ROS metabolism; antioxidant enzymes; defense inducers

**Citation:** Tripathi, R.; Vishunavat, K.; Tewari, R.; Kumar, S.; Minkina, T.; De Corato, U.; Keswani, C. Defense Inducers Mediated Mitigation of Bacterial Canker in Tomato through Alteration in Oxidative Stress Markers. *Microorganisms* **2022**, *10*, 2160. https://doi.org/10.3390/microorganisms10112160

Academic Editor: Dawn L. Arnold

Received: 12 September 2022
Accepted: 27 October 2022
Published: 31 October 2022

**Copyright:** © 2022 by the authors. Licensee MDPI, Basel, Switzerland. This article is an open access article distributed under the terms and conditions of the Creative Commons Attribution (CC BY) license (https://creativecommons.org/licenses/by/4.0/).

## 1. Introduction

Tomato (*Solanum lycopersicum* L.) is one of the most extensively cultivated vegetable crops [1]; however, its cultivation is heavily affected by several plant pathogens which affect the quantity and quality of the produce. The crop is generally affected by different fungal, bacterial, viral and nematode diseases in the hilly region of northern India. Among the bacterial diseases, bacterial wilt (*Ralstonia solanacearum*) and bacterial spot (*Xanthomonas campestris* pv. *vesicatoria*) have been reported [2], often adding to the grower's losses. While most of the tomato diseases are caused by Gram-negative bacteria, a bacterial disease causing leaf necrosis, wilting and splitting of stem and cankers in the fruits have been observed in recent years, causing damage to the crop in the tomato-growing areas of the northern hilly region. The disease was detected to be bacterial canker caused by the Gram-positive bacterium *Clavibacter michiganensis* ssp. *michiganensis* (Smith) Davis. The world-wide spread of this bacterium is facilitated by contaminated seed stocks, in which a single infected seed in 10,000 can initiate an epidemic [3]. The bacterium reduces the quality and quantity of the product, leading to substantial economic losses both in protected and open field conditions, which may sometimes even lead to complete yield loss [4–8].

Bacterial diseases, including bacterial canker, are expected to become more aggressive in the near future due to climate instability, with devastating effects on basic food-producing areas [9]. Disease management becomes difficult due to the unavailability of resistant cultivars and other effective measures. However, defense inducers or eustressors can play a potential role in reducing the disease's severity by alleviating the defense response and thus by providing efficient resistance in the plants toward the invading pathogen [10]. These eustressors are also known as elicitors and can be obtained either from a plant, microbe or derived synthetically [11]. The priming of susceptible plants with these elicitors leads to an improvement in plant growth and development [12], and also increases the resistance to biotic stress (plant pathogens), both locally and systemically, by inducing the systemic-acquired resistance (SAR) in plants [13–15]. The initiation of SAR is frequently linked with varied defense responses taking place within the cells, viz., pathogenesis-related (PR) proteins, phytoalexins synthesis, reactive oxygen species (ROS) accumulation, boosted action of defense-related enzymes [16] and lignin formation [17]. The pretreatment of tomato seedlings with acibenzolar-S-methyl (ASM) significantly reduces *Cmm* growth and disease severity during the course of the infection [18]. This ASM-mediated enhanced resistance is associated with increased activities of plant peroxidase and chitinase [19]. Additionally, DL-$\beta$-amino butyric acid (BABA) treatment remarkably suppresses symptom development caused by *Cmm* via the enhanced peroxidase and phenylalanine ammonia-lyase activities of the host plant [20]. Therefore, the purpose of this study was to evaluate the effective defense inducers so that an alternative to antibiotics can be explored for the management of bacterial canker disease in tomato.

## 2. Materials and Methods

### 2.1. Source of Cmm and Planting Material

The strain *Cmm10* (Genbank Accession No. MH321815) of *C. michiganensis* subsp. *michiganensis*, isolated from Himachal Pradesh, India, was routinely sub-cultured in nutrient broth–glucose–yeast medium (NGY: Nutrient Broth: 8.0 g, Yeast extract: 2.0 g, $K_2HPO_4$: 2.0 g, $KH_2PO_4$: 0.5 g, Glucose: 2.5 g, Agar: 15.0 g, in 1 L of distilled water followed by sterilization at 121 °C, at 15 psi for 15–20 min) was routinely used for the experiments.

Seedlings of susceptible tomato cv. Rohini and US 2853 were grown in 10 cm pots in a soil + sand + vermicompost mix (2: 1: 1) in the glasshouse at 28 ± 2 °C with 68–80% RH with alternate light and dark cycle (10 h: 14 h: dark: light). Leaves, disinfected with 70% ethanol and sterile distilled water (DW) were used for the study.

### 2.2. Preparation and Application of Defense Inducers

Salicylic acid (SA), isonicotinic acid (INA), benzothiadiazole (BTH) and lysozyme were procured from Sigma-Aldrich, Co. (Darmstadt, Germany) and were dissolved in sterile distilled water for the final concentrations of 200 ppm, 500 ppm and 800 ppm. Plants were grouped into five sets for each treatment with ten replicates. Set I: sprayed with SA, INA, BTH and lysozyme (200 ppm); Set II: sprayed with SA, INA, BTH and lysozyme (500 ppm); Set III; sprayed with SA, INA, BTH and lysozyme (800 ppm); Set IV; sprayed with *Cmm*; Set V: sprayed with water. About 200 µL of the prepared concentration was sprayed onto whole seedlings [18] and the seedlings were maintained in glasshouse in the aforementioned conditions. Leaves (1 gm) were collected after two days of treatment to evaluate various biochemical attributes.

### 2.3. Bacterial Strain Inoculation

The bacterial strain *Cmm10* was maintained in NGY at 4 °C. Inoculum was prepared from early log-phase cells of bacterial strain grown in nutrient yeast extract broth at 27 °C on an orbital shaker at 200 rpm for 24 h. Bacteria were then pelleted by centrifugation (twice, each at 3500 rpm for 5 min) and the pellet was rinsed twice by sterilized water and adjusted to the value of 0.06 at OD660 nm, which corresponds to $10^8$ cfu/mL for inoculation.

The two youngest leaves were then inoculated by cutting at the tips and dipping them into the bacterial suspension [21].

*2.4. Sample Preparation for Enzymatic Activity Determination*

2.4.1. Sample Collection

The samples were taken from the seedlings sprayed with four defense inducers at three different concentrations (200 ppm, 500 ppm and 800 ppm), at two different durations for biochemical analysis after 48 h in all the sets of experiment to assess the change in enzymatic activity. Leaf tissues were taken at the actual site of inoculation with *Cmm* in the case of inoculated plants and from sites similar to those on inoculated leaves in the case of the control plants. These leaf tissues were stored in a deep freezer ($-80\,^\circ$C) until used for biochemical analysis.

2.4.2. Biochemical Analysis

Polyphenol Oxidases (PPOs) Assay (EC 1.14.18.1)

Leaf samples (0.1 g) were homogenized in 2 mL of ice-cold phosphate buffer (0.1 M/L) at pH 6.5 and the homogenate was centrifuged at 16,000 rpm for 30 min at 4 $^\circ$C. The supernatant thus obtained was used directly in the enzyme assay. The reaction mixture for this assay contained 0.4 mL catechol, the substrate for PPOs estimation (1 mM/L) in 3 mL of (0.05 M/L) sodium phosphate buffer at pH 6.5 and 0.4 mL enzyme extract, whereas the reaction mixture without the enzyme extract served as control. The change in absorbance was recorded at 405 nm [22] and the PPO enzyme activity was expressed as change in OD min/mg/FW.

Phenylalanine Ammonia Lyases (PALs) Assay (EC 4.1.3.5)

For PALs assay, 0.1 g leaf samples from each treatment were homogenized in 2 mL of sodium borate buffer (0.1 M/L; pH 7.0; 4 $^\circ$C) containing (1.4 mM/L) 2-mercaptoethanol. Homogenate was centrifuged at 16,000 rpm at 4 $^\circ$C for 15 min and the supernatant was used as enzyme source. The reaction mixture containing 0.2 mL enzyme extract, 0.5 mL (0.2 M/L) borate buffer at pH 8.7 and 1.3 mL of water was prepared, and the reaction was initiated by the addition of 1 mL L-phenylalanine (0.1 M/L at pH 8.7) followed by an incubation at 32 $^\circ$C for 30 min. The termination of the ongoing reaction was achieved by pouring 0.5 mL of trichloroacetic acid (TCA, 1 M/L) into the reaction mixture. The measurement of PALs activity was conducted by estimating trans-cinnamic acid formation at 290 nm and was expressed as μmol/min/g fresh weight (FW) TCA [23].

Peroxidases (POs) Assay (EC 1.11.1.7)

Peroxidases assay was performed by homogenizing leaf samples (0.1 g) in 2 mL of ice-cold phosphate buffer (0.1 M/L), with a pH of 7.0 at 4 $^\circ$C, centrifuged at 16,000 rpm at 4 $^\circ$C for 15 min, and the supernatant was used as enzyme source. Pyrogallol (0.05 M/L) (1.5 mL), enzyme extract (0.05 mL) and 0.5 mL $H_2O_2$ (1% *v/v*) was used as reaction mixture. The changes in the absorbance at 420 nm were recorded after 30 s intervals for 3 min and the enzyme activity was expressed as change in the U/min/g FW [24].

Total Phenol Content (TPC) Assay

For estimation of total phenol content, the leaf tissues (0.1 g) dished in 5 mL ethanol (95%) were placed at 0 $^\circ$C for 48 h. Each sample was homogenized and centrifuged at 10,000 rpm for 10 min. Reaction mixture containing 1 mL of 95% ethanol, 5 mL of autoclaved distilled water and 0.5 mL of 50% Folin–Ciocalteu reagent was added to 1 mL of the supernatant and well shaken. Then, 1 mL of 5% sodium carbonate was added after 5 min and the reaction mixture was incubated at room temperature for an hour. The absorbance of the developed color was recorded at 725 nm [25]. Standard curves were prepared for each assay using different concentrations of gallic acid (GA) in 95% ethanol. Absorbance values were converted to mg GA equivalents (GAEs) $g^{-1}$ FW.

PR2 ($\beta$-1,3-Glucanase) Protein Assay

The $\beta$-1,3-glucanase activity was assayed by the laminarin-dinitrosalicylic acid method [26]. The reaction mixture consisted of 62.5 $\mu$L of 4% laminarin and 62.5 $\mu$L of enzyme extract. The reaction mixture was carried out at 40 °C for 10 min. The reaction was stopped, 375 $\mu$L of dinitrosalicylic acid was added and the reaction was heated for 5 min in boiling water, then vortexed and the absorbance was measured at 500 nm. The enzyme activity was measured as $\mu$g glucose released in $\text{min}^{-1}$ $\text{mg}^{-1}$ protein.

### 2.5. Histochemical Analysis

#### 2.5.1. Hydrogen Peroxide Production

The histochemical analysis of hydrogen peroxide ($H_2O_2$) was performed by DAB (3, 3-diaminobenzidine) resulting in a reddish-brown staining [27]. Leaf discs were immersed in DAB solution (1 mg $\text{ml}^{-1}$; pH 7.5) and incubated in the dark for 20 h at room temperature for qualitative estimation of $H_2O_2$. The leaf discs were then boiled in 15 mL solution containing absolute ethanol and lactophenol (2:1) for 5 min and then rinsed with ethanol (1 mL of 50%) twice and finally examined under light microscope (Nikon DS-fi1, Tokyo, Japan) to estimate $H_2O_2$ production in leaf tissues.

#### 2.5.2. Lignification

Transverse stem sections from each treatment were examined. The stem sections were fixed in 95% ($v/v$) ethanol mounted on a slide in a solution of saturated aqueous phloroglucinol in 20% hydrochloric acid and observed with light microscope (Leica (Wetzlar, Germany) DM2500). Positive lignin staining was indicated by red-violet coloration of the tissue [28].

### 2.6. Assessment of Defense Inducers Efficacy on Disease Incidence under Protected Conditions

Tomato seedlings (Cv. Rohini and US2853) were maintained in the glasshouse in the aforementioned conditions and were treated with defense inducers in two sets at the fifth week stage. In Set I: the curative spray of the defense inducers was given at 200 ppm, 500 ppm and 800 ppm, and in Set II: prophylactic spray of the defense inducers was given at equal concentrations to determine the most efficient concentration and spray duration for disease management. For the inoculation of healthy plants with *Cmm*, the two youngest leaves at the fifth week stage of the plants were inoculated by cutting them at the tips and dipping them into the bacterial suspension ($10^8$ cfu) [21]. The tomato plants treated only with water served as negative control and plants inoculated only with *Cmm* served as positive control. Disease progress was monitored until 21 days after inoculation. Disease incidence (percentage of plant infection) was recorded by using the standard formula. Fifty plants were used for each set of concentration and the experiments were replicated thrice.

Disease incidence (%) = (Number of diseased plants in the plot/Total number of plants in the plot) $\times$ 100.

### 2.7. Statistical Packages

Data obtained were analyzed by Statistical Package for Social Science (SPSS) version 26.0 for Windows (SPSS, Chicago IL, USA). Descriptive statistics (mean and standard deviation) analysis of variance with means separated using Tukey's test and level of significance were considered as $p \leq 0.05$. The figures were generated in Prism 9.0.1 (GraphPad Software, La Jolla, CA, USA).

## 3. Results

### 3.1. Effect on Peroxidase Activity

POx activity was significantly higher in the plants sprayed with defense inducers at 500 ppm in both varieties, followed by 200 ppm (Figure 1A). However, at an 800 ppm concentration, a phytotoxicity symptom was observed in some plants. The enzymatic activity in the plants sprayed only with the pathogen was near to that of a 200 ppm concentration and the least enzymatic activity was observed in the plants sprayed only with water.

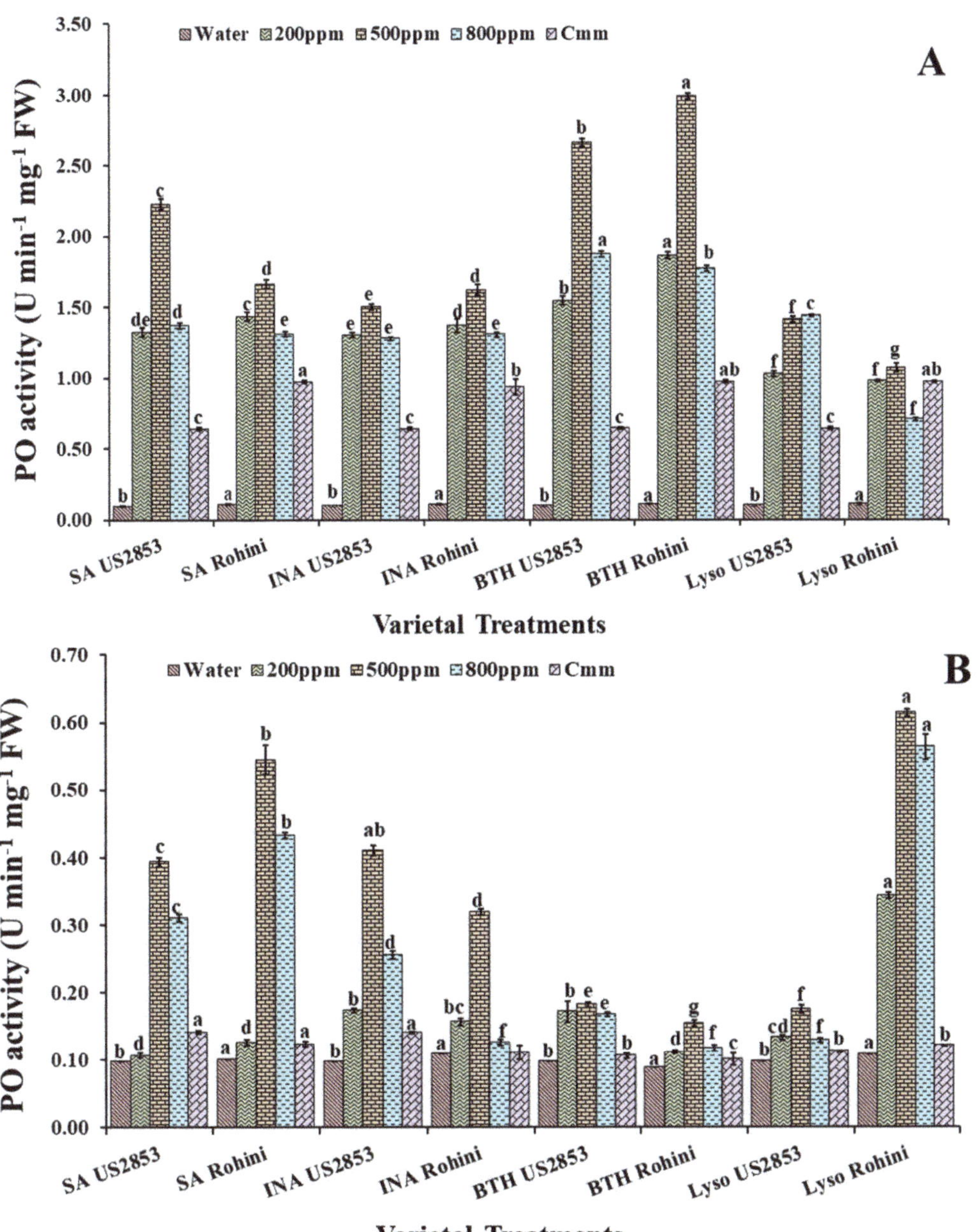



**Figure 1.** Peroxidase activity in tomato plants at 48 h after (**A**) prophylactic and (**B**) curative spray of SA, INA, BTH and lysozyme in cultivars US2853 and Rohini. Vertical bars indicate standard deviations of the means. Different letters indicate significant differences among treatment results taken at the same time interval according to Tukey's test at $p \leq 0.05$.

When the prophylactic spray of the defense inducers was applied, the maximum enzymatic activity was observed in the plant sprayed with BTH followed by SA, INA and lysozyme. Enzymatic activity was lesser in US2853 in comparison to Rohini. A significant elevation of the PO activity was observed in the varieties Rohini and US2853, which received the BTH foliar spray at a 500 ppm concentration (2.97, 2.68) ($p < 0.05$), followed by the plants sprayed with defense inducers at 200 ppm and 800 ppm. The least enzymatic activity was observed in cultivar US2853 (0.098), sprayed only with water, followed by the plants inoculated only with the pathogen (0.642) in cultivar US2853.

Enzymatic activity reached the maximum in the plants treated with SA in both the cultivars (1.78, 0.925; $p \leq 0.05$), respectively, at a 500 ppm concentration followed by 200 ppm and 800 ppm when the curative spray of defense inducers was applied (Figure 1B). However, a rise in enzymatic activity was also observed in the Rohini cultivar plants receiving a curative spray of lysozyme. The least enzymatic activity was observed in cultivar US2853 (0.098), sprayed only with water, followed by the plants only inoculated with the pathogen (0.642) in cultivar US2853.

### 3.2. Effect of Defense Inducers on Phenylpropanoid Activity

#### 3.2.1. PAL Activity

The maximum PAL activity was observed in the plants given the prophylactic spray with BTH in the cultivar Rohini followed by US2853 (12.02, 7.82, $p \leq 0.05$) at a 500 ppm concentration (Figure 2A). The second highest enzymatic activity was observed in the plants sprayed with SA (6.14, 4.13, $p \leq 0.05$), INA and lysozyme. Amongst the four treatments, the least effective treatment was that of lysozyme at 800 ppm (2.23, 2.17) ($p \leq 0.05$), in cultivars Rohini and US2853, respectively.

Within the plants receiving the spray of defense inducers after the inoculation of the pathogen, the highest enzymatic activity was observed in the plants treated with SA at 500 ppm (5.37, 2.46; $p \leq 0.05$) (Figure 2B) followed by BTH and lysozyme at 500 ppm concentrations. The least enzymatic activity amongst the treatment was observed in the plants treated with INA (3.71, 1.45; $p \leq 0.05$) in cultivars Rohini and US2853, respectively.

#### 3.2.2. Total Phenolic Content

The treatments with defense inducers (SA, INA, BTH and lysozyme) elevated the total phenolic content significantly in plants receiving the prophylactic spray of defense inducers in comparison to the control. However, phenolic content was significantly higher in the plants sprayed with BTH at a 500 ppm concentration in the cultivars Rohini and US2853 (5.18, 0.726, $p < 0.05$), respectively. An elevated phenolic content was observed in the SA-treated plants followed by the INA-treated plants, and the minimum phenolic content amongst the four treatments was observed in the plants treated with lysozyme (2.18, 0.515, $p \leq 0.05$) (Figure 3A) in cultivar Rohini.

It was observed that in the plants receiving the curative spray of defense inducers (SA, INA, BTH and lysozyme), at three varied concentrations (200 ppm, 500 ppm and 800 ppm), the phenolic content was higher in plants sprayed with SA in both the cultivars viz., Rohini and US2853 (3.92, 0.559, $p \leq 0.05$) (Figure 3B), followed by the plants treated with BTH and lysozyme. The minimum phenolic content amongst the four treatments was observed in the plants treated with INA (1.36, 0.417; $p \leq 0.05$).

#### 3.2.3. Polyphenol Oxidase Activity

The prophylactic spray of defense inducers showed a significant increase in the PPO activity in comparison to the positive and negative controls. The highest PPO activity was observed in the plants sprayed with BTH at 500 ppm in the cultivar Rohini followed by the cultivar US2853 (81.68, 61.30, $p < 0.05$) (Figure 4A). The least enzymatic activity was observed in the plants treated with lysozyme (31.82, 19.57, $p \leq 0.05$) at the same concentration. The minimum enzymatic activity was observed in the plants sprayed only with water.

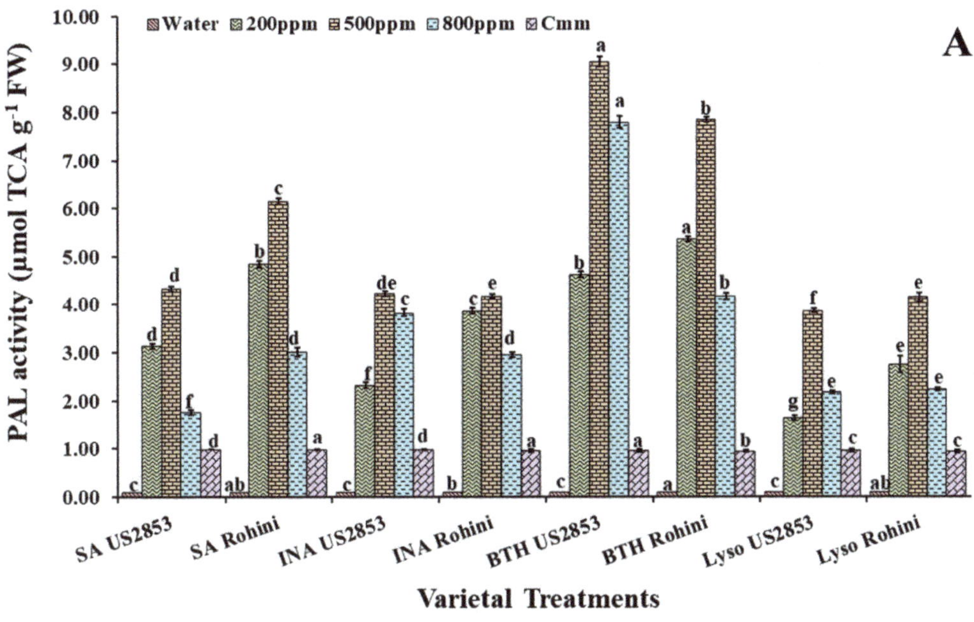


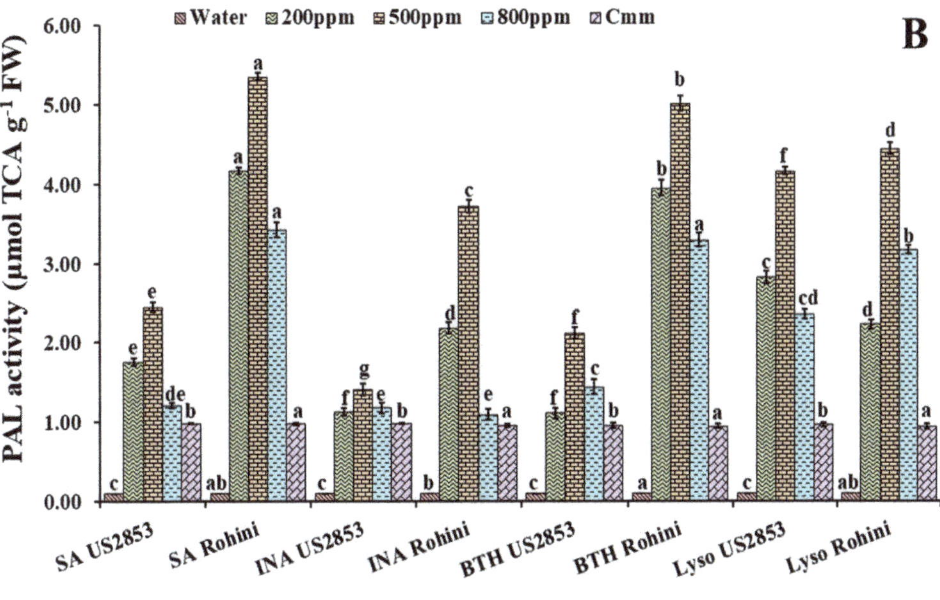


**Figure 2.** PAL activity in tomato plants at 48 h after (**A**) prophylactic and (**B**) curative spray of SA, INA, BTH and lysozyme in cultivars US2853 and Rohini. Vertical bars indicate standard deviations of the means. Different letters indicate significant differences among treatment results taken at the same time interval according to Tukey's test at $p \leq 0.05$.

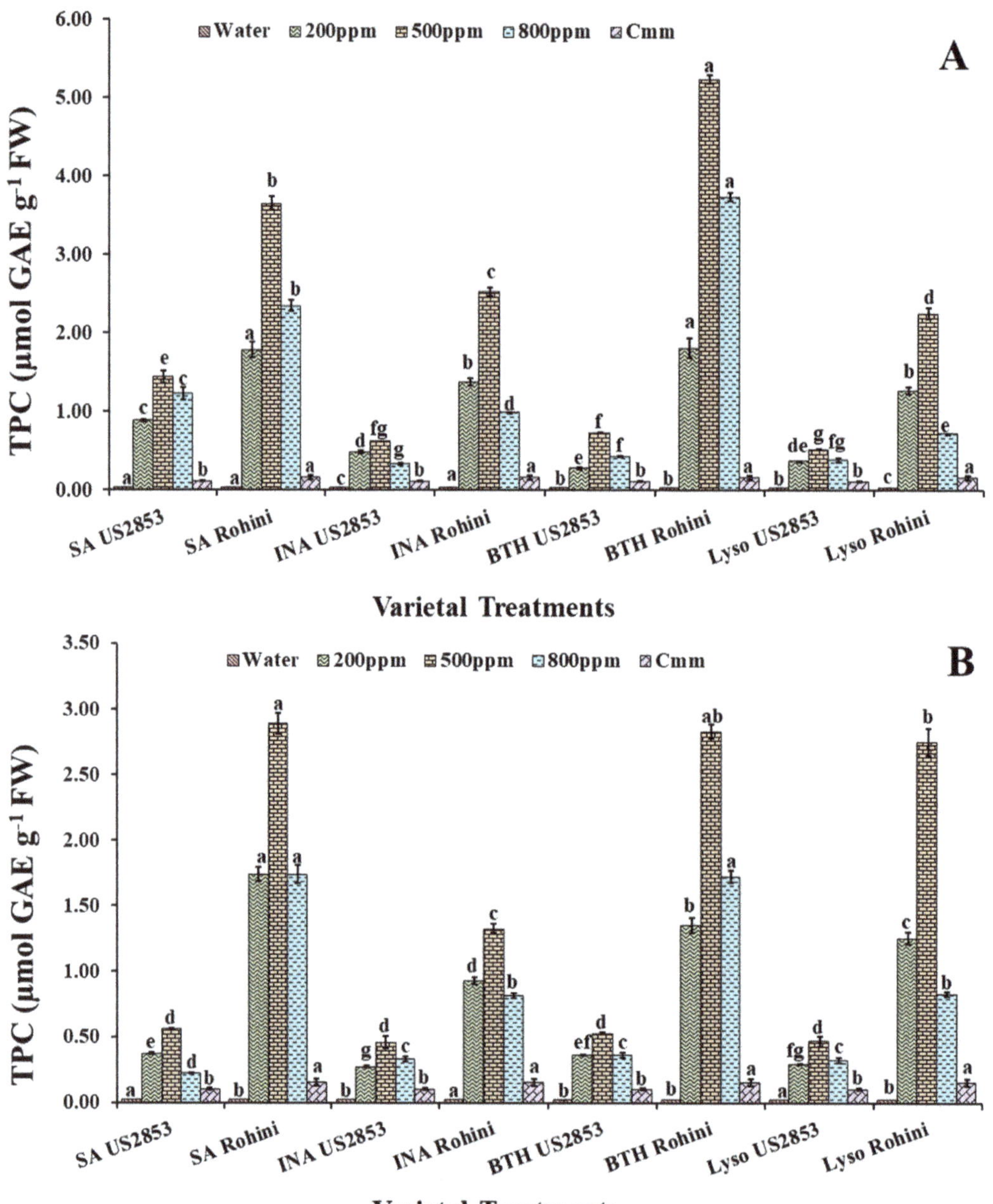



**Figure 3.** Total phenols in tomato plants at 48 h after (**A**) prophylactic and (**B**) curative spray of SA, INA, BTH and lysozyme in cultivars US2853 and Rohini. Vertical bars indicate standard deviations of the means. Different letters indicate significant differences among treatment results taken at the same time interval according to Tukey's test at $p \leq 0.05$.

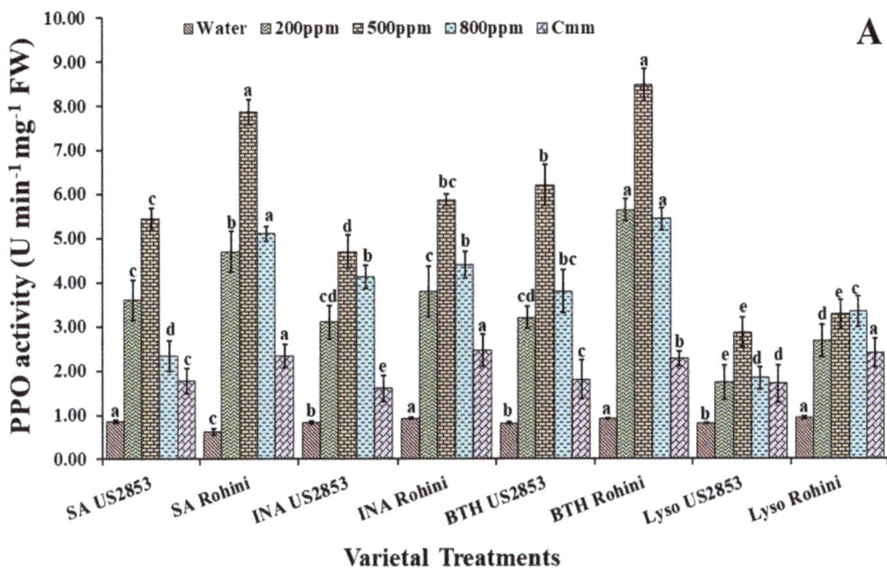


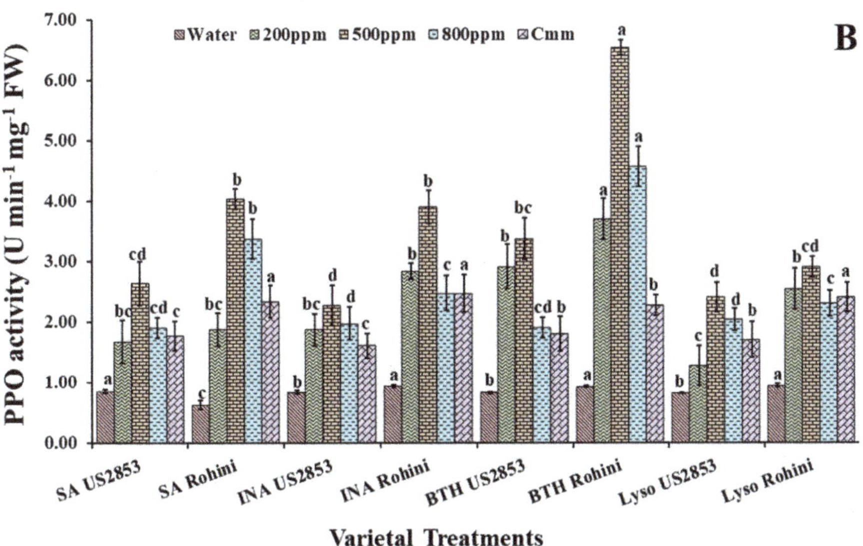


**Figure 4.** Polyphenol oxidase activity in tomato plants at 48 h after (**A**) prophylactic and (**B**) curative spray of SA, INA, BTH and lysozyme in cultivars US2853 and Rohini. Vertical bars indicate standard deviations of the means. Different letters indicate significant differences among treatment results taken at the same time interval according to Tukey's test at $p \leq 0.05$.

The findings indicate that, while receiving the treatment of defense inducers after inoculation with the pathogen, the maximum enzymatic activity was observed in the plants treated with SA in both the cultivars viz., Rohini and US2853 (67.39, 38.3, $p < 0.05$) (Figure 4B) at a 500 ppm concentration followed by the enzymatic activity at 200 ppm and 800 ppm. The second highest enzymatic activity was observed in the plants treated with BTH followed by INA and lysozyme with similar concentration patterns. The minimum enzymatic activity was observed in the plants sprayed only with water followed by the plants inoculated only with the pathogen.

### 3.3. Effect of Defense Inducers on β-1,3-Glucanases (PR-2 Protein) Activity

It is also observed that in the plants receiving the prophylactic spray of the defense inducers, the level of β-1,3-glucanases was significantly higher when treated with BTH at a 500 ppm concentration (3.28, 2.75, $p \leq 0.05$). The second highest enzymatic activity was observed in the plants treated with SA followed by the plants treated with INA. The least enzymatic activity was observed in the plants receiving the lysozyme treatment (1.09, 1.42, $p \leq 0.05$) (Figure 5A) in cultivars Rohini and US2853, respectively.

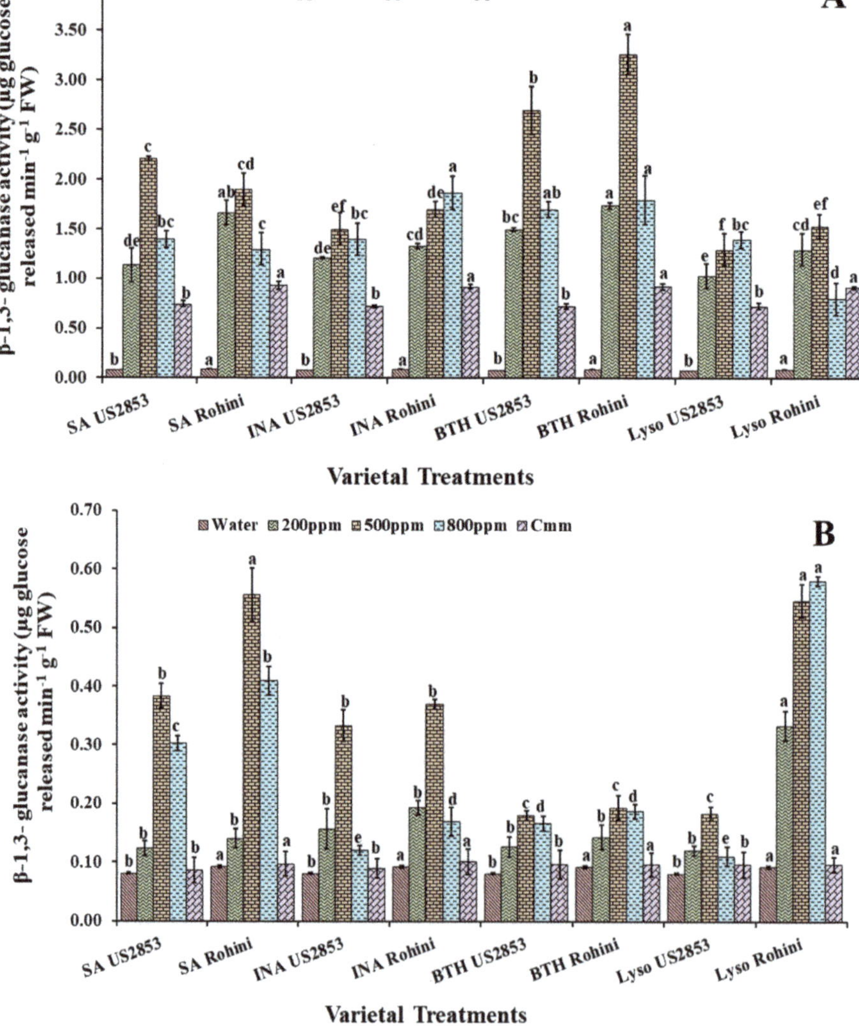

**Figure 5.** β-1,3-glucanases activity in tomato plants at 48 h after (**A**) prophylactic and (**B**) curative spray of SA, INA, BTH and lysozyme in cultivars US2853 and Rohini. Vertical bars indicate standard deviations of the means. Different letters indicate significant differences among treatment results taken at the same time interval according to Tukey's test at $p \leq 0.05$.

Studies have indicated that in the plants receiving the treatments of defense inducers at 200 ppm, 500 ppm and 800 ppm for the four-defense inducer, after the inoculation of the pathogen, the maximum activity in the cultivar Rohini was observed at the lysozyme 500 ppm concentration (0.672, $p \leq 0.05$) (Figure 5B), while in cultivar US2853 the maximum enzymatic activity was found at the SA 500 ppm concentration (0.393, $p \leq 0.05$). The minimum PR-2 protein activity was observed in the tomato plants treated with BTH (0.11, 0.12, $p \leq 0.05$) in cultivars Rohini and US2853, respectively.

### 3.4. Effect of Defense Inducers on Hydrogen Peroxide Generation

The formation of hydrogen peroxide in the leaves can be observed as the reddish-brown coloration of the leaf tissue easily visible to the naked eye. The leaf tissue sections undergo the DAB staining, which is distributed uniformly throughout the leaves. A simple test required for ensuring the peroxidase activity involves the exposure of the plant tissues to DAB and $H_2O_2$. DAB polymerization was studied in treated and untreated leaves after 48 h of incubation and the effect of the defense inducers and pathogen inoculation on peroxidase activity was determined on leaf tissues.

It was observed that in the tomato leaves, when given the prophylactic spray of the defense inducers (Figure 6), the peroxidase level was higher at 48 h after treatment in plants sprayed with BTH at a 500 ppm concentration followed by the plants sprayed with SA and INA. The minimum hydrogen peroxide production was observed in the plants treated with lysozyme before pathogen inoculation.

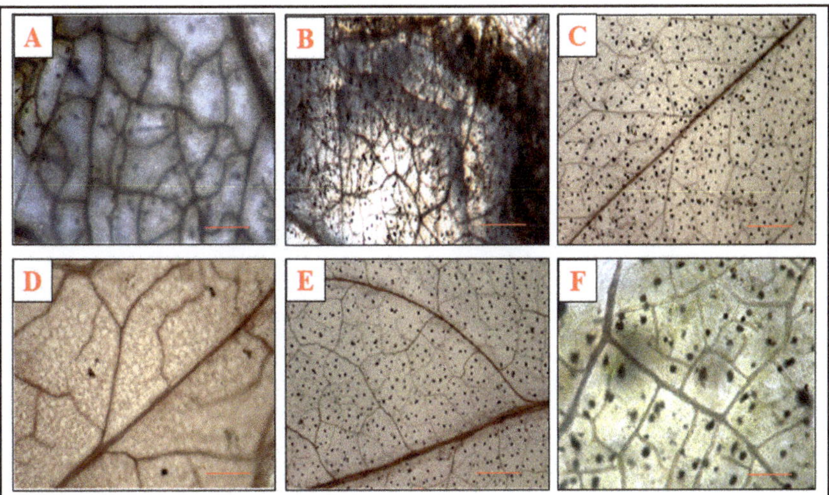

**Figure 6.** Visualization of hydrogen peroxide formation in the leaves of the plants receiving the prophylactic spray of defense inducers: (**A**) positive control (PC), (**B**) negative control (NC), (**C**) salicylic acid (SA), (**D**) isonicotinic acid (INA), (**E**) benzothiodiazole (BTH) and (**F**) lysozyme at 500 ppm after 48 h of treatment.

### 3.5. Effect of Defense Inducers on Lignin Deposition

A significant variation in the transverse section of the stem from different treatments was observed through histochemical staining. The tomato plants were given the prophylactic treatments of the defense inducers (Figure 7) and were then screened for lignin production at 48 h after the treatment by cutting the transverse section of the stem 1 cm above the point of inoculation. It was observed that lignin content was the highest in the plants sprayed with SA at 500 ppm concentrations, followed by the plants sprayed with BTH and INA. The least lignin content was observed in the plants treated with lysozyme before pathogen inoculation.

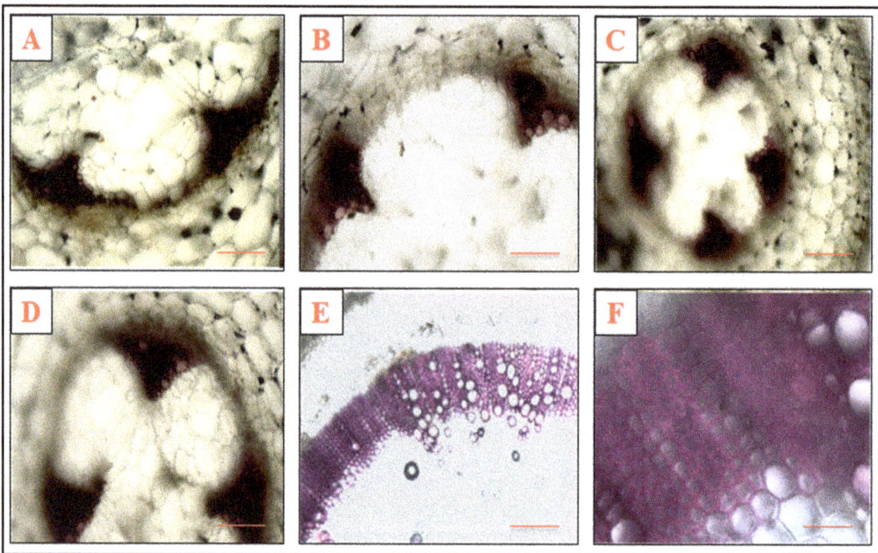

**Figure 7.** Lignin formation in the xylem vessels of the tomato plants' stem receiving the prophylactic treatment of defense inducers ((**A**) SA, (**B**) INA, (**C**) BTH, (**D**) lysozyme, (**E**) positive control and (**F**) negative control) at 500 ppm after 48 h of treatment.

*3.6. Assessment of Defense Inducers Efficacy on Disease Incidence Management and Disease Severity under Protected Conditions*

While comparing the timings of application and the most efficient doses of defense inducers for effective disease management, it was found that the disease incidence was at its minimum in BTH when applied before pathogen inoculation (27.88%) and it was at its highest in lysozyme (49.72%) with the prophylactic spray at 500 ppm (Figures 8 and 9).

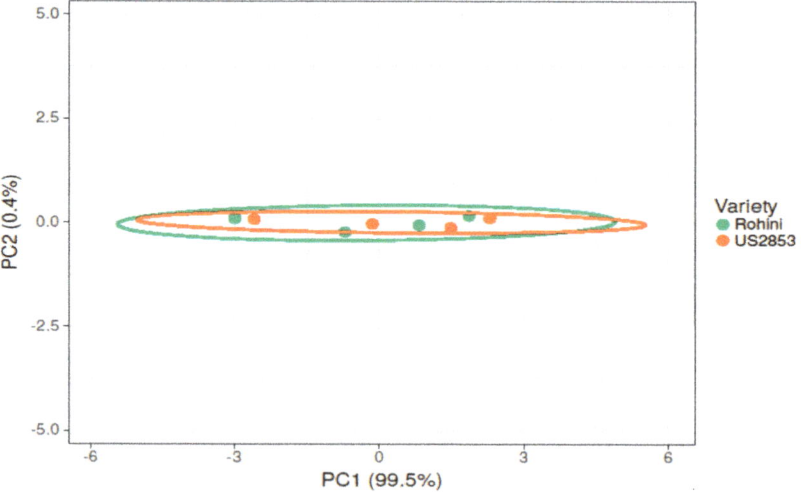



**Figure 8.** Principal component analysis depicting correlation and significance between defense inducers used at different concentrations and varieties Rohini (moderately resistant) and US 2853 (susceptible check) during prophylactic spray. The analysis was generated using the ClustVis website, https://biit.cs.ut.ee/clustvis/ (accessed on 9 July 2022).

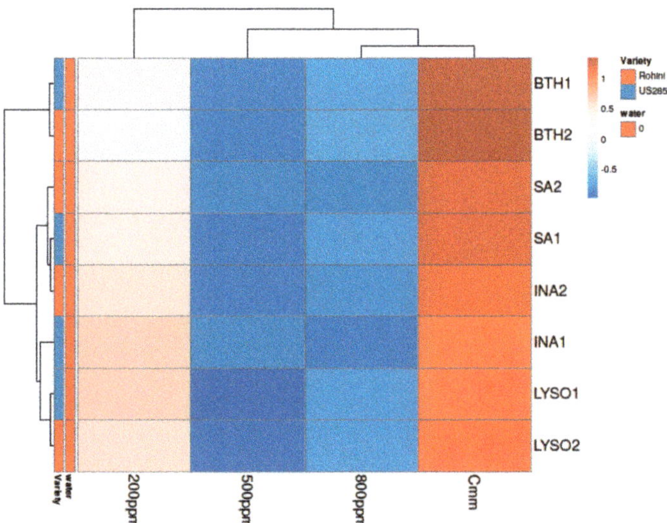



**Figure 9.** Clustering and heatmap analysis of effect of defense inducers on % disease incidence during prophylactic spray. The rows express the individual defense inducers concentrations and the columns indicate the defense inducers used in the study viz., lysozyme (LYSO), benzothiodiazole (BTH), isonicotinic acid (INA) and salicylic acid (SA). Here, 1 and 2 indicate Variety 1 (US2853) and Variety 2 (Rohini), respectively. Herein, untreated and uninoculated seedlings (water only) served as negative control while untreated and inoculated seedlings (*Cmm* only) served as positive control. Lower numerical values are the blue color, whereas higher numerical values are the red color. Map was generated by using the ClustVis website, https://biit.cs.ut.ee/clustvis/ (accessed on 9 July 2022). The results are expressed as averages of three replications.

When the plants received the treatments with defense inducers after the inoculation of the pathogen, the treatment exhibiting the lowest disease incidence was BTH (53.62%) and the treatment exhibiting the highest disease incidence was INA (62.94%) at 500 ppm (Figures 10 and 11).

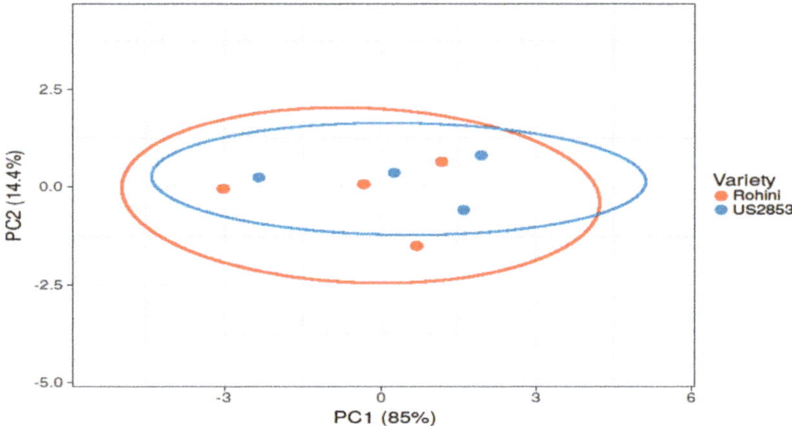



**Figure 10.** Principal component analysis depicting correlation and significance between defense inducers used at different concentrations and varieties Rohini (moderately resistant) and US 2853 (susceptible check) during curative spray. The analysis was generated by using the ClustVis website, https://biit.cs.ut.ee/clustvis/ (accessed on 9 July 2022).

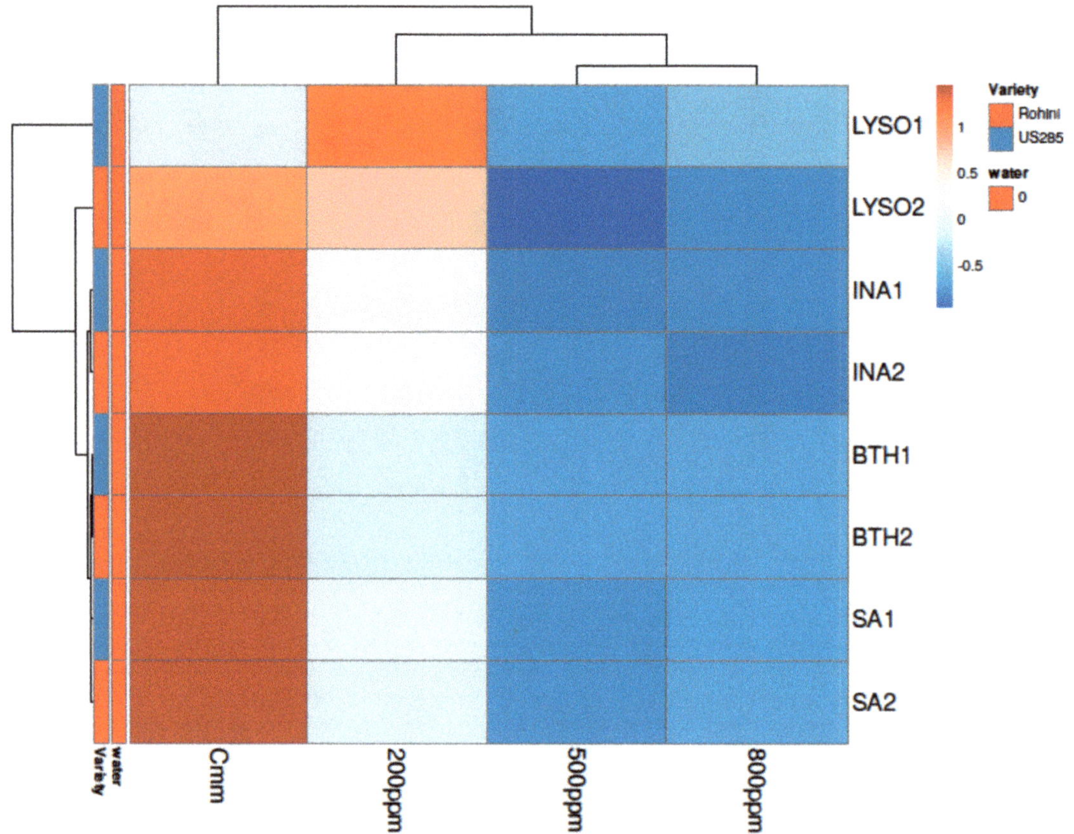



**Figure 11.** Clustering and heatmap analysis of effect of defense inducers on % diseases incidence during curative spray. The rows express the individual defense inducers concentrations and the columns indicate the defense inducers used in the study viz., lysozyme (LYSO), benzothiodiazole (BTH), isonicotinic acid (INA) and salicylic acid (SA). Here, 1 and 2 indicate Variety 1 (US2853) and Variety 2 (Rohini), respectively. Herein, untreated and uninoculated seedlings (water only) served as negative control while untreated and inoculated seedlings (*Cmm* only) served as positive control. Lower numerical values are the blue color, whereas higher numerical values are the red color. Map was generated by using the ClustVis website, https://biit.cs.ut.ee/clustvis/ (accessed on 9 July 2022). The results are expressed as averages of three replications.

The application of lysozyme at a 500 ppm concentration before or after the inoculation did not exhibit much difference in disease incidence; however, in the case of the other three defense inducers, the prophylactic spray was more effective in the management of the pathogen at 500 ppm. A decline in the percentage of disease incidence was observed at 800 ppm as well, but the treated plants also showed phytotoxic symptoms, making this concentration not suitable for disease management.

## 4. Discussion

The activation of the antioxidant network and the phenylpropanoid pathway are the primary steps taken by the plant as management of biotic stress [29]. SAR development is found to be connected with the varied cellular defense responses, such as PR protein synthesis, formation of phytoalexins and buildup of reactive oxygen species (ROS), fast changes in cell wall composition and significant increases in the concentration and activity of defense-related enzymes [16]. In this study, the dense inducer-treated pathogen-inoculated plants showed an elevation in their enzymatic and phenylpropanoid activities in comparison to the positive and negative controls. The studies exploring active oxygen species (AOS) during SAR expression showed ample evidence that AOS, $H_2O_2$ in particular, execute numerous significant functions in early defense responses of the plant against pathogens, and that these responses generally include mechanisms such as antimicrobial action, lignin formation, phytoalexin production and induction of SAR [17]. AOS may also render deleterious effects on cell health. $O_2$, OH and $H_2O_2$ are some of the ROS generally produced under stress conditions [30], acting as attacking oxidizing species that can rapidly affect all types of bio-molecules and easily damage them. Oxygen radical detoxifying enzymes such as catalase, peroxidase and superoxide dismutase (SOD), as well as nonenzymatic antioxidants such as ascorbate peroxidase and glutathione-S-transferase (GST) [31], play an important role in protecting the plant cells from the damage caused due to increased ROS generation [32].

The pre-treatment of plants with different biotic (pathogens and insect pests) and abiotic (chemicals) inducers induces plant resistance which helps defend the plants against subsequent attacks [33–35]. The plant phytohormones induce plant defense against many biotic and abiotic stresses. Salicylic acid is an important and well-studied endogenous plant growth regulator that generates a wide range of metabolic and physiological responses in plants involved in plant defense, in addition to their impact on plant growth and development [36,37]. In the current study, the leaf extract analysis showed an increase in the PO accumulation in all the treatments in comparison to the negative and positive controls. SA also activates the generation of ROS and other defensive processes such as hypersensitive response and cell death [38]. The simultaneous inclusion of phenolic compounds in the cell wall during incompatible plant–microbe/elicitor interactions can be associated with an increase in POX activity. The enzyme POXs are supposed to catalyze the last few steps of the lignification pathways in the tomato. A low activity of POX in plants treated with defense inducers at higher concentrations may be due to the phytotoxicity experienced by the plant at higher concentrations [39].

PAL plays an important role in plant defense; it is involved in the biosynthesis of salicylic acid, an essential signal involved in plant systemic resistance. The enhanced enzymatic activity of PAL is the foremost response in a number of plant species to pathogen challenges and is very much associated with resistance [40]. An alleviation in the PAL activity by addition of salicylic acid to *Saussurea medusa* cell cultures at a concentration of 20 µM [41] and a reduction in severity of *Lasidiploidia theobromae* [42] were reported, due to the priming of tea plants with BTH before inoculation with *L. theobromae* after an increase in PAL activity. In the present study, PAL activity was at its highest in the pathogen-inoculated defense-inducers-treated plant as compared to the negative and positive controls.

The stress posed on the plants by varied biotic and abiotic stress inducers can be defended by phenolic compounds [33,43,44]. The physiology of plants and their displayed metabolism can be altered by the oxidation of phenols that produce many defensive compounds, which help the plant in surviving against different stresses either directly or through diverse plant-signaling pathways [45]. Furthermore, ROS such as superoxide anion, hydroxide radicals, $H_2O_2$ and singlet oxygen produced by the oxidation of phenols activate plant defense enzymes [46,47]. PPOs can be considered as one of the most important enzymes involved in plant defense against many biotic and abiotic stresses [33]. Defense inducers such as jasmonic acid, SA and ethylene have been observed to stimulate such enzymes in plants [48]. An increase in the enzymatic activities and phenylpropanoid syntheses was also observed in the Bacillus cereus-treated, *Cmm*-infected tomato plant [49]. The result observed in the present study is in agreement with the findings of various researchers, wherein an elevation in the phenolic content and PPO activity was observed in the defense-inducer-treated pathogen-inoculated plant as compared to the positive and negative controls.

The PR proteins $\beta$-1,3-glucanases (PR-2) and chitinases (PR-3) have been recognized to possess enzymatic activities, including direct antimicrobial activity by degrading microbial cell wall components. The enzymatic activities of these proteins lead to the breakdown of the pathogen and/or plant cell wall components which act as elicitors to plant defense responses [50]. Chitinase enzymes have been reported to have lysozymal activity leading to bacterial cell wall hydrolyses in a few plants [51,52]. The PR genes expression and the connected accretion of the encoded PR proteins have often been considered as the molecular basis of induced resistance. SA, JA and ET stimulate the production of antimicrobial compounds such as phytoalexins and pathogenesis-related (PR) proteins [53] that eventually initiate a hypersensitive response (HR) that causes infected cell death and pathogen containment [52,54]. In the present study, an elevation in the $\beta$-1,3-glucanases activity was also observed in the defense-inducer-treated pathogen-inoculated plant as compared to the positive and negative controls.

Salicylic acid and its functional analogous have also been reported to increased $H_2O_2$ levels in treated tobacco leaves [55]. $H_2O_2$ plays a vital role not only in stimulating hypersensitive cell death, but also in restricting the spread of cell death by inducing the expression of cell-protecting genes in surrounding cells [31]. In the present study, a higher accumulation of $H_2O_2$ was also observed in plants treated with defense inducers, and similar histochemical studies carried out by [34] also clearly showed that defense inducers such as BTH and ASM induce $H_2O_2$ accumulation in bean plants. The elevated levels of $H_2O_2$, which result from the inhibition of ROS, eventually serve as a secondary messenger for the induction of defense responses [39].

A polymer of the phenylpropanoid compound, lignin, is present constitutively in plants. However, its composition and content can vary at alternate levels on the basis of the degree of stress to which the plants are exposed. Alleviation in lignification is often observed in plants as a response to the biotic stresses experienced by them and is considered as one of the mechanisms adopted by the plant in its defense [55]. The resistance of plants to the cell-wall-degrading enzymes produced by the pathogen can be enhanced by strengthening the plant cell wall by phenolics and lignin, thereby acting as a perfunctory blockade to toxin invasion and to physical penetration toward the protoplast [56]. The transverse section of stems, exhibiting the lignin deposition in stem cells, may block the pathogen penetration, as was also observed by some researchers [57,58]. A higher lignin deposition was also observed in the present study in the defense-inducer-treated pathogen-inoculated plant, thereby indicating that the priming of plants by defense inducers before pathogen inoculation activates a faster and more pronounced defense response in them [59,60].

### 5. Conclusions

The present study very clearly demonstrates an augmented defense response in tomato plants against *Cmm* by defense inducers. The defense inducers not only elicited the antioxidant activity that caused an increased $H_2O_2$ generation but also the phenylpropanoid activity that caused an increased PAL activity followed by alleviated phenolic accumulation and lignin deposition. Moreover, there is also evidence of ROS generation as indicated by higher POX and $\beta$-1,3-glucanases activities. The increased response of all these activities in the defense-inducer treatment compared to the negative and positive controls is positively correlated with reduced disease incidence. The defense inducers in tomato crops may thus be used to enhance the defense responses toward *Cmm*.

**Author Contributions:** Conceptualization, K.V. and R.T. (Ruchi Tripathi).; methodology, R.T. (Rashmi Tewari), C.K.; software, S.K.; validation, R.T. (Ruchi Tripathi)., R.T. (Rashmi Tewari) and C.K.; formal analysis, R.T. (Ruchi Tripathi), S.K.; investigation, R.T. (Ruchi Tripathi); resources, K.V.; data curation R.T. (Ruchi Tripathi); writing—original draft preparation, R.T. (Ruchi Tripathi); writing—review and editing, R.T. (Rashmi Tewari), C.K. and U.D.C.; visualization, K.V., R.T. (Ruchi Tripathi); supervision, K.V.; project administration, K.V.; funding acquisition, T.M., U.D.C., C.K. All authors have contributed equally. In addition, all authors have read and agreed to the published version of the manuscript.

**Funding:** C.K. gratefully acknowledges the financial support from the project of the Ministry of Science and Higher Education of the Russian Federation on the Young Scientist Laboratory within the framework of the Interregional Scientific and Educational Center of the South of Russia (no. LabNOTs-21-01AB, FENW-2021-0014) and the Strategic Academic Leadership Program of the Southern Federal University ("Priority 2030"). R.T. (Ruchi Tripathi) and K.V. acknowledges the financial support from the Indian Council of Agricultural Research.

**Institutional Review Board Statement:** Not applicable.

**Informed Consent Statement:** Not applicable.

**Data Availability Statement:** The data has been presented in the research article itself and any further data can be presented by the corresponding author on demand.

**Conflicts of Interest:** The authors declare no conflict of interest.

# References


1. Paduchuri, P.; Gohokar, S.; Thamke, B.; Subhas, M. Transgenic tomatoes. *Int. J. Adv. Biotechnol. Res.* **2010**, *2*, 69–72.
2. Gupta, S.K.; Thind, T.S. *Disease Problems in Vegetable Production*; Scientific Publishers: Jodhpur, India, 2006; p. 576.
3. de León, L.; Siverio, F.; López, M.M.; Rodríguez, A. Clavibacter michiganesis subsp. michiganensis, a Seedborne Tomato Pathogen: Healthy Seeds Are Still the Goal. *Plant Dis.* **2011**, *95*, 1328. [CrossRef] [PubMed]
4. Chang, R.J.; Ries, S.M.; Pataky, J.K. Local sources of Clavibacter michiganensis ssp. michiganensis in the development of bacterial canker on tomatoes. *Phytopathology* **1992**, *82*, 553–560. [CrossRef]
5. Gitaitis, R.D.; Beaver, R.W.; Voloudakis, A.E. Detection of Clavibacter michiganensis subsp. michiganensis in symptomless tomato transplants. *Plant Dis.* **1991**, *75*, 834–838. [CrossRef]
6. Hadas, R.; Kritzman, G.; Klietman, F.; Gefen, T.; Manulis, S. Comparison of extraction procedures and determination of the detection threshold for *Clavibacter michiganensis* ssp. michiganensis in tomato seeds. *Plant Pathol.* **2005**, *54*, 643–649. [CrossRef]
7. Kawaguchi, A.; Tanina, K.; Inoue, K. Molecular typing and spread of *Clavibacter michiganensis* subsp. *michiganensis* in greenhouses in Japan. *Plant Pathol.* **2010**, *59*, 76–83. [CrossRef]
8. Kawaguchi, A.; Tanina, K.; Inoue, K. Spatiotemporal distribution of tomato plants naturally infected with bacterial canker in greenhouses. *J. Gen. Plant Pathol.* **2013**, *79*, 46–50. [CrossRef]
9. Velásquez, A.C.; Castroverde, C.D.M.; He, S.Y. Plant-pathogen warfare under changing climate conditions. *Curr. Biol.* **2018**, *28*, 619–634. [CrossRef]
10. Wiesel, L.; Newton, A.C.; Elliot, I.; Booty, D.; Gilory, E.M.; Birch, P.R.J.; Hein, I. Molecular effects of resistance elicitors from biological origin and their potential for crop protection. *Front. Plant Sci.* **2014**, *5*, 655. [CrossRef]
11. Vázquez-Hernández, M.C.; Parola-Contreras, I.; Montoya-Gómez, L.M.; Torres-Pacheco, I.; Schwarz, D.; Guevara-González, R.G. Eustressors: Chemical and physical stress factors used to enhance vegetables production. *Sci. Hortic.* **2019**, *250*, 223–229. [CrossRef]
12. Yakhin, O.I.; Lubyanov, I.A.; Yakhin, I.A.; Brown, P.H. Biostimulants in Plant Science: A global perspective. *Front Plant Sci.* **2017**, *7*, 2049. [CrossRef] [PubMed]
13. Pilon-Smits, E.A.H.; Quinn, C.F.; Tapken, W.; Malagoli, M.; Schiavon, M. Physiological functions of beneficial elements. *Curr. Opin. Plant Biol.* **2009**, *12*, 267–274. [CrossRef] [PubMed]
14. Salazar, B.; Ortiz, A.; Keswani, C.; Minkina, T.; Mandzhieva, S.; Singh, S.P.; Rekadwad, B.; Borriss, R.; Jain, A.; Singh, H.B.; et al. *Bacillus* spp. as Bio-factories for Antifungal Secondary Metabolites: Innovation Beyond Whole Organism Formulations. *Microb. Ecol.* **2022**, 1–24. [CrossRef] [PubMed]
15. Keswani, C.; Singh, H.B.; García-Estrada, C.; Caradus, J.; He, Y.W.; Mezaache-Aichour, S.; Glare, T.R.; Borriss, R.; Sansinenea, E. Antimicrobial secondary metabolites from agriculturally important bacteria as next-generation pesticides. *Appl. Microbiol. Biotechnol.* **2020**, *104*, 1013–1034. [CrossRef]
16. Ryals, J.A.; Neuenschwander, U.H.; Willits, M.G.; Molina, A.; Steiner, H.S.; Hunt, M.D. Systemic Acquired Resistance. *Plant Cell* **1996**, *8*, 1809–1819. [CrossRef]
17. Mehdy, M.C.; Sharma, Y.K.; Sathasivan, K.; Bays, N.W. The role of activated oxygen species in plant disease resistance. *Physiol. Plant.* **1996**, *98*, 365–374. [CrossRef]
18. Baysal, O.; Soylu, E.M.; Soylu, S. Induction of defense related enzymes and resistance by the plant activator acibenzolar S methyl in tomato seedlings against bacterial canker caused by *Clavibacter michiganensis* ssp michiganensis. *Plant Pathol. J.* **2003**, *52*, 747–753. [CrossRef]
19. Soylu, S.; Baysal, O.; Soylu, E.M. Induction of disease resistance by the plant activator, acibencolar-Smethyl (ASM), against bacterial canker (*Clavibacter michiganensis* subsp. michiganensis) in tomato seedlings. *Plant Sci.* **2003**, *165*, 1069–1075. [CrossRef]
20. Baysal, O.; Gursoy, Y.; Ornek, H.; Duru, A. Induction of oxidants in tomato leaves treated with DL-beta-amino butyric acid (BABA) and infected with *Clavibacter michiganensis* ssp. michiganensis. *Eur. J. Plant. Pathol.* **2005**, *112*, 361–369. [CrossRef]
21. Van den Bulk, R.W.; Zevenhuizen, L.P.T.M.; Cordewener, J.H.G.; Dons, J.J.M. Characterization of the extracellular polysaccharide produced by *Clavibacter michiganensis* subsp. michiganensis. *Phytopathology* **1991**, *81*, 619–623. [CrossRef]

22. Gauillard, F.; Richard-Forget, F.; Nicolas, J. New spectrophotometric assay for polyphenol oxidase activity. *Ann. Biochem.* **1993**, *215*, 59–65. [CrossRef] [PubMed]

23. Brueske, C.H. Phenylalanine ammonia-lyase activity in Lycopersicon esculentum roots infected and resistant to the root knot nematode, Meloidogyne incognita. *Physiol. Plant Pathol.* **1980**, *16*, 409–414. [CrossRef]

24. Hammerschmidt, R.; Nuckles, E.M.; Kuc, J. Association of enhanced peroxidase activity with induced systemic resistance of cucumber to *Colletotrichum lagenarium*. *Physiol. Plant Pathol.* **1982**, *20*, 73–82. [CrossRef]

25. Zheng, Z.; Shetty, K. Solid-state bioconversion of phenolics from cranberry pomace and role of Lentinus edodes β-glucosidase. *J. Agric. Food Chem.* **2000**, *48*, 895–900. [CrossRef] [PubMed]

26. Pan, S.O.; Ye, X.S.; Kuc, J. A technique for detection of chitinase, β-1.3-glucanase, and protein patterns after a single separation using polyacrylamide gel electrophoresis or isoelectrofocusing. *Phytopathology* **1991**, *81*, 970–974. [CrossRef]

27. Thordal-Christensen, H.; Zhang, Z.; Wei, Y.; Collinge, D.B. Subcellular localization of $H_2O_2$ in plants: $H_2O_2$ accumulation in papillae and hypersensitive response during the barley-powdery mildew interaction. *Plant J.* **1997**, *11*, 1187–1194. [CrossRef]

28. Jensen, W. *Botanical Histochemistry: Principles and Practice*; W. H. Freeman: San Francisco, CA, USA, 1962.

29. Keswani, C.; Dilnashin, H.; Birla, H.; Singh, S.P. Unravelling efficient applications of agriculturally important microorganisms for alleviation of induced inter-cellular oxidative stress in crops. *Acta Agric. Slov.* **2019**, *114*, 121–130. [CrossRef]

30. Dat, J.; Vandenabeele, S.; Vranová, E.; Montagu, M.V.; Inze, D.; Breusegem, F.V. Dual action of the active oxygen species during plant stress responses. *Cell Mol. Life Sci.* **2000**, *57*, 779–795. [CrossRef]

31. Alscher, R.G.; Donahue, J.L.; Cramer, C.L. Reactive oxygen species and antioxidants: Relationships in green cells. *Physiol. Plant* **1997**, *100*, 224–233. [CrossRef]

32. Kuzniak, E.; Sklodowska, M. Ascorbate, glutathione andrelated enzymes in chloroplasts of tomato leaves infected by Botrytis cinerea. *Plant Sci.* **2001**, *160*, 723–731. [CrossRef]

33. War, A.R.; Paulraj, M.G.; War, M.Y.; Ignacimuthu, S. Differential defensive response of groundnut germplasms to *Helicoverpa armigera* (Hubner) (*Lepidoptera*: Noctuidae). *J. Plant Interact.* **2011**, *7*, 45–55. [CrossRef]

34. Hu, X.; Li, W.; Chen, Q.; Yang, Y. Early signal transduction linking the synthesis of jasmonic acid in plant. *Plant Signal. Behav.* **2009**, *4*, 696–697. [CrossRef] [PubMed]

35. Lu, H. Dissection of salicylic acid-mediated defense signaling networks. *Plant Signal. Behav.* **2009**, *4*, 712–717. [CrossRef] [PubMed]

36. Chen, H.; Xue, L.; Chintamanani, S.; Germain, H.; Lin, H.; Cui, C.; Cai, R.; Zuo, J.; Tang, X.; Li, X.; et al. Ethylene insensitive3 and ethylene insensitive3-like1 repress salicylic acid induction deficient2 expression to negatively regulate plant innate immunity in Arabidopsis. *Plant Cell* **2009**, *21*, 2527–2540. [CrossRef] [PubMed]

37. Zhao, L.Y.; Chen, J.L.; Cheng, D.F.; Sun, J.R.; Liu, Y.; Tian, Z. Biochemical and molecular characterizations of Sitobion avenae–induced wheat defense responses. *Crop Prot.* **2009**, *28*, 435–442. [CrossRef]

38. Hayat, Q.; Hayat, S.; Irfan, M.; Ahmad, A. Effect of exogenous salicylic acid under changing environment: A review. *Environ. Exp. Bot.* **2009**, *68*, 14–25. [CrossRef]

39. Rajjou, L.; Belghazi, M.; Huguet, R.; Robin, C.; Moreau, A.; Job, C.; Job, D. Proteomic investigation of the effect of salicylic acid on Arabidopsis seed germination and establishment of early defense mechanisms. *Plant Physiol.* **2006**, *141*, 910–923. [CrossRef]

40. Pallas, J.A.; Paiva, N.L.; Lamb, C.; Dixon, R.A. Tobacco plants epigenetically suppressed in phenylalanine ammonia-lyase expression do not develop systemic acquired resistance in response to infection by tobacco mosaic virus. *Plant J.* **1996**, *10*, 281–293. [CrossRef]

41. Yu, Z.; Fu, C.; Han, Y.; Li, Y.; Zhao, D. Salicylic acid enhances jaceosidin and syringin production in cell cultures of Saussurea medusa. *Biotechnol. Lett.* **2006**, *28*, 1027–1031. [CrossRef]

42. Das, S.; Chakraborty, P.; Mandal, P.; Saha, D.; Saha, A. Phenylalanine ammonia-lyase gene induction with benzothiadiazole elevates defence against *Lasiodiplodia theobromae* in tea in India. *J. Phytopathol.* **2017**, *165*, 755–761. [CrossRef]

43. Sharma, H.C.; Sujana, G.; Manohar Rao, D. Morphological and chemical components of resistance to pod borer, *Helicoverpa armigera* in wild relatives of Pigeonpea. *Arthropod-Plant Interact.* **2009**, *3*, 152–161. [CrossRef]

44. Tripathi, R.; Tewari, R.; Singh, K.P.; Keswani, C.; Minkina, T.; Srivastava, A.K.; De Corato, U.; Sansinenea, E. Plant Mineral Nutrition and Disease Resistance: A Significant Linkage for Sustainable Crop Protection. *Front. Plant Sci.* **2022**, *13*, 883970. [CrossRef]

45. Usha Rani, P.; Jyothsana, Y. Biochemical and enzymatic changes in rice plants as a mechanism of defense. *Acta Physiol. Plant.* **2010**, *32*, 695–701. [CrossRef]

46. Kawano, T. Roles of the reactive oxygen species-generating peroxidase reactions in plant defense and growth induction. *Plant Cell Rep.* **2003**, *21*, 829–837. [CrossRef] [PubMed]

47. Maffei, M.E.; Mithofer, A.; Boland, W. Insects feeding on plants: Rapid signals and responses preceding the induction of phytochemical release. *Phytochemistry* **2007**, *68*, 2946–2959. [CrossRef]

48. Noreen, Z.; Ashraf, M. Change in antioxidant enzymes and some key metabolites in some genetically diverse cultivars of radish (*Raphanus sativus* L.). *Environ. Exp. Bot.* **2009**, *67*, 395–402. [CrossRef]

49. Solano-Alvarez, N.; Valencia-Hernández, J.A.; Rico-García, E.; Torres-Pacheco, I.; Ocampo-Velázquez, R.V.; Escamilla-Silva, E.M.; Romero-García, A.L.; Alpuche-Solís, Á.G.; Guevara-González, R.G. Novel Isolate of Bacillus cereus Promotes Growth in Tomato and Inhibits *Clavibacter michiganensis* Infection under Greenhouse Conditions. *Plants* **2021**, *10*, 506. [CrossRef]

50. Van Loon, L. Induced resistance in plants and the role of pathogenensis-related proteins. *Eur. J. Plant Pathol.* **1997**, *103*, 753–765. [CrossRef]
51. Boller, T.; Gehri, A.; Mauch, F.; Vogeli, U. Chitinase in bean leaves: Induction by ethylene, purification, properties, and possible function. *Planta* **1983**, *157*, 22–31. [CrossRef]
52. Heitz, T.; Segond, S.; Kauffmann, S.; Geoffroy, P.; Prasad, V.; Brunner, F.; Fritig, B.; Legrand, M. Molecular characterization of a novel tobacco PR protein: A new plant chitinase/lysozyme. *Mol. Genet. Genom.* **1994**, *245*, 246–254. [CrossRef]
53. Lamb, C.; Dixon, R.A. The Oxidative Burst in Plant Disease Resistance. *Annu. Rev. Plant Physiol. Plant Mol. Biol.* **1997**, *48*, 251–275. [CrossRef] [PubMed]
54. Dangl, J.L.; Dietrich, R.A.; Richberg, M.H. Death don't have no mercy: Cell death programs in plant-microbe interactions. *Plant Cell* **1996**, *8*, 1793–1807. [CrossRef] [PubMed]
55. Rogers, L.A.; Campbel, M.M. The genetic control of lignin deposition during plant growth and development. *New Phytol.* **2004**, *164*, 17–30. [CrossRef]
56. Nicholson, R.L.; Hammerschmidt, R. Phenolic compounds and their role in disease resistance. *Annu. Rev. Phytopathol.* **1992**, *30*, 369–389. [CrossRef]
57. Luna, E.; Pastor, V.; Robert, J.; Flors, V.; Mauch-Mani, B.; Ton, J. Callose deposition: A multifaceted plant defense response. *Mol. Plant-Microbe Interact.* **2011**, *24*, 183–193. [CrossRef] [PubMed]
58. Park, P.; Ikeda, K.-I. Ultrastructural analysis of responses of host and fungal cells during plant infection. *J. Gen. Plant Pathol.* **2008**, *74*, 2–14. [CrossRef]
59. Conrath, U.; Thulke, O.; Katz, V.; Schwindling, S.; Kohler, A. Priming as a mechanism in induced systemic resistance of plants. *Eur. J. Plant Pathol.* **2001**, *107*, 113–119. [CrossRef]
60. Dehghanian, Z.; Habibi, K.; Dehghanian, M.; Aliyar, S.; Lajayer, B.A.; Astatkie, T.; Minkina, T.; Keswani, C. Reinforcing the bulwark: Unravelling the efficient applications of plant phenolics and tannins against environmental stresses. *Heliyon* **2022**, *8*, e09094. [CrossRef]

 *microorganisms*

*Article*

# *Bacillus velezensis* BV01 Has Broad-Spectrum Biocontrol Potential and the Ability to Promote Plant Growth

Ting Huang [1,2], Yi Zhang [1], Zhihe Yu [2], Wenying Zhuang [1] and Zhaoqing Zeng [1,*]

[1]  State Key Laboratory of Mycology, Institute of Microbiology, Chinese Academy of Sciences, Beijing 100101, China; h_ting17@163.com (T.H.); zhangyicrazy@163.com (Y.Z.); zhuangwy@im.ac.cn (W.Z.)
[2]  College of Life Sciences, Yangtze University, Jingzhou 434025, China; zhiheyu@hotmail.com
*   Correspondence: zengzq@im.ac.cn

**Abstract:** To evaluate the potential of a bacterial strain as a fungal disease control agent and plant growth promoter, its inhibitory effects on phytopathogens such as *Bipolaris sorokiniana*, *Botrytis cinerea*, *Colletotrichum capsici*, *Fusarium graminearum*, *F. oxysporum*, *Neocosmospora rubicola*, *Rhizoctonia solani*, and *Verticillium dahliae* were investigated. The results showed that the inhibitory rates in dual-culture and sterile filtrate assays against these eight phytopathogens ranged from 57% to 83% and from 36% to 92%. The strain was identified as *Bacillus velezensis* based on morphological and physiological characterization as well as phylogenetic analyses of *16S* rRNA and the gyrase subunit A protein (*gyrA*) regions. The results demonstrated that *B. velezensis* was able to produce fungal cell-wall-degrading enzymes, namely, protease, cellulase, and β-1,3-glucanase, and the growth-promotion substances indole-3-acetic acid (IAA) and siderophore. Furthermore, *B. velezensis* BV01 had significant control effects on wheat root rot and pepper *Fusarium* wilt in a greenhouse. Potted growth-promotion experiments displayed that BV01 significantly increased the height, stem diameter, and aboveground fresh and dry weights of wheat and pepper. The results imply that *B. velezensis* BV01, a broad-spectrum biocontrol bacterium, is worth further investigation regarding its practical applications in agriculture.

**Keywords:** *Bacillus*; antifungal activity; fungal phytopathogens; wheat root rot; *Fusarium* wilt; greenhouse pot experiment

**Citation:** Huang, T.; Zhang, Y.; Yu, Z.; Zhuang, W.; Zeng, Z. *Bacillus velezensis* BV01 Has Broad-Spectrum Biocontrol Potential and the Ability to Promote Plant Growth. *Microorganisms* **2023**, *11*, 2627. https://doi.org/10.3390/microorganisms11112627

Academic Editors: Chetan Keswani and Rainer Borriss

Received: 12 September 2023
Revised: 23 October 2023
Accepted: 23 October 2023
Published: 25 October 2023

**Copyright:** © 2023 by the authors. Licensee MDPI, Basel, Switzerland. This article is an open access article distributed under the terms and conditions of the Creative Commons Attribution (CC BY) license (https://creativecommons.org/licenses/by/4.0/).

## 1. Introduction

Crop diseases caused by phytopathogens have resulted in a decrease in agricultural yields and quality, leading to significant economic losses [1]. In particular, soil-borne fungal infections of important crops such as wheat, corn, rice, and pepper cause large economic losses [2]. The United Nations 2030 Sustainable Development Goals suggested that the world should ensure sustainable consumption and production patterns, promote sustainable agriculture, and reduce environmental pollution [3]. For a long time, synthetic chemical pesticides were commonly used in traditional agriculture to combat plant diseases, but they often caused environmental pollution and residual toxic effects in animals and humans [4]. Thus, discovery of eco-friendly, long-lasting, and effective methods are required for disease prevention and management in agriculture. The use of microbial and biochemical agents has been explored as a practical alternative approach [5].

The plant-growth-promoting rhizobacteria (PGPRs) are often used for the production of bioactive substances that can protect plants by suppressing pathogens, inducing systemic resistance, or improving resistance to environmental stresses, by facilitating nutrient acquisition and modulating phytohormone levels in plants [6,7]. In recent years, *Bacillus subtilis* and its closest relatives *B. amyloliquefaciens*, *B. velezensis*, *B. cereus*, and *B. licheniformis* have been widely used as biofertilizers and biofungicides [8,9]. *Bacillus velezensis* FZB42, the classical PGPR strain, was successfully used as a biocontrol agent in potato, strawberry, wheat, and cabbage [10–14]. The most prevalent plant fungal diseases, such as grey mold, *Fusarium* head blight, anthracnose, and root rot, etc., are mainly caused by species of *Botrytis*, *Fusarium*, *Colletotrichum*, and *Rhizoctonia* [15]. This can be attributed to their broad host range, genetic diversity, rapid adaptation to plant disease resistance, and production of toxins [16]. Previous studies have shown that *B. velezensis* is a promising agent for control of *Rhizoctonia solani* [17], *Gaeumannomyces graminis* var. *tritici* [18], *Fusarium oxysporum* f. sp. *niveum* [19], *Botrytis cinerea*, *Colletotrichum gloeosporioides*, and *Phytophthora infestans* [20], and it has attracted widespread attention in agricultural disease research. Nevertheless, studies on its biocontrol mechanism, screening of excellent strains, analyses of transcriptomics, proteomics, metabolomics, and research on industrial and commercial applications of *B. velezensis* are needed [21].

In this study, we aimed to assess the potential of a newly isolated bacterial strain, *B. velezensis* BV01, as a broad-spectrum biocontrol agent and investigate its capacity to control plant diseases and promote wheat and pepper development. The findings are of great significance for reducing the use of chemical fungicides to control soil-borne fungal diseases, thereby improving the ecological environment, and for providing technical support for food safety and sustainable development.

## 2. Materials and Methods

### 2.1. Tested Strains

Information on the 12 bacterial and fungal strains used is listed in Table 1. *Bacillus velezensis* BV01 was isolated from a contaminated potato dextrose agar (PDA) plate in the laboratory. *Bacillus velezensis* JDF and *B. subtilis* L01 and BS208 were isolated from three commercially available bacterial agents NongBaoShengWu®, LvLong®, and GuanLan®, respectively, and the eight fungal plant pathogen strains were provided by colleagues from Beijing Academy Agriculture and Forestry Sciences, China Academy Agricultural Sciences, Guangxi Academy Agricultural Sciences, Nanjing Agricultural University, and our institute (Table 1). All strains were deposited in the China General Microbiological Culture Collection Center (CGMCC) and the State Key Laboratory of Mycology, Institute of Microbiology, Chinese Academy of Sciences.

**Table 1.** The bacterial and fungal strains tested in this study.

| Strain | Characteristics Relevant to This Work | Source |
|---|---|---|
| *Bacillus velezensis* CGMCC 1.60184 | Isolated from the State Key Laboratory of Mycology, Institute of Microbiology, Chinese Academy of Sciences (116.38982° E, 40.01076° N) on 6 November 2019 | This work |
| *B. velezensis* JDF | Registration of broad-spectrum antagonism, stress resistance and growth-promoting ability, used as a positive control | Isolated from NongBaoShengWu® bacterial agent |
| *B. subtilis* L01 | Registration of antimicrobial activity and strong stress resistance, used as a positive control | Isolated from LvLong® bacterial agent |
| *B. subtilis* BS208 | Registration of prevention and control of gray mold and powdery mildew, used as a positive control | Isolated from GuanLan® bacterial agent |
| *Bipolaris sorokiniana* PP12 | Causes wheat root rot | Provided by Prof. Niu Yongchun of Chinese Academy of Agricultural Sciences |
| *Botrytis cinerea* PP1 | Causes tomato gray mold | Provided by Prof. Qiu Jiyan of Beijing Academy of Agricultural Sciences |
| *Colletotrichum capsici* PP6 | Causes ring rot disease | Agricultural Sciences |
| *Fusarium graminearum* PP15 | Causes *Fusarium* head blight | Provided by Prof. Ma Zhengqiang of Nanjing Agricultural University |

**Table 1.** *Cont.*

| Strain | Characteristics Relevant to This Work | Source |
|---|---|---|
| *F. oxysporum* F6 | Causes blight disease | Provided by Prof. Li Qili of Guangxi Academy of Agricultural Sciences |
| *Neocosmospora rubicola* PP23 | Causes root and stem rot | Provided by Dr. Fu Shenzhan of Institute of Microbiology, Chinese Academy of Sciences |
| *Rhizoctonia solani* PP11 | Causes wilt disease | Provided by Prof. Niu Yongchun of Chinese Academy of |
| *Verticillium dahliae* PP8 | Causes greensickness | Agricultural Sciences |

### 2.2. Evaluation of Antagonistic Abilities In Vitro

All bacterial strains were tested for their antagonistic abilities against eight phytopathogens in dual-culture assays [18]. The bacterial strains were grown in nutrient broth (NB: peptone 10.0 g, beef paste 3.0 g, NaCl 5.0 g, dH$_2$O 1000 mL) at 28 °C for 2 d, and the cell suspension was adjusted to $5 \times 10^8$ CFU/mL. The fungal strains were grown on PDA at 25 °C for 5 d. Mycelial plugs (5 mm in diam.) were placed in the center of each new PDA plate (90 mm in diam.); then, four sterile filter papers (5 mm in diam.), which contained 6 μL of a bacterial suspension, were placed evenly around individual mycelial plugs. Sterile water was used as the control. All plates were incubated at 25 °C for 5 d. Each treatment had three replicates. Measurements were calculated with the following formula: y = (A − B)/A × 100% (y: percentage of inhibition, A: colony diameter of pathogen on control plate, and B: colony diameter of pathogen in experimental group) [22].

### 2.3. Assessment of Antifungal Activity of Bacterial Sterile Filtrate

The antifungal activity of bacterial sterile filtrate was evaluated by measuring the diameter of the fungal colony [23]. Four bacterial strains were incubated in NB with shaking at 180 rpm at 28 °C for 2 d, and then the cultures were centrifuged at 8000 rpm at 4 °C for 20 min to collect the supernatant. The supernatant was filtered through a 0.22 μm filter and then mixed into molten PDA at 45–50 °C with a concentration of 20% (*v/v*). The mycelium plug (5 mm in diam.) of each pathogen was placed at the center of each PDA plate with bacterial filtrate and incubated for 5 d at 25 °C. The PDA plate without bacterial filtrate was used as the control. Three replicates were set up for each treatment. The inhibition percentage (y%) was calculated with the following formula: y = (A − C)/A × 100% (A: growth radius of pathogen in control and C: growth radius of pathogen in different treatments).

### 2.4. Morphological Observation and Molecular Identification of Strain BV01

The morphological characteristics of strain BV01 were recorded after incubation on a nutrient agar (NA: peptone 10.0 g, beef paste 3.0 g, NaCl 5.0 g, agar 18.0 g, dH$_2$O 1000 mL) plate at 28 °C for 2 d. A single colony was then taken and evenly spread onto a glass slide, dry fixed, and Gram stained for 1 min [24]. The microscopic photographs were taken with a Zeiss AxioCam MRc 5 digital camera (Carl Zeiss, Jena, Germany) attached to Zeiss Axioskop 2 plus microscope (Carl Zeiss, Göttingen, Germany).

Genomic DNA was extracted from fresh cultures using a bacterial genomic DNA kit (ZOMAN, Beijing, China) according to the manufacturer's instructions. The *16S* rRNA and the gyrase subunit A protein (*gyrA*) regions were amplified using the primer pairs of 27F (5′-AGAGTTTGATCCTGGCTCAG-3′) and 1492R (5′-GGTTACCTTGTTACGACTT-3′) and *gyrA*-F (5′-CAGTCAGGAAATGCGTACGTCTT-3′) and *gyrA*-R (5′-CAAGGTAAT GCTCCAGGCATTGCT-3′), respectively. PCR reactions were conducted using an ABI 2720 Thermal Cycler (Applied Biosciences, Foster City, CA, USA), and the products were then sequenced using an ABI3730XL DNA Sequencer (Applied Biosciences, Foster City, CA, USA). The newly obtained sequences and those of the ex-type strain as well as the related ones retrieved from GenBank were aligned using BioEdit 7.2 [25] and analyzed with the neighbor-joining (NJ) method [26] via MEGA X [27]. The topological confidence of the resulting tree and the statistical supports of the branches were tested using the neighbor-joining bootstrap proportion (NJBP) with 1000 replications, each with 10 replicates of random addition of taxa [28].

### 2.5. Determination of Enzyme Activity and Secondary Metabolites

Protease, cellulase, and $\beta$-1,3-glucanase were detected on skim milk, carboxymethyl cellulose (CMC), and glucose agar, respectively [29]. Siderophore and indole-3-acetic acid (IAA) were tested using modified chrome azurol S and Salkowski reagent agar, respectively [30]. Enzyme activity and production of metabolites were observed based on the presence of clear zones around bacterial colonies after incubation at 28 °C for 3 d. All treatments were repeated three times.

### 2.6. Biocontrol Activity of Strain BV01 In Vivo

After evaluation of the antagonistic abilities of the four bacterial strains, two strains, namely, JDF and BV01, were further tested for their biocontrol and plant-growth-promotion abilities in a pot experiment. Four germinated wheat seeds were sowed per pot containing a mixture of podsolic soil and vermiculite ($v$:$v$ = 1:1). On the 14th day, the roots of the plants were punctured and inoculated nearby the wounds with pathogen mycelium blocks (5 mm in diam.). After 24 h, the roots were irrigated with 30 mL ($2.5 \times 10^8$ CFU/mL) of bacterial suspensions per wheat plant, and the same amount of NB was used as the control. The plants were kept in a greenhouse with a 14 h/10 h photoperiod/dark period at $26 \pm 1$ °C. The incidence of disease of wheat treated with BV01, JDF, and the non-treated control was recorded and calculated at 20 d after *B. sorokiniana* inoculation. Infection types (ITs) of *B. sorokiniana* on wheat were evaluated based on the area of a brown or black lesion at the stem base, with scores varying from 0 to 4 (IT0: no lesion, IT1: coverage of necrotic lesion less than 1/4, IT2: lesion coverage between 1/4 and 1/2, IT3: coverage between 1/2 and 2/3, and IT4: coverage between 2/3 and total). The disease index = $\sum (d_i \times l_i) / (L \times d_{imax}) \times 100\%$ ($d_i$: infection type, li: number of plants with each infection type, and L: number of wheat plants investigated) [31].

Three true leaves at the age of 40 d were detached from pepper seedlings and punctured in two places with a sterile needle on both sides of the leaf. Mycelium blocks (5 mm in diam.) of *F. graminearum* were placed on the two wounds and then sprayed with 2 mL of either BV01 or JDF fermentation broth at a concentration of $2.5 \times 10^8$ CFU/mL, and the same volume of NB was used as the non-inoculated control at 24 h after *F. graminearum* inoculation. The treatments were kept in a greenhouse at $26 \pm 1$ °C with a 14 h/10 h photoperiod/dark period for 6 d; sterile water was added once to keep the treatment moist. There were 3 leaves in each pot and 3 replicates in each treatment. The diameters of diseased spots were recorded and calculated at the 6th day after *F. graminearum* inoculation. The inhibition rate was calculated as follows: $(D_{CK} - D_i)/D_{CK} \times 100\%$ ($D_{CK}$: the control group's average colony diameter and Di: the treatment group's average colony diameter).

### 2.7. Plant-Growth-Promoting Assays in a Greenhouse

Sterilized wheat and pepper seeds were soaked in suspensions of BV01 or JDF at a concentration of $2.5 \times 10^8$ CFU/mL and then in sterile water for 10 min. NB was used as the control. The plants were kept in a greenhouse with a 14 h/10 h photoperiod/dark period at $26 \pm 1$ °C. The wheat and peppers were harvested at the 21st and 35th days, respectively. Plant height, fresh weight, dry weight, and leaf width were recorded, and the strong seedling index (SSI) was calculated [32].

### 2.8. Statistical Analysis

Statistical analysis was performed using SPSS 21 (Armonk, NY, USA). ANOVA was performed, and mean values were compared using Duncan's multiple range test with $p < 0.05$ as the level of significance. All analyses were conducted using GraphPad Prism 8 (San Diego, CA, USA).

## 3. Results

### 3.1. Inhibitory Effects of Four Tested Bacterial Strains against Eight Fungal Phytopathogens

Strain BV01 exhibited varying degrees of antagonism against different phytopathogens, and the inhibition rates ranged from 57% to 83% (Table 2) with the highest potential inhibitory effects against *B. cinerea* PP1 (Figure 1, Treatment 1). The inhibition rates showed that BV01 had significantly higher inhibitory effects than JDF, L01, and BS208 on six of the eight tested fungal phytopathogens.

**Table 2.** Antifungal activities of four bacterial strains.

| Strains | Dual-Culture Inhibition Rate (%) | | | | Fermentation Broth Inhibition Rate (%) | | | |
|---|---|---|---|---|---|---|---|---|
| | CGMCC 1.60184 | JDF | L01 | BS208 | CGMCC 1.60184 | JDF | L01 | BS208 |
| *B. sorokiniana* PP12 | 80.68 ± 1.23 [a] | 79.55 ± 0.93 [a] | 77.27 ± 1.05 [b] | 27.27 ± 2.47 [c] | 92.26 ± 0.68 [a] | 76.77 ± 0.23 [b] | 70.32 ± 1.03 [c] | 8.39 ± 1.89 [d] |
| *B. cinerea* PP1 | 82.95 ± 1.23 [b] | 85.23 ± 2.23 [a] | 85.23 ± 2.14 [a] | 0.00 ± 0.83 [c] | 81.82 ± 0.82 [a] | 64.77 ± 1.00 [c] | 71.59 ± 0.58 [b] | 2.27 ± 1.44 [d] |
| *C. capsici* PP6 | 79.55 ± 0.64 [a] | 75.00 ± 1.25 [b] | 77.27 ± 1.07 [b] | 26.14 ± 1.68 [c] | 72.73 ± 0.75 [a] | 71.59 ± 0.86 [a] | 72.73 ± 0.42 [a] | 4.55 ± 1.45 [b] |
| *F. graminearum* PP15 | 61.36 ± 1.09 [b] | 65.91 ± 0.49 [a] | 62.50 ± 0.62 [b] | 0.00 ± 0.62 [c] | 59.09 ± 1.66 [a] | 43.18 ± 0.52 [c] | 47.53 ± 0.66 [b] | 13.64 ± 0.97 [d] |
| *F. oxysporum* F6 | 65.91 ± 0.66 [a] | 62.50 ± 1.03 [b] | 52.27 ± 0.71 [c] | 32.95 ± 1.27 [d] | 56.79 ± 1.23 [a] | 13.58 ± 1.06 [b] | 12.04 ± 0.69 [b] | 8.95 ± 0.46 [c] |
| *N. rubicola* PP23 | 56.52 ± 1.02 [a] | 50.72 ± 0.93 [b] | 50.72 ± 1.23 [b] | 42.03 ± 1.24 [c] | 36.13 ± 1.19 [a] | 12.18 ± 0.99 [b] | 12.18 ± 0.52 [b] | 4.19 ± 0.74 [c] |
| *R. solani* PP11 | 69.32 ± 0.71 [a] | 54.55 ± 0.86 [b] | 53.41 ± 1.23 [b] | 0.00 ± 1.7 [c] | 46.59 ± 1.23 [a] | 0.00 ± 1.08 [c] | 0.00 ± 1.21 [c] | 5.68 ± 0.56 [b] |
| *V. dahliae* PP8 | 70.15 ± 1.18 [a] | 58.21 ± 0.73 [b] | 55.22 ± 1.35 [c] | 32.84 ± 0.88 [d] | 71.72 ± 1.18 [a] | 59.07 ± 0.73 [b] | 61.03 ± 1.35 [b] | 33.95 ± 0.88 [c] |

The inhibition rates (%) ($n = 3$, mean ± SE). Different letters indicate significantly different groups ($p < 0.05$, ANOVA, Tukey HSD).

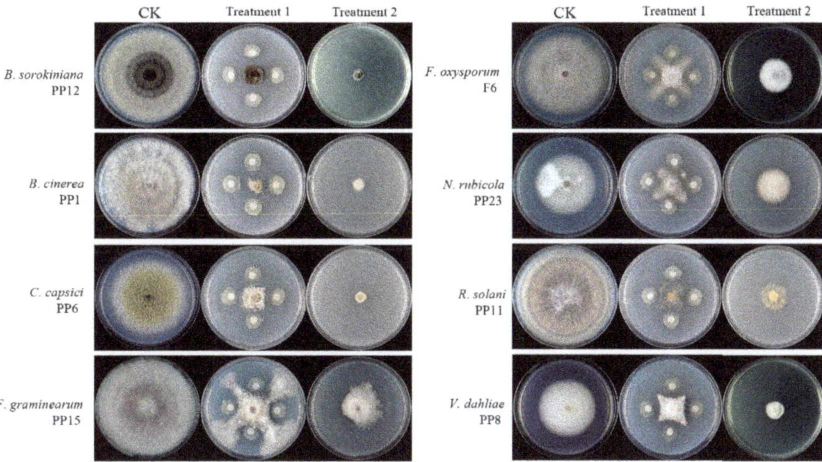


**Figure 1.** Inhibitory effects of BV01 against fungal phytopathogens. CK: only pathogen on PDA at 25 °C for 5 d; Treatment 1: dual culture of BV01 against pathogen on PDA at 25 °C for 5 d; Treatment 2: pathogen on PDA amended with fermentation broth of BV01 at 25 °C for 5 d.

Antifungal assay by fermentation broth test showed that BV01 had relatively high inhibitory effects against different pathogens (Table 2), and the highest inhibitory rate reached 92% (against *B. sorokiniana* PP12) (Figure 1, Treatment 2). Overall, the effects of BV01 were better than those of JDF and L01 and were significantly superior to those of BS208.

### 3.2. Identification of Strain BV01

The colony of BV01 was ivory white and non-transparent with a rough surface on NA medium (Figure 2A–C). The cells were Gram-positive (Figure 2D), rod-shaped, 1.43–2.53 µm long, 0.66–0.88 µm wide, and occurred singly, in pairs, or occasionally in short chains. The analysis of *16S* rRNA sequences showed that strain BV01 shared 99% identity with the type strain of *B. velezensis* (CR502) according to a BLAST search. The resulting NJ trees based on sequences of *16S* rRNA and *gyrA* (Figure 3) showed that BV01 clustered with *Bacillus* species and grouped with the type strain of *B. velezensis*, which confirmed its taxonomic position.

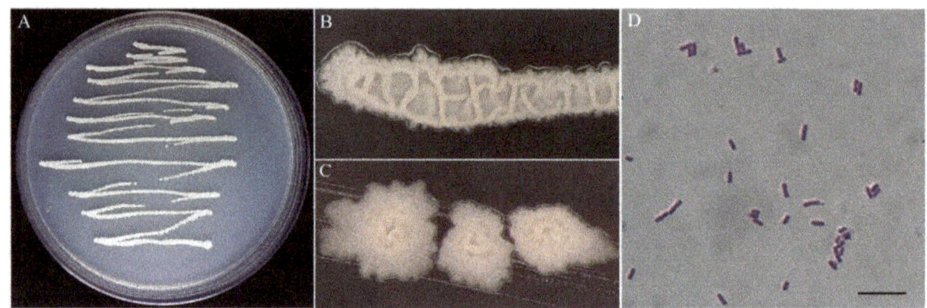

**Figure 2.** Colony and microscopic characteristics of BV01. (**A–C**) General colonies on nutrient agar; (**D**) Gram-stained cells. Bar: D = 10 μm.

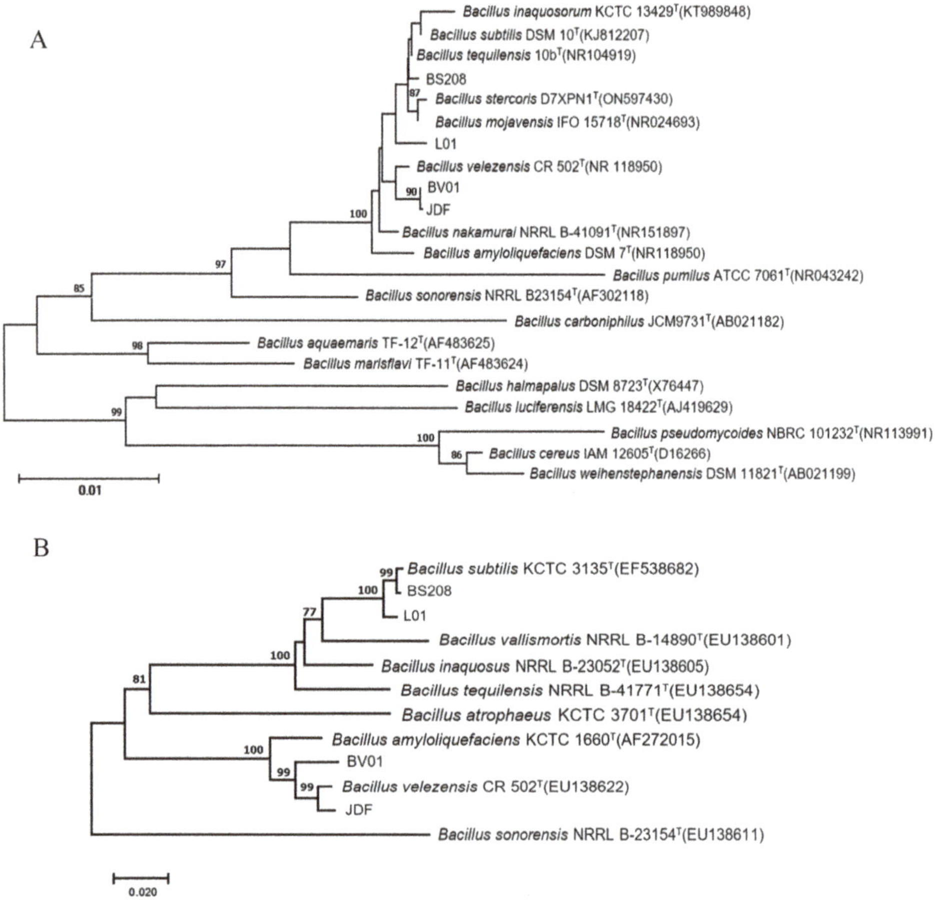

**Figure 3.** Phylogenetic trees generated based on sequences of *16S* rRNA (**A**) and *gyrA* (**B**) regions of *Bacillus* species. NJBP values greater than 75% are shown at the nodes.

### 3.3. Detection of Antagonism-Related Lytic Enzymes

Clear zones detected around the colony of BV01 indicated that the strain produced protease, cellulase, and β-1,3-glucanase (Figure 4A–C) as well as siderophore (Figure 4D) and IAA (Figure 4E), which suggested its high potential in biological control. The production of IAA reached 12.17 mg/mL after incubation for 6 d.

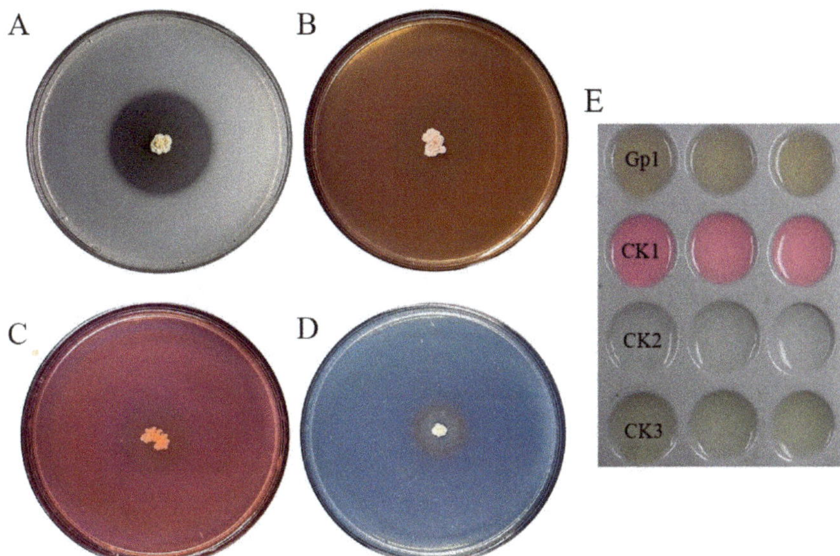

**Figure 4.** Detection of extracellular enzyme production and growth-promotion traits of BV01. (**A**) protease; (**B**) cellulase; (**C**) β-1,3-glucanase; (**D**) siderophore; (**E**) indole-3-acetic acid (IAA); Gp$_1$: BV01 suspension, CK$_1$: 10 mg/mL IAA, CK$_2$: sterilize distilled water, CK$_3$: NB.

### 3.4. Biocontrol Effects of Bacterial Strains BV01 and JDF on Wheat Root Rot

Lesions at the stem bases of wheat were obviously brown in the non-treated control, while those treated with BV01 and JDF were very slightly infected (Figure 5A). The disease indices of CK, BV01, and JDF were 76.4, 40.8, and 53.6, respectively (Figure 5B). In the BV01 treatment, infection with wheat root rot was significantly ($p < 0.05$) reduced, the relative control efficacy was 47% (Figure 5C), and the fresh and dry weights (Figure 5D) and plant height (Figure 5E) were increased by 24%, 91%, and 34%, respectively.

### 3.5. Biocontrol Effect of Strain BV01 on Fusarium Wilt

The symptoms on pepper leaves of the control were severe, on those treated with JDF were moderate, and on those treated with BV01 were weak (Figure 6A). The average diameter of a spot was 2.31, 0.99, and 1.76 cm in the CK, BV01, and JDF treatments (Figure 6B). The control effect reached 57% and 24% for the treatments with BV01 and JDF, respectively (Figure 6C).

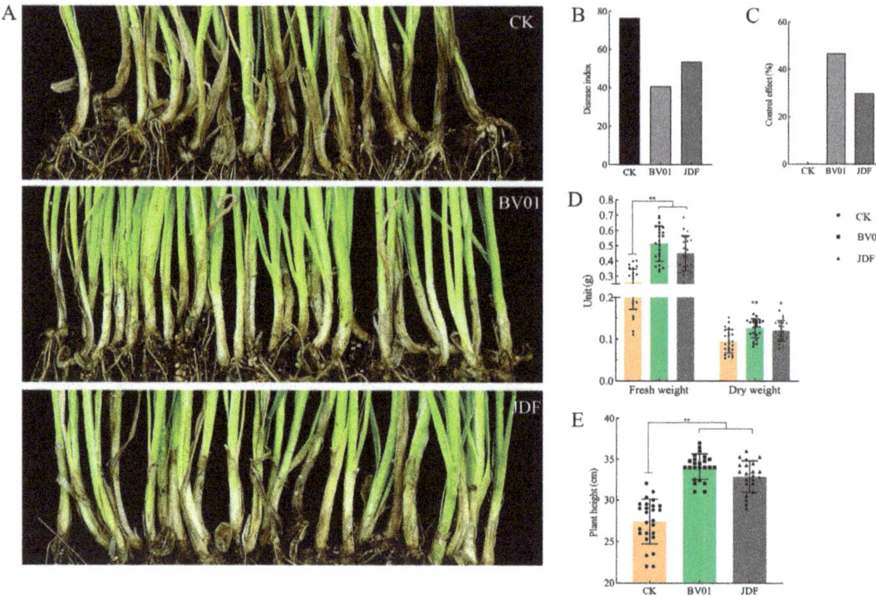

**Figure 5.** Inhibitory effect of *Bacillus* strains on wheat root rot disease caused by *B. sorokiniana* PP12. (**A**) Symptoms of *B. sorokiniana* on wheat roots with different treatments; (**B**) disease index of *B. sorokiniana* in the treatments with BV01 and JDF; (**C**) inhibition rates of BV01 and JDF; (**D**) fresh and dry weights of wheat; (**E**) shoot biomass (cm) measured by plant height of wheat. The results were observed after 40 d of incubation. Values are the means $\pm$ SEs, $n$ = 27 plants, ** $p < 0.001$, and * $p < 0.05$.

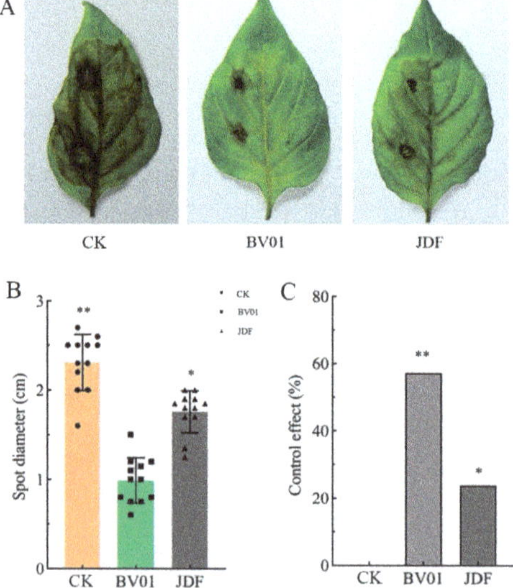

**Figure 6.** Effect of *Bacillus* strains on disease symptoms caused by *F. graminearum* PP15 on leaves. (**A**) Symptoms of *F. graminearum* on leaves with different treatments; (**B**) spot diameter after treatment with BV01 or JDF; (**C**) inhibition rates of BV01 and JDF. Values are the means $\pm$ SEs, $n$ = 9 leaves, ** $p < 0.001$, and * $p < 0.05$.

*3.6. Growth-Promotion Effects of Strain BV01 on Wheat and Pepper*

Wheat treated with BV01 exhibited an increase in height of 13% (Figure 7A), while the fresh weight and dry weight were improved by 10% and 5% (Figure 7B). Pepper treated with BV01 exhibited increases in the fresh weight, leaf width, and stem thickness of 20%, 9%, and 9%, respectively. The dry weight and plant height were improved by 12% and 2% (Figure 7C,D). The SSI for treatment with BV01 and JDF increased by 19% and 10%. These findings suggest that strain BV01 was more effective in promoting plant growth than JDF.

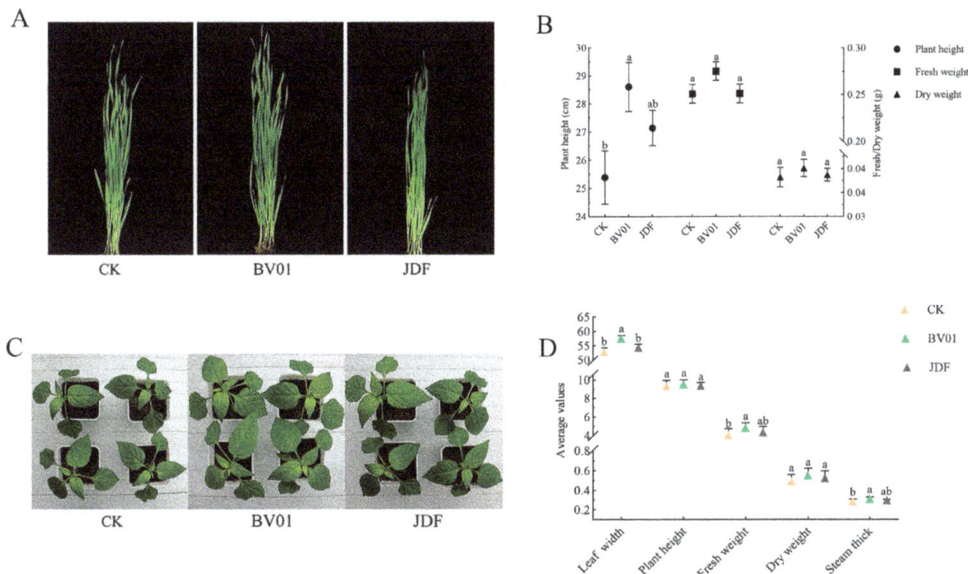

**Figure 7.** Growth-promotion effects of BV01 and JDF on wheat and pepper. Wheat growth (**A**) and experimental conditions (**B**) in CK, BV01, and JDF pot experiments, *n* = 12 plants. Pepper growth (**C**) and experimental conditions (**D**) in CK, BV01, and JDF pot experiments, *n* = 9 plants. Values are the means ± SEs, different letters show significant differences (Fisher's LSD, $p < 0.05$).

## 4. Discussion

For a long time, *Bacillus amyloliquefaciens* and *B. subtilis* were known to have biocontrol functions against various plant pathogens [33]. Recently, *B. velezensis* was reported as a biocontrol agent against many phytopathogens. For example, *B. velezensis* strain F21 can control *Fusarium* wilt on watermelon [19], and strain BR-01 has strong antagonistic effects on rice pathogens [34], while strain CE100 increases fruit yield of strawberries by controlling fungal diseases [35]. The star strain FZB42 was initially established in 1998, and successive studies on its antimicrobial substances, interactions between plants and bacteria, regulatory small RNAs, and biocontrol enzymes have been carried out [33]. In previous studies, antagonistic strains of *B. velezensis* were often isolated from water, soil, air, plant roots, plant surfaces, and animal intestines [7]. In the present study, strain BV01 was derived from a PDA plate in the laboratory and speculated to be an air source strain. Based on morphological characteristics and phylogenetic evidence, strain BV01 was identified as *B. velezensis*; further exploration of its biological control potential was then performed. Its dual-culture inhibition rates against different pathogens were greater than 56%, and the fermentation broth inhibition rates were reduced by more than 36% when compared to the control. The results indicate that BV01 produces a special antibacterial substance. Some lipopeptide extract components of *B. amyloliquefaciens* have been demonstrated as key substances in controlling the growth of *Xanthomonas citri* subsp. *citri* [36]. Zhou et al. [34]

proved that the relative inhibition rate of *B. velezensis* BR-01 against *F. fujikuroi* was 57%, while the strain showed no antagonistic ability against *R. solani*. The results of the current study revealed that strain BV01 possessed very strong antagonistic activity and broad-spectrum biological ability against *B. cinerea, F. oxysporum, C. capsici, V. dahliae, R. solani, B. sorokiniana, F. graminearum,* and *N. rubicola*.

Many *Bacillus* species produce a variety of hydrolytic enzymes, such as cellulase, β-1,3-glucanase, and protease, which are responsible for the degradation of diverse components of fungal pathogens [35,37]. The detection of cellulase, protease, and β-1,3-glucanase in BV01 supports its association with the growth suppression of several fungal phytopathogens. Our results also revealed that strain BV01 effective in vitro against fungal pathogens was also able to produce siderophores, which are related to indirect antagonistic processes such as plant defenses and growth promotion [30]. Moreover, some members of *Bacillus* invade the rhizosphere of plants and promote plant growth by producing plant hormones, such as IAA, cytokinins, and gibberellins, and chelating minerals and siderophores. Many plant-growth-promoting bacteria produce IAA, which promotes the development of plant roots, and are usually utilized as bioinoculants [38–45]. In a previous study, *B. velezensis* BY6 was reported to significantly increase the dry and fresh mass and plant height of Pdpap poplar seedlings [46]. In the present study, *B. velezensis* BV01 produced IAA during its growth. Moreover, our pot experiment results revealed that pepper and wheat treated with strain BV01 possessed higher fresh weight, dry weight, plant height, leaf width, stem thickness, and SSI than controls. Both the antifungal activity assay and greenhouse pot experiment indicated that the strain BV01 has biocontrol and plant-growth-promotion potential.

Wheat and pepper are two of the most commonly grown crops and vegetables in the world. Several pathogens cause severe diseases of them and thus reduce significantly their yields. For example, wheat root rot caused by *B. sorokiniana, Fusarium* spp., and other pathogens alone or in combination generally can lead to wheat yield reductions of 20%–30%, with severe cases of more than 50% [47,48]. Previous studies revealed that *B. subtilis* and *B. amyloliquefaciens* can prevent and control wheat root rot [47]. However, there are few studies on the effects of *B. velezensis* on wheat root rot caused by *B. sorokiniana. Bacillus velezensis* strains CC09 and NEAU-242-2 could be used as potential biocontrol agents to control wheat disease [49,50]. In this study, *B. velezensis* strain BV01 was able to effectively control wheat root rot caused by *B. sorokiniana* in a greenhouse, with a control rate of 47%. The occurrence of pepper wilt is increasing currently and seriously affects the quality of pepper. For example, the incidence of pepper wilt disease in China is generally 15%–30%, with severe cases decreasing quality by 70%–80% [51]. The main pathogen, *Fusarium graminearum*, is a highly destructive phytopathogen, not only lowering crop yields but also producing mycotoxins and affecting crop quality. Previous studies have confirmed that *B. velezensis* could control pepper root rot [52], wheat spikes [53], corn stalk rot [54], and corn head blight [55]. To our knowledge, the present study is the first report that *B. velezensis* can serve as a potential biocontrol agent for controlling pepper wilt induced by *F. graminearum. Bacillus velezensis* BV01 not only promotes the growth of wheat and pepper seedlings but also significantly controls wheat root rot and pepper wilt. In summary, *Bacillus velezensis* BV01 has good control effects in both dual-culture and fermentation broth tests against *B. sorokiniana* and *F. graminearum,* and it obviously reduced the disease symptoms and promoted the growth of wheat and pepper.

## 5. Conclusions

*Bacillus velezensis* BV01 showed protease, cellulase, and β-1,3-glucanase activities, which are related to phytopathogen cell wall degradation, and produced growth-promotion substances such as IAA and siderophore. This strain also suppressed the growth of eight phytopathogens both in dual-culture and sterile filtrate assays and significantly reduced the disease incidence of wheat root rot and *Fusarium* wilt in greenhouse settings. Moreover, it significantly promoted wheat and pepper growth. In conclusion, BV01 exhibits broad and effective antagonistic activity against several phytopathogens, promotes plant growth, and is worthy of further exploration of its biocontrol applications in eco-friendly agriculture practices.

**Author Contributions:** Conceptualization, Z.Z. and W.Z.; resources, W.Z. and Z.Z.; data curation, T.H.; writing—original draft preparation, T.H.; writing—review and editing, Y.Z., W.Z., Z.Y. and Z.Z.; visualization, T.H. All authors have read and agreed to the published version of the manuscript.

**Funding:** This research was supported by the National Natural Science Foundation of China (32270009, 31870012, 31750001), the Biological Resources Programme, Chinese Academy of Sciences (KFJ-BRP-017-082), and the Frontier Key Program of the Chinese Academy of Sciences (QYZDY-SSW-SMC029).

**Data Availability Statement:** All the data relevant to this manuscript are available on request from the corresponding author.

**Acknowledgments:** The authors would like to thank Yongchun Niu, Jiyan Qiu, Zhengqiang Ma, Qili Li, and Shenzhan Fu for providing the phytopathogens used in this study and Hongjun Chen for corrections to the language.

**Conflicts of Interest:** The authors declare no conflict of interest. The funders had no role in the design of the study; in the collection, analyses, or interpretation of data; in the writing of the manuscript; or in the decision to publish the results.

# References


1. Savary, S.; Willocquet, L.; Pethybridge, S.J.; Esker, P.; McRoberts, N.; Nelson, A. The global burden of pathogens and pests on major food crops. *Nat. Ecol. Evol.* **2019**, *3*, 430–439. [CrossRef]
2. Xu, S.; Liu, Y.X.; Cernava, T.; Wang, H.; Zhou, Y.; Xia, T.; Cao, S.; Berg, G.; Shen, X.X.; Wen, Z.; et al. Fusarium fruiting body microbiome member *Pantoea agglomerans* inhibits fungal pathogenesis by targeting lipid rafts. *Nat. Microbiol.* **2022**, *7*, 831–843. [CrossRef]
3. Wennersten, R.; Qie, S. United nations sustainable development goals for 2030 and resource use. In *Handbook of Sustainability Science and Research*; Filho, W.L., Ed.; Springer International Publishing: Cham, Switzerland, 2018; pp. 317–339.
4. Mesnage, R.; Defarge, N.; Spiroux de Vendômois, J.; Séralini, G. Major pesticides are more toxic to human cells than their declared active principles. *BioMed Res. Int.* **2014**, *2014*, 179691. [CrossRef] [PubMed]
5. Borriss, R. Use of plant-associated *Bacillus* strains as biofertilizers and biocontrol agents in agriculture. In *Bacteria in Agrobiology: Plant Growth Responses*; Maheshwari, D., Ed.; Springer: Berlin/Heidelberg, Germany, 2011; pp. 41–76.
6. Jing, J.Y.; Cong, W.F.; Bezemer, T.M. Legacies at work: Plant-soil-microbiome interactions underpinning agricultural sustainability. *Trends Plant Sci.* **2022**, *27*, 781–792. [CrossRef] [PubMed]
7. Chowdhury, S.P.; Hartmann, A.; Gao, X.W.; Borriss, R. Biocontrol mechanism by root-associated *Bacillus amyloliquefaciens* FZB42—A review. *Front. Microbiol.* **2015**, *6*, 780. [CrossRef] [PubMed]
8. Fan, B.; Blom, J.; Klenk, H.P.; Borriss, R. *Bacillus amyloliquefaciens, Bacillus velezensis*, and *Bacillus siamensis* form an "operational group B. amyloliquefaciens" within the *B. subtilis* species complex. *Front. Microbiol.* **2017**, *8*, 22. [CrossRef] [PubMed]
9. Solano-Alvarez, N.; Valencia-Hernández, J.A.; Rico-García, E.; Torres-Pacheco, I.; Ocampo-Velázquez, R.V.; Escamilla-Silva, E.M.; Romero-García, A.L.; Alpuche-Solís, Á.G.; Guevara-González, R.G. A novel isolate of *Bacillus cereus* promotes growth in tomato and inhibits *Clavibacter michiganensis* infection under greenhouse conditions. *Plants* **2021**, *10*, 506. [CrossRef]
10. Schmiedeknecht, G.; Bochow, H.; Junge, H. Use of *Bacillus subtilis* as biocontrol agent. II. Biological control of potato diseases. *J. Plant Dis. Protect.* **1998**, *105*, 376–386.
11. Sylla, J.; Alsanius, B.W.; Krüger, E.; Reineke, A.; Strohmeier, S.; Wohanka, W. Leaf microbiota of strawberries as affected by biological control agents. *Phytopathology* **2013**, *103*, 1001–1011. [CrossRef]
12. Talboys, P.J.; Owen, D.W.; Healey, J.R.; Withers, P.J.A.; Jones, D.L. Auxin secretion by *Bacillus amyloliquefaciens* FZB42 both stimulates root exudation and limits phosphorus uptake in *Triticum aestivum*. *BMC Plant Biol.* **2014**, *14*, 51. [CrossRef]
13. Chowdhury, S.P.; Dietel, K.; Rändler, M.; Schmid, M.; Junge, H.; Borriss, R.; Hartmann, A.; Grosch, R. Effects of *Bacillus amyloliquefaciens* FZB42 on lettuce growth and health under pathogen pressure and its impact on the rhizosphere bacterial community. *PLoS ONE* **2013**, *8*, 68818. [CrossRef]
14. Mácha, H.; Marešová, H.; Juříková, T.; Švecová, M.; Benada, O.; Škríba, A.; Baránek, M.; Novotný, Č.; Palyzová, A. Killing effect of *Bacillus velezensis* FZB42 on a *Xanthomonas campestris* pv. *campestris* (Xcc) strain newly isolated from cabbage *Brassica oleracea* Convar. *capitata* (L.): A metabolomic study. *Microorganisms* **2021**, *9*, 1410. [CrossRef] [PubMed]
15. Wang, H.; Guo, Y.J.; Luo, Z.; Gao, L.W.; Li, R.; Zhang, Y.X.; Kalaji, H.M.; Qiang, S.; Chen, S.G. Recent advances in alternaria phytotoxins: A review of their occurrence, structure, bioactivity, and biosynthesis. *J. Fungi* **2022**, *8*, 168. [CrossRef] [PubMed]
16. Dean, R.; Van Kan, J.A.; Pretorius, Z.A.; Hammond-Kosack, K.E.; Di Pietro, A.; Spanu, P.D.; Rudd, J.J.; Dickman, M.; Kahmann, R.; Ellis, J.; et al. The Top 10 fungal pathogens in molecular plant pathology. *Mol. Plant Pathol.* **2012**, *13*, 414–430. [CrossRef]

17. Ali, S.A.M.; Sayyed, R.Z.; Mir, M.I.; Khan, M.Y.; Hameeda, B.; Alkhanani, M.F.; Haque, S.; Mohammad Al Tawaha, A.R.; Poczai, P. Induction of systemic resistance in maize and antibiofilm activity of surfactin from *Bacillus velezensis* MS20. *Front. Microbiol.* **2022**, *13*, 879739. [CrossRef] [PubMed]

18. Kang, X.X.; Guo, Y.; Leng, S.; Xiao, L.; Wang, L.H.; Xue, Y.R.; Liu, C.H. Comparative transcriptome profiling of *Gaeumannomyces graminis* var. *tritici* in wheat roots in the absence and presence of biocontrol *Bacillus velezensis* CC09. *Front. Microbiol.* **2019**, *10*, 1474. [CrossRef]

19. Jiang, C.H.; Yao, X.F.; Mi, D.D.; Li, Z.J.; Yang, B.Y.; Zheng, Y.; Qi, Y.J.; Guo, J.H. Comparative transcriptome analysis reveals the biocontrol mechanism of *Bacillus velezensis* F21 against *Fusarium* wilt on watermelon. *Front. Microbiol.* **2019**, *10*, 652. [CrossRef]

20. Yan, H.H.; Qiu, Y.; Yang, S.; Wang, Y.Q.; Wang, K.Y.; Jiang, L.L.; Wang, H.Y. Antagonistic activity of *Bacillus velezensis* SDTB038 against *Phytophthora infestans* in potato. *Plant Dis.* **2021**, *105*, 1738–1747. [CrossRef]

21. Zhang, D.F.; Gao, Y.X.; Wang, Y.J.; Liu, C.; Shi, C.B. Advances in taxonomy, antagonistic function and application of *Bacillus velezensis*. *Microbiol. China* **2020**, *47*, 3634–3649.

22. Xu, W.; Zhang, L.Y.; Goodwin, P.H.; Xia, M.C.; Zhang, J.; Wang, Q.; Liang, J.; Sun, R.H.; Wu, C.; Yang, L.R. Isolation, identification, and complete genome assembly of an endophytic *Bacillus velezensis* YB-130, potential biocontrol agent against *Fusarium graminearum*. *Front. Microbiol.* **2020**, *11*, 598285. [CrossRef]

23. Li, Z.; Guo, B.; Wan, K.; Cong, M.; Huang, H.; Ge, Y. Effects of bacteria-free filtrate from *Bacillus megaterium* strain L2 on the mycelium growth and spore germination of *Alternaria alternata*. *Biotechnol. Biotechnol. Equip.* **2015**, *29*, 1062–1068. [CrossRef]

24. Tripathi, N.; Sapra, A. Gram staining. In *StatPearls*; StatPearls Publishing: Treasure Island, FL, USA, 2022.

25. Tippmann, H.F. Analysis for free: Comparing programs for sequence analysis. *Brief. Bioinform.* **2004**, *5*, 82–87. [CrossRef] [PubMed]

26. Saitou, N.; Nei, M. The neighbor-joining method: A new method for reconstructing phylogenetic trees. *Mol. Biol. Evol.* **1987**, *4*, 406–425. [PubMed]

27. Kumar, S.; Stecher, G.; Li, M.; Knyaz, C.; Tamura, K. MEGA X: Molecular evolutionary genetics analysis across computing platforms. *Mol. Biol. Evol.* **2018**, *35*, 1547–1549. [CrossRef]

28. Felsenstein, J. Confidence limits on phylogenies: An approach using the bootstrap. *Evolution* **1985**, *39*, 783–791. [CrossRef] [PubMed]

29. Syed-Ab-Rahman, S.F.; Carvalhais, L.C.; Chua, E.; Xiao, Y.W.; Wass, T.J.; Schenk, P.M. Identification of soil bacterial isolates suppressing different *Phytophthora* spp. and promoting plant growth. *Front. Plant Sci.* **2018**, *9*, 1502. [CrossRef]

30. Demange, P.; Bateman, A.; Mertz, C.; Dell, A.; Piémont, Y.; Abdallah, M.A. Bacterial siderophores: Structures of pyoverdins Pt, siderophores of *Pseudomonas tolaasii* NCPPB 2192, and pyoverdins Pf, siderophores of *Pseudomonas fluorescens* CCM 2798. Identification of an unusual natural amino acid. *Biochemistry* **1990**, *29*, 11041–11051. [CrossRef]

31. Dong, N.; Liu, X.; Lu, Y.; Du, L.; Xu, H.; Liu, H.; Xin, Z.; Zhang, Z. Overexpression of TaPIEP1, a pathogen-induced ERF gene of wheat, confers host-enhanced resistance to fungal pathogen *Bipolaris sorokiniana*. *Funct. Integr. Genom.* **2010**, *10*, 215–226. [CrossRef]

32. Anwar, A.; Yan, Y.; Liu, Y.M.; Li, Y.S.; Yu, X.C. 5-aminolevulinic acid improves nutrient uptake and endogenous hormone accumulation, enhancing low-temperature stress tolerance in cucumbers. *Int. J. Mol. Sci.* **2018**, *19*, 3379. [CrossRef]

33. Fan, B.; Wang, C.; Song, X.F.; Ding, X.L.; Wu, L.M.; Wu, H.J.; Gao, X.W.; Borriss, R. *Bacillus velezensis* FZB42 in 2018: The gram-positive model strain for plant growth promotion and biocontrol. *Front. Microbiol.* **2018**, *9*, 2491. [CrossRef]

34. Zhou, J.P.; Xie, Y.Q.; Liao, Y.H.; Li, X.Y.; Li, Y.M.; Li, S.P.; Ma, X.G.; Lei, S.M.; Lin, F.; Jiang, W.; et al. Characterization of a *Bacillus velezensis* strain isolated from *Bolbostemmatis rhizoma* displaying strong antagonistic activities against a variety of rice pathogens. *Front. Microbiol.* **2022**, *13*, 983781. [CrossRef]

35. Hong, S.; Kim, T.Y.; Won, S.J.; Moon, J.H.; Ajuna, H.B.; Kim, K.Y.; Ahn, Y.S. Control of fungal diseases and fruit yield improvement of strawberry using *Bacillus velezensis* CE 100. *Microorganisms* **2022**, *10*, 365. [CrossRef] [PubMed]

36. Wang, X.; Liang, L.; Shao, H.; Ye, X.; Yang, X.; Chen, X.; Shi, Y.; Zhang, L.; Xu, L.; Wang, J. Isolation of the novel strain *Bacillus amyloliquefaciens* F9 and identification of lipopeptide extract components responsible for activity against *Xanthomonas citri* subsp. *citri*. *Plants* **2022**, *11*, 457. [CrossRef] [PubMed]

37. Jing, R.X.; Li, N.; Wang, W.P.; Liu, Y. An endophytic strain JK of genus *Bacillus* isolated from the seeds of super hybrid rice (*Oryza sativa* L., Shenliangyou 5814) has antagonistic activity against rice blast pathogen. *Microb. Pathog.* **2020**, *147*, 104422. [CrossRef] [PubMed]

38. Idris, E.E.; Iglesias, D.J.; Talon, M.; Borriss, R. Tryptophan-dependent production of indole-3-acetic acid (IAA) affects level of plant growth promotion by *Bacillus amyloliquefaciens* FZB42. *Mol. Plant Microbe Interact.* **2007**, *20*, 619–626. [CrossRef] [PubMed]

39. Kloepper, J.W.; Ryu, C.M.; Zhang, S.A. Induced systemic resistance and promotion of plant growth by *Bacillus* spp. *Phytopathology* **2004**, *94*, 1259–1266. [CrossRef]

40. Yao, A.V.; Bochow, H.; Karimov, S.A.; Boturov, U.; Sanginboy, S.; Sharipov, A. Effect of FZB24® *Bacillus subtilis* as a biofertilizer on cotton yields in field tests. *Arch. Phytopathol. Plant Prot.* **2006**, *39*, 323–328. [CrossRef]

41. Bloemberg, G.V.; Lugtenberg, B.J.J. Molecular basis of plant growth promotion and biocontrol by rhizobacteria. *Curr. Opin. Plant Biol.* **2001**, *4*, 343–350. [CrossRef]

42. Compant, S.; Duffy, B.; Nowak, J.; Clement, C.; Barka, E.A. Use of plant growth-promoting bacteria for biocontrol of plant diseases: Principles, mechanisms of action, and future prospects. *Appl. Environ. Microb.* **2005**, *71*, 4951–4959. [CrossRef]

43. Yi, Y.J.; Luan, P.Y.; Wang, K.; Li, G.L.; Yin, Y.A.; Yang, Y.H.; Zhang, Q.Y.; Liu, Y. Antifungal activity and plant growth-promoting properties of *Bacillus mojovensis* B1302 against *Rhizoctonia cerealis*. *Microorganisms* **2022**, *10*, 1682. [CrossRef]

44. Balderas-Ruíz, K.A.; Bustos, P.; Santamaria, R.I.; González, V.; Cristiano-Fajardo, S.A.; Barrera-Ortíz, S.; Mezo-Villalobos, M.; Aranda-Ocampo, S.; Guevara-García, Á.A.; Galindo, E.; et al. *Bacillus velezensis* 83 a bacterial strain from mango phyllosphere, useful for biological control and plant growth promotion. *AMB Express* **2020**, *10*, 163. [CrossRef] [PubMed]

45. Hernandez, J.; Tamez-Guerra, P.; Gomez-Flores, R.; Delgado, C.; Robles-Hernandez, L.; Gonzalez-Franco, A.C.; Infante-Ramirez, R. Pepper growth promotion and biocontrol against *Xanthomonas euvesicatoria* by *Bacillus cereus* and *Bacillus thuringiensis* formulations. *Peer J.* **2023**, *11*, e14633. [CrossRef] [PubMed]

46. Zhang, P.; Xie, G.; Wang, L.; Xing, Y. *Bacillus velezensis* BY6 promotes growth of poplar and improves resistance contributing to the biocontrol of *Armillaria solidipes*. *Microorganisms* **2022**, *10*, 2472. [CrossRef]

47. Al-Sadi, A.M. *Bipolaris sorokiniana*-induced black point, common root rot, and spot blotch diseases of wheat: A review. *Front. Cell. Infect. Microbiol.* **2021**, *11*, 584899. [CrossRef] [PubMed]

48. Zhou, H.; Ren, Z.H.; Zu, X.; Yu, X.Y.; Zhu, H.J.; Li, X.J.; Zhong, J.; Liu, E.M. Efficacy of plant growth-promoting bacteria *Bacillus cereus* YN917 for biocontrol of rice blast. *Front. Microbiol.* **2021**, *12*, 684888. [CrossRef]

49. Kang, X.; Zhang, W.; Cai, X.; Zhu, T.; Xue, Y.; Liu, C. *Bacillus velezensis* CC09: A potential 'Vaccine' for controlling wheat diseases. *Mol. Plant Microbe Interact.* **2018**, *31*, 623–632. [CrossRef] [PubMed]

50. Zhao, T.; Zhang, L.; Qi, C.; Bing, H.; Ling, L.; Cai, Y.; Guo, L.; Wang, X.; Zhao, J.; Xiang, W. A seed-endophytic bacterium NEAU-242-2: Isolation, identification, and potential as a biocontrol agent against *Bipolaris sorokiniana*. *Biol. Control* **2023**, *185*, 105312. [CrossRef]

51. Zhang, J.Q.; Zheng, A.K.; Gao, L.H.; Sun, Y.M.; Jiang, Z.Y. Isolation and identification of the main pathogens causing *Capsicum* wilt and screening of agricultural antagonisms. *J. Anhui Agric. Sci.* **2021**, *49*, 134–137.

52. Pei, D.; Zhang, Q.; Zhu, X.; Yao, X.; Zhang, L. The complete genome sequence resource of rhizospheric soil-derived *Bacillus velezensis* Yao, with biocontrol potential against *Fusarium solani*-induced pepper root rot. *Phytopathology* **2023**, *113*, 580–583. [CrossRef]

53. Cantoro, R.; Palazzini, J.M.; Yerkovich, N.; Miralles, D.J.; Chulze, S.N. *Bacillus velezensis* RC 218 as a biocontrol agent against *Fusarium graminearum*: Effect on penetration, growth and TRI5 expression in wheat spikes. *BioControl* **2020**, *66*, 259–270. [CrossRef]

54. Wang, S.; Sun, L.; Zhang, W.; Chi, F.; Hao, X.; Bian, J.Y.; Li, Y. *Bacillus velezensis* BM21, a potential and efficient biocontrol agent in control of corn stalk rot caused by *Fusarium graminearum*. *Egypt. J. Biol. Pest Control* **2020**, *30*, 9. [CrossRef]

55. Kim, J.A.; Song, J.S.; Kim, P.I.; Kim, D.H.; Kim, Y. *Bacillus velezensis* TSA32-1 as a promising agent for biocontrol of plant pathogenic fungi. *J. Fungi* **2022**, *8*, 1053. [CrossRef] [PubMed]

**Disclaimer/Publisher's Note:** The statements, opinions and data contained in all publications are solely those of the individual author(s) and contributor(s) and not of MDPI and/or the editor(s). MDPI and/or the editor(s) disclaim responsibility for any injury to people or property resulting from any ideas, methods, instructions or products referred to in the content.

*microorganisms*

MDPI

*Article*

# Plant-Associated Representatives of the *Bacillus cereus* Group Are a Rich Source of Antimicrobial Compounds

Joachim Vater [1], Le Thi Thanh Tam [2], Jennifer Jähne [1], Stefanie Herfort [1], Christian Blumenscheit [1], Andy Schneider [1], Pham Thi Luong [2], Le Thi Phuong Thao [2], Jochen Blom [3], Silke R. Klee [4], Thomas Schweder [5,6], Peter Lasch [1] and Rainer Borriss [5,7,*]

[1] Proteomics and Spectroscopy Unit (ZBS6), Center for Biological Threats and Special Pathogens, Robert Koch Institute, 13353 Berlin, Germany; vatermj@web.de (J.V.); jennifer.jaehne@web.de (J.J.); herforts@rki.de (S.H.); blumenscheitc@rki.de (C.B.); schneidera@rki.de (A.S.); laschp@rki.de (P.L.)

[2] Division of Pathology and Phyto-Immunology, Plant Protection Research Institute (PPRI), Duc Thang, Bac Tu Liem, Hanoi, Vietnam; thanhtamle10_2012@yahoo.co.uk (L.T.T.T.); luong3tb@gmail.com (P.T.L.); lethao0602@gmail.com (L.T.P.T.)

[3] Bioinformatics and Systems Biology, Justus-Liebig Universität Giessen, 35392 Giessen, Germany; jochen.blom@computational.bio.uni-giessen.de

[4] Highly Pathogenic Microorganisms Unit (ZBS2), Center for Biological Threats and Special Pathogens, Robert Koch Institute, 13353 Berlin, Germany; klees@rki.de

[5] Institute of Marine Biotechnology e.V. (IMaB), 17489 Greifswald, Germany; schweder@uni-greifswald.de

[6] Pharmaceutical Biotechnology, University of Greifswald, 17489 Greifswald, Germany

[7] Institute of Biology, Humboldt University Berlin, 10115 Berlin, Germany

[*] Correspondence: rainer.borriss@rz.hu-berlin.de

Citation: Vater, J.; Tam, L.T.T.; Jähne, J.; Herfort, S.; Blumenscheit, C.; Schneider, A.; Luong, P.T.; Thao, L.T.P.; Blom, J.; Klee, S.R.; et al. Plant-Associated Representatives of the *Bacillus cereus* Group Are a Rich Source of Antimicrobial Compounds. *Microorganisms* **2023**, *11*, 2677. https://doi.org/10.3390/microorganisms11112677

Academic Editor: Francesco Celandroni

Received: 8 October 2023
Revised: 26 October 2023
Accepted: 27 October 2023
Published: 31 October 2023

**Copyright:** © 2023 by the authors. Licensee MDPI, Basel, Switzerland. This article is an open access article distributed under the terms and conditions of the Creative Commons Attribution (CC BY) license (https://creativecommons.org/licenses/by/4.0/).

**Abstract:** Seventeen bacterial strains able to suppress plant pathogens have been isolated from healthy Vietnamese crop plants and taxonomically assigned as members of the *Bacillus cereus* group. In order to prove their potential as biocontrol agents, we perform a comprehensive analysis that included the whole-genome sequencing of selected strains and the mining for genes and gene clusters involved in the synthesis of endo- and exotoxins and secondary metabolites, such as antimicrobial peptides (AMPs). Kurstakin, thumolycin, and other AMPs were detected and characterized by different mass spectrometric methods, such as MALDI-TOF-MS and LIFT-MALDI-TOF/TOF fragment analysis. Based on their whole-genome sequences, the plant-associated isolates were assigned to the following species and subspecies: *B. cereus* subsp. *cereus* (6), *B. cereus* subsp. *bombysepticus* (5), *Bacillus tropicus* (2), and *Bacillus pacificus*. These three isolates represent novel genomospecies. Genes encoding entomopathogenic crystal and vegetative proteins were detected in *B. cereus* subsp. *bombysepticus* TK1. The in vitro assays revealed that many plant-associated isolates enhanced plant growth and suppressed plant pathogens. Our findings indicate that the plant-associated representatives of the *B. cereus* group are a rich source of putative antimicrobial compounds with potential in sustainable agriculture. However, the presence of virulence genes might restrict their application as biologicals in agriculture.

**Keywords:** *Bacillus cereus*; phylogenomics; DNA–DNA hybridization (ddH); biocontrol; plant growth promotion (PGP); biosynthesis gene cluster (BGC); kurstakin; thumolycin

## 1. Introduction

At present, the replacement of harmful chemical pesticides by environmentally friendly biological means is a pressing need in agriculture worldwide. Microbes, such as bacteria and fungi, have been proven to be promising candidates for the development of efficient agents useful in sustainable agriculture. At present, endospore-forming *Bacillus* spp. and Gram-negative *Pseudomonas* spp. are the most used constituents of bioformulations applied in biological plant protection. The main advantage of bioformulations based on *Bacillus* endospores is their longevity, which makes their stability comparable with that of chemical fungicides [1].

During a survey of plant-beneficial bacteria as part of the microbiome of different plant-associated sites, such as the rhizosphere, the tissues of the inner root, and the attached insect larvae of Vietnamese crop plants (black pepper, coffee and orange trees, brown mustard, and tomato), a number of Gram-positive, endospore-forming bacteria, able to suppress common plant pathogens, were isolated. Based on their draft genome sequences, the isolates were taxonomically assigned as being members of the *Bacillaceae* family, representing four main taxonomic groups: *Lysinibacillus* spp., *Brevibacillus* spp., the *Bacillus subtilis* species complex, and the *Bacillus cereus* group [2]. Our further studies revealed that, in contrast to *Lysinibacillus* sp., the plant-associated *Brevibacilli* harbored a multitude of interesting antimicrobial peptides with a strong potential to suppress phytopathogenic bacteria, fungi, and nematodes [3]. The *Bacillus velezensis* isolates TL7 and S1, members of the *B. subtilis* species complex, were identified in large-scale trials as the most promising candidates for developing efficient biocontrol agents [4]. In this study, we focus on the plant-associated isolates belonging to the *B. cereus* group in order to investigate their potential for biocontrol and plant growth promotion.

The *B. cereus* group, also known as *B. cereus sensu lato* (*s.l.*), comprises a steadily increasing number of species, but is still plagued by taxonomic inconsistencies. Several phenotypic traits important for taxonomic assignment, such as the synthesis of the anthrax toxin and capsule, entomopathogenic crystal proteins, and the synthesis of emetic toxins (cereulide), are plasmid-encoded and can be lost during strain evolution [5]. Well-known members of the *B. cereus* group are the human-pathogenic *B. anthracis*, the entomopathogenic *B. thuringiensis*, and the opportunistic pathogen *B. cereus sensu stricto* (*s.s.*). The three species are able to cause human diseases with different severity [6]. They are closely related and harbor very similar genome sequences, which do not necessarily justify their delineation in different species. Traditionally, they have been discriminated due to properties mainly encoded by extrachromosomal elements.

*B. anthracis* (risk group 3) was identified in 1876 by the German physician Robert Koch as the causative agent of anthrax [7], the first disease that was linked to a microbe. Its virulence is based on the ability to form exotoxins and a capsule, which are encoded by the plasmids pXO1 and pXO2 [8].

The plasmid-encoded production of the highly toxic cereulide is restricted to rare emetic *B. cereus* strains occurring in some foods, whilst the production of the diarrheal-inducing enterotoxins (hemolysin BL, HBL; non-hemolytic enterotoxin, NHE; and cytotoxin K, CytK) is common in *B. cereus s.l.* [9]. Due to the extreme stability of the cyclic dodecadepsipeptide celeuride, which withstands current food processing techniques, their emetic *B. cereus* producer strains are of particular concern for human health [9].

*B. thuringiensis* (Bt), isolated 1901 by Ishikawa as "B. sotto" from the silkworm, and some years later as *B. thuringiensis* by Berliner from the meal moth [10], is an insect pathogen that is successfully used in agriculture as a biopesticide based on the production of diverse crystal toxins, also known as δ-endotoxins [11].

However, the *B. cereus* taxonomy solely based on the presence of virulence plasmids with a specific function becomes increasingly questionable in light of the recent phylogenomic data. The occurrence of *B. cereus* strains containing pXO1-like plasmids [12] and of crystal protein-harboring *B. thuringiensis* strains, which are phylogenetically related to *B. anthracis* [13], make the identification of these species a difficult task. Moreover, the occurrence of virulence genes in the *B. cereus s.l.* species cannot be excluded. Therefore, *B. cereus s.l.* strains with potential for use in sustainable agriculture can be a risk for public health and need to be carefully checked for their genomic content, also in the case of their taxonomic delineation suggests them as a "safe" species.

In this study, we aim to elucidate the potential of plant-associated members of the *B. cereus* species complex as biological plant protection. In the utilization of these strains as biocontrol agents, their potential due to their rich biosynthetic potential, but also the risks connected with the presence of virulence genes, need to be considered. Genome-based phylogenetic analyses revealed that most of the isolates were clustered within two subspecies of the *B. cereus s.s.* species. Interestingly, the isolate *B. cereus* TK1 harbored genes encoding two different crystal proteins and one vegetative insectopathogenic toxin (Vip3). Genome mining for biosynthetic gene clusters probably involved in the synthesis of antimicrobial peptides (AMPs) and direct mass-spectrometric investigation of the synthesized AMPs revealed that the isolates are promising candidates for use in sustainable agriculture. In this context, the lipopeptides kurstakin and thumolycin seem to be of special importance. A special highlight of our research is resolving the primary structure of the plasmid-encoded thumolycin pentapeptide by LIFT-MALDI-TOF/TOF fragment analysis. The inhibiting action of the *B. cereus s.l.* isolates against plant pathogens was corroborated in direct assays performed with pathogenic oomycetes, fungi, and nematodes. Finally, the plant-growth-promoting activity of some of the isolates is demonstrated. Regardless of these promising results, we have to consider the risk for public health when the *B. cereus s.l.* isolates are applied as biological means in agriculture.

## 2. Materials and Methods

### 2.1. Strain Isolation and Cultivation

Isolation from Vietnamese healthy crop plants, and insects attached on plant surfaces (Table 1), and the purification of the strains were performed as described previously [2,4]. In order to exclude vegetative cells, the samples were heat-treated at 80 °C for 20 min. Only isolates able to suppress fungal plant pathogens were selected for further characterization [2]. The cultivation of the bacterial strains and DNA isolation have been previously described [14]. The *B. cereus* group strains were cultivated on Cereus Ident agar and Cereus-selective agar, as described in Section 2.4.

### 2.2. Reconstruction of the Complete Genomes

The genome sequences of *Bacillus cereus* A22, *B. cereus* A24, *B. cereus* HD1.4B, and *B. cereus* HD2.4 were reconstructed using a combined approach of two sequencing technologies that generated short paired-end reads and long reads. The resulting sequences were then used for hybrid assembly. Short-read sequencing has been previously described [14]. Long-read sequencing was conducted in house with the Oxford Nanopore MinION with the flowcell (R9.4.1), as described previously [3]. The quality of assemblies was assessed by determining the ratio of falsely trimmed proteins by using Ideel (https://github.com/phiweger/ideel, accessed on 1 November 2021). The genome coverage of the obtained contigs was 50× in average. Genome annotation and visualization was performed as described previously [3].

### 2.3. Screening of the Virulence Genes

The screening of virulence genes in whole-genome shotgun sequences (WGS) and complete genomes was performed by using a combined analysis of the PATRIC annotation system [15] and tblastN in the 17 genomes. The most characteristic genes from *B. anthracis* Vollum, including four genes of the pXO1 plasmid (*cya*, *lef*, *pag*A, and *rep*X) and six genes of the pXO2 plasmid (*cap*A, *cap*B, *cap*C, *cap*D, *cap*E, and *rep*S), were used as the reference sequences. The tblastN threshold for both similarity and coverage was >30%, and all BLAST results were cross-checked against the PATRIC annotation, available at the Bacterial and Viral Bioinformatics Resource Center, BV-BRC, https://www.bv-brc.org/, accessed on 1 November 2021.

The criteria for the presence of virulence plasmids were established as described by Liu et al. [16]. Sequences representing the different types of δ-endotoxins (Supplementary Table S1) were extracted from the NCBI data bank. Searches for the presence of genes encoding crystal proteins and toxins in the 17 plant-associated *B. cereus* genome sequences were performed with tblastN using the respective protein sequences as query.

**Table 1.** The plant-associated *B. cereus* group isolates and their collection sites. Two *B. cereus* genomosubspecies, A, and B, were distinguished (Figure 1). The crop plants used for isolating the strains were black pepper (*Piper nigrum*) trees, tomato (*Lycopersicon esculentum*) plants, orange (*Citrus sinensis*) trees, maize (*Zea mays*), and brown mustard (*Brassica juncea*). Samples obtained from the inner root tissues were obtained after the sterilization of the root surface [2].

| Accession | Size | G + C | Genes | | Sample | GTDB | Collection | | | |
|---|---|---|---|---|---|---|---|---|---|---|
| | bp | % | Coding | RNA | Name | Genomospecies | Plant | Organ | Site | Date |
| CP085501.1 | 5,268,018 | 35.7 | 5533 | 149 | A24 | *B. cereus* ssp. *A* | black pepper | root | Vietnam | 24 May 2018 |
| CP085506.1 | 5,183,312 | 34.9 | 5450 | 154 | HD2.4B | *B. cereus* ssp. *A* | tomato plants | rhizosphere | Vietnam, Hoai Duc, Ha Noi | 23 April 2019 |
| CP085510.1 | 5,183,870 | 34.9 | 5450 | 157 | HD1.4B | *B. cereus* ssp. *A* | tomato plant | rhizosphere | Vietnam, Hoai Duc, Ha Noi | 23 April 2019 |
| JABSVB000000000.1 | 5,796,358 | 34.8 | 5664 | 80 | HB3.1 | *B. cereus* ssp. *A* | orange tree | rhizosphere | Vietnam, Cao Phong, Hoa Binh | 17 April 2019 |
| VDDR00000000.1 | 5,071,716 | 35.4 | 5619 | 63 | A8 | *B. cereus* ssp. *A* | coffee tree | root | Vietnam | 9 May 2018 |
| VEPT00000000.1 | 5,319,678 | 35.3 | 5332 | 63 | A31 | *B. cereus* ssp. *A* | black pepper | root | Vietnam | 22 May 2018 |
| VEPS00000000.2 | 6,195,299 | 34.7 | 6254 | 78 | TK1 | *B. cereus* ssp. *B* | black pepper | rhizosphere | Vietnam | 9 May 2018 |
| CP085498.1 | 5,310,791 | 35.2 | 5640 | 143 | A22 | *B. cereus* ssp. *B* | coffee tree | root | Vietnam | 9 May 2018 |
| JABSVF000000000.1 | 5,869,336 | 34.8 | 5640 | 80 | M2.1B | *B. cereus* ssp. *B* | maize | rhizosphere | Vietnam, Phu An, Thanh Da | 6 December 2019 |
| VEPQ00000000.1 | 5,604,011 | 35.6 | 5466 | 63 | A42 | *B. cereus* ssp. *B* | black pepper | root | Vietnam | 12 May 2018 |
| VEPR00000000.1 | 5,594,617 | 35.6 | 5480 | 67 | SN4-3 | *B. cereus* ssp. *B* | maize | dead insect | Vietnam | 28 May 2018 |
| VEPV00000000.1 | 5,335,513 | 35.2 | 5339 | 58 | SN1 | *B. tropicus* ssp. *B* | *Ostrinia nubilalis* | | Vietnam | 28 May 2018 |
| VEPW00000000.1 | 5,958,606 | 35.8 | 5843 | 93 | CD3-2 | *B. tropicus* ssp. | brown mustard | rhizosphere | Vietnam | 28 May 2018 |
| VEPU00000000.1 | 5,443,801 | 35.2 | 5389 | 72 | SN4.1 | *B. pacificus* ssp. *B* | *Ostrinia nubilalis* | | Vietnam | 28 May 2018 |
| JABSVD000000000.1 | 5,695,940 | 35.1 | 5534 | 87 | HD1.3 | *Bacillus* sp. | tomato plant | rhizosphere | Vietnam, Hoai Duc, Ha Noi | 23 April 2019 |
| VEPX00000000.1 | 5,695,940 | 35.3 | 5136 | 65 | CD3-5 | *Bacillus* sp. | brown mustard | rhizosphere | Vietnam | 28 May 2018 |
| VEPY00000000.1 | 5,150,560 | 35.2 | 5175 | 61 | CD3-1a | *Bacillus* sp. | brown mustard | rhizosphere | Vietnam | 28 May 2018 |

### 2.4. Genotypic and Phenotypic Characterization of the Isolate B. cereus CD3-1a

In addition to the *B. anthracis* virulence genes mentioned above, the genome of strain *B. cereus* CD3-1a was also screened for presence of the four *B. anthracis*-specific prophage regions (dhp) described by Radnedge et al. [17]. These in silico analyses were complemented by real-time PCR assays targeting *pag*A, *cap*B, *rpo*B, and dhp61.183. Colony morphology was examined on Columbia blood agar, blood trimethoprim agar, Cereus Ident agar, and Cereus-selective agar [18].

### 2.5. Taxonomical Phylogeny Assessment

Species and subspecies delineation were performed using the Type (Strain) Genome Server (TYGS) platform [4]. Information on nomenclature was provided by the List of Prokaryotic names with Standing in Nomenclature (LPSN), available at https://lpsn. dsmz.de (accessed on 1 November 2021) [19]. The EDGAR3.0 pipeline [20] was used for elucidating taxonomic relationships as described previously [4].

### 2.6. Genome Mining

The in silico prediction of gene clusters involved in secondary metabolite synthesis was performed using the antiSMASH pipeline version 6 [21], the bioinformatic tool described by Bachmann and Ravel [22], and BAGEL4 [23].

### 2.7. Sample Preparation and Mass-Spectrometric Detection of the Bioactive Peptides

The bioactive compounds of the investigated *B. cereus s.l.* strains were detected and identified by MALDI-TOF MS, as outlined previously [24,25]. A Bruker Autoflex Speed TOF/TOF mass spectrometer (Bruker Daltonics; Bremen, Germany) was used with Smartbeam laser technology applying a 1 kHz frequency-triple Nd-YAG laser ($\lambda_{ex}$ = 355 nm). Samples (2 µL) of the colony surface extracts and culture supernatants were mixed with a 2 µL matrix solution (a saturated solution of $\alpha$-hydroxy-cinnamic acid in 50% aqueous ACN containing 0.1% TFA) spotted on the target, air-dried, and measured. Mass spectra were obtained by positive-ion detection in the reflector mode. The monoisotopic masses were observed. Parent ions were detected with a resolution of 10.000. The sequence analysis of peptide products was performed by MALDI-LIFT-TOF/TOF mass spectrometry in the laser induction decay (LID) mode [26]. The product ions in the LIFT-TOF/TOF fragment spectra were obtained with a resolution of 1000.

### 2.8. Antifungal, Nematocidal, and Plant-Growth-Promoting Activity Assays

Assays for activity against plant pathogens (oomycetes, fungi, and nematodes) were performed as previously described [4]. In brief, antifungal activities were assayed by placing agar plugs containing the respective fungi onto potato dextrose agar (PDA). The test bacteria were then streaked between the plugs, and the diameter of the fungal colonies as indicative for direct growth inhibition was recorded daily.

The bioassays for nematocidal activity were performed with *Caenorhabditis elegans* N2 and *Meloidogyne* sp., as described previously [4]. In the slow killing test, the nematodes were added to an agar-solidified growth medium containing the test bacteria and incubated for a period from 3 to 5 days at 25 °C. In the liquid fast killing test, overnight cultures of the test bacteria were transferred into 12-well plates containing the nematodes in a liquid M9 medium. The mortality of nematodes was defined as the ratio of dead (non-motile) nematodes to the total number of nematodes [4].

The root-knot nematode *Meloidogyne* sp. was isolated from the roots of infested pepper plants, according to Hooper et al. [27]. Tomato plantlets were grown in pots with sterilized alluvial Red River soil under subtropical climate conditions in the local greenhouse [4]. Test bacteria and second-stage juvenile (J2) nematodes were added to the pots two weeks after transplanting. Ten weeks after infesting with the nematodes, the number of knots in tomato plants was estimated [28].

Plant growth promotion assays were performed with *Arabidopsis thaliana* seedlings, as described previously [29]. Seven-day-old seedlings were dipped into a spore suspension of the test bacteria and transferred into a square Petri dish with a half-strength Murashige–Skoog medium solidified with 1% agar. After three weeks of incubation at 22 °C and a daily photoperiod of 14 h, the fresh weight of the plants was measured.

*2.9. Data Analysis*

The data obtained from the biocontrol and plant growth promotion experiments were analyzed using a one-factorial analysis of variance (ANOVA). The mean values were calculated from the results of the replicates ($n \geq 3$). The Fisher´s least significant difference (LSD) test was conducted as a post hoc test for estimating significant differences ($p \leq 0.05$) between the mean values as described previously [4].

*2.10. Gene Bank Accession Numbers of the Complete Genome Sequences*

*Bacillus cereus* A22 chromosome: CP085498.1, *Bacillus cereus* A22 plasmid P1: CP085499.1, *Bacillus cereus* A22 plasmid P2: CP085500.1, *Bacillus cereus* A24 chromosome: CP085501.1, *Bacillus cereus* A24 plasmid P1: CP085502.1, *Bacillus cereus* A24 plasmid P2: CP085503.1, *Bacillus* cereus HD1.4B chromosome: CP0855510.1, *Bacillus cereus* HD1.4B plasmid P1: CP085511.1, *Bacillus cereus* HD1.4B plasmid P2: CP085512.1, *Bacillus cereus* HD1.4B plasmid P3: CP085513.1, *Bacillus cereus* HD2.4 chromosome: CP0855506.1, *Bacillus cereus* HD2.4 plasmid P1: CP085507.1, *Bacillus cereus* HD2.4 plasmid P2: CP085508.1, *Bacillus cereus* HD2.4 plasmid P3: CP085509.1.

## 3. Results and Discussion

*3.1. Comparative Genome Analysis of the Isolates from Vietnamese Crop Plants Representing the Bacillus cereus s.l. Complex*

3.1.1. Genome-Based Species and Subspecies Delineation of the Plant-Associated Isolates Belonging to the *B. cereus* Group

Seventeen of the endospore-forming bacterial strains, isolated from Vietnamese crop plants and insects attached at their surface (Table 1), were previously assigned to the *Bacillus cereus s.l.* group [2]. All isolates displayed the typical features of *B. cereus*: they developed phospholipase C and hemolytic activity when cultivated on Cereus Ident and sheep blood agar plates. The phylogenetic tree obtained from the 16S rRNA sequences supported their previous taxonomic assignment as members of the *B. cereus s.l.* group, but possessed an average branch support of only 28.5% (Supplementary Figure S1), which is not sufficient for robust species delineation.

We used a whole-genome-based approach for the robust delineation of the taxonomic position of the plant-associated *B. cereus s.l.* isolates. The phylogenomic tree containing a total of 128 *B. cereus s.l.* genomes, mainly extracted from the NCBI data bank, yielded three main branches (1–3). Branch 3 was subdivided into clusters 3A and 3B. All of our isolates were distributed within the cluster 3B and were to be found related to the clusters formed by the type strains of *B. cereus*, *B. anthracis*, *B. tropicus*, and *B. pacificus* (Figure 1A).

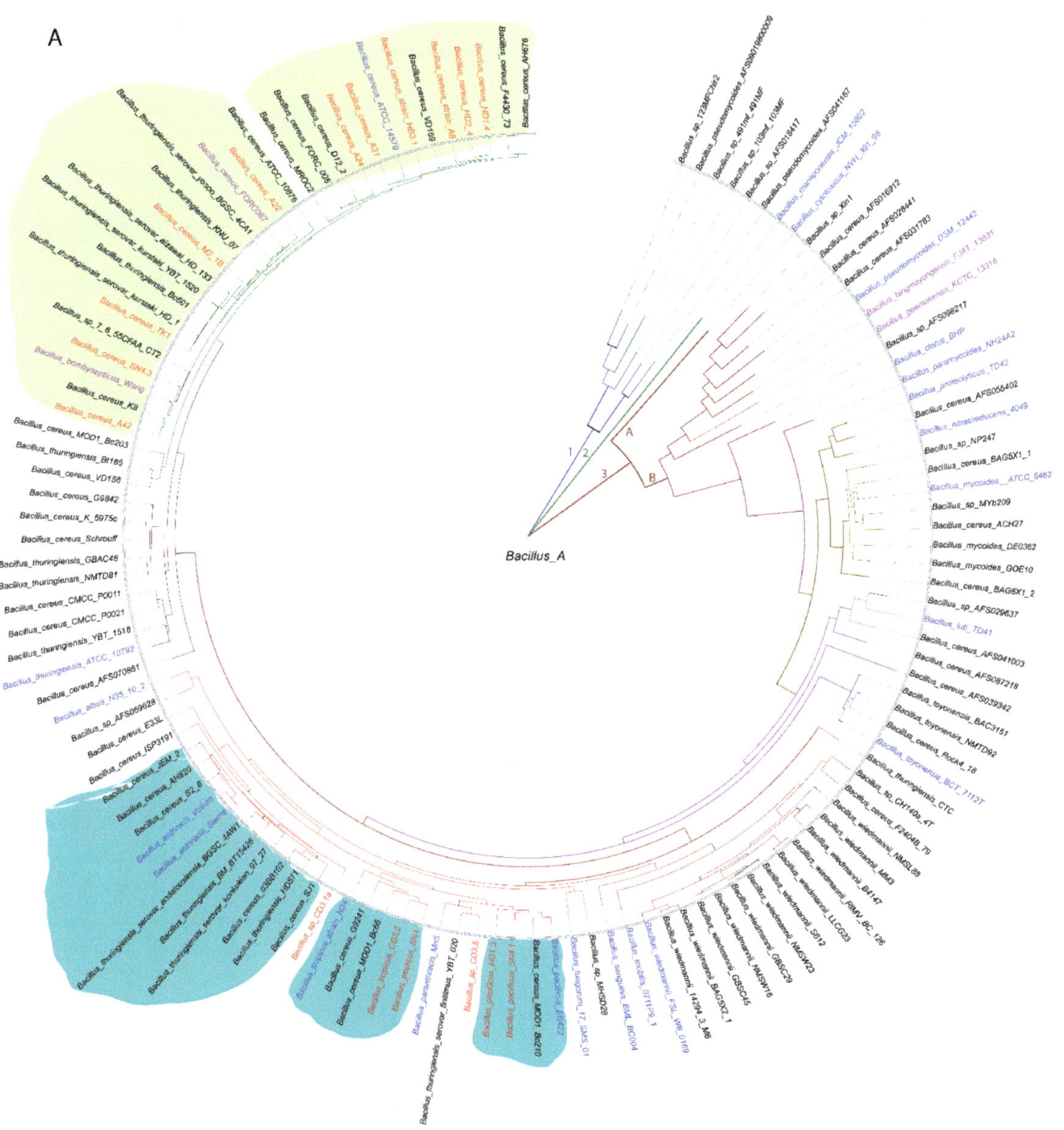

**Figure 1.** *Cont.*

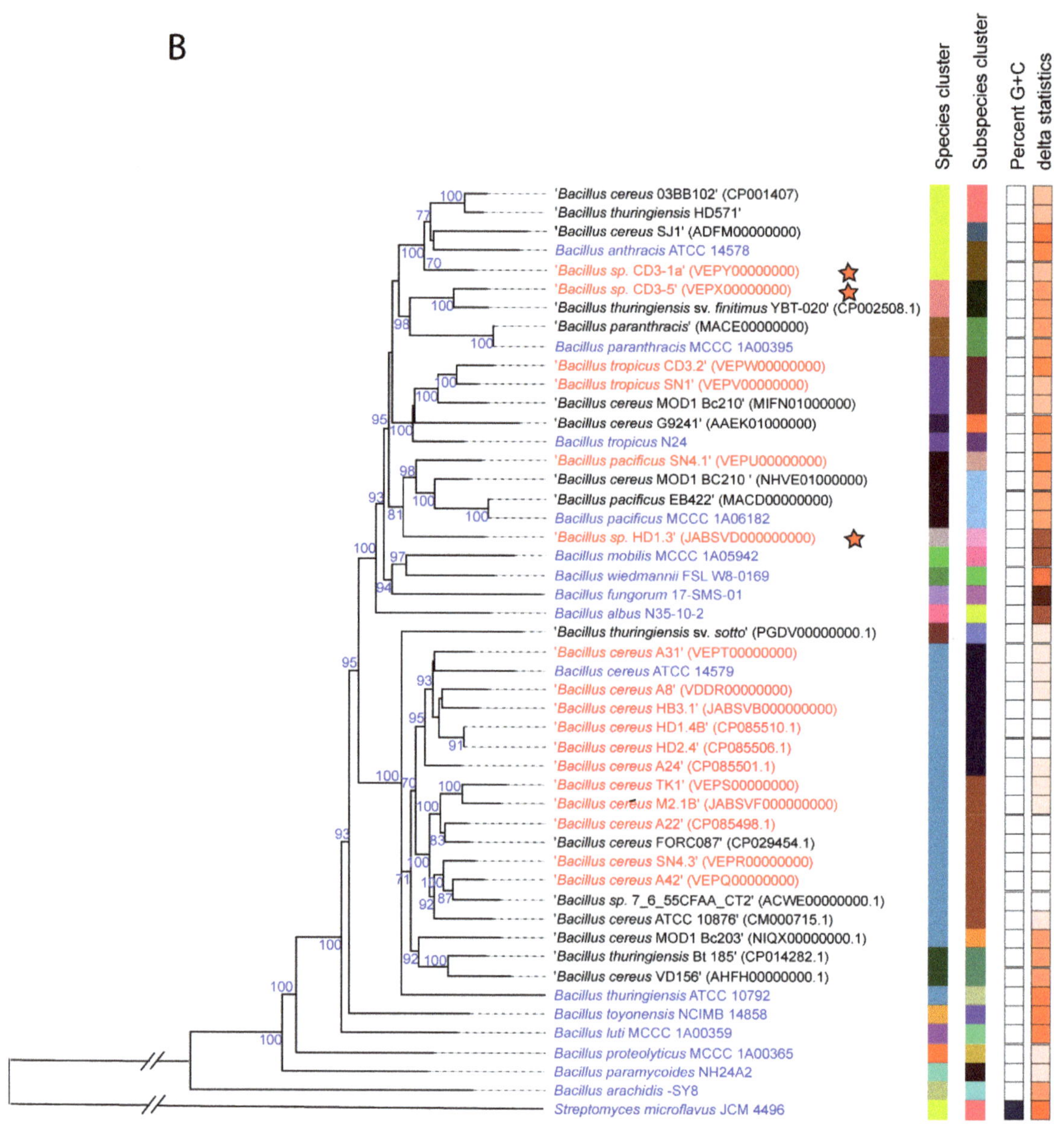

**Figure 1.** *Cont.*

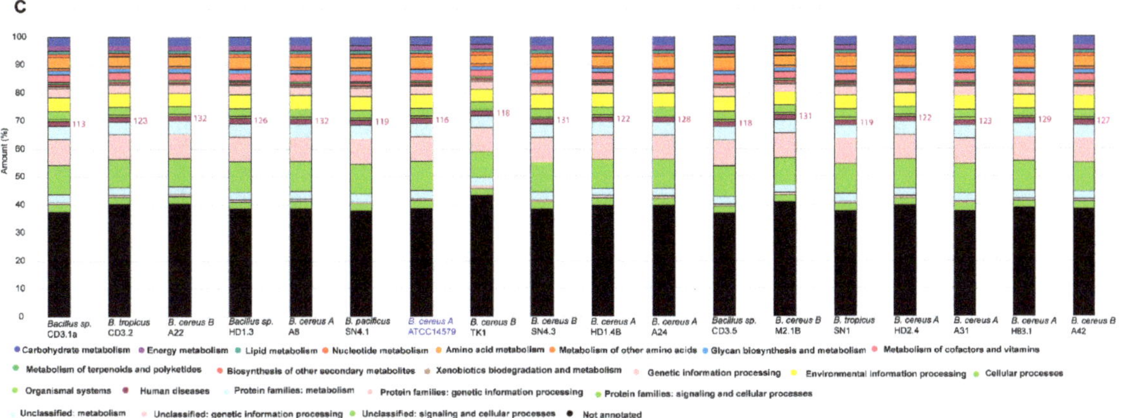



**Figure 1.** (**A**) Approximately maximum likelihood phylogenetic tree for 128 *Bacillus cereus* group genomes, calculated by EDGAR3.0 using the FastTree software (http://www.microbesonline.org/fasttree/ (accessed on 1 November 2021)). The unrooted tree was built out of a core of 1054 genes per genome. The core has 280,349 AA residues/bp per genome and 35,884,672 in total. *Bacillus_A* is a term used from GTDB (https://gtdb.ecogenomic.org (accessed on 1 November 2021)) for the genomospecies belonging to the *B. cereus* group. The tree is divided into three main branches (1–3). Cluster 3 is further divided into the subclusters 3A and 3B. The 21 type strains are labelled in blue letters. The isolates investigated in this study are indicated by red letters and all belong to the subcluster 3B. Related subclusters containing the plant-associated Vietnamese isolates are marked by irregular-colored fields. The ANI values are documented in Supplementary Figure S3. (**B**) *Bacillus cereus* tree inferred with FastMe 2.01 [30] from GBDP distances calculated from whole-proteome data using the Type (Strain) Genome Server TYGS (https://tygs.dsmz.de (accessed on 1 November 2021)). Analysis was performed using both maximum likelihood and maximum parsimony, with 16 type strains (blue letters) and 17 genome sequences obtained from the *Bacillus cereus* strains isolated from Vietnamese crop plants (red letters). In addition, 16 strains with similar proteomes obtained from the NCBI data bank were included, yielding a total of 49 proteomes. The branch lengths are scaled in terms of the GBDP distance formula $d_5$. Putative novel genomospecies were indicated by red stars. The numbers above the branches are GBDP pseudo-bootstrap support values > 60% from the replications, with an average branch support of 87.5%. The first two colored columns to the right of each name refer to the genome-based species and subspecies clusters, specified by dDDH cutoff values of 70% and 79%, respectively. (**C**) Functional KEGG category analysis of plant-associated *B. cereus* group isolates. The type strain *B. cereus* ATCC 14,579 was included in the analysis. The number of genes associated with human diseases is indicated. A total of 125,216 KEGG functional categories (including non-annotated sequences) were found in the selected 18 contigs.

Similar, but more detailed, results were obtained when using the Type (Strain) Genome server TYGS [31]. Our survey resulted in assigning six species and seven subspecies clusters (Figure 1B, Supplementary Figure S2). A total of 15 of the isolates were assigned to four valid species, *B. cereus* (11), *Bacillus pacificus* (1), *Bacillus tropicus* (2), and *Bacillus anthracis* (1). In the case of *B. cereus*, the dDDH values obtained after the comparison of 11 isolates with the type strain ATCC14579 exceeded the species cut off (>70%, Supplementary Table S2). At the genomic level, two subclusters were distinguished: six isolates, yielding dDDH values above the subspecies cutoff (>79%), represented the subspecies 'A' (*B. cereus* subsp. *cereus*), whilst five isolates showed dDDH values ranging from 72% to 74%, when compared with ATCC14579. The latter cluster formed a second subcluster 'B' together with *B. cereus* FORC087.1, which was clearly distinguished from subcluster 'A' (Supplementary Table S2). When the genomes of the members of the *B. cereus* subcluster 'B' were compared with the genome of *Bacillus bombysepticus* Wang [32], their dDDH values exceeded the subspecies cutoff (>79%, Supplementary Table S2). The direct comparison of FORC087 with *B. bombysepticus* Wang yielded a dDDH value of 87.8%, suggesting their close relationship. In the Genome Taxonomy Database, GTDB [33], the genomospecies *Bacillus_A bombysepticus* harbored 667 members, whilst the *B. cereus* genomospecies represented by *B. cereus* ATCC 14,579 harbored only 310 genomes (GTDB release 08-RS214, 28th April 2023).

Although *B. bombysepticus* is still not listed as a valid species in the List of Prokaryotic names with Standing in Nomenclature, LPSN [34], we propose to designate the *B. cereus* subcluster 'B' as genomosubspecies *B. cereus* subsp. *bombysepticus*, taking into account that the members of the '*bombysepticus* group' shared dDDH values above the species cutoff with *B. cereus* ATCC 14579.

The genomes of two other isolates, SN1 and CD3-2, were assigned, according to their dDDH and Fast ANI values, to *Bacillus tropicus*. However, when compared with the *B. tropicus* type strain N24, their dDDH and ANI values were below the subspecies cutoff, indicating that these isolates form the subcluster 'B' together with *B. cereus* MOD1 Bc210, distinct from the *B. tropicus* type strain (Figure 1B, Supplementary Table S2).

The isolate *Bacillus* sp. CD3-1a clustered together with the *B. anthracis* type strains. However, this species delineation appeared to be questionable, since we did not detect the *B. anthracis* virulence plasmids pXO1 and pXO2 in the draft genome of CD3-1a (see next section).

Two isolates, *Bacillus* sp. CD3-5 and *Bacillus* sp. HD1.3, although distantly related to *B. tropicus* and *B. pacificus*, could not be assigned to any species present in the TYGS database (17 April 2023) and might represent novel genomospecies (Figure 1B).

3.1.2. Occurrence of Virulence Genes Might Restrict the Application of *B. cereus s.l.* Isolates

The functional KEGG analysis revealed the presence of genes possibly involved in human disease in the genomes of all plant-associated *B. cereus s.l.* isolates (Figure 1C).

Within the *B. cereus* group, the occurrence of virulence factors, which are closely linked to disease symptoms [35,36], and of entomopathogenic Cry toxins [37] have been reported. In the past, these elements have been widely applied to the assignation of *B. anthracis*, *B. cereus*, and *B. thuringiensis*.

Since the genome sequence of the isolate CD3-1a formed a cluster together with the *B. anthracis* type strain ATCC 14,578 (Figure 1), we checked the CD3-1a draft genome for the presence of sequences of the characteristic anthrax toxin plasmids pXO1 and pXO2. No sequences similar to the genes encoding the Rep proteins RepX (pXO1) and RepS (pXO2) were detected in CD3-1a. Moreover, no sequences exhibiting a significant similarity with the anthrax genes of pXO1 (*cya*, *pagA*, and *lef*) and pXO2 (*capABCDE*) were found in CD3-1a and in the other plant-associated *B. cereus s.l.* isolates, excluding their taxonomic delineation as representative of the human-pathogenic *B. anthracis* species (Supplementary Table S3).

To distinguish "true" *B. anthracis* isolates from non-anthrax-causing representatives of the *B. cereus* group, a tblastN search within the genome sequence of the dhp chromosomal marker sequences, which indicate the presence of *B. anthracis*-specific prophages, was proposed [17]. None of the *B. anthracis*-specific dhp fragments could be detected in the CD3-1a genome. In addition, the real-time PCR amplification of the protective antigen *pagA* gene, the capsule *capB* gene, and the dhp61.183 gene (one of the prophage regions) using CD3-1a DNA was not achieved. A delayed amplification signal was observed for the *B. anthracis*-specific *rpoB* gene, which is known for non-anthrax strains of the *B. cereus* group [17]. In contrast to *B. anthracis*, but similar to the other isolates, CD3-1a was hemolytic when cultivated in Columbia blood agar or blood trimethoprim agar, and the genes encoding the hemolysin BL (HBL) toxin were present on the chromosome (Supplementary Table S3). The isolate also displayed phospholipase C and lecithinase activity, like the typical strains of the *B. cereus* group.

Interestingly, the isolate *B. pacificus* SN4.1 harbored a pXO1-like *repX* gene, and the isolate *B. tropicus* CD3.2 harbored a sequence resembling the pXO2-like *repS* gene in their draft genomes (Figure 2, Supplementary Table S3), suggesting that the *rep* genes characteristic for the pOX plasmids can occur in other members of the *B. cereus s.l.* species complex. This is in line with the previous findings of Liu et al. [16].

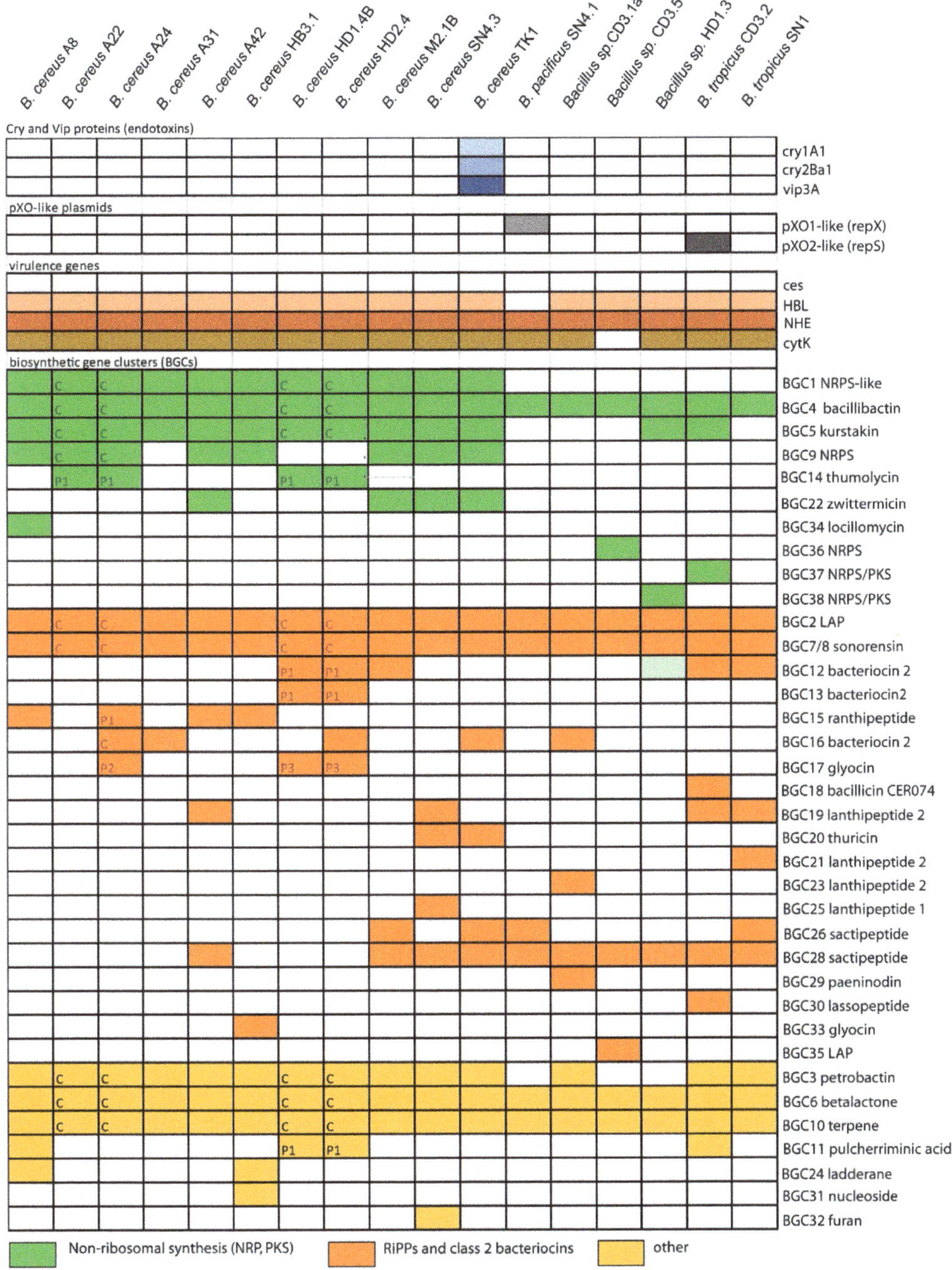


**Figure 2.** Occurrence of genes encoding entomopathogenic crystal (light blue) and vegetative (blue) proteins, the replication proteins RepS (grey) and RepX (dark grey) of *B. anthracis* pXO plasmids, and

virulence genes (HBL/NHE, *cyt*K). The gene cluster for synthesizing celeuride (ces) was not detected in any of the isolates. The 36 biosynthetic gene clusters (BGCs) encoding secondary metabolites in the 17 *B. cereus s.l.* isolates were identified by AntiSMASH6.0 and BAGEL4. The location of the BGC on either the chromosome (C) or the plasmids (P1, P3) is indicated when available. Further information is presented in Supplementary Tables S3–S8.

Next, we probed the *B. cereus s.l.* isolates for the presence of other virulence genes involved in the production of toxins responsible for foodborne diseases in human beings. Cereulide, the causative agent of the emetic syndrome [9], is known to be non-ribosomally synthesized by giant peptide synthetases encoded by the *ces* gene cluster. None of our isolates harbored this gene cluster (Figure 2), suggesting that the plant-associated isolates did not represent emetic *B. cereus* strains.

By contrast, the HBL/NHE enterotoxin operons encoding the non-hemolytic entero-toxin A (NHE) and the hemolysin component BL (HBL) [38] occurred in nearly all isolates, with one exception. *B. pacificus* SN4.1 harbored the genes responsible for synthesis of NHE enterotoxin, but not the genes for hemolysin synthesis (Supplementary Table S3). HBL and NHE are the causative agents of the diarrheal syndrome in human beings, which is caused by ingestion of vegetative cells and spores that produce enterotoxins in the small intestine [39].

Due to these findings, we cannot exclude that the plant-associated *B. cereus s.l.* isolates can cause the diarrheal syndrome in human beings. The application of plant-associated *B. cereus s.l.* strains in crop protection agents represents a possible risk for public health and should be considered with care.

### 3.1.3. Genes Encoding Insecticidal Proteins in *B. cereus* subsp. Bombysepticus TK1

Furthermore, we proved the occurrence of *cry* genes encoding entomocidal proteins (δ-endotoxins). Sequences that completely matched the crystal proteins Cry1A1 and Cry2Ba1 were detected in *B. cereus* subsp. *bombysepticus* TK1. The gene encoding the cytolytic CytK protein was detected in all *B. cereus* group isolates (Figure 2). The synthesis of δ-endotoxins is considered as a typical feature of *B. thuringiensis* [40]. However, in line with our results, Liu et al. [16] found that the ability to synthesize δ-endotoxins is widespread in different members of the *B. cereus* species complex. Thus, the presence or absence of *cry* genes cannot be considered to discriminate between the *B. cereus* and *B. thuringiensis* species.

In addition to Cry proteins, TK1 harbored a gene for the synthesis of the vegetative insecticidal protein, Vip3. Vip proteins are referred as second-generation insecticidal proteins. Vip3 proteins have insecticidal activity against Lepidopteran pests [41] and can be used for the management of various detrimental pests.

### 3.1.4. Plasmid-Encoded Virulence Genes and Biosynthetic Gene Clusters in *B. cereus* Isolates A22, A24, HD1.4B, and HD2.4

A first survey of the draft genome sequences for the presence of gene clusters encoding lipopeptides revealed that *B. cereus* ssp. *bombysepticus* A22 and the *B. cereus* ssp. *cereus* strains A24, HD1.4B, and HD2.4 harbored gene clusters similar to the thumolycin gene cluster, previously detected in *B. thuringiensi* BMB171 [42]. This finding prompted us to sequence completely the four strains using the nanopore sequencing technology (see Materials and Methods). The complete genomes consisted of one single chromosomal DNA molecule and extrachromosomal DNA elements, bearing the features of plasmid DNA (Figure 3). The chromosomes of all four isolates contained more than 5000 kb. The large P1 plasmids of A22 and A24 contained 480,744 bps and 471,669 bps, respectively. The smaller P2 plasmid of A22 contained 93,778 bps. Small plasmids not exceeding 12 kb were detected in A24 (P2), HD1.4B (P3), and HD2.4 (P3). The DNA elements found in HD1.4B and HD2.4 were nearly identical, suggesting that both isolates represented clones of the same strain. Both harbored one chromosome and three plasmids of nearly identical size and gene content (Supplementary Table S4). The presence of plasmid-specific Rep proteins in all extrachromosomal elements was corroborated by using the SEED and the RAST annotation system [43] (Supplementary Figure S4).

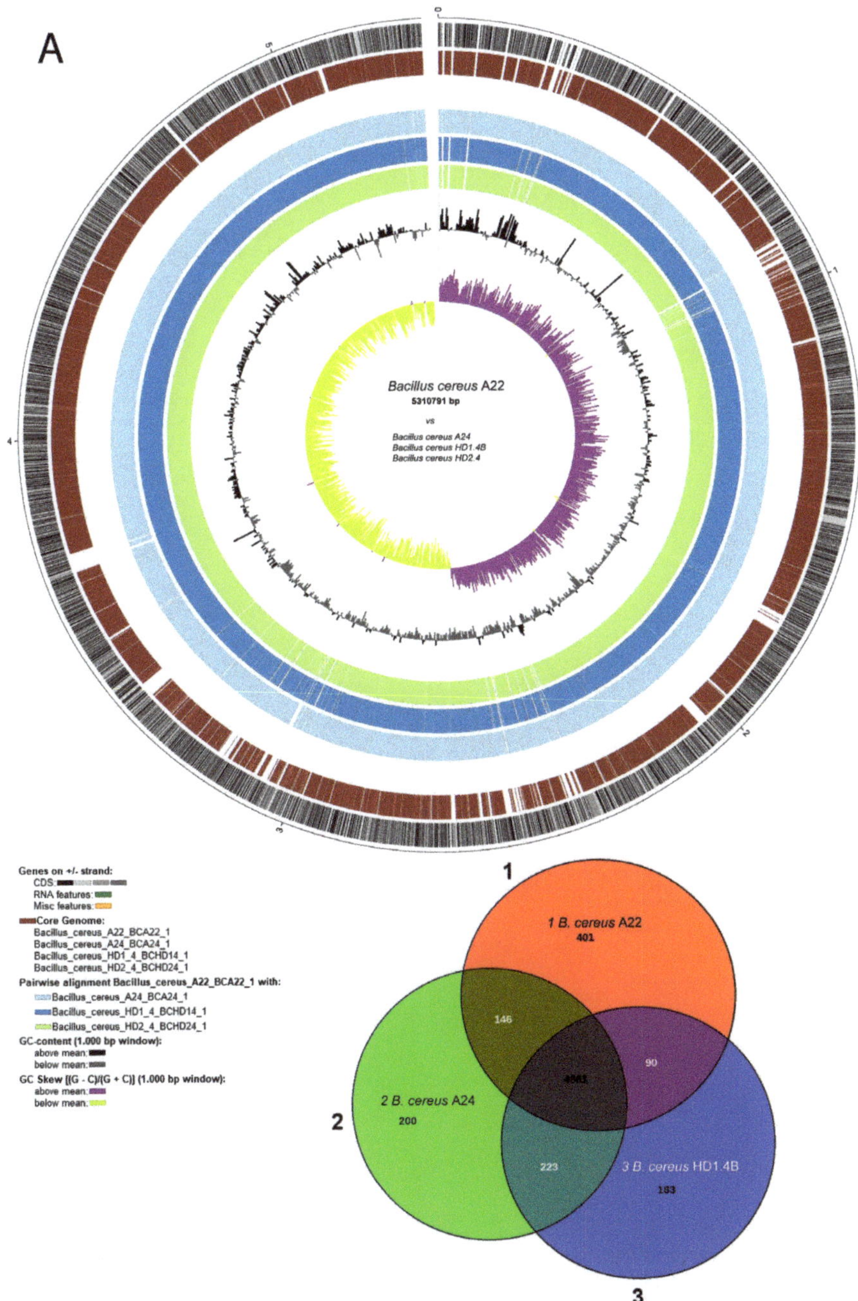


**Figure 3.** *Cont.*

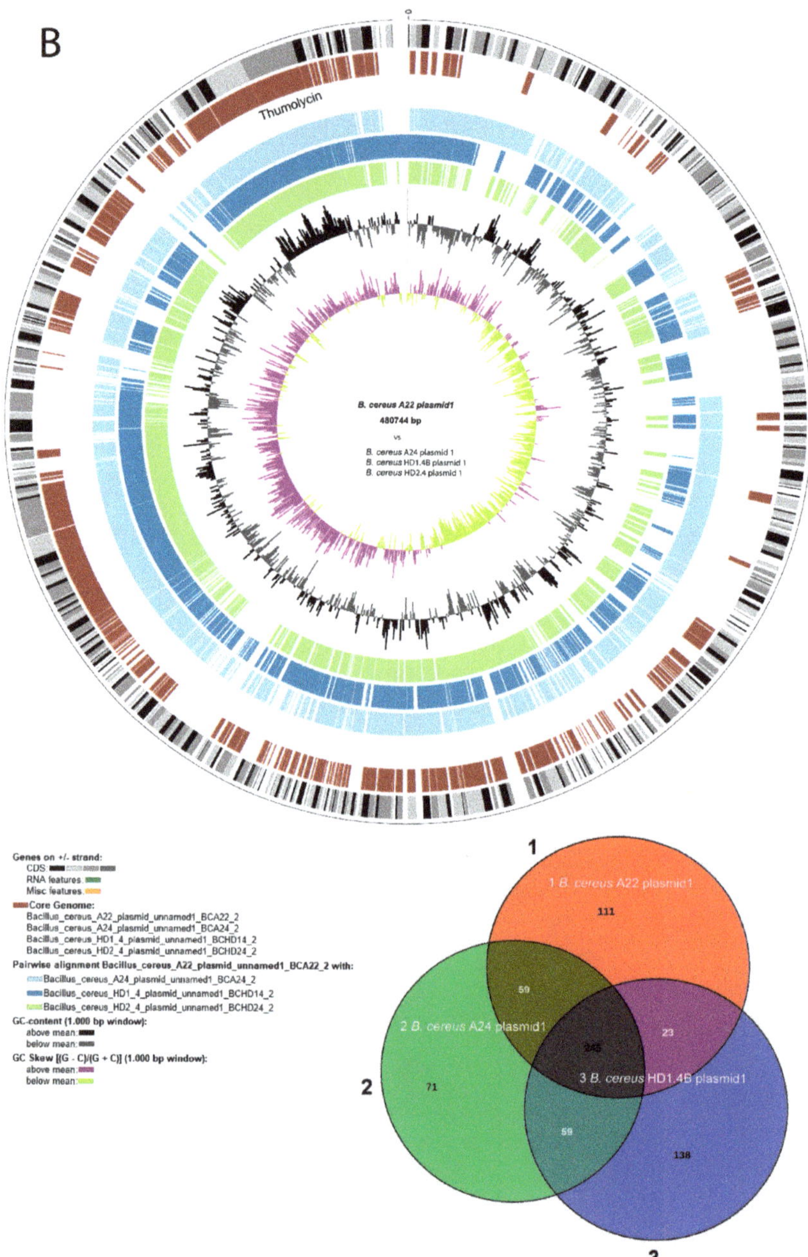



**Figure 3.** Circular plots of the *Bacillus cereus* A22 chromosome (**A**) and plasmid P1 (**B**) generated with Biocircos. The Venn diagrams below show the comparison of A22 with the chromosomes and plasmid P1 of *B. cereus* A24 and HD1.4B. From outer to inner circle: Genes (CDS) on +(1)/− strand (2); core genome, brown (3); GC content (1000 bp window) above the mean: black, below mean: grey (4); GC skew ((G-C)/(G+C)) (1000 bp window), above mean: purple, below mean: light green (5). The grey line within the inner circle shows deviations of the average GC content. The 30 kb thumolycin gene cluster is part of the core genome in all four plasmid P1 species.

Comparing *B. cereus* ssp. *cereus* A22 with the representatives of the *B. cereus bombysepticus* clade (*B. bombysepticus* Wang, FORC087, ATCC10876, A42, M2.1B, SN4.3, and TK1) yielded 113 singletons, including two catalases, 5-methylcytosine-specific restriction enzyme A, and other restriction enzymes. Comparing *B. cereus* ssp. *bombysepticus* A24 with the representatives of the *B. cereus cereus* clade (ATCC14579, A22, HD1.4B, HD2.4, A8, HB3.1, and A31) revealed 367 singletons, including proteins involved in conjugative transfer, the AlwI family type II restriction endonuclease, and urease subunits and associated proteins.

The potential virulence factor phosphatidylinositol-specific phospholipase C (PI-PLC), a characteristic marker of the *B. cereus* group [38], was encoded by the large P1 plasmids of A22, A24, HD1.4B, and HD2.4. PI-PLCs catalyze the cleavage of the membrane lipid phosphatidylinositol (PI), or its phosphorylated derivatives, to produce diacylglycerol (DAG) and the water-soluble head group, phosphorylated *myo*-inositol [44].

The annotation of the chromosomal elements detected in A22, A24, HD1.4B, and HD2.4 is summarized in Supplementary Table S4. Surprisingly, the plasmid P1 sequences from HD1.4B and HD2.4 harbored three genes with similarity to the NHE/HBL enterotoxin operons. These genes are known to be located on the chromosome. In fact, the chromosomes of the four isolates, including HB1.4 and HD2.4B, harbored the complete NHE/HBL gene set (Supplementary Figure S5).

Many metabolic features were found to be encoded by the large P1 plasmids harboring more than 500 coding genes. In addition to the thumolycin gene cluster, present in the large plasmids of all four isolates, two other BGCs encoding pulcheriminic acid and the bacteriocin cerein 7B precursor were found to be located in the large plasmids of HD1.4B and HD2.4.

Interestingly, in addition to the chromosomal-encoded type 1 restriction modification systems (RM) [45], type 1 RM gene clusters encoding the subunits M, S, and R were present in the 481 kb plasmid P1 of A24 and in the 94 kb plasmid P2 of A22. A fragmentary type III RM system consisting of RMIII helicase and the methylation subunit flanked by UvrD helicase and a transposase was detected in the P1 plasmid of HD2.4

A gene cluster detected in plasmid P1 of the *B. cereus* strains was similar to the anthrose BGC, previously described in *B. anthracis* Sterne [46]. The anthrose-containing oligosaccharide attached at the surface of the exosporium might contribute to enhanced survival rates under multiple stress conditions. Our results are in line with previous results of Dong et al. [47] demonstrating that the presence of anthrose-containing exosporia is not restricted to *B. anthracis*.

The complete operon for *myo*-inositol catabolism was detected in the large plasmids of all the four *B. cereus* isolates. The gene cluster was found to harbor the genes encoding the same enzymes as the *myo*-inositol operon previously detected in the chromosome of *B. subtilis* [48]. The presence of repeats and mobile elements in the flanking regions suggested that the operon might be acquired by horizontal gene transfer (Supplementary Figure S6).

*3.2. Genome Mining for Biosynthetic Gene Clusters (BGCs) Encoding Secondary Metabolites*

Antimicrobial compounds belong to structurally diverse groups of molecules, such as non-ribosomal peptides (NRPs) and polyketides (PKs), ribosomally synthesized and post-translationally modified (RiPPs) and unmodified (class 2 bacteriocins) peptides [49,50]. Genome mining using the software pipelines of antiSMASH6.0 [20], PKS/NRPS Analysis [21], and BAGEL4 [22] was performed with the genomes of all the Vietnamese isolates of the *B. cereus sensu lato* complex. The results were subsequently compared with the MIBiG database [51] in order to distinguish between characterized and uncharacterized BGCs. Our survey yielded a total of 209 BGCs representing 36 different gene clusters involved in the biosynthesis of secondary metabolites. Only a few, such as the siderophores petrobactin (BGC0000942) and bacillibactin (BGC0000309), zwittermicin (BGC0001059), locillomycin (BGC0001005), and pulcherrimic acid (BGC0002103), were listed in the MIBiG data bank. Two BGCs, kurstakin and thumolycin, although not listed in the MIBiG repository, were identified due to their similarity to genes already deposited in the NCBI data bank. Most of the BGCs exhibited no or only low similarity to the known BGCs present in the MIBiG data bank. Five BGCs encoding bacillibactin, RiPPs (2), betalactone (1), and terpene (1) were found conserved in all *B. cereus s.l.* isolates (Figure 2). An overview about the BGC species detected in the plant-associated *B. cereus* isolates is presented in Supplementary Table S5.

### 3.2.1. Non-Ribosomally Synthesized Antimicrobial Peptides (NRPs) and Polyketides (PKs)

NRPs are secondary metabolites that are synthesized through giant multi-modular peptide synthetases [52]. The complete *krs* gene cluster (BGC5) encoding the cyclic lipoheptapeptide kurstakin [53] was found to be widely distributed and occurs in the genomes of most *B. cereus sensu lato* isolates, except *B. pacificus* SN4.1, *B. tropicus* SN1, and *Bacillus* sp. CD3-1a and CD3.5. Kurstakin is responsible for biofilm formation [54]. Although kurstakin was present in the majority of the investigated isolates (13/17 strains), the kurstakin gene cluster, containing the genes *krsE, krsA, krsB, krsC, sfp,* and *krsD* (Figure 4A), is not listed in the MIBiG data bank.

The lipopeptide thumolycin, recently detected in *B. thuringiensis* BMB171, enabled the bacterium to develop a broad spectrum of antimicrobial and nematocidal activities [42]. Unfortunately, the structure of the lipopeptide is still not resolved. We detected the thumolycin (*tho*) gene cluster (BGC14) in plasmids of the *B. cereus* strains A22, A24, HD1.4B, and HD2.4. The genes of the *tho* cluster spanned around 30 kb. Two multimodular non-ribosomal peptide synthetases (ThoH and ThoI) synthesized a putative pentapeptide Orn-D-X-Leu/Ile-XS-Leu (Figure 4B). The *tho*C-, *tho*D-, and *tho*E-encoded proteins are probably involved in the synthesis of the fatty acid chain [42].

Fragments of the locillomycin gene cluster [55] (BGC34) were detected in *B. cereus* A8 (Supplementary Table S5). To the best of our knowledge, to date, the locillomycin gene cluster has been detected only in members of the *B.subtilis* species complex. The gene cluster for the synthesis of the catecholic iron siderophore bacillibactin, 2,3-dihydroxybenzoyl-Gly-Thr trimeric ester, has been previously reported in the genomes of *B. subtilis* [56] (BGC0000309) and *B. velezensis* FZB42 [57] (BGC0001185). Its non-ribosomal synthesis was found to be dependent on Sfp (phosphopantetheinyl transferase) [58]. BGCs with a similar structure as that of BGC0000309 and BGC0001185 were detected in all 17 *B. cereus sensu lato* isolates investigated in this study. Whilst the core structure of the bacillibactin transcription unit was well conserved, a *sfp* gene in the flanking region was identified as a unique feature for the *B. cereus* bacillibactin operon (Figure 5A). This is in contrast to the operon structure in the *B. subtilis* species complex, where the *sfp* gene is located in a more remote location, downstream flanking the surfactin operon [58].

The gene cluster for the synthesis of the aminopolyol antibiotic zwittermicin was detected in four *B. cereus* genomes. The highly polar zwittermicin A (ZmA) possesses antiprotist and antibacterial activities, and consists of numerous ethanolamine and glycolyl moieties flanked by N-terminal D-serine and an unusual amide generated from ß-ureidoalanine. The aminopolyol structure of the final product results from different processing events of the NRPS/PK hybrid precursor molecule (Figure 5B), in which a multitude of gene products of the *zma* gene cluster are involved [59].

In addition, numerous uncharacterized NRPs, PKS, and NRP/PKS hybrids were found (Supplementary Tables S5 and S6). A unique gene in *B. tropicus* CD3.2, located downstream of an uncharacterized NRP + PK cluster (BGC37, Supplementary Figure S7), encoded a putative necrose-inducing protein (NPP1 family) [60].

### 3.2.2. Gene Clusters Representing RiPPs and Bacteriocins

In contrast to polyketides and peptides, which are synthesized independently from ribosomes, numerous peptides with antimicrobial activity (bacteriocins) are synthesized by a ribosome-dependent mechanism. According to Zhao and Kuipers [49], several groups of ribosomally synthesized peptides (RiPPs) can be identified:

- Class I: post-translationally modified peptides smaller than 10 kDa.

- Class II: small (<10 Da), unmodified peptides with or without a leader sequence.
- Class III: peptides larger than 10 kDa.

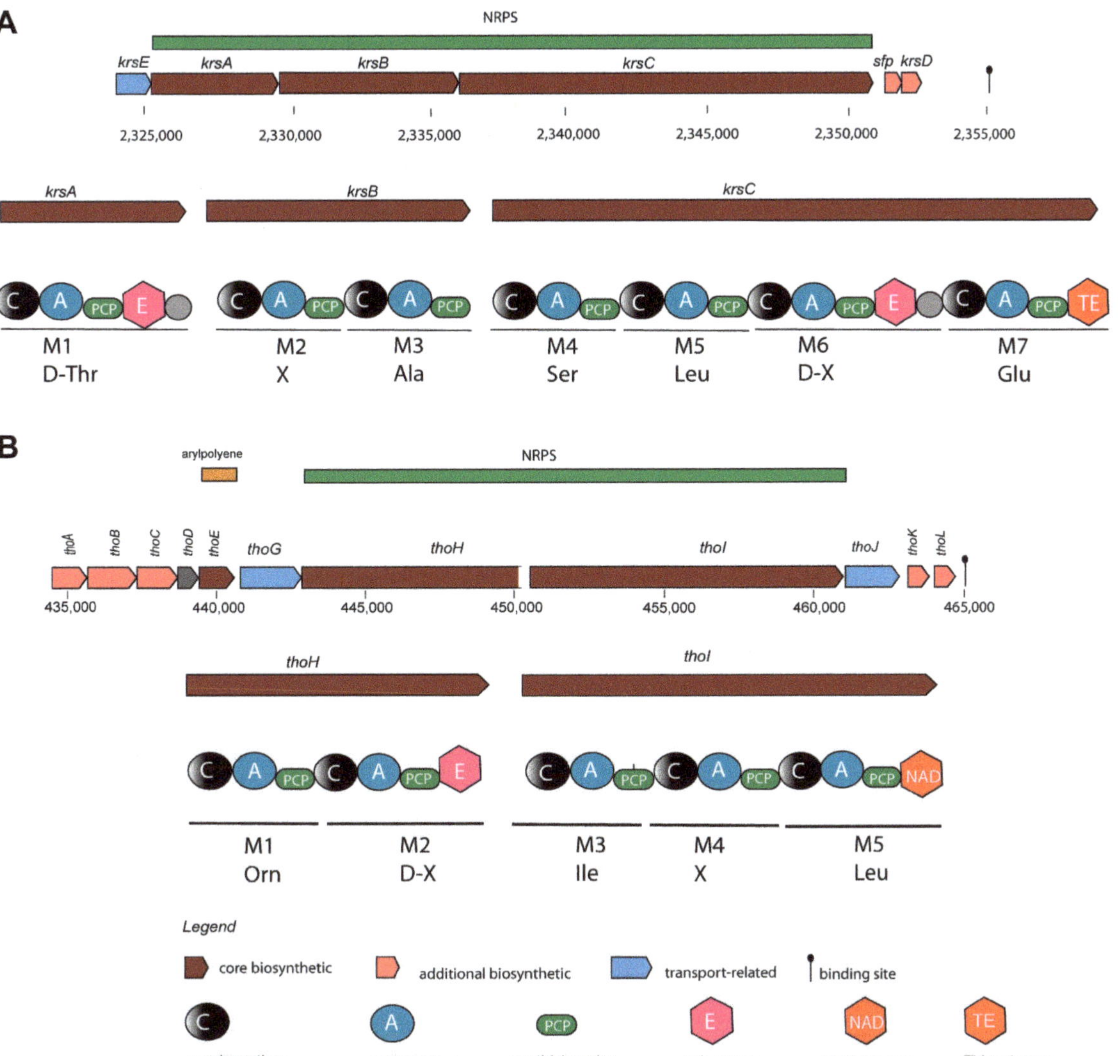



**Figure 4.** Gene cluster and domain structure involved in the non-ribosomal synthesis of cyclic lipopeptides in *B. cereus* A22. (**A**) The kurstakin (*krs*) gene cluster is located on the chromosome of A22 in the range of 2325–2355 kb. The amino acid sequence deduced from the adenylation domains was experimentally corrected and completed by LIFT-MALDI-TOF/TOF MS (see Table 2). (**B**) The thumolycin (*tho*) gene cluster resides in the *B. cereus* A22 plasmid 1 between 435 kb and 465 kb. The domain structure of ThoH and ThoI, including the amino acids deduced from their adenylation domains, is shown. The complete amino acid sequence determined by LIFT-MALDI -TOF/TOF MS is shown in Table 3. Further domains were detected in ThoC (A), ThoE (KS), ThoK (TE), and ThoL (ACPS).

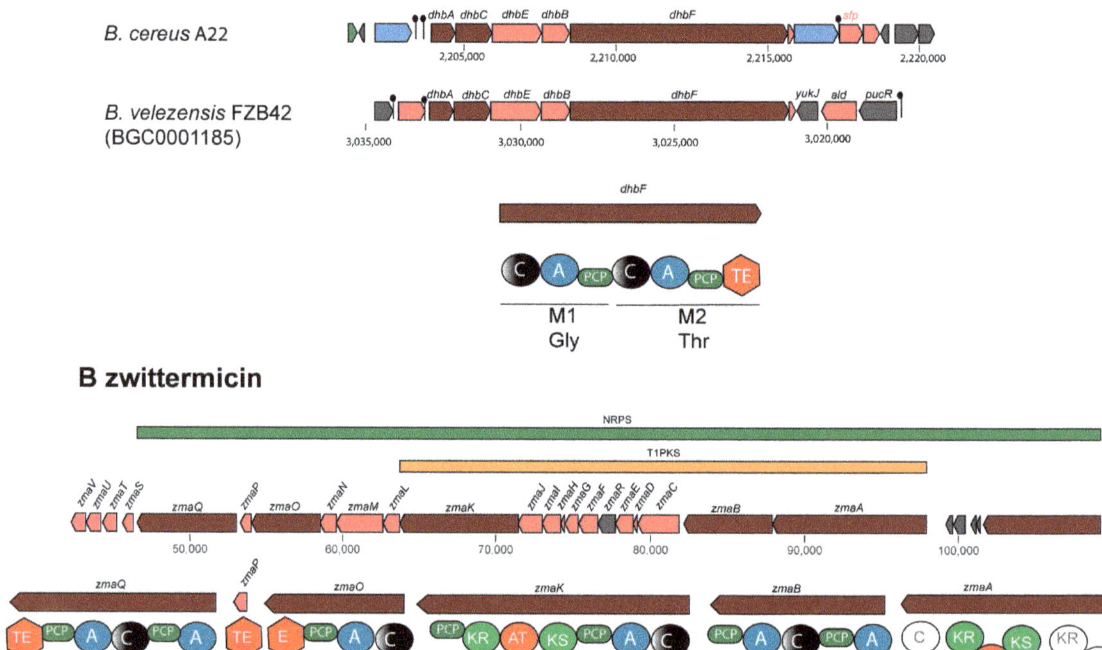


**Figure 5.** Gene clusters involved in the non-ribosomal synthesis of bacillibactin and zwittermicin. The siderophore gene cluster in the *B. cereus* A22 chromosome. (**A**) Comparison of the corresponding *B. velezensis* FZB42 gene cluster revealed the presence of the *sfp* gene downstream of the ballibactin transcription unit as a unique feature. (**B**) The gene cluster for the synthesis of the PK/NRP hybrid zwittermicin in *B. cereus* SN4-3.

RiPPs, such as lanthipeptides (class1 and class2), linear azol(in)e-containing peptides (LAPs), lassopeptides, sactipeptides, thiopeptides, and representatives of the class II unmodified bacteriocins, such as UviB peptides (holin-like proteins), were detected in the *B. cereus* group isolates applying the antiSMASH and BAGEL4 toolkits (Supplementary Figures S8 and S9). Many RiPP biosynthetic proteins recognize and bind their cognate precursor peptide through a domain known as the RiPP recognition element (RRE) [61]. The detection of RRE domains using antiSMASH-supported genome mining was helpful in identifying the BGCs involved in the synthesis of RiPPs, which did not contain known core peptide-encoding sequences [62].

A total of 19 different BGCs encoding RiPPs and unmodified class 2 bacteriocins were detected. Only six encoded precursor peptides with an apparent similarity to known

RiPPs: BGC20 (thuricin), BGC17 and BGC18 (sublancin/CER074), BGC29 (paeninodin), and BGC26 (thurincin H) (Supplementary Tables S5 and S7).

Antimicrobial lanthipeptides (lantibiotics) are post-translationally highly modified and contain the thioether amino acid lanthionine as well as several other modified amino acids [63]. LanA precursor peptides consist of an N-terminal leader peptide and a C-terminal core region. The first step in post-translational modification is the activation and elimination of water from the Ser and Thr residues forming dehydroalanine (DhA) and dehydrobutyrine (DhB), respectively. Then, ß-thioether cross-links are generated between the DhA, DhB, and the Cys residues. The modifying enzymes involved in the formation of the thioether link in class AI lanthipeptides are the dehydratase LanB and the cyclase LanC. The modification of A2 lanthipeptides is accomplished by LanM, containing the dehydratase and the cyclase domain in one protein. Classes A3 and A4 lanthipeptides are modified by LanKC and LanL, respectively [64]. We detected a BGC encoding a representative of the A1 lanthipeptides in *B. cereus* SN4.3 (BGC25). Four genes encoding precursor peptides similar to paenibacillin and subtilomycin were identified within BGC25. BGCs encoding A2 lantibiotics similar to plantaricin (BGC19), thuricin (BGC20), lichenicidin (BGC21), salivaricin (BGC23), and paenibacillin (BGC25) occurred in several *B. cereus* isolates (Supplementary Tables S5 and S7).

Gene clusters encoding LAPS were found to be widely distributed in *B. cereus* and related species. LAPS are characterized by the post-translational modification of the precursor peptide, yielding thiazol(in)e and (methyl)oxazol(in)e heterocycles. Modifying enzymes are the FMN-dependent dehydrogenase (SagB) and cyclodehdratase (SagC and YcaO) [65]. BGC2 (Supplementary Table S7) was identified as being member of the TOMM class (thiazole/oxazole-modified microcins), characterized by a gene cluster consisting of a cyclodehydratase gene and associated genes encoding dehydrogenase and a maturation protein. The core region of the TOMM precursor leader peptide contained a region enriched with Cys residues (BGC7/8), which is typically for the hetero-cycloanthracin/sonorensin family [66].

A glycocin-encoding gene cluster (BGC17) was detected in the plasmid P2 of *B. cereus* A24. Glyocins are defined as post-translationally glycosylated RiPPs with antimicrobial activity [65,67]. BGC17 resembled sublancin, which is an S-linked glycopeptide coding a SunS family peptide S-glcosyltransferase and a bacillicin CER074 peptide (BGC0001863) containing a glucose attached to a cysteine residue [68]. A second gene cluster (BGC33) harboring genes encoding a glycosyltransferase and a putative 75 aa precursor peptide was detected in *B. cereus* HB3.1 (Supplementary Table S5).

Lassopeptides are characterized by an N-terminal macrolactam ring threaded by the C-terminal tail. A cysteine protease B and a lactam synthetase C are necessary for the post-translational modification of the precursor peptide [69]. Two gene clusters, probably encoding lassopeptides, were identified. BGC29 harbored, in addition to a structural gene (*paeA*) for the synthesis of paeninodin lasso peptide [70], the genes for the synthesis of the essential components of post-translational modification, *paeB1* (PQQD family protein), *paeB2* (cysteine protease), and *paeC* (lactam ring closing cyclase). Four copies of the lasso precursor gene and split B1 and B2 genes were detected in BGC30 (Supplementary Table S7).

Two gene clusters (BGC26 and BGC28), encoding the radical S-adenosylmethionine (rSAM) enzyme, necessary for the post-translational modification of sactipeptides, occurred in the representatives of the *B. cereus* group. A well-known representative of sactipeptides is subtilosin A (SboA), synthesized by *Bacillus subtilis*. The rSAM enzyme (AlbA) catalyzes the linkage of a thiol with an $\alpha$-carbon of a functional amino acid residue [71]. BGC26 harbored genes involved in the synthesis and rSAM-dependent modification of a thurincin H-like precursor peptide. A gene encoding a protein containing an N-terminal radical SAM domain (pfam04055) and a C-terminal pfam08756 domain with a CxCxxxxC motif (BmbF) was detected in BGC28. In contrast to *B. subtilis*, the YfkA and YfkB regions, originally reported as separate ORFs in *B. subtilis*, were found fused in the *B. cereus* gene cluster (Supplementary Table S7).

Like the structurally related sactipeptides, the thioether linkage in ranthipeptides is generated via a radical-initiated mechanism. However, ranthipeptides do not contain $\alpha$-carbon links and were recently designated as non-$\alpha$ thioether peptides [72]. The ranthipeptide gene cluster (BGC15) detected in four *B. cereus* isolates harbored a gene encoding the rSAM protein belonging to the MoaA/NifB/PqqE/SkfB superfamily (Supplementary Table S7).

A gene cluster (BGC7/8), involved in the synthesis and modification of an 82 aa precursor thiopeptide belonging to the heterocycloanthracin/sonorensin family [73], occurred in all *B. cereus* group isolates. Its C-terminal region contained an extended repeat region with Cys at every third residue (Supplementary Table S7).

Two different subclasses of bacteriocin class II peptides were detected: the holin-like BhlA encoding genes (BGC12 and BGC16) and a cluster (BGC13) harboring a gene encoding a cerein-like prepeptide belonging to the Blp family. Similar as in lanthipeptides, the Blp family prepeptides are characterized by a conserved GlyGly processing site between the N-terminal leader and the C-terminal core peptide region [74]. The BhlA holin of *Bacillus pumilus* causes bacterial death by cell membrane disruption [75]. In addition to the genes encoding leaderless BhlA peptides, the holin gene clusters harbored genes encoding muramidases (GH25 glycosyl hydrolases) that hydrolyze the peptidoglycan cell wall (Supplementary Table S7, Supplementary Figures S8 and S9).

### 3.2.3. Other Antimicrobial Secondary Metabolites

Seven BGCs encoding other antimicrobial secondary metabolites were identified in the plant-associated *B. cereus* isolates (Supplementary Figure S7, Supplementary Table S8). Three of them, BGC3 (=BGC0000942, petrobactin), BGC11 (=BGCBGC0002103, pulcherriminic acid), and BGC32 (=BGC0000914, furan), were similar to the known BGCs deposited in the MiBIG data bank.

The *asbABCDEF* gene cluster is responsible for the biosynthesis of petrobactin, a catecholate siderophore that functions in both iron acquisition and virulence [76]. We detected the petrobactin gene cluster in the genomes of 14 isolates. Only *B. pacificus* SN4-1 and HD1-3 and *Bacillus* sp. CD3.5 did not harbor the BGC0000942 cluster, which is common in most representatives of the *B. cereus* group [77].

Pulcherriminic acid is a cyclic dipeptide able to chelate $Fe^{3+}$ [78]. Due to its high affinity to Fe ions, *Bacillus* strains producing pulcherriminic acid compete successfully with other microorganisms in low iron environments. A gene cluster similar to the pulcherriminic acid synthesis cluster in *B. subtilis* (BGC0000914) was detected in the *B. cereus* HD1.4B and HD2.4 plasmid sequences (Supplementary Table S4) and in the draft genomes of *B. cereus* A8 and *B. tropicus* CD3-2.

Several genes of BGC32, possibly involved in the synthesis of a furan-like metabolite, showed a striking similarity to the methylenomycin A gene cluster in *Streptomyces coelicolor* [79].

Four BGCs did not show similarity to any characterized biosynthetic gene clusters. BGC6 possibly encoded ß-lactone-harbored genes with similarity to the genes flanking the plipastatin BGC in *B. subtilis*. BGC24 contained several genes of the carbohydrate metabolism, probably involved in the synthesis of ladderane. BGC10 and BGC31 encoded enzymes for the synthesis of terpene and nucleoside metabolites, respectively.

### 3.3. Detection of Bioactive Peptides by MALDI-TOF Mass Spectrometry

The genome mining data summarized in Figure 2 indicate the presence of BGCs at the genomic level. However, the real biosynthetic capacity of the investigated isolates can only be verified by the isolation and structural analysis of the compounds actually produced. In Figure 6, we demonstrate the production of the non-ribosomally formed secondary metabolites of strain A22 as a representative for *B. cereus* detected by MALDI-TOF MS. As an overview, Figure 6A–C show mass spectra for the compounds found in a surface extract of A22 taken from cell materials grown on agar plates in the Landy medium for 48 h. Two prominent products were observed. Figure 6B shows the mass peaks for two kurstakins with chain lengths of their fatty acid component of 12 and 13 carbon atoms, respectively. The following mass data were found: C12-kurstakins: $[M + H,Na,K]^+$ = 892,5/914,5/930.5 Da; C13-kurstakins: $[M + H,Na,K]^+$ = 906.5/928.5/944.5 Da. In addition, as yet unknown compounds with the mass numbers of 1051.8 and 1065.9 were found, which dominate the MALDI-TOF mass spectrum of the surface extract in Figure 6A. Presumably, there are two isomers that differ by a methylene group (Figure 6C). This compound cannot be correlated to any of the BGCs of the antiSMASH profile of strain A22.

Figure 6D–F show the mass spectra of the products of A22 released into the culture broth for growth in liquid cultures for 48 in the Landy medium. Here, the arylpolyene lipopeptide thumolycin and both siderophores bacillibactin and petrobactin were detected. Figure 6E exhibits the mass peaks of thumolycin ($m/z$ = 697.8) and petrobactin ($m/z$ = 720.2). A relatively small part of the kurstakins was released into the culture filtrate, while the main part remained attached at the outer surface of A22. In Figure 6F, kurstakin mass peaks ($m/z$ = 906.8/930.8 and 944.8) overlap with those of the siderophore bacillibactin $[M + H,Na,K]+$ = 883.6/905.6/921 Da. These results demonstrate that kurstakins were predominantly found attached to the outer surface of *B. cereus* cells, while thumolycin and both siderophores bacillibacin and petrobactin were released into the culture medium. Similar profiles were obtained for strains A24, HD1.4B, and HD2.4. All other investigated *B, cereus* isolates did not produce thumolycin.

The structure of both lipopeptide products of strain A22, kurstakin and thumolycin, were investigated in detail by LIFT-MALDI-TOF/TOF fragment analysis [26]. Table 2 shows the sequence determination of C13 kurstakin with a parent ion $[M + H]+$ = 906.504 Da derived from product ions obtained by LIFT-MALDI-TOF/TOF fragment ion spectra. In Table 3, the structure of this compound is modelled from the nearest neighbor relationships using di-, tri-, and tetrapeptide fragments.

**Table 2.** Mass spectrometric sequence determination of C13-kurstakin produced by *B. cereus* A22 with a parent ion $[M + H]^+$ = 906.504 derived from product ion patterns obtained by LIFT-MALDI-TOF/TOF fragment ion spectra. FA: fatty acid component.

| $b_n$-$H_2O$ (*found*) | - | 280.123 | 337.149 | 408.107 | 495.159 | 632.284 | 760.445 | - |
|---|---|---|---|---|---|---|---|---|
| $b_n$ (*found*) | 197.030 | 298.133 | 355.112 | - | - | - | 778.446 | 906.504 |
| $b_n$ (*calc.*) | 197.190 | 298.238 | 355.259 | 426.296 | 513.328 | 650.387 | 778.446 | 906.504 |
| | **C13-FA** | **Thr** (1) | **Gly** (2) | **Ala** (3) | **Ser** (4) | **His** (5) | **Gln** (6) | **Gln** (7) |
| $y_n$ (*calc.*) | 906.502 | 710.322 | 609.275 | 552.253 | 481.216 | 394.184 | 257.125 | 129.066 |
| $y_n$ (*found*) | 906.504 | 710.266 | 609.163 | 552.133 | 481.106 | 394.084 | 257.050 | 129.066 |
| $y_n$-$H_2O$ (*found*) | - | 692.185 | - | - | 463.103 | - | - | - |

Using the same technique, we investigated thumolycin, which is a combination of a pentapeptide attached to a yet unknown arylpolyene lipidic residue. By LIFT-MALDI-TOF/TOF fragment analysis, we obtained the complete sequence of the pentapeptide part for the first time, which is compatible with the initial results from Zheng et al. [42] and the module organization of the corresponding BGC derived from antiSMASH 6.0 genome mining. The structure of this pentapeptide is shown in Table 4.

In summary, by MALDI-TOF mass spectrometry, we detected all compounds produced by *B. cereus* strains non-ribosomally. The investigation of the RiPPs, such as lanthipeptides, sactipeptides, and bacteriocins, is in still progress.

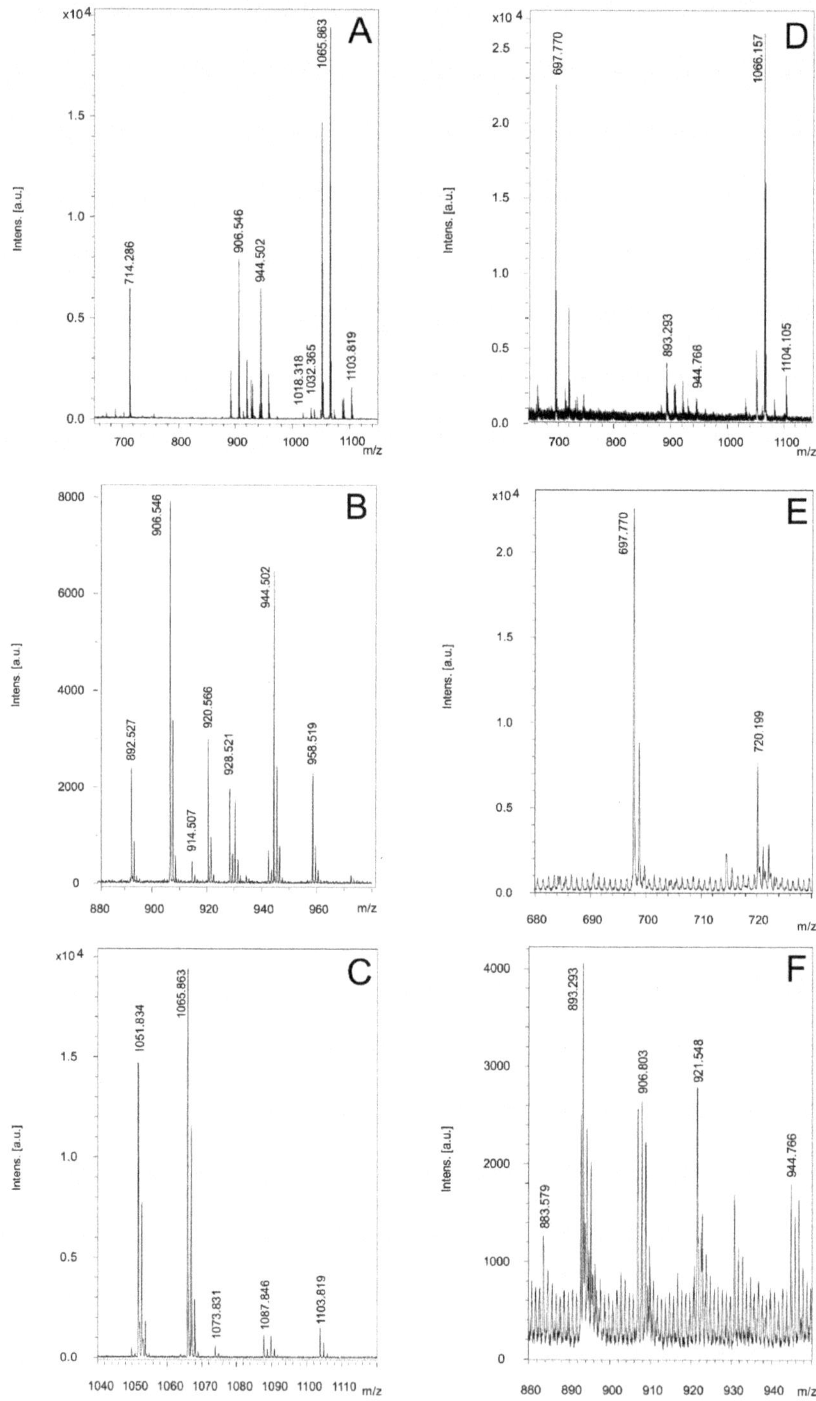

**Figure 6.** Bioactive compounds produced by *B. cereus* A22. (**A–C**) Compounds detected on the surface extracts of this strain. (**A**) MALDI-TOF mass spectrum of a surface extract of strain A22 grown on agar

plates in the Landy medium in the mass range of *m/z* = 650–1150. (**B**) Kurstakins observed in the range of *m/z* = 880–980. (**C**) Detection of a yet unknown prominent product with mass numbers of 1051.83 and 1065.86 Da. (**D–F**) Compounds found in the culture filtrate of strain A22 grown in the Landy medium for 48 h. (**D**) MALDI-TOF mass spectrum of a culture filtrate of strain A22 grown in the Landy medium for 48 h in the mass range of *m/z* = 690–1120. (**E**) Detection of the arylpolyene-lipopeptide thumolycin and the siderophore petrobactin with mass numbers of 697.77 and 720.20. Da. (**F**) Kurstakins and the siderophore bacillibactin ([M + H;K] $^{+}$ = 883.58 and 921.55 Da) found in the mass range of *m/z* = 880–950.

**Table 3.** Modeling of the structure of C13-kurstakin (*m/z* = 906.5) by nearest-neighbor relationships obtained by MALDI-TOF MS.

| (A) Dipeptide fragments | *m/z* | *m/z* |
|---|---|---|
| | Calc. | found |
| C13-FA-Thr | 298.238 | 298.141/280.129 |
| Thr-Gly | 159.077 | 159.007/141.022 |
| Gly-Ala | 129.066 | 129.016 |
| Ala-Ser | 159.077 | 159.007/141.022 |
| Ser-His | 225.099 | 225.034/207.019 |
| His-Gln | 266.125 | 266.060 |
| Gln-Gln | 257.125 | 257.050 |
| (B) Tripeptide fragments | | |
| C13-FA-Thr-Gly | 341.244 | 341.170/323.167 |
| Thr-Gly-Ala | 230.114 | 230.026/212.025 |
| Gly-Ala-Ser | 216.098 | 216.024/198.022 |
| Ala-Ser-His | 296.136 | -/278.050 |
| Ser-His-Gln | 353.157 | 353.084/335.070 |
| His-Gln-Gln | 394.184 | 394.101/335.070 |
| (C) Tetrapeptide fragments | | |
| C13-FA-Thr-Gly-Ala | 426.296 | -/408.121 |
| Thr-Gly-Ala-Ser | 317.146 | 317.057/299.070 |
| Gly-Ala-Ser-His | 353.157 | 353.084/335.070 |
| Ala-Ser-His-Gln | 424.194 | - |
| Ser-His-Gln-Gln | 481.216 | 481.120/463.120 |

*3.4. B. cereus s.l. Strains Suppressed Plant Pathogens and Promoted Plant Growth*

3.4.1. Antifungal and Nematocidal Activity

Our in vitro bioassays revealed that only few of the *B. cereus* group isolates efficiently inhibited phytopathogenic fungi and nematodes. Antifungal activity was examined in vitro using *Fusarium oxysporum*, known for causing fusarium wilt disease [80], and *Phytophthora palmivora*, one of the most detrimental plant pathogens in Vietnam [81]. Whilst most of the isolates did not significantly suppress the growth of the pathogenic fungi and oomycetes, a strong antimicrobial activity was exerted by *B. cereus* HB3.1 and *Bacillus* sp. HD.3 (Figure 7).

**Table 4.** Sequence of the pentapeptide part of the lipopeptide thumolycin derived from the product ion pattern obtained by LIFT-MALDI-TOF/TOF fragment ion spectra. The yet unknown arylpolyene lipid part of thumolycin of unknown length is linked to the Orn residue. The amino acids Orn, Ile, and Leu were also predicted by their adenylation domain sequences from genome mining using antiSMASH 6.0 (Figure 3B).

| b ions → | | | | | |
|---|---|---|---|---|---|
| $b_n$ *(found)* | - | 216.17 | 328.25 | 457.38 | (p)?? |
| $b_n$ *(calc.)* | 115.09 | 216.14 | 329.22 | 457.28 | (P)570.36 |
| | **Orn** | **Thr** | **Ile** | **Gln** | **Leu** |
| | (1) | (2) | (3) | (4) | (5) |
| $y_n$ *(calc.)* | (p)570.36 | 456.28 | 355.24 | 242.15 | 114.09 |
| $y_n$ *(found)* | **(p)??** | **456.39** | **355.25** | **242.17** | - |
| | | | | | <-- y ions |

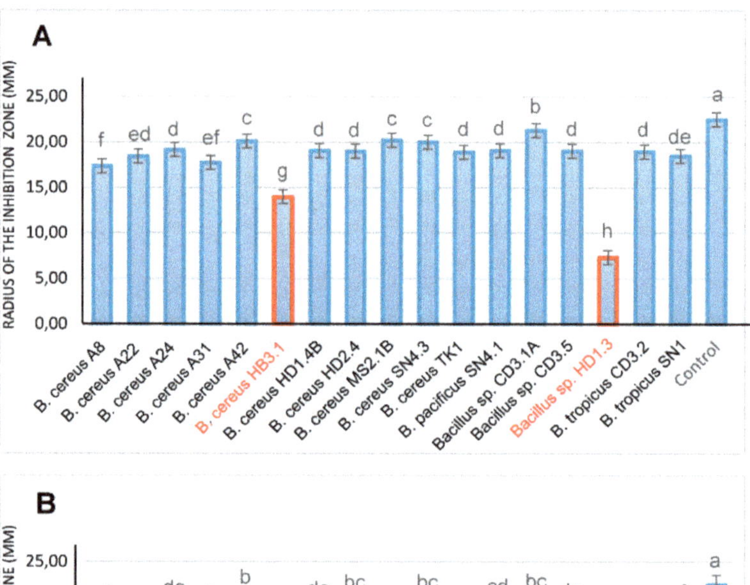

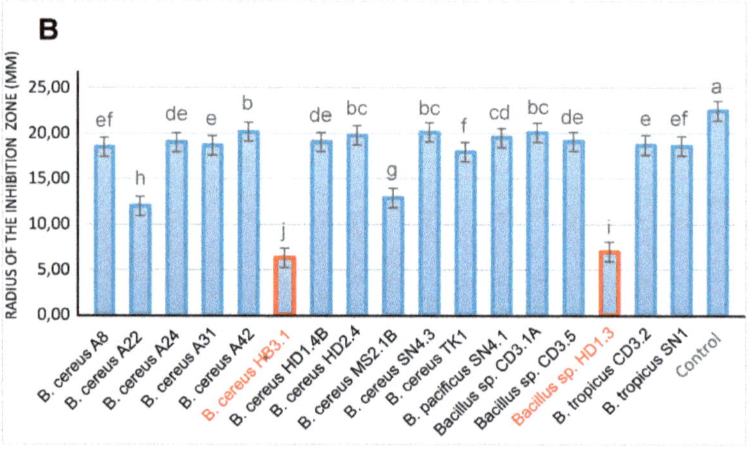

**Figure 7.** In vitro assay of the antifungal activity of *B. cereus* group strains isolated from Vietnamese crop plants. (**A**) Suppression of *Phytophthora palmivora*. (**B**) Suppression of *Fusarium oxysporum*. Strains with enhanced antimicrobial action are indicated in red. All diagrams show the means of at least three replicates (n ≥ 3). Negative controls were performed without treatment with the bacteria. Columns with superscripts with the same letter are not significantly different according to the Fisher's least significance difference (LSD) test ($p \leq 0.05$).

Root-knot nematodes, such as *Meloidogyne spp.*, are one of the most important plant pathogens in tropical and temperate agriculture, and are responsible for the significant harvest losses of main Vietnamese crops, such as coffee trees (*Coffea arabica* and *Coffea canephora*) and black pepper plants [82]. In order to analyze the antagonistic activity of the *B. cereus* isolates, we tested at first their suppressing effect against the model nematode *Caenorhabditis elegans*. Fast and slow death rates were estimated in a bioassay under laboratory conditions. Most of the investigated *B. cereus* isolates were able to kill considerable amounts of the nematodes, as revealed in both test systems (Figure 8).

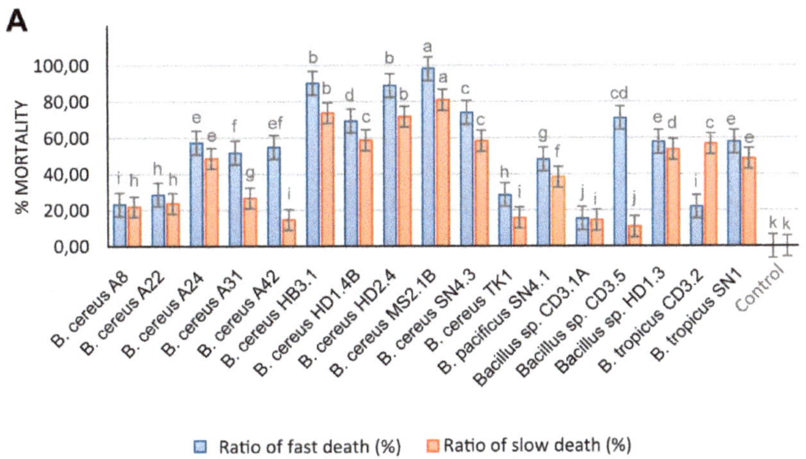

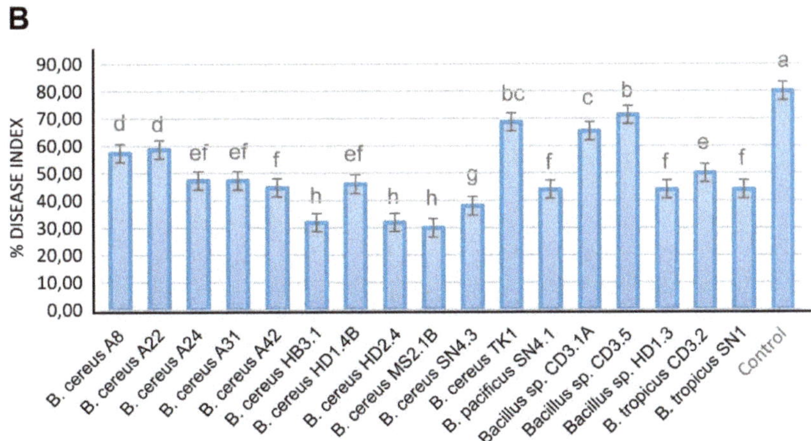

**Figure 8.** Nematocidal activity of *B. cereus* isolates. (**A**) Bioassay with *Caenorhabditis elegans*. Slow killing activity was determined in NGM plates and fast killing activity in a liquid medium as described previously [3]. (**B**) Determination of the biocontrol action of the *B. cereus* isolates on the root-knot nematode *Meloidogyne* sp. in greenhouse experiments. Tomato plants infested with *Meloidogyne* sp. were used for the test (counting of "knots" in the tomato roots). The increase compared to the control without adding with the *Bacillus* isolates is shown. All experiments were conducted with three independent repetitions and a randomized design. The bars above the columns indicate the standard error (SE). Different letters at each treatment indicate significance between inoculated and uninoculated conditions, at a $p \leq 0.05$ level after the *t*-test.

In order to examine the inhibiting effects against phytopathogenic nematodes more directly, we isolated a representative strain of *Meloidogyne sp.* from the galls of infested black pepper plant roots according to the hypochlorite procedure [83]. The inhibiting effect exerted by the test bacteria on disease development was examined in a greenhouse experiment. Ten weeks after the transplanting of the tomato plantlets into the soil, the formation of root knots was visually registered and used as a measure for calculating the disease index according to Bridge and Page [28]. All *B. cereus* isolates were found to be efficient against nematodes. In the presence of HB3.1, HD1.4B, HD2.4, MS2.1B, and SN4.3, the killing rates (estimated as slow and fast killing rates) of *Caenorhabditis elegans* exceeded 60% (Figure 8A). The *Melodoine* sp.-caused disease index of tomato plants was reduced by more than 50% after the application of *B. cereus* HB3.1, HD2.4, MS2.1B, and SN4.3 (Figure 8B). Similar rates were previously detected in representatives of *B. velezensis* [4] and *Brevibacillus* spp. [3].

### 3.4.2. Plant Growth Promotion

We examined the effect of the Vietnamese *B. cereus* isolates on the *Arabidopsis thaliana* biotest system [29]. *B. cereus* HD1.4B and *B. cereus* HD2.4 enhanced the growth of the *Arabidopsis* seedlings by more than 20% (Figure 9). However, using the same biotest system, the increase rates observed for some plant-associated *Brevibacillus* and *B. velezensis* strains isolated during the same survey [2] were higher and estimated to range from 30 to 40% [3,4].

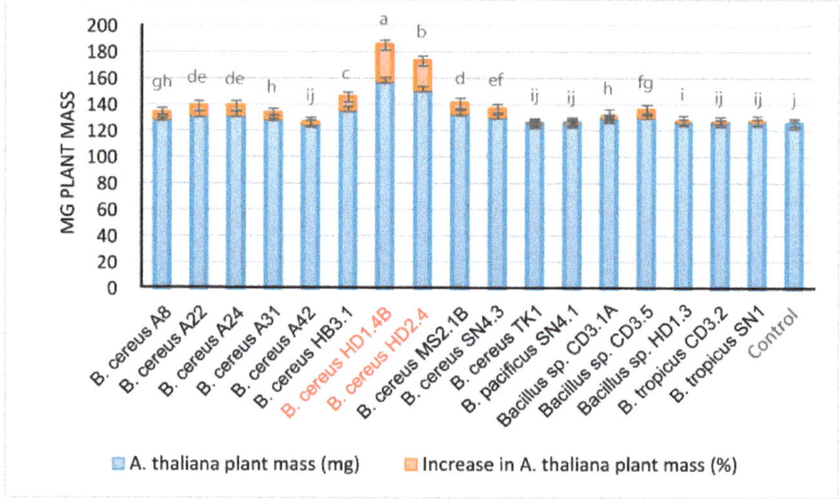


**Figure 9.** Growth-promoting effects of *B. cereus* isolates on *Arabidopsis thaliana* seedlings. The blue columns in the diagram represent the fresh weight obtained after 21 days under controlled conditions in the growth chamber. The % increase compared to the untreated control (red columns) is indicated on top of the columns. Each treatment value is presented as the means of three replications (n = 3) with the standard error. Different letters at each treatment indicate significance between inoculated and uninoculated conditions, at a $p \leq 0.05$ level after the *t*-test.

## 4. Conclusions

In this study, we showed that plant-associated representatives of the *B. cereus* group were able to suppress important plant pathogens, such as fungi (*Fusarium oxysporum*), oomycetes (*Phytophthora palmivora*), and root-knot forming nematodes (*Meloidogyne* sp.) The plant-growth-promoting activity of some of the isolates could also be demonstrated.

Genome mining revealed that the members of the *B. cereus* group are rich in gene clusters probably involved in the synthesis of antimicrobial peptides that are efficient in inhibiting plant pathogens and triggering plant-induced systemic resistance [84]. A total of 36 different biosynthetic gene clusters (BGCs), many of them not listed in the MiBIG data bank, were detected in the 17 isolates obtained from Vietnames crop plants. Mass-spectrometric analysis revealed that, in addition to some hitherto unknown compounds, several species of the antimicrobial lipopeptides kurstakin and thumolycin and the siderophores bacillibactin and petrobactin were expressed in many of the isolates. The arylpolyene lipopeptide thumolycin was reported to possess interesting antimicrobial and nematocidal activities, but its primary structure was not resolved [42]. Here, the primary structure of the pentapeptide part was resolved, and the Orn residue was identified as being linked with the yet unknown arylpolyene lipid part.

In addition to antimicrobial peptides, the biocontrol of plant pathogens can be exerted by δ-endotoxins (parasporal inclusion proteins), traditionally known to be produced by *B. thuringiensis*, a close relative of *B. cereus*. *Cry* genes encoding the entomocidal crystal proteins Cry1Aa1 and Cry2Ba1 were detected in the genome of *B. cereus* ssp. *Bombysepticus* TK1. In the same strain, another gene encoding the vegetative insecticidal protein Vip3 was also detected, suggesting that the presence of insecticidal proteins is not restricted to *B. thuringiensis*. The encouraging results described above might lead to the development of the selected representatives of the *B. cereus* group as biocontrol agents. However, as known food poisoning organisms and members of the risk group 2, the potential of the *B. cereus* group isolates to produce toxins needs to be carefully examined before they can be applied in sustainable agriculture.

In this context, we showed that no gene clusters encoding *B. anthracis* pXO plasmid-related toxins were present in all the *B. cereus s.l.* genomes investigated here. Furthermore, we could rule out the synthesis of the heat-stable cereulide toxin, the causative agent of the emetic syndrome. However, the regular appearance of chromosomally localized virulence genes, encoding the heat-labile enterotoxins HBL and NHE, might restrict the direct application of *B. cereus s.l.* strains in biological plant protection.

In order to avoid such conflicts, the utilization of interesting *B. cereus* BGCs and genes encoding entomocidal proteins can be achieved by their heterologous expression in safe plant-beneficial host strains, which has been already demonstrated with *B. velezensis* FZB42 [85,86].

**Supplementary Materials:** The following supporting information can be downloaded at: https://www.mdpi.com/article/10.3390/microorganisms11112677/s1. Figure S1: Tree inferred with FastMe 2.1.6.1 from GBDP distances calculated from 16S rDNA gene sequences. Figure S2: GBDP tree (whole-genome-sequence-based) inferred with FastMe 2.1.6.1 from GBDP distances calculated from genome sequences. Figure S3: Heat map of the FastANI values of 128 *B. cereus* group genomes calculated and drawn by the EDGAR3.0 software package. Figure S4: Environment of the Rep protein genes in the plasmid sequences of A22, A24, HD1.4B, and HD2.4; Figure S5: Localization of the virulence genes and gene clusters on the chromosomes and plasmids of *B. cereus* isolates. The *cytK* gene and the NHE/HBL gene clusters were chromosomally localized. The complete set of NHE and HBL genes was chromosomally localized in all four completely sequenced strains (A22, A24, HD1.4B, and HD2.4). The P1 plasmid sequences of HD1.4B and HD2.4 harbored genes with similarity to the NHE/HBL enterotoxin family. Figure S6: Plasmid-encoded catabolic operons, biosynthetic gene clusters (BGCs), and restriction/modification systems. Figure S7: BGCs in the *B. cereus* group isolates encoding NRPS/PKS and other secondary metabolites. Figure S8: RiPP gene clusters detected by applying the BAGEL4 software (http://bagel4.molgenrug.nl/ (accessed on 1 November 2021)) in the genomes of the Vietnamese *Bacillus cereus* group genomes. Figure S9: RiPP gene clusters detected by applying the antiSMASH version 6 software in the Vietnamese *Bacillus cereus* group genomes. Table S1: Types of entomocidal cry toxins. Table S2: Species and subspecies delineation of the 17 Vietnamese *B. cereus* sensu lato isolates according to their dDDH and Fast ANI values. Table S3: List of the 17 *B. cereus* group strains used in this study with detailed annotation. Table S4: Genome annotation of the four completely sequenced B. cereus strains by RASTk. Table S5: Gene clusters (BGCs) of the Vietnamese Bacillus cereus group isolates. Table S6: Gene clusters (BGCs) of the Vietnamese Bacillus cereus group isolates involved in the non-ribosomal synthesis of antimicrobial peptides

(RiPPs and bacteriocins). Table S7: Gene clusters (BGCs) of the Vietnamese Bacillus cereus group isolates involved in the ribosomal synthesis of antimicrobial peptides (RiPPs and bacteriocins). Table S8: Gene clusters (BGCs) of the Vietnamese Bacillus cereus group isolates involved in the synthesis of other secondary metabolites.

**Author Contributions:** Conceptualization, R.B., L.T.T.T., T.S. and P.L.; software, J.B.; investigation, J.V., J.J., S.H., L.T.T.T., A.S., P.T.L., L.T.P.T., S.R.K. and R.B.; resources, T.S. and P.L.; data curation, J.J., S.H., J.V. and C.B.; writing—original draft preparation, R.B., J.V. and L.T.T.T.; writing—review and editing, J.V., S.R.K., T.S., P.L., S.R.K. and R.B.; project administration, R.B., T.S. and P.L.; funding acquisition, R.B., T.S., P.L. and L.T.T.T. All authors have read and agreed to the published version of the manuscript.

**Funding:** This research was supported through the project ENDOBICA by the Bundesministerium für Bildung und Forschung (BMBF) (grant no. 031B0582A/031B0582B), the National Foundation for Science and Technology Development (NAFOSTED; code no. 106.03-2017.28), and the Ministry of Science and Technology (MOST) in Vietnam (code no. NDT.40.GER/18).

**Data Availability Statement:** Gene bank accession numbers of the complete genome sequences available in the NCBI data bank are listed in Section 2, Materials and Methods.

**Acknowledgments:** Silke Becker and Petra Lochau are thanked for their excellent technical support.

**Conflicts of Interest:** The authors declare no conflict of interest.

# References


1. Borriss, R. Use of plant-associated *Bacillus* strains as biofertilizers and biocontrol agents. In *Bacteria in Agrobiology: Plant Growth Responses*; Maheshwari, D.K., Ed.; Springer: Berlin/Heidelberg, Germany, 2011; pp. 41–76. [CrossRef]
2. Tam, L.T.T.; Jähne, J.; Luong, P.T.; Thao, L.T.P.; Chung, L.T.K.; Schneider, A.; Blumenscheit, C.; Lasch, P.; Schweder, T.; Borriss, R. Draft genome sequences of 59 endospore-forming Gram-positive bacteria associated with crop plants grown in Vietnam. *Microbiol. Resour. Announc.* **2020**, *9*, e01154-20. [CrossRef]
3. Jähne, J.; Le Thi, T.T.; Blumenscheit, C.; Schneider, A.; Pham, T.L.; Le Thi, P.T.; Blom, J.; Vater, J.; Schweder, T.; Lasch, P.; et al. Novel Plant-Associated Brevibacillus and Lysinibacillus Genomospecies Harbor a Rich Biosynthetic Potential of Antimicrobial Compounds. *Microorganisms* **2023**, *11*, 168. [CrossRef]
4. Thanh Tam, L.T.; Jähne, J.; Luong, P.T.; Phuong Thao, L.T.; Nhat, L.M.; Blumenscheit, C.; Schneider, A.; Blom, J.; Kim Chung, L.T.; Anh Minh, P.L.; et al. Two plant-associated *Bacillus velezensis* strains selected after genome analysis, metabolite profiling, and with proved biocontrol potential, were enhancing harvest yield of coffee and black pepper in large field trials. *Front. Plant. Sci.* **2023**, *14*, 1194887. [CrossRef] [PubMed]
5. Carroll, L.M.; Cheng, R.A.; Wiedmann, M.; Kovac, J. Keeping up with the *Bacillus cereus* group: Taxonomy through the genomics era and beyond. *Crit. Rev. Food Sci. Nutr.* **2022**, *62*, 7677–7702. [CrossRef]
6. Ehling-Schulz, M.; Lereclus, D.; Koehler, T.M. The *Bacillus cereus* Group: *Bacillus* Species with Pathogenic Potential. *Microbiol. Spectr.* **2019**, *7*. [CrossRef]
7. Koch, R. Die Ätiologie der Milzbrand-Krankheit, begründet auf die Entwicklungsgeschichte des Bacillus Anthracis. *Cohns beiträge ur Biol. Der Pflanz.* **1876**, *2*, 277.
8. Moayeri, M.; Leppla, S.H.; Vrentas, C.; Pomerantsev, A.P.; Liu, S. Anthrax Pathogenesis. *Annu. Rev. Microbiol.* **2015**, *69*, 185–208. [CrossRef] [PubMed]
9. Yang, S.; Wang, Y.; Liu, Y.; Jia, K.; Zhang, Z.; Dong, Q. Cereulide and Emetic Bacillus cereus: Characterizations, Impacts and Public Precautions. *Foods* **2023**, *12*, 833. [CrossRef]
10. Berliner, E. Über die Schlafsucht der Mehlmottenraupe (Ephestia kühniella Zell) und ihren Erreger *Bacillus thuringiensis* n. sp. *Z. fAngew. Entomol.* **1915**, *2*, 29–56. [CrossRef]
11. Palma, L.; Muñoz, D.; Berry, C.; Murillo, J.; Caballero, P. Bacillus thuringiensis toxins: An overview of their biocidal activity. *Toxins* **2014**, *6*, 3296–3325. [CrossRef]
12. Baldwin, V.M. You Can't *B. cereus*—A Review of *Bacillus cereus* Strains That Cause Anthrax-Like Disease. *Front. Microbiol.* **2020**, *11*, 1731. [CrossRef]
13. Kolstø, A.B.; Tourasse, N.J.; Økstad, O.A. What sets Bacillus anthracis apart from other Bacillus species? *Annu. Rev. Microbiol.* **2009**, *63*, 451–476. [CrossRef] [PubMed]
14. Blumenscheit, C.; Jähne, J.; Schneider, A.; Blom, J.; Schweder, T.; Lasch, P.; Borriss, R. Genome sequence data of *Bacillus velezensis* BP1.2A and BT2.4. *Data Brief.* **2022**, *41*, 107978. [CrossRef] [PubMed]
15. Olson, R.D.; Assaf, R.; Brettin, T.; Conrad, N.; Cucinell, C.; Davis, J.J.; Dempsey, D.M.; Dickerman, A.; Dietrich, E.M.; Kenyon, R.W.; et al. Introducing the Bacterial and Viral Bioinformatics Resource Center (BV-BRC): A resource combining PATRIC, IRD and ViPR. *Nucleic Acids Res.* **2023**, *51*, D678–D689. [CrossRef] [PubMed]

16. Liu, Y.; Lai, Q.; Göker, M.; Meier-Kolthoff, J.P.; Wang, M.; Sun, Y.; Wang, L.; Shao, Z. Genomic insights into the taxonomic status of the Bacillus cereus group. *Sci. Rep.* **2015**, *5*, 14082. [CrossRef]
17. Radnedge, L.; Agron, P.G.; Hill, K.K.; Jackson, P.J.; Ticknor, L.O.; Keim, P.; Andersen, G.L. Genome differences that distinguish Bacillus anthracis from Bacillus cereus and Bacillus thuringiensis. *Appl. Environ. Microbiol.* **2003**, *69*, 2755–2764. [CrossRef]
18. Klee, S.R.; Ozel, M.; Appel, B.; Boesch, C.; Ellerbrok, H.; Jacob, D.; Holland, G.; Leendertz, F.H.; Pauli, G.; Grunow, R.; et al. Characterization of Bacillus anthracis-like bacteria isolated from wild great apes from Cote d'Ivoire and Cameroon. *J. Bacteriol.* **2006**, *188*, 5333–5344. [CrossRef]
19. Meier-Kolthoff, J.P.; Sardà Carbasse, J.; Peinado-Olarte, R.L.; Göker, M. TYGS and LPSN: A database tandem for fast and reliable genome-based classification and nomenclature of prokaryotes. *Nucleic Acid Res.* **2022**, *50*, D801–D807. [CrossRef]
20. Dieckmann, M.A.; Beyvers, S.; Nkouamedjo-Fankep, R.C.; Hanel, P.H.G.; Jelonek, L.; Blom, J.; Goesmann, A. EDGAR3.0, Comparative genomics and phylogenomics on a scalable infrastructure. *Nucleic Acids Res.* **2021**, *49*, W185–W192. [CrossRef]
21. Blin, K.; Shaw, S.; Kloosterman, A.M.; Charlop-Powers, Z.; van Wezel, G.P.; Medema, M.H.; Weber, T. antiSMASH 6.0, Improving cluster detection and comparison capabilities. *Nucleic Acids Res.* **2021**, *49*, W29–W35. [CrossRef]
22. Bachmann, B.O.; Ravel, J. Chapter 8. Methods for in silico prediction of microbial polyketide and nonribosomal peptide biosynthetic pathways from DNA sequence data. *Methods Enzymol.* **2009**, *458*, 181–217. [CrossRef] [PubMed]
23. van Heel, A.J.; de Jong, A.; Song, C.; Viel, J.H.; Kok, J.; Kuipers, O.P. BAGEL4, A user-friendly web server to thoroughly mine RiPPs and bacteriocins. *Nucleic Acids Res.* **2018**, *46*, W278–W281. [CrossRef] [PubMed]
24. Vater, J.; Herfort, S.; Doellinger, J.; Weydmann, M.; Borriss, R.; Lasch, P. Genome Mining of the Lipopeptide Biosynthesis of Paenibacillus polymyxa E681 in Combination with Mass Spectrometry: Discovery of the Lipoheptapeptide Paenilipoheptin. *Chembiochem.* **2018**, *19*, 744–753. [CrossRef]
25. Mülner, P.; Schwarz, E.; Dietel, K.; Junge, H.; Herfort, S.; Weydmann, M.; Lasch, P.; Cernava, T.; Berg, G.; Vater, J. Profiling for Bioactive Peptides and Volatiles of Plant Growth Promoting Strains of the *Bacillus subtilis* Complex of Industrial Relevance. *Front. Microbiol.* **2020**, *11*, 1432. [CrossRef] [PubMed]
26. Suckau, D.; Resemann, A.; Schuerenberg, M.; Hufnagel, P.; Franzen, J.; Holle, A. A novel MALDI LIFT-TOF/TOF mass spectrometer for proteomics. *Anal. Bioanal. Chem.* **2003**, *376*, 952–956. [CrossRef] [PubMed]
27. Hooper, D.J.; Hallmann, J.; Subbotin, S.A. Methods for extraction, processing and detection of plant and soil nematodes. In *Plant Parasitic Nematodes in Subtropical and Tropical Agriculture*; Luc, M., Sikora, R., Bridge, J., Eds.; CAB International: Wallingford, UK, 2005; pp. 53–86.
28. Bridge, J.; Page, S.L.J. Estimation of Root Knot Nematode Infestation Levels in Roots Using a Rating Chart. *Trop. Pest Manag.* **1980**, *26*, 296–298. [CrossRef]
29. Budiharjo, A.; Chowdhury, S.P.; Dietel, K.; Beator, B.; Dolgova, O.; Fan, B.; Bleiss, W.; Ziegler, J.; Schmid, M.; Hartmann, A.; et al. Transposon mutagenesis of the plant-associated Bacillus amyloliquefaciens ssp. plantarum FZB42 revealed that the nfrA and RBAM17410 genes are involved in plant-microbe-interactions. *PLoS ONE* **2014**, *9*, e98267. [CrossRef]
30. Lefort, V.; Desper, R.; Gascuel, O. FastME 2.0, A comprehensive, accurate, and fast distance-based phylogeny inference program. *Mol. Biol. Evol.* **2015**, *32*, 2798–2800. [CrossRef]
31. Meier-Kolthoff, J.P.; Göker, M. TYGS is an automated high-throughput platform for state-of-the-art genome-based taxonomy. *Nat. Commun.* **2019**, *10*, 2182. [CrossRef]
32. Cheng, T.; Lin, P.; Jin, S.; Wu, Y.; Fu, B.; Long, R.; Liu, D.; Guo, Y.; Peng, L.; Xia, Q. Complete Genome Sequence of Bacillus bombysepticus, a Pathogen Leading to Bombyx mori Black Chest Septicemia. *Genome Announc.* **2014**, *2*, e00312-14. [CrossRef]
33. Parks, D.H.; Chuvochina, M.; Rinke, C.; Mussig, A.J.; Chaumeil, P.A.; Hugenholtz, P. GTDB: An ongoing census of bacterial and archaeal diversity through a phylogenetically consistent, rank normalized and complete genome-based taxonomy. *Nucleic Acids Res.* **2022**, *50*, D785–D794. [CrossRef] [PubMed]
34. Parte, A.C.; Sardà Carbasse, J.; Meier-Kolthoff, J.P.; Reimer, L.C.; Göker, M. List of Prokaryotic names with Standing in Nomenclature (LPSN) moves to the DSMZ. *Int. J. Syst. Evol. Microbiol.* **2020**, *70*, 5607–5612. [CrossRef]
35. Agata, N.; Ohta, M.; Mori, M. Production of an emetic toxin, cereulide, is associated with a specific class of *Bacillus cereus*. *Curr. Microbiol.* **1996**, *33*, 67–69. [CrossRef] [PubMed]
36. Baillie, L.; Read, T.D. *Bacillus anthracis*, a bug with attitude! *Curr. Opin. Microbiol.* **2001**, *4*, 78–81. [CrossRef] [PubMed]
37. Bravo, A.; Gómez, I.; Porta, H.; García-Gómez, B.I.; Rodriguez-Almazan, C.; Pardo, L.; Soberón, M. Evolution of *Bacillus thuringiensis* Cry toxins insecticidal activity. *Microb. Biotechnol.* **2013**, *6*, 17–26. [CrossRef]
38. Beecher, D.J.; Wong, A.C.L. Cooperative, synergistic and antagonistic haemolytic interactions between haemolysin BL, phosphatidylcholine phospholipase C and sphingomyelinase from Bacillus cereus. *Microbiology* **2000**, *146 Pt 12*, 3033–3039. [CrossRef]
39. Dietrich, R.; Jessberger, N.; Ehling-Schulz, M.; Märtlbauer, E.; Granum, P.E. The Food Poisoning Toxins of *Bacillus cereus*. *Toxins* **2021**, *13*, 98. [CrossRef]
40. Sanahuja, G.; Banakar, R.; Twyman, R.M.; Capell, T.; Christou, P. Bacillus thuringiensis: A century of research, development and commercial applications. *Plant Biotechnol. J.* **2011**, *9*, 283–300. [CrossRef]
41. Gupta, M.; Kumar, H.; Kaur, S. Vegetative Insecticidal Protein (Vip): A Potential Contender From *Bacillus thuringiensis* for Efficient Management of Various Detrimental Agricultural Pests. *Front. Microbiol.* **2021**, *12*, 659736. [CrossRef]
42. Zheng, D.; Zeng, Z.; Xue, B.; Deng, Y.; Sun, M.; Tang, Y.J.; Ruan, L. Bacillus thuringiensis produces the lipopeptide thumolycin to antagonize microbes and nematodes. *Microbiol. Res.* **2018**, *215*, 22–28. [CrossRef]

43. Overbeek, R.; Olson, R.; Pusch, G.D.; Olsen, G.J.; Davis, J.J.; Disz, T.; Edwards, R.A.; Gerdes, S.; Parrello, B.; Shukla, M.; et al. The SEED and the Rapid Annotation of microbial genomes using Subsystems Technology (RAST). *Nucleic Acids Res.* **2014**, *42*, D206-14. [CrossRef] [PubMed]

44. Griffith, O.H.; Ryan, M. Bacterial phosphatidylinositol-specific phospholipase C: Structure, function, and interaction with lipids. *Biochim. Biophys. Acta* **1999**, *1441*, 237–254. [CrossRef] [PubMed]

45. Gao, Y.; Cao, D.; Zhu, J.; Feng, H.; Luo, X.; Liu, S.; Yan, X.X.; Zhang, X.; Gao, P. Structural insights into assembly, operation and inhibition of a type I restriction-modification system. *Nat. Microbiol.* **2020**, *5*, 1107–1118. [CrossRef] [PubMed]

46. Dong, S.; McPherson, S.A.; Wang, Y.; Li, M.; Wang, P.; Turnbough, C.L., Jr.; Pritchard, D.G. Characterization of the enzymes encoded by the anthrose biosynthetic operon of Bacillus anthracis. *J. Bacteriol.* **2010**, *192*, 5053–5062. [CrossRef]

47. Dong, S.; McPherson, S.A.; Tan, L.; Chesnokova, O.N.; Turnbough, C.L., Jr.; Pritchard, D.G. Anthrose biosynthetic operon of Bacillus anthracis. *J. Bacteriol.* **2008**, *190*, 2350–2359. [CrossRef]

48. Yoshida, K.; Yamaguchi, M.; Morinaga, T.; Kinehara, M.; Ikeuchi, M.; Ashida, H.; Fujita, Y. myo-Inositol catabolism in Bacillus subtilis. *J. Biol. Chem.* **2008**, *283*, 10415–10424. [CrossRef]

49. Zhao, X.; Kuipers, O.P. Identification and classification of known and putative antimicrobial compounds produced by a wide variety of Bacillales species. *BMC Genom.* **2016**, *17*, 882. [CrossRef] [PubMed]

50. Tracanna, V.; de Jong, A.; Medema, M.H.; Kuipers, O.P. Mining prokaryotes for antimicrobial compounds: From diversity to function. *FEMS Microbiol. Rev.* **2017**, *41*, 417–429. [CrossRef] [PubMed]

51. Terlouw, B.R.; Blin, K.; Navarro-Muñoz, J.C.; Avalon, N.E.; Chevrette, M.G.; Egbert, S.; Lee, S.; Meijer, D.; Recchia, M.J.J.; Reitz, Z.L.; et al. MIBiG 3.0, a community-driven effort to annotate experimentally validated biosynthetic gene clusters. *Nucleic Acids Res.* **2023**, *51*, D603–D610. [CrossRef]

52. Finking, R.; Marahiel, M.A. Biosynthesis of nonribosomal peptides. *Annu. Rev. Microbiol.* **2004**, *58*, 453–488. [CrossRef]

53. Béchet, M.; Caradec, T.; Hussein, W.; Abderrahmani, A.; Chollet, M.; Leclère, V.; Dubois, T.; Lereclus, D.; Pupin, M.; Jacques, P. Structure, biosynthesis, and properties of kurstakins, nonribosomal lipopeptides from *Bacillus* spp. *Appl. Microbiol. Biotechnol.* **2012**, *95*, 593–600. [CrossRef] [PubMed]

54. Yu, Y.Y.; Zhang, Y.Y.; Wang, T.; Huang, T.X.; Tang, S.Y.; Jin, Y.; Mi, D.D.; Zheng, Y.; Niu, D.D.; Guo, J.H.; et al. Kurstakin Triggers Multicellular Behaviors in *Bacillus cereus* AR156 and Enhances Disease Control Efficacy Against Rice Sheath Blight. *Plant Dis.* **2023**, *107*, 1463–1470. [CrossRef] [PubMed]

55. Luo, C.; Liu, X.; Zhou, X.; Guo, J.; Truong, J.; Wang, X.; Zhou, H.; Li, X.; Chen, Z. Unusual Biosynthesis and Structure of Locillomycins from Bacillus subtilis 916. *Appl. Environ. Microbiol.* **2015**, *81*, 6601–6609. [CrossRef]

56. May, J.J.; Wendrich, T.M.; Marahiel, M.A. The dhb operon of Bacillus subtilis encodes the biosynthetic template for the catecholic siderophore 2,3-dihydroxybenzoate-glycine-threonine trimeric ester bacillibactin. *J. Biol. Chem.* **2001**, *276*, 7209–7217. [CrossRef] [PubMed]

57. Chen, X.H.; Koumoutsi, A.; Scholz, R.; Borriss, R. More than anticipated—Production of antibiotics and other secondary metabolites by Bacillus amyloliquefaciens FZB42. *J. Mol. Microbiol. Biotechnol.* **2009**, *16*, 14–24. [CrossRef] [PubMed]

58. Chen, X.H.; Koumoutsi, A.; Scholz, R.; Eisenreich, A.; Schneider, K.; Heinemeyer, I.; Morgenstern, B.; Voss, B.; Hess, W.R.; Reva, O.; et al. Comparative analysis of the complete genome sequence of the plant growth-promoting bacterium Bacillus amyloliquefaciens FZB42. *Nat. Biotechnol.* **2007**, *25*, 1007–1014. [CrossRef] [PubMed]

59. Kevany, B.M.; Rasko, D.A.; Thomas, M.G. Characterization of the complete zwittermicin A biosynthesis gene cluster from Bacillus cereus. *Appl. Environ. Microbiol.* **2009**, *75*, 1144–1155. [CrossRef]

60. Fellbrich, G.; Romanski, A.; Varet, A.; Blume, B.; Brunner, F.; Engelhardt, S.; Felix, G.; Kemmerling, B.; Krzymowska, M.; Nürnberger, T. NPP1, a Phytophthora-associated trigger of plant defense in parsley and Arabidopsis. *Plant J.* **2002**, *32*, 375–390. [CrossRef]

61. Burkhart, B.J.; Hudson, G.A.; Dunbar, K.L.; Mitchell, D.A. A prevalent peptide-binding domain guides ribosomal natural product biosynthesis. *Nat. Chem. Biol.* **2015**, *11*, 564–570. [CrossRef]

62. Kloosterman, A.M.; Shelton, K.E.; van Wezel, G.P.; Medema, M.H.; Mitchell, D.A. RRE-Finder: A Genome-Mining Tool for Class-Independent RiPP Discovery. *mSystems* **2020**, *5*, e00267. [CrossRef]

63. Sahl, H.G.; Jack, R.W.; Bierbaum, G. Biosynthesis and biological activities of lantibiotics with unique post-translational modifications. *Eur. J. Biochem.* **1995**, *230*, 827–853. [CrossRef]

64. Wang, J.; Ma, H.; Ge, X.; Zhang, J.; Teng, K.; Sun, Z.; Zhong, J. Bovicin HJ50-like lantibiotics, a novel subgroup of lantibiotics featured by an indispensable disulfide bridge. *PLoS ONE* **2014**, *9*, e97121. [CrossRef]

65. Walker, M.C.; Eslami, S.M.; Hetrick, K.J.; Ackenhusen, S.E.; Mitchell, D.A.; van der Donk, W.A. Precursor peptide-targeted mining of more than one hundred thousand genomes expands the lanthipeptide natural product family. *BMC Genom.* **2020**, *21*, 387. [CrossRef]

66. Arnison, P.G.; Bibb, M.J.; Bierbaum, G.; Bowers, A.A.; Bugni, T.S.; Bulaj, G.; Camarero, J.A.; Campopiano, D.J.; Challis, G.L.; Clardy, J.; et al. Ribosomally synthesized and post-translationally modified peptide natural products: Overview and recommendations for a universal nomenclature. *Nat. Prod. Rep.* **2013**, *30*, 108–160. [CrossRef]

67. Ren, H.; Biswas, S.; Ho, S.; van der Donk, W.A.; Zhao, H. Rapid Discovery of Glycocins through Pathway Refactoring in Escherichia coli. *ACS Chem. Biol.* **2018**, *13*, 2966–2972. [CrossRef] [PubMed]

68. Oman, T.J.; Boettcher, J.M.; Wang, H.; Okalibe, X.N.; van der Donk, W.A. Sublancin is not a lantibiotic but an S-linked glycopeptide. *Nat. Chem. Biol.* **2011**, *7*, 78–80. [CrossRef] [PubMed]
69. Hegemann, J.D.; Zimmermann, M.; Xie, X.; Marahiel, M.A. Lasso peptides: An intriguing class of bacterial natural products. *Acc. Chem. Res.* **2015**, *48*, 1909–1919. [CrossRef] [PubMed]
70. Zhu, S.; Hegemann, J.D.; Fage, C.D.; Zimmermann, M.; Xie, X.; Linne, U.; Marahiel, M.A. Insights into the Unique Phosphorylation of the Lasso Peptide Paeninodin. *J. Biol. Chem.* **2016**, *291*, 13662–13678. [CrossRef] [PubMed]
71. Flühe, L.; Knappe, T.A.; Gattner, M.J.; Schäfer, A.; Burghaus, O.; Linne, U.; Marahiel, M.A. The radical SAM enzyme AlbA catalyzes thioether bond formation in subtilosin A. *Nat. Chem. Biol.* **2012**, *8*, 350–357. [CrossRef]
72. Hudson, G.A.; Burkhart, B.J.; DiCaprio, A.J.; Schwalen, C.J.; Kille, B.; Pogorelov, T.V.; Mitchell, D.A. Bioinformatic Mapping of Radical S-Adenosylmethionine-Dependent Ribosomally Synthesized and Post-Translationally Modified Peptides Identifies New Cα, Cβ, and Cγ-Linked Thioether-Containing Peptides. *J. Am. Chem. Soc.* **2019**, *141*, 8228–8238. [CrossRef]
73. Chopra, L.; Singh, G.; Choudhary, V.; Sahoo, D.K. Sonorensin: An antimicrobial peptide, belonging to the heterocycloanthracin subfamily of bacteriocins, from a new marine isolate, Bacillus sonorensis MT93. *Appl. Environ. Microbiol.* **2014**, *80*, 2981–2990. [CrossRef] [PubMed]
74. Franz, C.M.; van Belkum, M.J.; Holzapfel, W.H.; Abriouel, H.; Gálvez, A. Diversity of enterococcal bacteriocins and their grouping in a new classification scheme. *FEMS Microbiol. Rev.* **2007**, *31*, 293–310. [CrossRef]
75. Aunpad, R.; Panbangred, W. Evidence for two putative holin-like peptides encoding genes of Bacillus pumilus strain WAPB4. *Curr. Microbiol.* **2012**, *64*, 343–348. [CrossRef] [PubMed]
76. Lee, J.Y.; Janes, B.K.; Passalacqua, K.D.; Pfleger, B.F.; Bergman, N.H.; Liu, H.; Håkansson, K.; Somu, R.V.; Aldrich, C.C.; Cendrowski, S.; et al. Biosynthetic analysis of the petrobactin siderophore pathway from Bacillus anthracis. *J. Bacteriol.* **2007**, *189*, 1698–1710. [CrossRef] [PubMed]
77. Koppisch, A.T.; Dhungana, S.; Hill, K.K.; Boukhalfa, H.; Heine, H.S.; Colip, L.A.; Romero, R.B.; Shou, Y.; Ticknor, L.O.; Marrone, B.L.; et al. Petrobactin is produced by both pathogenic and non-pathogenic isolates of the Bacillus cereus group of bacteria. *Biometals.* **2008**, *21*, 581–589. [CrossRef]
78. Yuan, S.; Yong, X.; Zhao, T.; Li, Y.; Liu, J. Research Progress of the Biosynthesis of Natural Bio-Antibacterial Agent Pulcherriminic Acid in *Bacillus*. *Molecules* **2020**, *25*, 5611. [CrossRef]
79. Corre, C.; Song, L.; O'Rourke, S.; Chater, K.F.; Challis, G.L. 2-Alkyl-4-hydroxymethylfuran-3-carboxylic acids, antibiotic production inducers discovered by Streptomyces coelicolor genome mining. *Proc. Natl. Acad. Sci. USA* **2008**, *105*, 17510–17515. [CrossRef]
80. Arie, T. *Fusarium* diseases of cultivated plants, control, diagnosis, and molecular and genetic studies. *J. Pestic. Sci.* **2019**, *44*, 275–281. [CrossRef]
81. Chi, N.M.; Thu, P.Q.; Nam, H.B.; Quang, D.Q.; Phong, L.V.; Van, N.D.; Trang, T.T.; Kien, T.T.; Tam, T.T.T.; Dell, B. Management of *Phytophthora palmivora* disease in *Citrus reticulata* with chemical fungicides. *J. Gen. Plant Pathol.* **2020**, *86*, 494–502. [CrossRef]
82. Eisenback, J.D.; and Triantaphyllou, H.H. Root-knot nematodes: Meloidogyne species and races. In *Manual of Agricultural Nematology*; Nickle, W.R., Ed.; Marcell Dekker: New York, NY, USA, 1991; pp. 191–274.
83. Hussey, P.S.; Barker, K.R. A comparison of methods of collecting inocula of Meloidogyne spp., including a new technique. *Plant Dis. Report.* **1973**, *57*, 1025–1028.
84. Chowdhury, S.P.; Hartmann, A.; Gao, X.; Borriss, R. Biocontrol mechanism by root-associated Bacillus amyloliquefaciens FZB42—A review. *Front. Microbiol.* **2015**, *6*, 780. [CrossRef] [PubMed]
85. Qiao, J.-Q.; Wu, H.-J.; Huo, R.; Gao, X.-W.; Borriss, R. Stimulation and biocontrol by *Bacillus amyloliquefaciens* subsp. plantarum FZB42 engineered for improved action. *Chem. Biol. Technol. Agric.* **2014**, *1*, 12. [CrossRef]
86. Wu, L.; Wu, H.J.; Qiao, J.; Gao, X.; Borriss, R. Novel Routes for Improving Biocontrol Activity of Bacillus Based Bioinoculants. *Front. Microbiol.* **2015**, *6*, 1395. [CrossRef] [PubMed]

**Disclaimer/Publisher's Note:** The statements, opinions and data contained in all publications are solely those of the individual author(s) and contributor(s) and not of MDPI and/or the editor(s). MDPI and/or the editor(s) disclaim responsibility for any injury to people or property resulting from any ideas, methods, instructions or products referred to in the content.

 *microorganisms*

Article

# Novel Plant-Associated *Brevibacillus* and *Lysinibacillus* Genomospecies Harbor a Rich Biosynthetic Potential of Antimicrobial Compounds

Jennifer Jähne [1,†], Thanh Tam Le Thi [2,†], Christian Blumenscheit [1,†], Andy Schneider [1], Thi Luong Pham [2], Phuong Thao Le Thi [2], Jochen Blom [3], Joachim Vater [1], Thomas Schweder [4,5], Peter Lasch [1] and Rainer Borriss [4,6,*]

1    Proteomics and Spectroscopy Unit (ZBS6), Center for Biological Threats and Special Pathogens, Robert Koch Institute, 13353 Berlin, Germany
2    Division of Pathology and Phyto-Immunology, Plant Protection Research Institute (PPRI), Duc Thang, Bac Tu Liem, Ha Noi, Vietnam
3    Bioinformatics and Systems Biology, Faculty of Biology and Chemistry, Justus-Liebig Universität Giessen, 35392 Giessen, Germany
4    Institute of Marine Biotechnology e.V. (IMaB), 17489 Greifswald, Germany
5    Pharmaceutical Biotechnology, University of Greifswald, 17489 Greifswald, Germany
6    Institute of Biology, Humboldt University Berlin, 10115 Berlin, Germany
*    Correspondence: rainer.borriss@rz.hu-berlin.de
†    These authors contributed equally to this work.

Citation: Jähne, J.; Le Thi, T.T.; Blumenscheit, C.; Schneider, A.; Pham, T.L.; Le Thi, P.T.; Blom, J.; Vater, J.; Schweder, T.; Lasch, P.; et al. Novel Plant-Associated *Brevibacillus* and *Lysinibacillus* Genomospecies Harbor a Rich Biosynthetic Potential of Antimicrobial Compounds. *Microorganisms* 2023, 11, 168. https://doi.org/10.3390/microorganisms11010168

Academic Editor: Nicole Hugouvieux-Cotte-Pattat

Received: 3 December 2022
Revised: 3 January 2023
Accepted: 6 January 2023
Published: 9 January 2023

Copyright: © 2023 by the authors. Licensee MDPI, Basel, Switzerland. This article is an open access article distributed under the terms and conditions of the Creative Commons Attribution (CC BY) license (https://creativecommons.org/licenses/by/4.0/).

**Abstract:** We have previously reported the draft genome sequences of 59 endospore-forming Gram-positive bacterial strains isolated from Vietnamese crop plants due to their ability to suppress plant pathogens. Based on their draft genome sequence, eleven of them were assigned to the *Brevibacillus* and one to the *Lysinibacillus* genus. Further analysis including full genome sequencing revealed that several of these strains represent novel genomospecies. In vitro and in vivo assays demonstrated their ability to promote plant growth, as well as the strong biocontrol potential of *Brevibacilli* directed against phytopathogenic bacteria, fungi, and nematodes. Genome mining identified 157 natural product biosynthesis gene clusters (BGCs), including 36 novel BGCs not present in the MIBiG data bank. Our findings indicate that plant-associated *Brevibacilli* are a rich source of putative antimicrobial compounds and might serve as a valuable starting point for the development of novel biocontrol agents.

**Keywords:** *Lysinibacillus*; *Brevibacillus*; genomic islands; phylogenomics; taxonomy; ANI; average nucleotide identity; dDDH; DNA-DNA hybridization: biocontrol; nematicidal activity; plant growth promotion; secondary metabolites

## 1. Introduction

Coffee and pepper are valuable products of Vietnamese agriculture. Their coffee production worldwide ranks second, after Brazil. Black pepper is of similar importance in Vietnam, and has been exported to more than 110 countries. However, during the last years, there has been a tendency of harvest yield reduction in the case of coffee of about 20% compared to the yield of 2014. Harvest losses are caused by plant pathogens such as fungi (e.g., *Fusarium oxysporum*), oomycetes (*Phytophthora palmivora*) and root-knot nematodes (e.g., *Meloidogyne incognita*). In the past, chemical pesticides were used to control the phytopathogens, but their use is no longer permitted due to their toxic remnants. The epidemic occurrence of fast death disease, which is damaging black pepper plantations, has led to a rural exodus of farmers. For this reason, the sustainable development of agriculture, which includes the applying of highly efficient and reliable biocontrol agents, useful for preventing and suppressing pests on black pepper and coffee, is an urgent need.

We recently isolated endospore-forming Gram-positive bacteria strains with putative biocontrol functions from healthy Vietnamese crop plants (e.g., coffee and black pepper) grown in fields infested with plant pathogens [1]. Based on their draft genome sequences, 47 isolates were assigned as being representative of the genus *Bacillus* belonging either to the *B. subtilis* or the *B. cereus* group [2]. Twelve additional isolates were either representatives of the *Brevibacillus* (11) or the *Lysinibacillus* (1) genus [1]. The majority of these strains could not be assigned down to species level. Preliminary experiments revealed that the *Brevibacillus* strains exert a strong nematicidal activity, and were able to suppress the growth of fungal plant pathogens.

The genus *Brevibacillus* encompassing the historical *Bacillus brevis* cluster [3] belongs to the family Bacillaceae of the phylum Firmicutes. The genus currently comprises 30 validated species according to the List of Prokaryotic Names with Standing in Nomenclature (LPSN; [4]). The members of the genus are characterized by the formation of spherical swollen sporangia, containing L-lysine in their peptidoglycan cell wall, and by their inability to utilize traditional carbohydrates. Although reports about the biocontrol activity of *Brevibacilli* against plant pathogens are rare [5–7], the application of antimicrobial compounds with medical importance produced by *Brevibacillus* spp. has a long tradition. Members of the *Brevibacillus brevis* cluster are a rich source of antimicrobial peptides and lipopeptides (AMPs), including gramicidin A and gramicidin S. The cyclic decapeptide gramicidin S (Soviet gramicidin) was discovered as early as 1942 [8]. Linear gramicidin A is the first AMP used clinically, and has still medical importance in topical medications. Until today, many novel AMPs such as the non-ribosomally synthesized cyclic decapeptide tyrocidine, the linear lipo-tri-decapeptide brevibacillin, the atypical cationic peptide edeine, and the bacteriocins laterosporulin and laterosporulin10 have been described [8]. *Brevibacillus* Leaf182 has previously been characterized as the most powerful inhibitor in the *Agrostemma thaliana* phyllosphere, and harbors a large number of gene clusters putatively involved in the synthesis of AMPs and other antimicrobial compounds. Marthiapeptide A and the previously unknown polyketide macrobrevin were identified in Leaf182 [9].

The genus *Lysinibacillus* [10], with the type species *Lysinibacillus boronitolerans*, at present comprises 21 validly published species (24-10-2022 LPSN). *L. sphaericus*, formerly *Bacillus sphaericus*, is known as an entomopathogen able to act as a biological insecticide that is efficient against mosquitos. In addition, other beneficial abilities such as plant growth promotion, the biocontrol of plant pathogens, and bioremediation have been reported [11].

In this study, we have fully characterized a new *Lysinibacillus* isolate and three novel *Brevibacillus* strains which were isolated from Vietnamese crop plants. We could assign their genomes to novel genomospecies, and were able to identify more than 150 biosynthetic gene clusters (BGCs) probably involved in the synthesis of a multitude of structurally diverse antimicrobial compounds. Biocontrol experiments performed with phytopathogenic bacteria, fungi, and nematodes revealed the high potential of the *Brevibacilli* isolates to suppress plant pathogens.

## 2. Materials and Methods

### 2.1. Strain Isolation, Growth Conditions and DNA Isolation

The plant-associated bacteria were isolated from healthy crop plants, such as coffee and pepper, from fields located at Dak Nong, Dak Lac provinces, Vietnam, and infested with plant pathogens such as *Phytophthora palmivora*, *Fusarium oxysporum*, and nematode *Meloidogyne* sp. [1]. Routinely, the leaf, stem and root of healthy plants were selected and taken to the laboratory for further processing. The plant organs were cut into $5 \times 5$ mm pieces and washed twice with sterile water. Afterwards, the plant parts were dipped into 75% ethanol for one min, and then into 0.1% mercury dichloride ($HgCl_2$) for two min. The cuts were then washed three times with sterile water, and taken up in 10 mL sterile water. The suspension was grounded in a sterile and chilled mortar. After 30 min. incubation, 0.1 mL of the solution was transferred to LB plates and allowed to grow for 72 h at 28 °C. Finally, single colonies were purified, transferred to 50 mL fresh LB medium, and cultured in a shaker at 28 °C and 180 rpm for 24 h. Due to their slow growth, *Brevibacillus* spp. were found to be enriched when soil samples adherent to plant roots were incubated under shaking for two weeks, diluted to $10^{-9}$, and then plated either onto R2A agar (Oxoid Lim., Basingstoke, UK) or onto 1/5 diluted NPT agar consisting of 0.4 g/L nutrient broth, 1 g/L potato dextrose, 1.2 g/L tryptic soy broth, 2 g/L MES hydrate (2-morpholino-ethan sulfonic acid), and 15 g/L agar, pH 5.75. Colonies appeared after three to seven days incubation at 28 °C. The purified strains were maintained as glycerol stocks (20%, *w/v*) at −80 °C. The cultivating of the bacterial strains and DNA isolation have been described previously [1,2].

### *2.2. Genome Sequencing, Assembly and Annotation*

The genome sequences of *Lysinibacillus* sp. CD3-6, *Brevibacillus parabrevis* HD3.3A, *Brevibacillus* sp. DP1.3A, and *Brevibacillus* sp. M2.1A were reconstructed using a combined approach of two sequencing technologies, which generated short paired-end reads and long reads. The resulting sequences were then used for hybrid assembly. Short-read sequencing was conducted in LGC Genomics (Berlin, Germany) using an Illumina HiSeqq in a paired 150 bp manner, as described previously [2]. Long read sequencing was done in house with the Oxford Nanopore MinION with the Flow Cell (R9.4.1) and prepared with the Ligation Sequencing Kit (SQK-LSK109). The samples were sequenced over 48 h and base called afterwards by Guppy v3.1.5. Long reads were trimmed using Porechop (https://github.com/rrwick/Porechop, v0.2.4, accessed on 3 December 2022) and filtered using Filtlong (https://github.com/rrwick/Filtlong, v0.2.0, accessed on 3 December 2022) on default settings. De-novo assemblies were generated by using the hybrid-assembler Unicycler (https://github.com/rrwick/Unicycler, v0.4.8, accessed on 3 December 2022) [11]. The quality of assemblies was assessed by determining the ratio of falsely trimmed protein by using Ideel (https://github.com/phiweger/ideel, accessed on 3 December 2022). Genome coverage of the obtained contigs was $50\times$ on average.

Automatic genome annotation was performed using the RAST (Rapid A using Subsystem Technology) server [12,13] implemented with RASTk [14], and with the NCBI Genome Automatic Annotation Pipeline (PGAP6.2, [7]) for the general genome annotation provided by NCBI RefSeq. Functional annotation was done by using PATRIC web resources [15]. Core- and pan-genome analysis was performed within the EDGAR3.0 pipeline [16]. Genomic islands (GI) were predicted with the webserver IslandViewer 4 (http://www.pathogenomics.sfu.ca/islandviewer/, accessed on 3 December 2022) [17]. Circular plots of genome and plasmid sequences were visualized with BioCircos [18].

For assessment of genome similarity and phylogeny, the genome sequence data were uploaded to the Type (Strain) Genome Server (TYGS), available at https://tygs.dsmz.de, accessed on 3 December 2022 [19]. Information on nomenclature was provided by the List of Prokaryotic names with Standing in Nomenclature (LPSN, available at https://lpsn.dsmz.de, accessed on 3 December 2022) [20]. All user genomes were compared with all type strain genomes available in the TYGS database via the MASH algorithm [21], and the ten strains with the smallest MASH distances were chosen per user genome. Using the Genome BLAST Distance Phylogeny approach (GBDP), the ten closest type strain genomes for each of the user genomes were calculated. GTDB sp. were calculated with GTDB-Tk and the FASTANI calculator (https://gtdb.ecogenomic.org/, accessed on 3 December 2022). GTDB-Tk is a toolkit for assigning objective classifications to bacterial and archaeal genomes based on the Genome Database Taxonomy (GTDB). In silico DNA-DNA hybridization (dDDH) values were calculated in the TYGS platform using formula $d_4$, which is the sum of all identities found in the high score segment pairs (HSPs) divided by the total length of all HSPs. A pan-genome analysis was performed using the Bacteria Pan Genome Analysis pipeline (BPGA) [22] with the amino acid sequences and default parameter settings (50% identity). The tree files provided by the BPGA pipeline and TYGS were visualized with iTOL (https://itol.embl.de/#, accessed on 3 December 2022).

In addition, the EDGAR3.0 pipeline [16] was used for elucidating taxonomic relationships based on genome sequences. To construct a phylogenetic tree for a project, the core genes of these genomes were computed. In a following step, the alignments of each core gene set are generated using MUSCLE, and the alignments are concatenated to one huge alignment. This alignment is the input for the FastTree software (http://www.microbesonline.org/fasttree/, accessed on 3 December 2022) to generate approximately-maximum-likelihood phylogenetic trees. The values at the branches of FastTree trees are _NOT_ bootstrapping values, but local support values computed by FastTree using the Shimodaira-Hasegawa test. FastANI [23] was used to calculate ANI heatmaps within the EDGAR pipeline for a selected set of genomes.

### 2.3. Biocontrol Activity against Plant Pathogens and Plant Growth Promotion

Antibacterial activity was examined as follows: 0.5 mL of a stationary culture (around $10^9$ cells) of the phytopathogenic indicator bacteria (*Clavibacter michiganensis, Dickeya solani, Erwinia amylovora*) were mixed with 2 mL liquid soft agar (0.7%) at 40 °C, and then added to Petri dishes with 1.5% LB-Agar. The test bacteria (*Brevibacillus* and *Lysinibacillus* strains) were grown in liquid culture for 48 h. under continuous shaking. 10 µL of the culture was allowed to soak into filter paper discs (2 mm in diameter), and were then placed onto the surface of the soft agar containing the indicator bacteria. The cultures were examined every day and allowed to grow for six days at 27 °C.

The antifungal activity of the isolates was determined as follows: Plugs (5 mm in diameter) with the pathogenic fungi were placed onto potato dextrose agar (PDA). Paper discs with the test bacteria were then added 20 mm away from the fungi. The cultures were incubated for six days at 27 °C and examined daily.

A bioassay of nematicidal activity was performed with *Caenorhabditis elegans* N2 (Carolina, U.S.A., https://www.carolina.com, accessed on 3 December 2022) fed with *Escherichia coli* OP50 cells. The culture and synchronization of the worms was performed as previously described [24]. The L4 stage was used for two different bioassays performed as described previously [25]. In the slow killing assay, around 40 L4 *C. elegans* worms were added to the nematode growth medium (NGM) agar plate containing the test bacteria. The mixture was incubated for 3–5 days at 25 °C and inspected daily. In the liquid fast killing assay, the test bacteria were grown overnight under shaking (200 rpm) at 37 °C in 3 mL liquid assay medium. 100 µL of the bacterial culture was diluted with 500 µL M9 medium, and transferred into 12 well plates. Each well was seeded with 40–60 L4 stage N2 nematodes, and the assay was performed at 25 °C for 24 h. Mortalities of nematodes were defined as the ratio of dead nematodes over the tested nematodes.

Root-knot nematode *Meloidogyne* sp. was isolated from roots of infested pepper plants according to [26]. Tomato plantlets were grown in pots with natural soil under controlled conditions in a greenhouse. Test bacteria and second stage juvenile (J2) nematodes were added to the pots two weeks after transplanting. The number of knots in tomato plants was estimated ten weeks after infestation with the nematodes [27].

A plant growth promotion assay was performed with wild type *Arabidopsis thaliana* (EDVOTEK, USA https://www.edvotek.com/, accessed on 3 December 2022) according to [28]. The surface sterilized seeds were pre-germinated on Petri dishes containing half-strength Murashige-Skoog medium semi-solidified with 0.6% agar and incubated at 22 °C under long daylight conditions (16 h light/8 h dark) for seven days. The roots of *Arabidopsis* seedlings were then dipped into a diluted spore suspension of the test bacteria ($10^5$ CFU/mL) for five min., and five seedlings were transferred into a square Petri dish containing half-strength MS-medium solidified with 1% agar. The square Petri dishes were incubated in a growth chamber at 22 °C at a daily photoperiod of 14 h. The fresh weight of the plants was measured 21 days after transplanting for estimation of the ability of bacterial strains for growth promotion. All experiments were performed in triplicate and the standard deviation SD was indicated as bars of the column diagrams.

*2.4. Identification of Gene Clusters Involved in the Synthesis of Secondary Metabolites*

Gene clusters for secondary metabolite synthesis were mined using antiSMASH pipeline version 6 [29] under settings of all features and BAGEL4 [30].

*2.5. Data Analysis*

The data obtained from biocontrol and plant growth promotion experiments were analyzed by Statistic Analysis Systems (SAS) software. The results of three replicates (n = 3) were expressed as SD (standard deviations). Significance was calculated with *t*-tests using the ANOVA procedure according to Duncan at LSD = 0.05. Every experiment was conducted using a completely random design. Excel software was used to create the graphical representations.

*2.6. Gene Bank Accession Numbers of DNA Sequences*

*Lysinibacillus* sp. CD3-6 chromosome: CP085880.1, *Lysinibacillus* sp. CD3-6 extra DNA (2485 bp): CP085881.1, *Brevibacillus parabrevis* HD3.3A chromosome: CP085874.1, *Brevibacillus parabrevis* HD3.3A plasmid (61,606 bp): CP085875.1, *Brevibacillus* sp. DP1.3A chromosome: CP085876.1, *Brevibacillus* sp. DP1.3A plasmid 1 (44,610 bp): CP085877.1, *Brevibacillus* sp. DP1.3A plasmid 2 (10,996 bp): CP085878.1, *Brevibacillus* sp. DP1.3A plasmid 3 (6349 bp): CP085879.1, *Brevibacillus* sp. M2.1A chromosome: NZ_JABSUY020000001.1, NZ_JABSUY020000002.1, NZ_JABSUY020000004.1, *Brevibacillus* sp. M2.1A plasmid (19,434 bp): NZ_JABSUY020000003.1.

## 3. Results and Discussion

*3.1. Resequencing of Selected Lysinibacillus and Brevibacillus Strains Revealed the Presence of Genomic Islands and Extrachromosomal Elements*

*Lysinibacillus* sp. CD3-6, *Brevibacillus parabrevis* HD3.3A, *Brevibacillus* sp. DP1.3A, and *Brevibacillus* sp. M2.1A were sequenced using nanopore sequencing technology. The complete genome of *Lysinibacillus* sp. CD3-6 consisted of two DNA elements: a single circular chromosome with 4,810,966 bps (CP085880.1), and a small DNA with 2485 bps (CP085881.1). The total size of both DNA elements was 4,813,451 bps harboring 4833 coding sequences. No genes with similarity to plasmid replication proteins were detected in the small DNA element, excluding its definition as a plasmid. A circular plot of the CD3-6 chromosome (4,810,966 bps) computed against the most related genome (*Lysinibacillus* sp. JNUCC-52) is shown in Supplementary Figure S1. CD3-6 was used as reference for computing the core genome against a set of 15 *Lysinibacillus* genomes representing the most related taxonomic groups (clusters A1-4, see Section 3.2). The core genome consisted of 2733 coding sequences (CDS). The pan genome consisted of 9100 CDS. A total of 187 singletons were detected in CD3-6 when compared with the other 15 genomes (Table 1). Genomic island (GI) prediction using the IslandViewer 4 webserver [17] revealed that the CD3-6 genome was rich in putative genomic islands, mainly characterized by the presence of site-specific integrases, HNH endonucleases, and phage proteins (Figure 1A). GI 1 (618,752–660,952) contained 31 genes including *tnpA* (IS200/IS605 family transposase) and a tyrosine-type recombinase/integrase encoding gene. GI 4 (1,776,323–1,805,601) contained 39 genes including several phage proteins and a gene encoding FAD dependent thymidylate synthase. A predicted class-ii lassopeptide was detected using the antiSMASH pipeline version 6 [29] in GI 5. The largest GI predicted in CD3-6 was GI 8 (2,649,327–2,738,124) harboring 119 genes including the type II toxin-antitoxin system (UED78437.1, UED78438.1) encoded by *hicA* and *hicB*. HicAB modules appear to be highly prone to horizontal gene transfer genes [31]. A complete list of the genes predicted in the CD3-6 GIs is given in Supplementary Table S1.

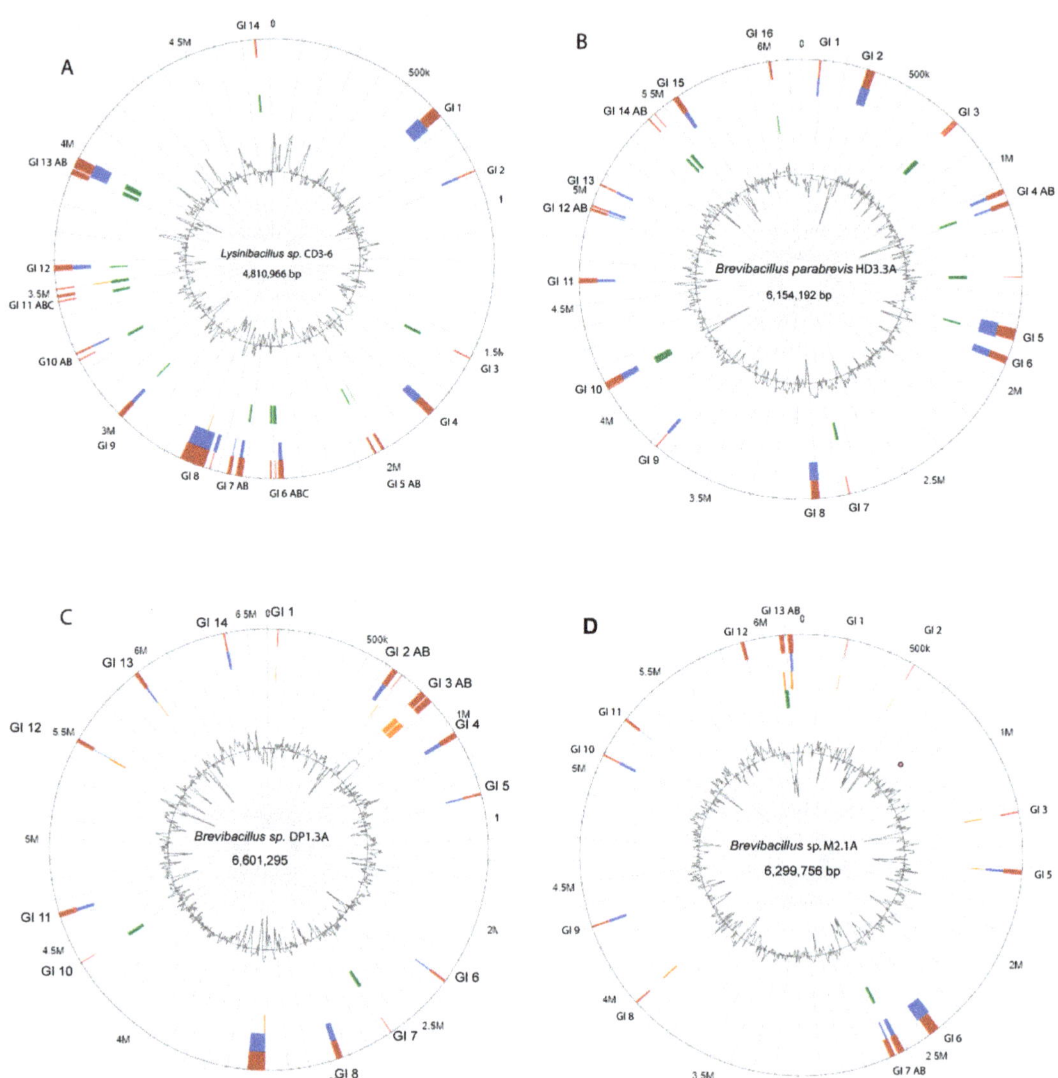

**Figure 1.** Circular plots of the chromosomes of *Lysinibacillus* sp. CD3-6 (**A**), *B. parabrevis* HD3.3A (**B**), *Brevibacillus* sp. DP1.3A (**C**), and *Brevibacillus* sp. M2.1A (**D**). Predicted GIs are shown as blocks colored according to the prediction method; Island Pick (green), Island-Path-DIMOB (blue), SIGI-HMM (orange), as well as the integrated results (dark red). The grey line within the inner circle shows deviations of the average GC-content.

The chromosome of the isolate *B. parabrevis* HD3.3A (6,154,192 bp) is presented in Supplementary Figure S2. HD3.3A was used as reference for computing the core genome against representatives of the *Brevibacillus*-A5 branch (clusters 25–27, see Section 3.2). The core genome consisted of 3044 CDS. The pan-genome consisted of 9438 CDS. A total of 166 singletons were detected in HD3.3A (Table 1).

Some 16 GIs were found to be distributed within the HD3.3A chromosome (Figure 1B). They were enriched with transposases of different families, such as IS3, IS21, IS110, IS256, Mu, TnsA, TnsD, Tns7, and site-specific integrases probably involved in horizontal gene transfer. An *agrB* gene probably involved in processing of cyclic lactone autoinducer (AIP) was detected in GI 5. Phage genes were detected in several GIs of this strain. GI 15 (5,552,099–5,579,031) harbored a number of flagellar proteins, possibly affecting the motility behavior of HD3.3A. (Supplementary Table S2).

*B. parabrevis* HD3.3A isolate harbored an 61 kb extrachromosomal element (GC:47.3%) containing the gene for the plasmid segregation protein ParM (WP_173600040.1) with similarity (41% identity) to the ParM protein (QYY44717.1) from *Aneurinibacillus thermoaerophilus* plasmid pAT1, and ParM proteins common in the *Bacillus cereus* group [32]. In addition, replicative DnaB helicase (WP_173600059.1), site-specific integrase (WP_229050018.1), and several phage proteins were detected (Figure 2A).

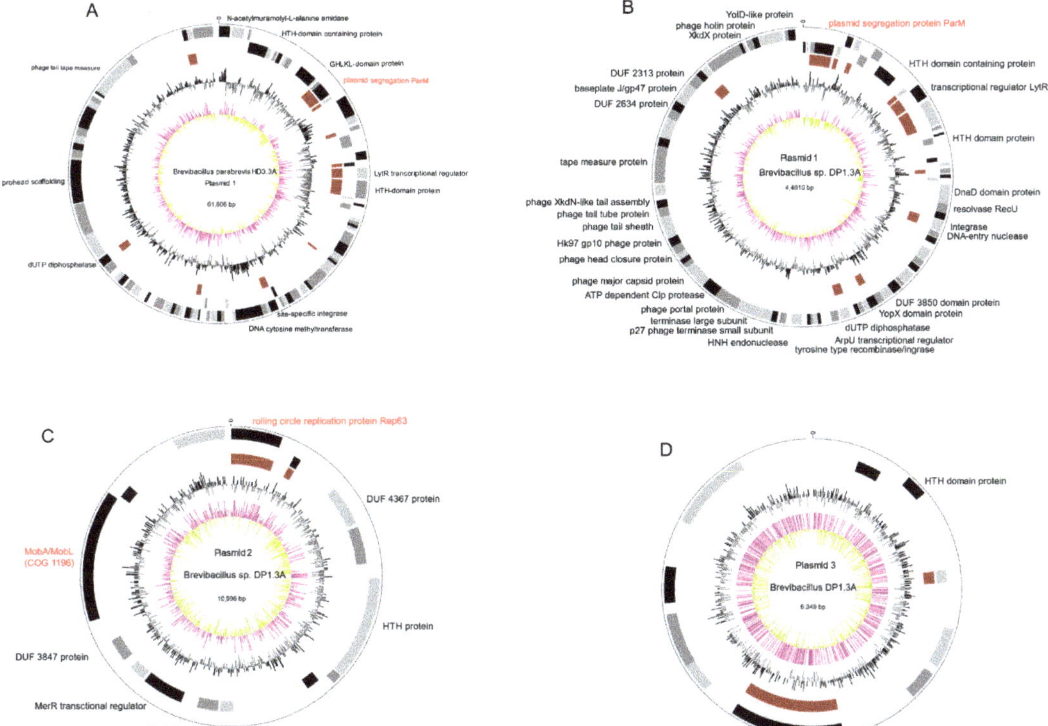

**Figure 2.** Circular plot of *Brevibacillus* plasmids generated with Biocircos. From outer to inner circle: Genes (CDS) on + (1)/− strand (2); core genome, brown (3); GC-content (1000 bp window) above mean: black, below mean: grey (4); GC Skew [(G-C)/(+C9] (1000 bp window), above mean: purple, below mean: light green (5). (**A**): Plasmid 1 (*B. parabrevis* HD3.3A) computed against plasmid 1 (*Brevibacillus* sp. DP1.3A). (**B**): Plasmid 1 (*Brevibacillus* sp. DP1.3A) computed against plasmid 1 (*B. parabrevis* HD3.3A). (**C**): Plasmid 2 (*Brevibacillus* sp. DP1.3A) computed against plasmid 3 (*Brevibacillus* sp. DP1.3A). (**D**): Plasmid 3 (*Brevibacillus* sp. DP1.3A) computed against plasmid 2 (*Brevibacillus* sp. DP1.3A).

The *Brevibacillus* sp. DP1.3A chromosome consisted of 6601 kb and displayed high similarity with *Brevibacillus* Leaf182 (Supplementary Figure S3). For estimating the core and pan genome, DP1.3A was computed against representatives of the *Brevibacillus*-A6 branch (clusters 28–35, see Section 3.2). The core genome consisted of 2607 CDs. The pan-genome consisted of 15,934 CDs. A total of 365 singletons were detected in DP1.3A (Table 1). Fourteen GIs were predicted in the chromosome of DP1.3A (Figure 1C). Three GIs contained gene clusters probably involved in the synthesis of secondary metabolites. GI 1 (50,050–55,329) harbored two genes involved in the synthesis of lanthipeptide class-ii peptides. GI 3 (805,782–878,612) harbored a gene cluster probably involved in the synthesis of macrobrevin [9]. Genes involved in the synthesis of a cyclic lactone autoinducer peptide (TIGR04223) were detected in GI 6 (2,311,687–2,324,264). An unknown NRP-PK hybrid scaffold was probably synthesized by genes present in GI 9 (3,310,924–3,397,894). In addition, GI 9 harbored a complete type I DNA restriction-modification system (Supplementary Table S3).

*Brevibacillus* sp. DP1.3A harbored three extrachromosomal elements (ECEs). The circular 44 kb DNA element (GC content: 43.9%, Figure 2B) displayed similarity with the HD3.3A plasmid, and also contained the *parM* gene (67.5% identity). In total, both *parM* containing plasmids shared 13 CDS including site-specific integrase, and dUTP diphosphatase. The circular 11 kb DP1.3A plasmid DNA (Figure 2C) encodes the replication initiator protein Rep (WP_173621461.1) involved in DNA-binding and the rolling circle replication of high-copy plasmids [33]. Notably, WP_173621461.1 shared high similarity (88.43% identity) with the Rep63 protein (HAJ4019592.1) from the *E. coli* isolate LA106_16-0310. Present in this plasmid was also a gene encoding the MobA/MobL mobilization protein UED78114.1, which is essential for conjugative plasmid transfer [34]. The protein resembled (74.7% identity) the MobA/MobL family protein of the gamma proteobacterium *Xanthomonas arbicola* (WP_080960787.1). A third ECE, 6349 bp in size, harbored a similar Rep protein gene as detected in plasmid 2 (Figure 2D).

The *Brevibacillus* sp. M2.1A chromosome (6300 kb) was found to be closely related to the *Brevibacillus brevis* strain 12B3 (Supplementary Figure S4). Computing with the other members of the *Brevibacillus*-A6 branch yielded 3295 core genes. The pan genome was formed by 13,428 genes. The M2.1A genome contained 113 singletons (Table 1). In contrast with DP1.3A, the GIs detected in the M2.1A chromosome (Figure 1D) did not harbor genes involved in the synthesis of secondary metabolites. A complete thioredoxin system with thioredoxin, thioredoxin-disulfide reductase, and thiol peroxidase was detected in GI 13 (Supplementary Table S4). This system might enable M2.1A to respond efficiently against oxidative stress [35].

M2.1A harbored low-copy plasmid DNA (size: 19,434 bps, GC content: 42.0%), whose partition might be governed by the plasmid partition protein A [36] (Table 1). The ParA family protein (MCC8438707.1) shared the highest similarity with the AAA family ATPase from *Brevibacillus borstelensis* (WP_251238174.1, 93.49% identity). In addition, the plasmid also shared partial sequence similarity (16% of their total length) with the low-copy plasmid from *Brevibacillus* sp. DP1.3A (CP085878).

**Table 1.** Extrachromosomal elements and general genomic features of *Lysinibacillus* sp. CD3-6, *B. parabrevis* HD3.3A, *Brevibacillus* sp. DP1.3A, and *Brevibacillus* sp. M2.1A. Methods used for generating the data are set in brackets (PGAP = RefSeq, EDGAR). The origin of replication (oriC) was estimated with Ori-Finder 2022 (http://tubic.tju.edu.cn/Ori-Finder2/, accessed on 3 December 2022) [37]. Calculation of core and pan genomes was performed as described in the text (Section 3.1). The protein features estimated with RAST are presented in Supplementary Table S5.

| Genus | *Lysini-bacillus* | *Brevibacillus* | | |
|---|---|---|---|---|
| Strain | CD3-6 | HD3.3A | DP1.3A | M2.1A |
| *Extrachrosomal elements (ECE)* | | | | |
| | CP085881 2485 bp | CP085875 61,606 bp<br>ParM-like segregation | CP085877 44,610 bp<br>ParM-like segregation | JABSUY020000003<br>19,434 bp ParA |
| | | | CP085878 10,996 bp<br>Rep, RC replication,<br>MobA/L | |
| | | | CP085879 6349 bp | |
| *Genomic features* | | | | |

**Table 1.** *Cont.*

| Genus | Lysini-bacillus | | Brevibacillus | |
|---|---|---|---|---|
| **Strain** | **CD3-6** | **HD3.3A** | **DP1.3A** | **M2.1A** |
| chromosome | CP085880 | CP085874 | CP085876 | JABSUY020000001 |
| Genome size (bp) | 4,810,966 | 6,154,192 | 6,601,295 | 6,172,625 |
| Replication origin (oriC) | 4,810,413 … 4,810,966 | 6,153,310 … 6,154,192 | 6,600,712 … 6,601,295 | 570,707 … 571,289 |
| G+C % | 37.1 | 52.1 | 47.4 | 47.4 |
| Number of genes (PGAP) | 4794 | 5819 | 6203 | 5956 |
| Genes coding (PGAP) | 4589 | 5570 | 5960 | 5728 |
| CDSs total (PGAP) | 5331 | 5661 | 6030 | 5783 |
| CDS core genome (EDGAR) | 2733 | 3044 | 2607 | 3295 |
| CDS pan genome (EDGAR) | 9100 | 9438 | 15,934 | 13,428 |
| CDS singletons (EDGAR) | 187 | 166 | 365 | 113 |
| Number of RNAs (PGAP) | 149 | 173 | 168 | 166 |
| rRNAs (5S, 16S, 23S, PGAP) | 37 | 38 | 41 | 41 |
| tRNAs (PGAP) | 112 | 130 | 127 | 106 |
| ncRNAs (PGAP) | 5 | 5 | 5 | 5 |
| Pseudo genes (PGAP) | 51 | 76 | 70 | 55 |

*3.2. Taxonomic Evaluation of Lysinibacillus and Brevibacillus Strains Revealed Novel Genomospecies*

3.2.1. Lysinibacillus CD3-6 Forms a Distinct Genomospecies Together with *Lysinibacillus* JNUCC-52

The 16S rRNA gene sequence was extracted using the TYGS server from the whole genome sequence of CD3-6 and used for phylogenetic analysis. We have also directly sequenced the 16S rRNA of CD3-6, which was deposited in the NCBI data base as MW820197.1. Both sequences differ by only one nucleotide. The resulting tree indicated a single species cluster formed by CD3-6 with *Lysinibacillus sphaericus* as the closest related species (Supplementary Figure S5). In order to exclude the possibility that the type of strains without known genome sequences, but more related to CD3-6, escaped our analysis, we performed a BLASTN-supported search (https://blast.ncbi.nlm.nih.gov, accessed on 3 December 2022) for related 16S rRNA sequences. However, no 16S rDNA sequences with more similarity than *L. sphaericus* (99.61%) were detected.

Since16S rRNA sequences are often not sufficient for species discrimination, we used the genome sequences for taxonomical strain identification. We started our analysis with the CD3-6 genome (CP085880) and 112 *Lysinibacillus* genomes obtained from the NCBI data bank. The phylogenetic tree (Figure 3) suggested that the members of the *Lysinibacillus* genus can be divided into two major groups, A and B. Group B members were not always representatives of the *Lysinibacillus* genus, but were often classified as representatives of different *Ureibacillus* species. Thus, group B seemed to be heterogenous, and to contain different genera. Group A contained *Lysinibacillus* genomes related to the type strain *Lysinibacillus sphaericus* DSM 28 [10], formerly *Bacillus sphaericus*. Based on the cut-off values for species delineation using ANI (96%) and dDDH (70%) [38], 24 clusters (A1–A24) were distinguished (Supplementary Table S6). Ten of the clusters contained type strains of recognized *Lysinibacillus* species, but most of the clusters (14) did not contain type strains, and represented according to the definition given by EZBioCloud (https://help.ezbiocloud.net/genomospecies/, accessed on 3 December 2022) for unnamed genomospecies consisting of so far unclassified or wrongly labelled *Lysinibacillus* genomes.

The type strain *Lysinibacillus sphaericus* DSM 28 was classified as being a member of cluster A4. Due to its genome sequence, *Lysinibacillus* sp. CD3-6 was assigned to belong to cluster A2 when using a 70% dDDH radius around each of the 13 *Lysinibacillus* type strains (Figure 3B).

The type strains most related to CD3-6, *Lysinibacillus sphaericus* KCTCC 3346, *Lysinibacillus tabacifolii* K3514, and *Lysinibacillus mangiferihumi*, possessed dDDH values (d4) far below of the species cut off (dDDH < 70, Supplementary Table S6). We conclude that CD3-6 represents, together with *Lysinibacillus* JNUCC-52, a novel genomospecies distinguished from the *L. sphaericus* species cluster. It should be noted that *L. mangiferihumi*, *L. tabacifolii*, and *L. varians* were recently characterized as later heterotrophic synonyms of *L. sphaericus* [39], and do not represent valid species.

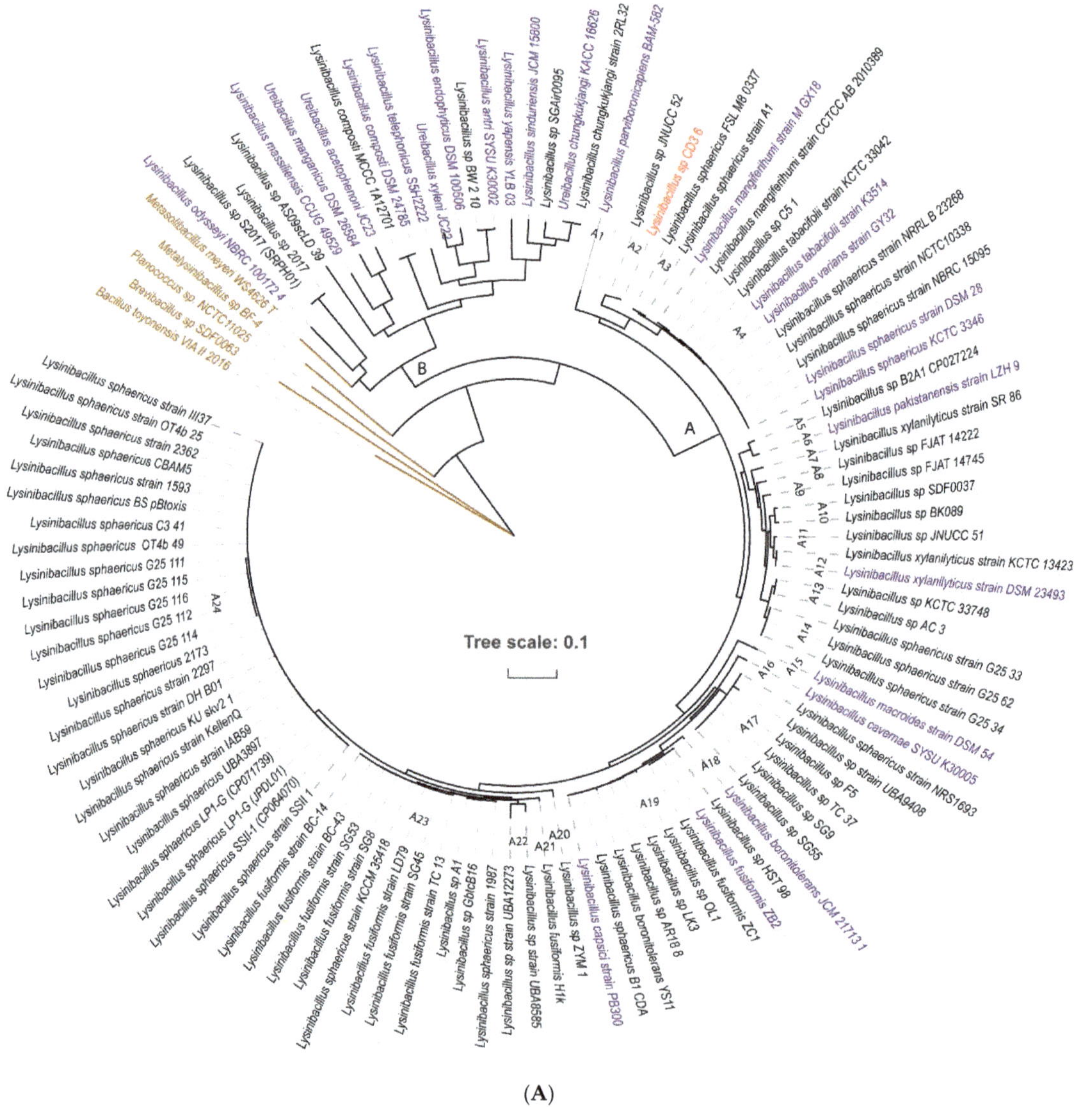

**(A)**

**Figure 3.** *Cont.*

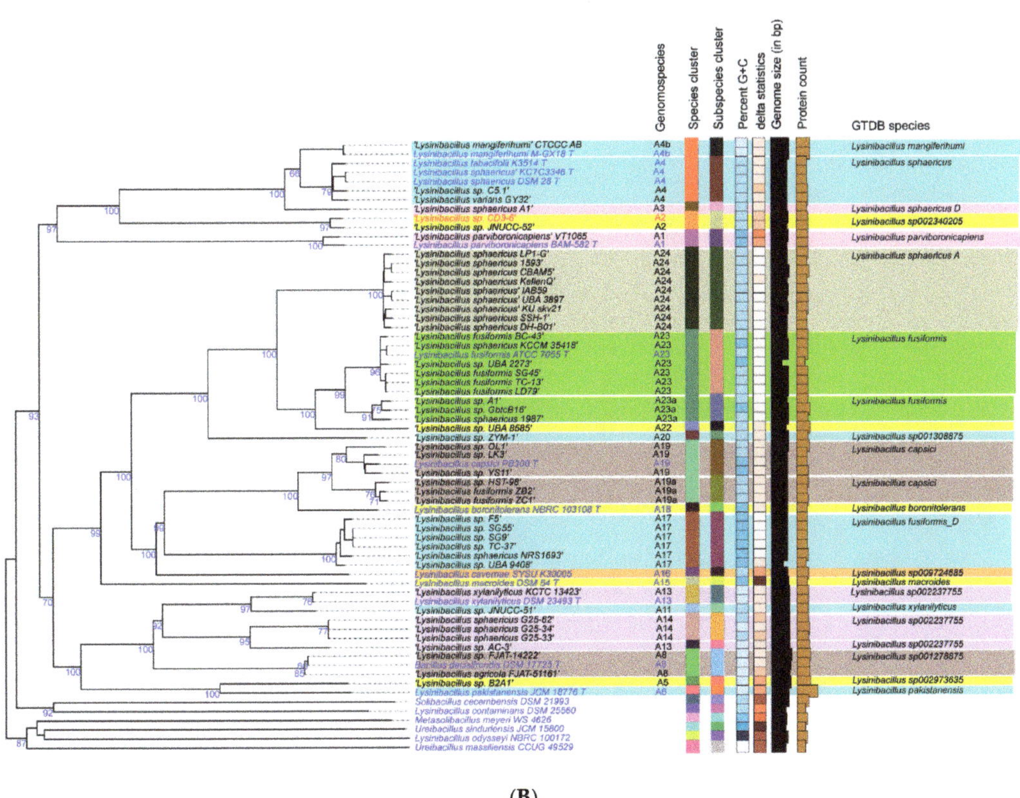

**(B)**

**Figure 3.** (**A**). Approximately-maximum-likelihood phylogenetic tree for 113 *Lysinibacillus* genomes, calculated by EDGAR using the Fast Tree software (http://www.microbesonline.org/fasttree/, accessed on 3 December 2022) and drawn by iTOL. Type strains are labelled in blue letters. Strains, previously misidentified as *Lysinibacillus* and now reclassified as representatives of other species, are labelled in brown. The tree harbored two main branches. Branch A contains the species related to *Lysinibacillus sphaericus*, whilst branch B contains the species which are more remote from *L. sphaericus*. The tree was built out of a core of 156 genes per genome, for a total of 17,628. The core has 58,525 AA-residues/bp per genome, for a total of 6,613,325. (**B**). *Lysinibacillus* tree inferred with FastMe 2.1.6.1 from GBDP distances calculated from whole genome sequences using the Type (Strain) Genome Server TYGS (https://tygs.dsmz.de, accessed on 3 December 2022). The tree consisted of 26 species and 28 subspecies clusters. Analysis was performed using both Maximum Likelihood and Maximum Parsimony. The numbers above the branches are GBDP pseudo-bootstrap support values >60% from replications, with an average branch support of 66.0%. Genomospecies according to (**A**) are indicated. The first two colored columns to the right of each name refer to the genome-based species and subspecies clusters, respectively, as determined by dDDH cut-off of 70 and 79%, respectively. The GTDB species are indicated at the right. The clustering yielded 16 species clusters and strain CD3-6 (labelled in red) was assigned together with JNUCC-52 as novel genomospecies (A2, *Lysinibacillus sp002340205*). Type strains are labelled in blue letters. The tree was rooted at the midpoint.

The extended analysis of whole-genome similarity by their Average Nucleotide Identity (ANI) corroborated the division of the *Lysinibacillus* group A into 24 species clusters (Supplementary Table S6, Supplementary Figure S7). Except clusters A21 and A22, all remaining clusters could be assigned to the genomospecies described in the genome taxonomy database (GTDB) release 07-RS207 (8 April 2022, [40]. The clusterA2 with *Lysini-*

*bacillus* CD3-6 together with *Lysinibacillus* JNUCC-52 was assigned as being *Lysinibacillus* sp002340205 following the GTDB taxonomy [41]. Average nucleotide identity and aligned nucleotides [%] were determined using JSpecies (https://jspecies.ribohost.com/, accessed on 3 December 2022) [42]. The method was based on a BLASTN comparison of the genome sequences [43].

### 3.2.2. Novel GS Were Assigned for Plant-Associated *Brevibacillus* Isolates

The phylogenomic tree constructed with 134 *Brevibacillus* genomes, including the eleven *Brevibacillus* isolates obtained from Vietnamese crop plants, demonstrated that three different major clusters (A, B, C) can be distinguished (Figure 4). Group B consisted of *Brevibacillus laterosporus* strains, whilst group A was formed by representatives of *Brevibacillus brevis* and their closest relatives. Within group A, six major branches with a total of 23 clusters were distinguished. Only 14 of those clusters were covered by type strains, whilst nine clusters represent novel genomospecies without validly recognized type strains. The majority of strains obtained from Vietnamese crop plants clustered within the *Brevibacillus* A group at branch 6. Only the putative *B. parabrevis* strains HD3.3A and HD1.4A clustered within branch 5 (Supplementary Table S7). A third cluster, group C, appeared as a heterogenous group of Brevibacilli, displaying a distant relationship to each other and to the other representatives of the *Brevibacillus* taxon (Figure 4A).

According to the analysis routine performed by the Type (Strain) Genome Server (TYGS), five of the Vietnamese *Brevibacillus* isolates clustered together with valid recognized type strains. The *Brevibacillus* strains HD1.4A, and HD3.3A were assigned as *B. parabrevis*, and HB1.1, HB1.2, HB1.4B formed a cluster with the *B. porteri* type strain. The remaining six isolates did not cluster together with validly published type strains and most likely represent four novel GS. *Brevibacillus* sp. RS1.1, HB2.2, and HB1.3 clustered together within one putative novel species cluster, but split into two different subspecies. Two distinguished novel species cluster were formed by *Brevibacillus* sp.: DP1.3A and M2.1A. Another novel GS was formed by *Brevibacillus* MS2.2 together with *Brevibacillus* sp. Leaf182, isolated from *Arabidopsis* phyllosphere [9] (Figure 4A).

The ANI was proposed to replace classical DNA-DNA hybridization (DDH) as the method for prokaryotic species circumscription in 2009 [45]. The FastANI heatmap constructed with the *Brevibacillus* group A genomes (Supplementary Figure S8) corroborated that the clusters presented in the phylogenomic tree (Supplementary Figure S9) correctly reflected their taxonomic relationship down to species level when using the cut-off level (95–96%) recommended for interspecies identity [43]. In addition, we have aligned our data with the most recent release of the Genome Taxonomy Database (GTDB) [40] (Release 07-RS207 (8 April 2022), and also found appropriate designations used in Supplementary Figure S8 in case of clusters not covered by recognized type strains. According to the classification given by ANI, dDDH, and GTDB, the query *Brevibacillus* isolate genomes were clustered as follows:

1. Strains HD1.4A and HD3.3A represented the *Brevibacillus parabrevis* species cluster (GS A5-25). They only shared ANIb values of ≤85% with the other group A *Brevibacillus* clusters.
2. The *Brevibacillus porteri* species cluster (GS A6-31) was formed by HB1.1, HB 1.2, and HB1.4B. Their ANIb-values (92–93%), when compared with the other *"brevis"* group strains, were below the species cut-off level (95–96%) recommended as the ANI criterion for interspecies identity.
3. The same was true for HB1.3, HB2.2, and RS1.1 forming together with other genomes GS A6-33. The cluster was designated according to the GTDB classification, as *Brevibacillus brevis* D and did not contain a type strain.
4. Strain MS2.2 formed together with Leaf182 the GS A6-29, designated as *Brevibacillus brevis* C. Their ANI values were found below 93% when compared with the most related clusters of the A6-branch.

5.  *Brevibacillus* sp. DP1.3A shared a common cluster (A6-30) with the genome of *Brevibacillus* sp. BC25. The GTDB classification of this cluster was s_*Brevibacillus* sp. 000282075.

*Brevibacillus* sp. M2.1A formed the unique GS A6-34 as a single strain, which was found to be most related to the *Brevibacillus formosus* cluster (GS A6-35). However, ANIb values estimated when compared with this cluster were found to be below the species cut-off (<96%). The GTDB classification of this cluster was s_*Brevibacillus* sp. 013284355.

(**A**)

**Figure 4.** *Cont.*

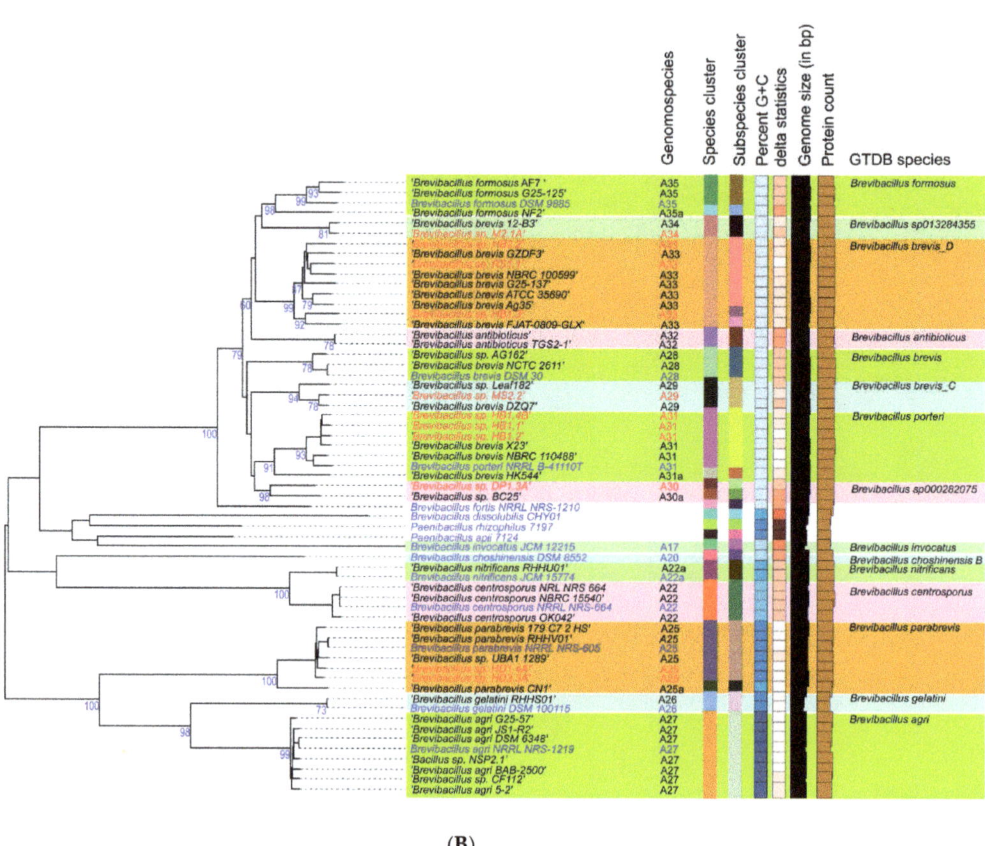

(B)

**Figure 4.** (**A**). Approximately-maximum-likelihood phylogenetic tree for 134 *Brevibacillus* genomes using the FastTree software accessible within the EDGAR package 3.0 [16]). Brevibacilli investigated in this study are labelled in red letters. The tree was built out of a core of 495 genes per genome, for 66,330 in total. The core has 167,683 AA-residues/ bp per genome, 15,461,980 in total. To construct a phylogenetic tree for a project, the core genes of these genomes are computed. In a following step, alignments of each core gene set are generated using MUSCLE, and the alignments are concatenated to one huge alignment. (**B**). *Brevibacillus* tree inferred with FastMe 2.1.6.1 [44] from GBDP distances calculated from whole genome sequences using the Type (Strain) Genome Server TYGS (https://tygs.dsmz.de, accessed on 3 December 2022). Analysis was performed using both Maximum Likelihood and Maximum Parsimony, with 14 type strains (labelled in blue letters) and 48 additional genome sequences including the *Brevibacillus* strains isolated from Vietnamese crop plants (labelled by red letters). The numbers above branches are GBDP pseudo-bootstrap support values >60% from replications, with an average branch support of 85.1%. The first column to the right of each name refers to the genome-based species according to the nomenclature given in (**A**). Species and Subspecies cluster (columns 2 and 3) are characterized by dDDH cut-off values of 70 and 79%, and ANI values of 96 and 98%, respectively. A total of 15 GTDB species clusters could be distinguished. The tree was rooted at the midpoint.

The strains including their corresponding GS are summarized in Supplementary Table S7 and Supplementary Figure S9, and were used for bioassays of their ability to control plant pathogens and to promote plant growth, as described in the next section.

*3.3. Plant-Associated Brevibacilli Promote Plant Growth and Suppress Plant Pathogens*

Our in vitro bioassays demonstrated that the *Brevibacillus* strains did efficiently inhibit the growth of phytopathogenic bacteria, fungi, and nematodes (Supplementary Table S8).

3.3.1. Antagonistic Activity against Gram-Positive and Gram-Negative Bacteria

Antagonistic activity against bacteria was indicated by inhibition zones around filter discs containing the test bacteria. The filter discs were placed onto soft agar mixed with the Gram-positive bacterium *Clavibacter michiganensis*, the causative agent of ring-rot in potato tubers. Antibiosis was demonstrated with all tested representatives of the *Brevibacillus* taxon independent of whether cells or supernatants were used. Inhibition zones appeared after one day and increased steadily during the whole period of observation. *Brevibacillus* sp. HB2.2, *Brevibacillus* sp. RS1.1, *B. porteri* HB1.4B, and *Brevibacillus* sp. DP1.3A were among the strains with the highest antagonistic activity against *Clavibacter michiganensis*. However, the antagonistic activity of *Lysinibacillus* CD3-6 was hard to detect (Figure 5).

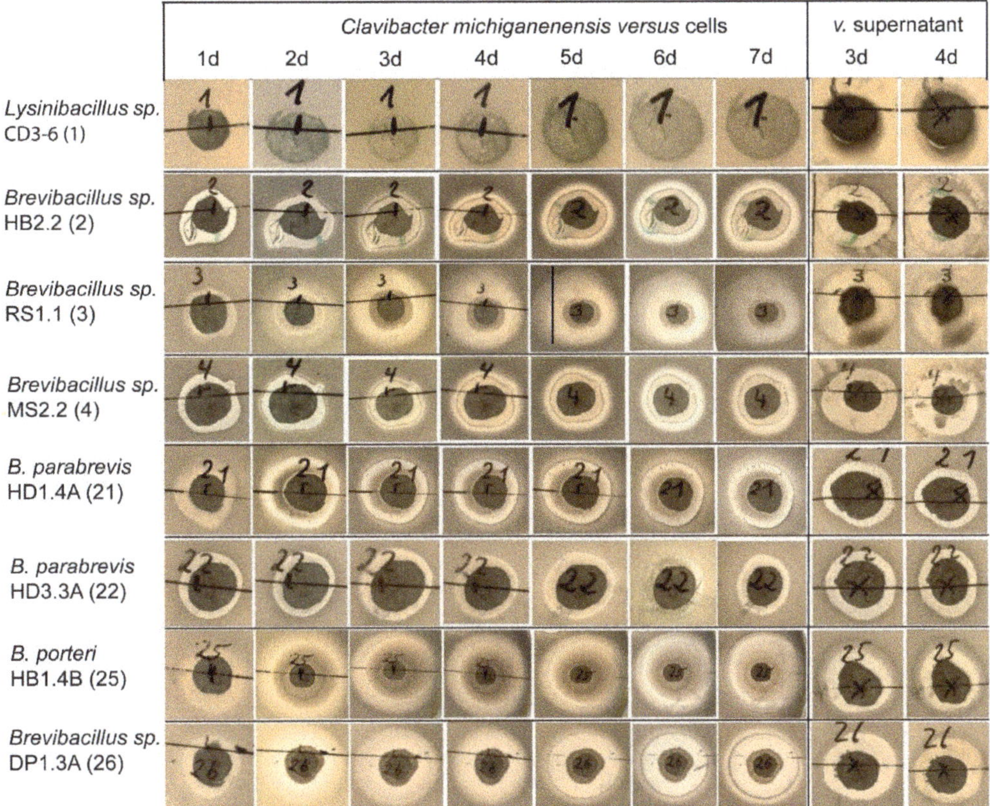

**Figure 5.** Inhibition of *Clavibacter michiganensis* by *Lysinibacillus* and *Brevibacillus* cells and supernatants indicated by clearance zones around the test bacteria.

*Xanthomonas campestris*, the causative agent of bacterial leaf spot disease in pepper, was used as an indicator of antagonistic activity. The largest growth inhibition showed *B. porteri* HB1.1, *Brevibacillus* DP1.3A, and *Brevibacillus* M2.1A. Corresponding to the results obtained with *C. michiganensis*, *Lysinibacillus* CD3-6 did not suppress the bacterial plant pathogen (Supplementary Table S8).

The Gram-negative plant pathogens *Dickeya solani* and *Erwinia amylovora* (causative agent of fire blight disease at orchard trees) were also used as indicators for antagonistic activity. This ruled out the possibility that the Gram-negative bacteria were less inhibited by the *Brevibacillus* strains as the Gram-positive bacterium *C. michiganensis*. *Brevibacillus* sp. RS1.1, *Brevibacillus* sp. MS2.2, *B. porteri* HB1.4B, and *Brevibacillus* DP1.3A inhibited the growth of *Dickeya solani* and *Erwinia amylovora* after an incubation period of two or three days, but their inhibition zones were relatively small. *Brevibacillus* sp. HB2.2, *B. parabrevis* HB2.2, and HD1.4A did not apparently inhibit the growth of both Gram-negative pathogens, suggesting that Brevibacilli were more efficient against Gram-positive bacteria (Supplementary Figure S10).

### 3.3.2. Antifungal Activity

Antifungal activity was examined in vitro using four different *Fusarium* species (*F. oxysporum*, *F. culmorum*, *F. poae*, *F.graminearum*) known for causing fusarium wilt disease [46], and *Aspergillus niger* ('black mold'). *Lysinibacillus* sp. CD3-6 was found to be inefficient against most of the *Fusarium* strains, but the *Brevibacillus* strains did suppress the growth of all phytopathogenic fungi and the oomycete *P. palmivora*, one of the most detrimental plant pathogens in Vietnam [47]. *B. porteri* HB1.4B, *Brevibacillus* sp. DP1.3A, *Brevibacillus* sp. RS1.1, and *Brevibacillus* sp. MS2.2 displayed strong inhibiting effects against *F. poae*, and to a minor degree against *F. culmorum*. *F. graminearum* was less sensitive, but was still inhibited by *Brevibacilllus* sp. HB2.2 and *Brevibacilllus* sp. RS1.1 (Supplementary Figure S11).

### 3.3.3. Nematicidal Activity

Root-knot nematodes, such as *Meloidogyne* spp., are one of the most important plant pathogens in tropical and temperate agriculture, and are responsible for significant harvest losses of main Vietnamese crops, such as coffee and black pepper [48]. In order to analyze the antagonistic activity of the *Brevibacillus* strains and *Lysinibacillus* CD3-6, we first tested their suppressing effect against the model nematode *Caenorhabditis elegans*. Fast and slow death rates were estimated in a bioassay under laboratory conditions. This ruled out the possibility that the *Brevibacillus* strains were much more efficient than *Lysinibacillus* sp. CD3-6. *Brevibacillus* sp. M2.1A displayed the highest killing effect against *C. elegans* (Figure 6A). In order to examine the suppressing effects against phytopathogenic nematodes more directly, we isolated a representative strain of *Meloidogyne* sp. directly from the galls of infested black pepper plant roots according to the hypochlorite procedure [49]. The suppressing effect exerted by the test bacteria on disease development was examined in a greenhouse experiment. Ten weeks after the transplanting of the tomato plantlets in soil, the formation of root knots was visually registered and used as a measure for calculating the disease index according to [27]. The *Brevibacillus* strains were found to be efficient in reducing the disease severity to around 50% and less compared to the untreated control. Again, *Brevibacillus* M2.1A performed the best, whilst *Lysinibacilllus* sp. CD3-6 was found to be less efficient than all tested *Brevibacillus* strains (Figure 6B).

Whilst reports about the biocontrol action of *Bacillus* ssp., such as *B. firmus* [50], *B. velezensis* former *B. amyloliquefaciens* [25], *B. cereus*, *B. thuringiensis*, and *B. subtilis* [51] are increasing, to the best of our knowledge we present here the first report on the nematicidal activity of Brevibacilli against root-knot nematodes.

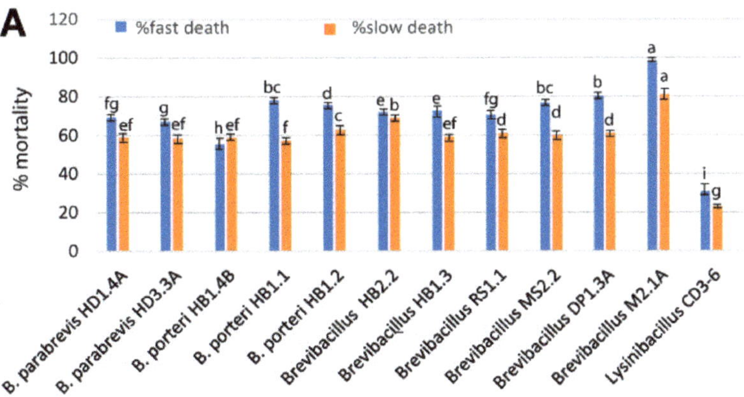

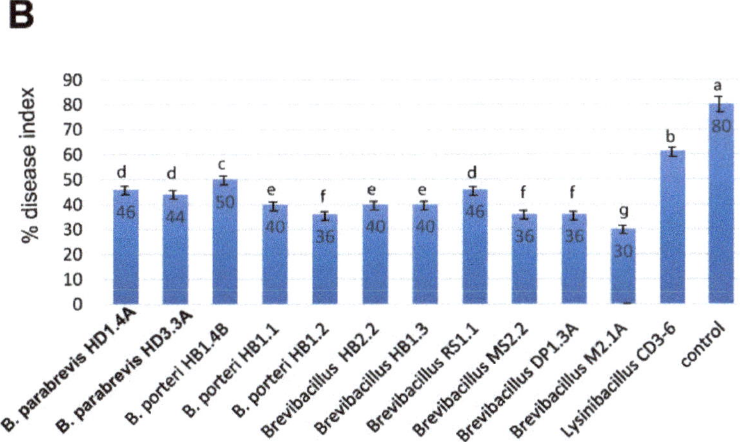

**Figure 6.** Nematicidal activity of *Brevibacillus* strains and *Lysinibacillus* CD3-6. (**A**): Bioassay with *Caenorhabditis elegans*. Slow killing activity was determined on NGM plates, and fast killing activity in liquid medium. (**B**): Tomato plants were infested with the root-knot nematode *Meloidogyne* sp. and the root-knot-forming rate in the presence and absence of the bacterial strains was estimated after 10 weeks growth of the tomato plants under controlled conditions in a greenhouse. All experiments were carried out with three independent repetitions and a completely randomized design. Different letters at each treatment indicate the significance between inoculated and uninoculated conditions at the $p \leq 0.05$ level after the *t*-test.

### 3.3.4. Plant Growth Promotion

We examined the effect of the *Brevibacillus* strains and *Lysinibacillus* CD3-6 in the *Arabidopsis thaliana* biotest system [28]. Several *Brevibacillus* strains enhanced the growth of the *Arabidopsis* seedlings in a considerable manner (Figure 7, Supplementary Table S9).

The highest increase in plant biomass was obtained for *B. porteri* HB1.2 (38.83%), *B. parabrevis* HD3.3A (34.93%), *B. porteri* HB1.1 (30%), and *B. parabrevis* HD1.4A (27.69%).

The above phenotypic experiment proved that plant-associated *Brevibacillus* strains isolated from Vietnamese crop plants are able to positively interact with *Arabidopsis* plants, and to stimulate their growth in the range previously reported for the prototype of the Gram-positive plant-growth-promoting rhizobacteria, *Bacillus velezensis* FZB42 [52].

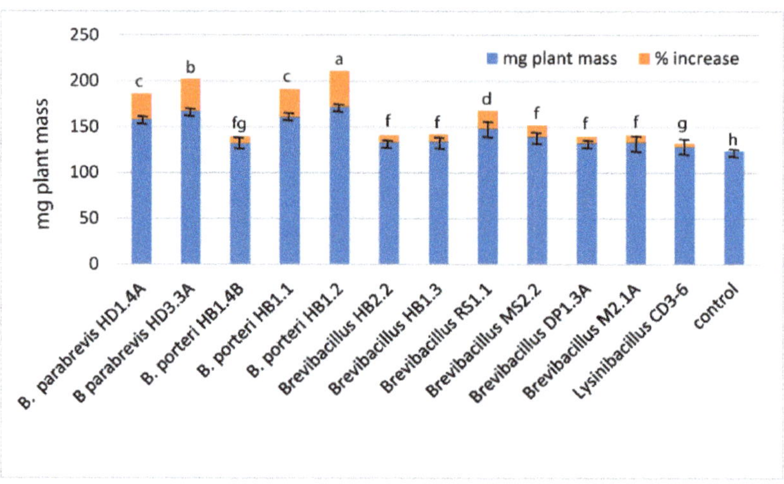



**Figure 7.** Growth promoting effects of *Brevibacillus* strains and *Lysinibacillus* CD3-6 on *Arabidopsis thaliana* seedlings. The % increase compared to the untreated control is indicated on top of the columns. Each treatment value is presented as the means of three replications (n = 3) with SE. Different letters at each treatment indicate the significance between inoculated and uninoculated conditions at the $p \leq 0.05$ level after the *t*-test.

*3.4. Genome Mining for Putative Natural Product Biosynthesis Gene Clusters*

The identification of biocontrol actions directed against phytopathogenic bacteria, fungi and nematodes prompted us to investigate the biosynthetic potential of the isolates using either their draft or their full genome sequences. Antimicrobial compounds belong to structurally diverse groups of molecules, such as nonribosomal peptides (NRP) and polyketides (PK), and ribosomally synthesized and posttranslationally modified peptides [53]. The antiSMASH 6.0 version [29] was used for the prediction and annotation of the biosynthetic gene clusters (BGCs). The antiSMASH results were subsequently compared against the MIBiG database [54] in order to identify characterized and uncharacterized BGCs. In total, 151 BGCs in the 11 *Brevibacillus* isolates, and 6 BGCs in *Lysinibacillus* sp. CD3-6 were identified and separated into nine distinct classes (Supplementary Table S10). A total of 36 of them were not detected in the MIBiG database and might encode the biosynthesis of uncharacterized secondary metabolites. The edeine gene cluster, previously detected in *Brevibacillus brevis* Vm4 and X23 [55], but not listed in the MIBiG databank, was found to be widely distributed in the *Brevibacillus* spp. strains. A variant of another modular PK-NRP hybrid, Paenilipoheptin, recently described in *Paenibacillus polymyxa* E681 [56], was detected for the first time in the genus *Brevibacillus*.

3.4.1. Gene Clusters Encoding Modular and Nonmodular Polyketides

Genes with more than 90% similarity to the gene cluster encoding the non-ribosomal synthesized polyketide macrobrevin [9] were detected in five of the Vietnam *Brevibacillus* strains (Supplementary Table S11). Generally, polyketides are synthesized by giant polyketide synthases in a non-ribosomal manner [57]. Similar to the gene cluster in *Brevibacillus* Leaf182 (BGC0001470), *Brevibacillus* sp. strains DP1.3A and HB1.3 contained the complete set of 15 modules, whilst strains MS2.2, HB2.2, and RS1.1, harboring only 12 modules, did probably encode truncated versions of the polyketide (Supplementary Figure S12). Macrobrevin displays a unique polyketide structure, and was shown to be active against leaf colonizing *Bacillus* strains [9].

Gene clusters with weak similarity to the nonmodular Type III PKS gene cluster (BGC0001964) were detected in most *Brevibacilllus* strains, and also in *Lysinibacillus* sp. CD3-

6 (Supplementary Table S11). Type III PKSs produce a large number of aromatic natural compounds in plants, fungi and bacteria. They catalyze the condensation of coenzyme activated starter units with (methyl)malonyl-CoA extender units by decarboxylative Claisen condensations [58]. A type III PKS chalcone synthase has been described in *Streptomyces griseus* [59] and other Gram-positive bacteria such as *B. subtilis* [60].

3.4.2. Non-Ribosomal-Synthesized Antimicrobial Peptides (NRP)

Gene clusters exhibiting similarity with the genes encoding the well-known NRPs tyrocidine (BGC0000452), gramicidin (BGC0000367), and marthiapeptide (BGC0001469) were detected in different *Brevibacillus* strains isolated from Vietnam crop plants (Supplementary Table S11). NRPs are secondary metabolites which are synthesized through giant multimodular peptide synthetases [61].

The three genes *tyc*A, *tyc*B, and *tyc*C encoding the cyclic ß-sheet decapeptide tyrocidine ($C_{66}H_{87}N_{13}O_{13}$) were detected in the 11 *Brevibacillus* strains representing the species *B. parabrevis*, *B.porteri* and four novel genomospecies (Supplementary Table S11). The gene clusters widely resembled the tyrocidine gene cluster from *Brevibacillus brevis* ATCC8185 [62]. Except for the adenylation domains present in modules 3 and 4, the other eight modules were found highly conserved in all *Brevibacillus* strains (Supplementary Figure S13). This corresponds to the structures reported for tyrocidine A, B and C bearing residues $Phe^{3/4}$, $Phe^{3}/Trp^{4}$, and $Trp^{3/4}$, respectively. The membrane effective antibiotics kill Gram-positive bacteria such as *Bacillus subtilis* and *Staphylococcus aureus* [63].

Gene clusters, similar to the four gene cluster (*lgrA-lgrB-lgrC-lgrD*) described by [64] as being involved in non-ribosomal synthesis of the linear gramicidin peptides, were detected in *B. parabrevis* HD1.4A and HD3.3A, *Brevibacillus* sp. DP1.3A, HB2.2, RS1.1, and M2.1A (Supplementary Table S11). The complete gene clusters present in *B. brevis* (AJ566197), *B. parabrevis* HD3.3A (CP085874.1) HD1.4A (JABSUW010000001), and *Brevibacillus* sp. DP1.3A (CP085876) were characterized by a formylation (F) domain in the first module (*lgrA*) and a final reductase (TD) domain in the last module (*lgrD*). Two different types were distinguished (Figure 8, Supplementary Figure S14). The gramicidin gene clusters in *B. parabrevis* HD1.4A and HD3.3A containing 16 modules were identical with the corresponding 74 kb *lgr* gene cluster in *B. brevis* ATCC 8185 (AJ566197). The product of the *lgr* gene cluster in ATCC 8185 is characterized as a linear gramicidin pentadecapeptide consisting of 15 hydrophobic amino acid residues with an alternating L- and D-configuration forming a β-helix-like structure. Module 16 is predicted to activate alanine (according to our antiSMASH prediction) or glycine (according to [64]), and subsequently to reduce alanine/glycine. Finally, N-formyl-pentadecapeptide-ethanolamine is released from the LgrD enzyme [64].

Surprisingly, we detected a novel variant of the gramicidin gene cluster in *Brevibacillus* sp. DP1.3A (GS A6-30) containing two additional modules in the 3′-region of the *lgrB* gene (Figure 8). Their adenylation domains were coding for Ala and D-Val, and the modules seem to be caused by a partial duplication in the 3′-region. We hypothesize that the 18-module gene cluster is coding for a linear heptadecapeptide (N-formyl-heptadecapeptide-ethanolamine) containing two additional amino acid residues, Ala and D-Val, in position 7 and 8. Other *Brevibacillus* strains harbored truncated forms of both gramicidin gene cluster variants. A truncated pentadecapeptide variant was detected in *Brevibacillus* sp. RS1.1, whilst gene clusters encoding truncated heptadecapeptides were detected in *Brevibacillus* sp. HB2.2, and MS2.2, belonging to genomospecies A6-33, and A6-34, respectively (Supplementary Table S11). Since the truncated gramicidin gene clusters were predicted from draft genomes (WGS), we cannot exclude the possibility that the deletions are due to incomplete sequences. The gramicidin gene cluster present in the MIBiG data bank, BGC0000367 (AP008955.1), was identified as a 13-module gene cluster bearing a 3′ deletion in the *lgrD* gene, probably encoding a truncated variant of the heptadecapeptide (Supplementary Table S11, Supplementary Figure S14). We propose to replace BGC0000367 by the gramicidin gene clusters from *B. brevis* ATCC8185, encoding the complete gramicidin pentadecapeptide, and *Brevibacillus* sp. DP1.3A, encoding the putative gramicidin heptadecapeptide.

*Brevibacillus* sp. MS2.2. harbored a gene cluster with high similarity to marthiapeptide (BGC0001469), a gene cluster recently detected in *Brevibacillus* Leaf182 [9] (Supplementary Table S11, Supplementary Figure S15). The polythiazole cyclopeptide marthiapeptide was first described in *Marinactinospora thermotolerans* and shown to inhibit Gram-positive bacteria and cancer cells [65].

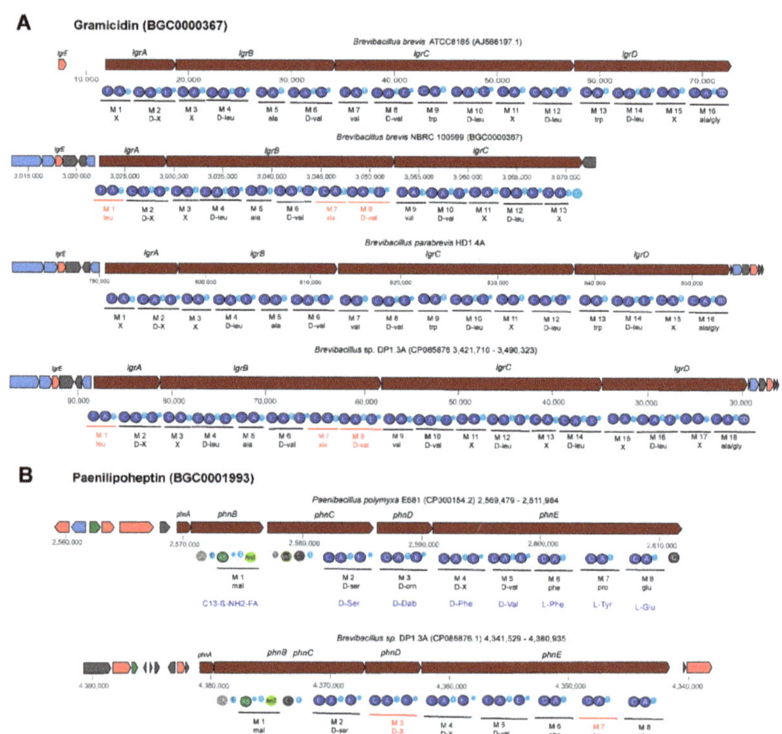

**Figure 8.** Gene clusters involved in non-ribosomal synthesis of peptides and lipopeptides detected in *Brevibacillus* genomes by antiSMASH. (**A**): Two types of gramicidin gene clusters were detected in the *Brevibacillus* isolates: a 16-module gene cluster in *B. parabrevis* HD1.4A encoding a putative *N*-formyl-pentadecapeptide-ethanolamine, and an 18-module gene cluster in *Brevibacillus* sp. DP1.3A encoding a putative *N*-formyl-heptadecapeptide-ethanolamine. (**B**): The gene cluster located in region 15 (4,380,935–4,341,529) of *Brevibacillus* sp. DP1.3A resembled BGC0001728, and is possibly involved in the non-ribosomal synthesis of paenilipoheptin in *Paenibacillus polymyxa* E681. Different modules are labelled in red. NRPS/PKS domains are indicated by filled circles. A: adenylation, C: condensation, CAL: co-enzyme A ligase, E: epimerization domain, KS: ketosynthase domain, T: thiolase.

### 3.4.3. Gene Clusters Encoding PK-NRP Hybrids

Gene clusters, exhibiting high similarity to the edeine gene cluster in *Brevibacillus brevis* Vm4 (Supplementary Figure S16B), were detected in nearly all the *Brevibacillus* strains representing *B. porteri* (HB1.1, HB1.2, HB1.4B), and novel genomospecies such as A6-29 (MS2.2), A6-30 (DP1.3A), A6-33 (HB1.3, RS1.1), and A6-34 (M2.1A), but not in the *B. parabrevis* strains (Supplementary Table S11). The six genes *ede*P, *ede*N, *ede*L, *ede*K, *ede*J, and *ede*I encode NRPS and PKS-NRPS hybrids involved in the non-ribosomal synthesis of the atypical cationic edeine peptides (Supplementary Figure S16A). Edeines contain a ß-Tyr or a ß-Phe residue at the N-terminus and a spermidine-polyamine structure at the

C-terminus, flanking five non-proteinogenic amino acids in the central part (Supplementary Figure S14A). Edeine is a broad-spectrum antimicrobial agent acting against bacteria and fungi. DNA synthesis is inhibited at low concentrations <15 µg/mL [66]. Despite the fact that edeines have been discovered as early as 1959 [67], corresponding gene clusters have not been deposited in the MIBiG data bank until now.

Interestingly, we found a gene cluster in *Brevibacillus* sp. DP1.3A, which strongly resembled the *phn* gene cluster in *Paenibacillus polymyxa* E681 (BGC0001728). The *phn* gene cluster is possibly involved in the non-ribosomal synthesis of the cyclic lipoheptapeptide paenilipoheptin [56]. Both gene clusters contain eight modules (Supplementary Table S11). The first module is involved in fatty acid synthesis, while the remaining modules, containing seven adenylation domains, were predicted to be responsible for the synthesis of a seven-member peptide chain containing D-ser- (1), D-phe- (5), tyr- (6), and glu-residues (7) (Figure 8B). The structure of paenilipoheptin synthesized by *P. polymyxa* was elucidated by MALDI-LIFT TOF/TOF MS as being C13-ß-NH2-FA—Ser (1)—Dab (2)—Trp (3)—Val (4)—Phe (5)—Tyr (6)—Glu (7) [56], which is compatible with the predicted structure of the gene cluster detected in *Brevibacillus* DP1.3A (Supplementary Figure S17). A thioesterase domain at the 3′ end of the phnE gene was missing in both gene clusters, indicating that cyclization might be accomplished by a free-standing TE enzyme as proposed by Vater et al. [56].

### 3.4.4. Gene Clusters Representing RiPPs, and Bacteriocins

In contrast to polyketides and peptides, which are synthesized independently from ribosomes, numerous secondary metabolites with antimicrobial activity such as RiPPs (ribosomally synthesized and posttranslationally modified peptides) and (unmodified) bacteriocins are synthesized by a ribosome-dependent mechanism. Several groups are distinguished [68]. Some of them, such as lassopeptides, LAPs, lanthipeptides, UviB peptides, and sactipeptides, were detected applying the antiSMASH and BAGEL4 [30] toolkits in the *Brevibacillus* isolates and/or the *Lysinibacillus* sp. CD3-6 strain (Figure 9).

Many RiPP biosynthetic proteins recognize and bind their cognate precursor peptide through a domain known as RiPP recognition element (RRE) [69]. The detection of RRE domains can be helpful in identifying gene clusters involved in the synthesis of novel classes of RiPPs [70], but does not necessarily identify a specific category of RiPPs. RRE containing domains with weak similarity to Pantocin-Microcin RRE (BGC 0000585) were found in nearly all *Brevibacillus* strains, but they were not associated with other genes involved in RiPP synthesis. In contrast, the gene for the lassopeptide RRE domain protein in *Lysinibacillus* sp. CD3-6 was associated with the genes encoding the lassopeptide class ii core (leader) peptide with a putative macrolactam sequence, and the lasso peptide biosynthesis B2 protein (Supplementary Table S12).

Gene clusters encoding linear azol(in)e-containing peptides (LAPs) were detected in all 11 *Brevibacillus* isolates, and *Lysinibacillus* CD3-6 (Supplementary Table S12). Most of them were identified as being members of the TOMM class (thiazole/oxazole-modified microcins) characterized by a gene cluster consisting of a cyclodehydratase gene and associated genes encoding dehydrogenase and a maturation protein. A gene encoding a TOMM precursor leader peptide was detected upstream of these three genes (Supplementary Figure S16). Typically, the TOMM precursor leader peptides were characterized by a homologous leader region and then a region enriched with Cys residues, a feature of the hetero-cycloanthracin/sonorensin family [68]. This type of thiopeptide encoding gene was detected in representatives of the *Brevibacillus* genomospecies 25, 29, 31, and 33 (Supplementary Table S12). Different sequences encoding TOMM leader peptides enriched with Val-Ala or Ser were detected in *Brevibacillus parabrevis* (GS 25), *Brevibacillus* sp. DP1.3A (GS 30), and *Brevibacillus* sp. M2.1A (GS34) (Supplementary Table S12). The LAP gene cluster in *Lysinibacillus* CD3-6 encoded a SagB/ThcOx family dehydrogenase, and a YcaO-like family protein, but did not possess genes encoding maturase and precursor peptide proteins.

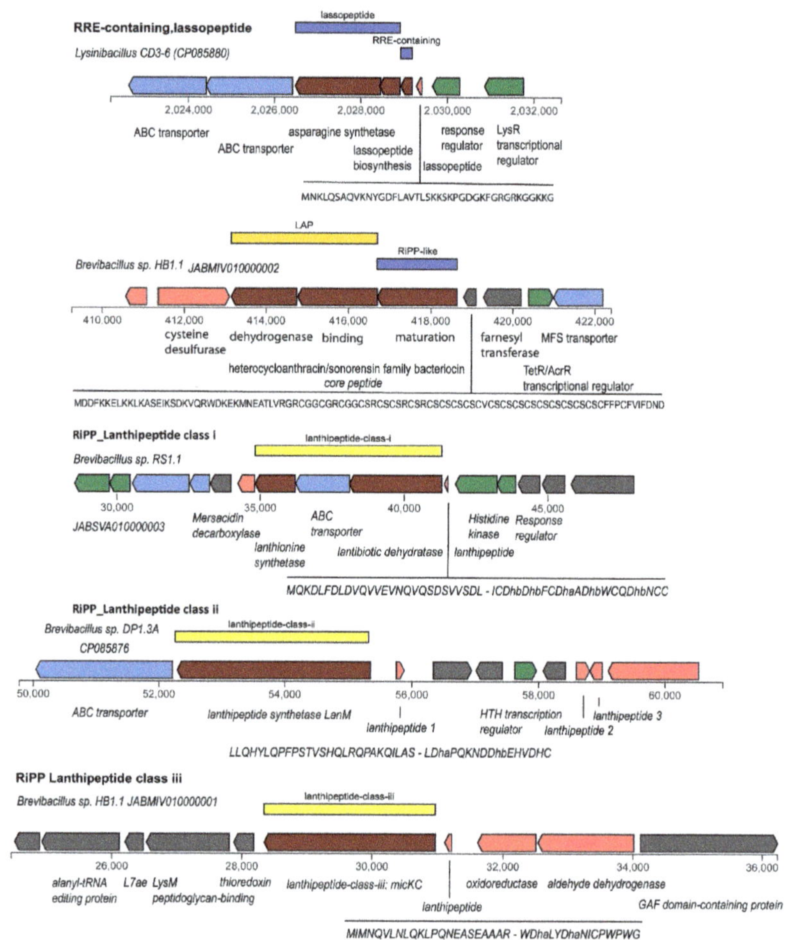



**Figure 9.** Different classes of RiPPs occurring in *Lysinibacillus* CD3-6 and in the plant-associated *Brevibacillus* strains.

Ripps similar to Linocin M18 were found in all *Brevibacillus* GS (Supplementary Table S12). The Linocin_M18 bacteriocin, first isolated from *Brevibacterium linens* M18 by Valdez-Stauber and Scherer [71], inhibits Gram-positive bacteria. These widely distributed proteins, referred to as encapsulins, form nano-compartments within the bacterium which contain ferritin-like proteins or peroxidaseenzymes. Lanthipeptides are defined by the presence of ß-thioether cross-links, which are generated by the posttranslational modification of Ser/Thr and Cys residues [72]. Best studied are class I and class II lanthipeptides, which are modified by different dehydratases (LanB or LanM) and cyclases. Gene clusters encoding class I lanthipeptides were detected in two representatives of *Brevibacillus* GS A6-33 (HB1.3, RS1.1) and GS A6-34 (M2.1A) (Supplementary Table S12, Supplementary Figure S19). Genes involved in synthesis and posttranslational modification of class II lanthipeptides were only detected in *Brevibacillus* DP1.3A, a representative of GS A6-30 (Supplementary Figure S20). Gene clusters for biosynthesis of class III lanthipeptides were detected in *B. porteri* strains HB1.1 and HB1.2, and in *Lysinibacillus* CD3-6 (Supplementary Figure S21, Supplementary Table S12).

A representative of UV inducible peptides (UviB) was detected in *Brevibacillus parabrevis* using BAGEL4 supported genome mining (Supplementary Table S12, Supplementary Figure S22). The Phage_holin_BhlA family is a family of holin-like proteins from both bacteriophages and bacterial chromosomes. BhlA, a putative holin-like protein of *Bacillus licheniformis* AnBa9 [73], and *Bacillus pumilus* WAPB4 [74], showed antibacterial activity against several Gram-positive bacteria.

### 3.4.5. Gene Cluster Involved in Synthesis of Siderophores and Other BGCs

The *asbABCDEF* gene cluster (BGC0000942) from *Bacillus anthracis, the causative agent of anthrax*, is responsible for the biosynthesis of petrobactin, a catecholate siderophore that functions in both iron acquisition and virulence [75]. It has been argued that the iron-siderophore petrobactin contributes to *B. anthracis* pathogenesis, which requires maintaining sufficient iron concentration during growth in the host tissue [76]. However, it has been documented that petrobactin synthesis is also common in non-pathogenic members of the *B. cereus senso latu* group [77]. Except for two representatives of *B. parabrevis* (HD1.4A, HD3.3A), we detected gene clusters with apparent similarity to BGC0000942 in representatives of *B. porteri* (HB1.1, HB1.2, HB1.4B) and the novel GS A6-29 (MS2.2), A6-30 (DP1.3A), and A6-33 (HB2.2, RS1.1, HB1.3) (Supplementary Table S12). By contrast, the gene clusters encoding non-ribosomal synthesis of the second siderophore bacillibactin, common in the *B. subtilis*, and *B. cereus* group, were not detected in the investigated *Brevibacillus* strains.

Other BGCs such as gene clusters involved in synthesis of terpenes, and the cyclic lactone autoinducer with a putative role in quorum sensing (AgrB precursorpeptides) were detected in most *Brevibacillus* isolates. *Lysinibacillus* sp. CD3-6 also harbored terpene encoding genes, and a gene cluster possibly involved in the synthesis of a betalactone containing protease inhibitor (Supplementary Table S12).

### 3.4.6. Uncharacterized NRP and PK-NRP Hybrid Gene Clusters in *Brevibacilli*

In addition to the characterized BGCs mentioned above, we detected 35 hitherto unknown NRPs and NRP-PK hybrid scaffolds in the *Brevibacillus* genomes, and one partial NRPS gene cluster containing a tyrosine module in *Lysinibacillus* CD3-6 (Supplementary Table S13).

A putative PK-NRP hybrid consisting of four modules (PK-orn-x-phe) occurred in all nine *Brevibacillus* strains belonging to the *Brevibacillus* A6-branch. In contrast, a three-module hybrid (asn-gly-pk) occurred only in *Brevibacillus* M2.2 (Supplementary Figure S23).

Giant PK-NRP hybrids consisting of either 16, 20, or 25 modules (Supplementary Figure S24) were detected in several *Brevibacillus* strains. The 16 M scaffold consisted of 14 PK modules (mal, ccmal, ohmal), and two modules involved in the non-ribosomal synthesis of amino acids (M6:X, M16: Ser). The three *B. porteri* strains, and *Brevibacillus* HB1.3, RS1.1, HB2.2, and M2.1.A harbored all 16 modules of the hybrid, whilst *Brevibacillus* DP1.3A harbored a deleted gene cluster, in which the modules 9–16 were missing. The largest gene cluster harboring 25 modules was detected in *B. parabrevis* HD3.3A. Some 14 modules were responsible for the non-ribosomal synthesis of amino acids, whilst the 11 pk-modules were involved in the synthesis of mal, ohmal, and ccmal. A deleted form of this scaffold harboring only 20 modules was detected in *B. parabrevis* HD1.4A.

Decapeptide scaffolds were detected adjacent to the *B. parabrevis* giant M25/20 PK-NRP hybrid scaffolds (Supplementary Table S13). Experimental proof is needed to corroborate their stand-alone state. Alternatively, it is also possible that they are directly connected with the giant hybrids.

Slightly modified NRP scaffolds consisting of either six or seven modules, and always starting with Glu-Ser at their N-terminus, were present in nearly all representatives of the A6 branch (Supplementary Figure S25). The six-module variant (glu-ser-ile-x-phe-D-orn) occurred in the genomes of *Brevibacillus* M2.2, and HB2.2. One of the two seven-module scaffolds contained the predicted sequence Glu-Ser-Val-Val-X-Phe-D-Orn, and occurred in *Brevibacillus* sp. DP1.3A, HB1.1, HB1.2, HB1.3. *Brevibacillus* sp. M2.1A harbored a second variant characterized by the predicted sequence Glu-Ser-Val-X-X-Phe-D-Orn.

Preliminary experiments did not reveal hexa- or heptapeptides in the *Brevibacillus* strains, but pentapeptides with partial corresponding sequences were detected. At present, we cannot exclude the possibility that the scaffolds are responsible for the synthesis of the pentapeptides, but one or two of the modules might be overread or not expressed.

An interesting NRP gene cluster encoding for three cysteine residues was detected in *B. parabrevis* HD1.4A, HD3.3A, *Brevibacillus* DP1.3A, and *B. porteri* HB1.1, HB1.2, HB1.4B (Supplementary Table S13). Due to the presence of predicted methyltransferase domains, it can be assumed that the cysteine residues are methylated (Supplementary Figure S26). The cryptic *nrs* gene cluster of the model biocontrol bacterium *Bacillus velezensis* FZB42 also harbors three modules encoding for cysteine. Recently, trithiazole was identified as the product of the *nrs* gene cluster, and NrsB as the oxidizing enzyme of the thiazoline precursor [78]. However, a more careful comparison of the domain structure of both gene clusters revealed significant differences between both BGCs, excluding the possibility that the final product can be a polythiazole (Figure 10). Although no o Ox domain was detected, two genes involved in dihydroxybenzoate (dhb) synthesis were (Figure 10A). The 2,3-dhb-AMP ligase and an oxido-reductase, predicted to be involved in dhb synthesis, possess counterparts in the *dhb* gene cluster of FZB42 known to be responsible for the non-ribosomal synthesis of the siderophore bacillibactin (BGC0000616.1, Figure 10C). For this reason, we assume that the unknown gene cluster might be involved in the biosynthesis of a novel siderophore a with function in iron acquisition.

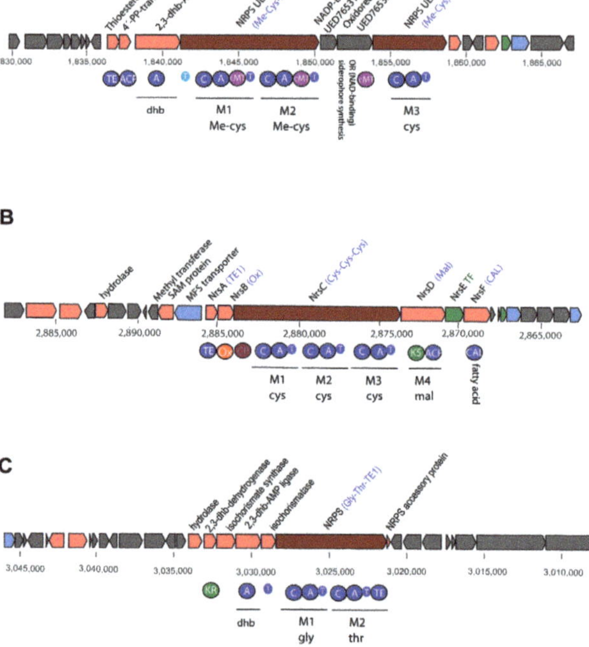

**Figure 10.** The uncharacterized NRPS gene cluster in *Brevibacillus* sp. DP1.3A (**A**) compared with the Bacillothiazol synthesizing *nrs* gene cluster (**B**) and the siderophore bacillibactin *dhb* gene cluster in *Bacillus velezensis* FZB42 (**C**). NRPS/PKS domains are indicated by filled circles. A: adenylation, ACS: 4′-phosphopantetheinyl transferase, C: condensation, CAL: co-enzyme A ligase, cMT: carbon methyltransferase, KR: ketoreductase, KS: ketosynthase, T: thiolase, TE: thioesterase.

## 4. Conclusions

Recently a total of 59 endospore-forming Gram-positive bacteria were isolated from healthy crop plants within fields in Vietnam that were infested with plant-pathogenic nematodes and fungi. According to their draft genome sequences, the majority of the strains were classified as being members of the *B. subtilis* group, mainly *B. velezensis*, and the *B.cereus* group. The remaining 12 strains were provisionally classified as *Brevibacillus* sp. and *Lysinibacillus* sp., respectively [1]. In this study we have focused on the members of the *Brevibacillus* and *Lysinibacillus* taxon, and a special procedure for enrichment of *Brevibacillus* strains was developed. The sequencing of the whole genomes of *Lysinibacillus* sp. CD3-6 and of three *Brevibacillus* strains allowed for a detailed genome analysis of the selected strains, and the identification of several extrachromosomal elements. The novel isolate *Brevibacillus* sp. DP1.3A, for example, harbored three different plasmids, representing either the low-copy type, whose segregation was directed by ParM, or the high copy type, replicating according to the rolling circle model. The high plasticity of the genomes was also indicated by the presence of numerous genomic islands within the *Lysinibacillus* and *Brevibacillus* chromosomes, and the high number of genes present in the pan-genomes.

Furthermore, we have elucidated the taxonomic position of the eleven *Brevibacillus* strains, and of the *Lysinibacillus* strain on the base of their complete or draft genome sequences. Five of the *Brevibacillus* strains were classified as being members of the known species, *B. parabrevis* (2), and *B. porteri* (3). The other six *Brevibacillus* strains, and *Lysinibacillus* CD3-6 were, according to their ANI and dDDH values, classified as being members of five novel genomospecies.

The main outcome of this work was characterizing the *Brevibacillus* isolates as potent antagonists of important plant pathogens. *Brevibacillus* strains were found to be efficient in directly inhibiting the growth of bacterial phytopathogens, such as *Clavibacter michiganensis*, *Xanthomonas campestris*, *Erwinia amylovora*, *Dickeya solani*, and fungal pathogens, such as *Fusarium oxysporum*, and other representatives of the *Fusarium* genus. The direct antagonistic action of the *Brevibacillus* strains against the phytopathogenic oomycete *Phytophthora palmivora* was also observed. Brevibacilli displayed high nematicidal activity, and significantly suppressed the formation of root-knots in tomato plants infested with *Meloidogyne* sp. in greenhouse experiments under controlled conditions.

The *Brevibacillus* genomes harbored a rich arsenal of known and hitherto uncharacterized BGCs predicted to synthesize ribosomally and non-ribosomally diverse classes of secondary metabolites with potential antagonistic action against plant pathogens. A total of 151 BGCs including 35 uncharacterized BGCs were identified by genome mining in the *Brevibacillus* genomes, whilst *Lysinibacillus* sp. CD3-6 harbored only six BGCs. The occurrence of important BGCs in the representatives of different genomospecies is shown in Figure 11. In several cases, it seems that the distribution of BGCs in Brevibacilli is dependent on their taxonomical position. Gene clusters for the synthesis of edeine and petrobactin, for example, were not detected in *B. parabrevis*, but were present in the species cluster A29, A30, A31, A33, and A34. The gene clusters devoted to tyrocidine synthesis were identified in all isolates. In contrast, gene clusters involved in the synthesis of marthiapeptide and paenilipoheptin were only detected in single strains representing the species cluster A29 and A30, respectively. The paenilipoheptin gene cluster exhibited striking similarity to a corresponding gene cluster in *Paenibacillus polymyxa* [56] Our results are in line with the previous finding that *Brevibacillus* sp. Leaf182 is the most potent antagonist within the *Arabidopsis* phyllosphere-microbiome consisting of 224 strains [9]. The BGC responsible for non-ribosomal synthesis of the antibacterial polyketide macrobrevin, originally described in Leaf182, was detected in three novel genomospecies (A29, A30, and A33). Its presence on the DP1.3A genomic islands suggested that the cluster could be transferred by horizontal gene transfer. Two types of gramicidin gene clusters were detected. One of them which harbored 18 modules has been not described previously. In summary, our genome mining results indicated that Brevibacilli are a treasure box of widely unexplored AMPs and other secondary metabolites. Biocontrol agents developed from plant-associated *Brevibacillus*

strains should extend our present arsenal of bio-based plant protection agents that are
necessary for a more sustainable agriculture.

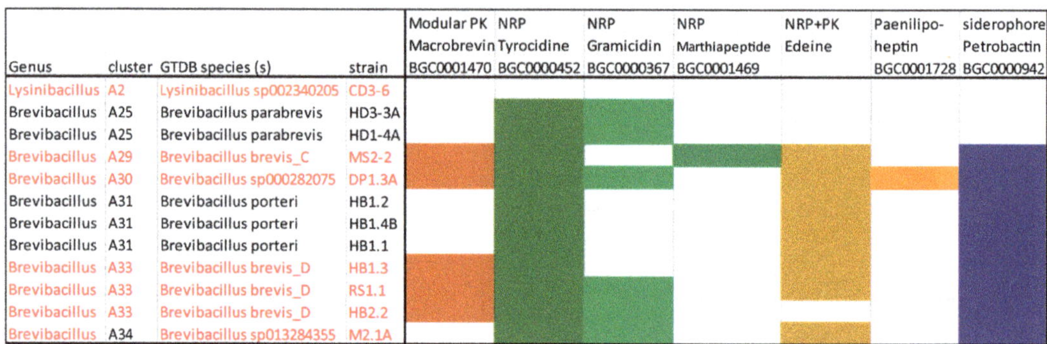


**Figure 11.** Occurrence of important BGCs in the different genomospecies investigated in this study.
Novel genomospecies without type strain are indicated by red letters.

**Supplementary Materials:** The following supporting information can be downloaded at The following are available online at https://www.mdpi.com/article/10.3390/microorganisms1101016 8/s1, Figure S1, Circular plot of the genome of *Lysinibacillus* sp. CD3-6 generated with BioCircos (Cui et al. 2016).; Figure S2, Circular plot of the genome of *Brevibacillus parabrevis* HD3.3A generated with BioCircos (Cui et al. 2016); Figure S3, Circular plot of the genome of *Brevibacillus* sp. DP1.3A generated with BioCircos (Cui et al. 2016); Figure S4, Circular plot of the genome of *Brevibacillus* sp. M2.1A generated with BioCircos; Figure S5, GBDP tree (16S rDNA gene sequence based) with *Lysinibacillus* sp. CD3-6; Figure S6, *Lysinibacillus* tree inferred with FastME 2.1.6.1 from GBDP distances calculated from 37 entries; Figure S7, Heat Map of the FastANI-Matrix [23] of group A *Lysinibacillus* genomes; Figure S8, Heat Map of the FastANI-Matrix [23] of *Brevibacillus* strains isolated from Vietnamese crop plants; Figure S9, Phylogenetic tree constructed with the Vietnamese *Brevibacillus* isolates; Figure S10, In vitro assay of biococontrol action of *Lysinibacillus* and *Brevibacillus* strains against *Dickeya solani* and *Erwinia amylovora*; Figure S11, Antagonistic action of *Brevibacillus* sp. HB2.2 (2), *Brevibacillus* sp. RS1.1 (3), *Brevibacillus* sp. MS2.2 (4), *B. parabrevis* HD1.4A (21), *Brevibacillus parabrevis* (22), *Brevibacillus porteri* HB1.4B (25), and *Brevibacillus* DP1.3A (26) against *Fusarium culmorum* (A), *Fusarium poae* (B), and *Fusarium graminearum* (C) after 4 days incubation at 27 °C; Figure S12, Modular PKS in Brevibacilli: The Macrobrevin variants harbored slight differences compared to *Brevibacillus* sp. leaf 182 (BGC0001470); Figure S13, NRP: Gene clusters similar to Tyrocidine (BGC0000452) present in *Brevibacillus parabrevis* HD1.4A (GS 25), *Brevibacillus* sp. MS2.2, DP1.3A, HB1.1, and M2.1A; Figure S14, NRP: Gene clusters similar to gramicidin (BGC0000367) and the gramicidin gene cluster in *Brevibacillus brevis* (AJ566197) were detected in *Brevibacillus parabrevis* HD1.4A (GS 25), and *Brevibacillus* sp. DP1.3A; Figure S15, NRP: Gene cluster similar to marthiapeptide A (BGC0001469) was detected in *Brevibacillus* sp. MS2.2; Figure S16, Modular PKS-NRP hybrids. A: Structure of edeine isoforms according to Westman et al. 2013. B: The edeine gene cluster of *Brevibacillus brevis* Vm4. C: Prediction of the domain structure in the *Brevibacillus* M2.1A gene cluster similar to the edeine gene cluster of *Brevibacillus brevis* Vm4 in Vietnamese *Brevibacillus* genomes by antiSMASH 6.0; Figure S17, Modular PKS-NRP hybrids. Gene cluster similar to the gene cluster encoding the synthesis of the cyclic lipoheptapeptide paenilipoheptin from *Paenibacillus polymyxa* E681 (BGC0001728) was detected in *Brevibacillus* sp. DP1.3A by antiSMASH 6.0 analysis; Figure S18, RiPP-LAP in *Brevibacillus* sp. HB1.1, DP1.3A, andHD3.3A; Figure S19, Class 1 lanthipeptides detected in *Brevibacillus* sp. HB1.3, *Brevibacillus* sp. RS1.1, and *Brevibacillus* sp. M2.1A; Figure S20, RiPP-Lanthipeptides class 2 in *Brevibacillus* sp. DP1.3A; Figure S21, RiPP lanthipeptides class 3 in *Brevibacillus porteri* HB1.1, HB1.2, and *Lysinibacillus* CD3-6; Figure S22, UviB and sactipeptides were detected in Brevibacillus HD1.4A by BAGEL4; Figure S23, Unknown PK-NRP hybrids pk-orn-x-phe and asn-gly-pk; Figure S24, Unknown PK-NRP hybrids: 16M PK-NRP and 20/25M PK-NRP; Figure S25, Unknown NRPS-like: glu-ser-ile-x-phe-orn/glu-ser-val-val-x-phe-orn/glu-ser-val-X-X-phe-D-orn (Brevipentin); Figure S26, Unknown NRPS-like: cys-cys-cys (Brevitriazol); Table S1, Genomic islands and their genes in *Lysinibacillus* sp. CD3-6 predicted with IslandViewer; Table S2, Genomic islands and their genes in *Brevibacillus parabrevis* HD3.3A predicted with IslandViewer; Table S3, Genomic islands and their genes in *Brevibacillus* sp. DP1.3A predicted with IslandViewer; Table S4, Genomic islands and their genes in *Brevibacillus* sp. M2.1A predicted by IslandViewer; Table S5, Protein features of *Lysinibacillus* CD3-6, *B. parabrevis* HD3.3A, *Brevibacillus* sp. DP1.3A, and *Brevibacillus* sp. M2.1A; Table S6, Genomes of *Lysinibacillus* group A used in this study; Table S7, *Brevibacillus* genomes used in this study and their classification; Table S8, In vitro assay of antagonistic activity of the 11 *Brevibacillus* strains and *Lysinibacillus* CD3-6 against pathogenic bacteria, fungi and nematodes (*C. elegans*); Table S9, Growth promotion of Arabidopsis thaliana by 11 *Brevibacillus* strains isolated from Vietnam and *Lysinibacillus* CD3-6; Table S10, Numbers of known and unknown BGCs detected in *Brevibacillus* spp. and *Lysinibacillus* spp. Isolated from Vietnamese crop plants; Table S11, Association of known BGC classes and antiSMSH annotations detected in the plant-associated Vietnamese *Brevibacillus* and *Lysinibacillus* strains; Table S12, Detection of gene clusters encoding RiPPs, terpenes, and Quorum sensing in the Vietnamese Brevibacillus and *Lysinibacillus* isolates; Table S13, Detection of gene clusters encoding uncharacterized NRPs and NRP-PK hybrids in Vietnamese *Brevibacillus* isolates.

**Author Contributions:** Conceptualization, R.B., T.T.L.T., T.S. and P.L.; investigation, J.J., T.L.P., P.T.L.T., J.B., A.S., J.V., R.B. and T.T.L.T.; resources, P.L.; data curation, J.J. and C.B.; writing, T.T.L.T. and R.B.; review and editing, J.V., T.S., P.L. and R.B.; project administration, R.B., T.S. and P.L.; funding acquisition, R.B. and T.T.L.T. All authors have read and agreed to the published version of the manuscript.

**Funding:** This research was supported through the project ENDOBICA by the Bundesministerium für Bildung und Forschung (BMBF) (grant no 031B0582A/031B0582B), the National Foundation for Science and Technology Development (NAFOSTED); code no. 106.03-2017.28), and the Ministry of Science and Technology (MOST) in Vietnam (code no. NDT.40.GER/18).

**Data Availability Statement:** Accession numbers of DNA sequences are given in Section 2.6.

**Acknowledgments:** We thank Stefanie Herfort, Robert Koch Institute Berlin, Germany for technical support. We are grateful the ABiTEP GmbH, Berlin for supplying phytopathogenic bacteria and fungi used in this study.

**Conflicts of Interest:** The authors declare that they have no conflict of interest.

# References

1. Tam, L.T.T.; Jähne, J.; Luong, P.T.; Thao, L.T.P.; Chung, L.T.K.; Schneider, A.; Blumenscheit, C.; Lasch, P.; Schweder, T.; Borriss, R. Draft genome sequences of 59 endospore-forming Gram-positive bacteria associated with crop plants grown in Vietnam. *Microbiol. Resour. Announc.* **2020**, *9*, e01154-20. [CrossRef] [PubMed]
2. Blumenscheit, C.; Jähne, J.; Schneider, A.; Blom, J.; Schweder, T.; Lasch, P.; Borriss, R. Genome sequence data of *Bacillus velezensis* BP1.2A and BT2.4. *Data Brief* **2022**, *41*, 107978. [CrossRef] [PubMed]
3. Shida, O.; Takagi, H.; Kadowaki, K.; Komagata, K. Proposal for two new genera, *Brevibacillus* gen. nov. and *Aneurinibacillus* gen. nov. *Int. J. Syst. Bacteriol.* **1996**, *46*, 939–946, Erratum in *Int. J. Syst. Bacteriol.* **1997**, *47*, 248. [CrossRef] [PubMed]
4. Parte, A.C.; Sardà Carbasse, J.; Meier-Kolthoff, J.P.; Reimer, L.C.; Göker, M. List of Prokaryotic names with Standing in Nomenclature (LPSN) moves to the DSMZ. *Int. J. Syst. Evol. Microbiol.* **2020**, *70*, 5607–5612. [CrossRef]
5. Edwards, S.G.; Seddon, B. Mode of antagonism of *Brevibacillus brevis* against *Botrytis cinerea* in vitro. *J. Appl. Microbiol.* **2001**, *91*, 652–659. [CrossRef]
6. Zhang, X.; Zhang, B.X.; Zhang, Z.; Shen, W.F.; Yang, C.H.; Yu, J.Q.; Zhao, Y.H. Survival of the biocontrol agents *Brevibacillus brevis* ZJY-1 and *Bacillus subtilis* ZJY-116 on the spikes of barley in the field. *J. Zhejiang Univ. Sci. B* **2005**, *8*, 770–777. [CrossRef]
7. Li, C.P.; Shi, W.C.; Wu, D.; Tian, R.M.; Wang, B.; Lin, R.S.; Zhou, B.; Gao, Z. Biocontrol of potato common scab by *Brevibacillus laterosporus* BL12 is related to the reduction of pathogen and changes in soil bacterial community. *Biol. Control* **2021**, *153*, 104496. [CrossRef]
8. Yang, X.; Yousef, A.E. Antimicrobial peptides produced by *Brevibacillus* spp.: Structure, classification and bioactivity: A mini review. *World J. Microbiol. Biotechnol.* **2018**, *34*, 57. [CrossRef]
9. Helfrich, E.J.N.; Vogel, C.M.; Ueoka, R.; Schäfer, M.; Ryffel, F.; Müller, D.B.; Probst, S.; Kreuzer, M.; Piel, J.; Vorholt, J.A. Bipartite interactions, antibiotic production and biosynthetic potential of the Arabidopsis leaf microbiome. *Nat. Microbiol.* **2018**, *3*, 909–919. [CrossRef]

10. Ahmed, I.; Yokota, A.; Yamazoe, A.; Fujiwara, T. Proposal of *Lysinibacillus boronitolerans* gen. nov. sp. nov., and transfer of *Bacillus fusiformis* to *Lysinibacillus fusiformis* comb. nov. and *Bacillus sphaericus* to *Lysinibacillus sphaericus* comb. nov. *Int. J. Syst. Evol. Microbiol.* **2007**, *57*, 1117–1125. [CrossRef]

11. Chen, S.; Zhou, Y.; Chen, Y.; Gu, J. fastp: An ultra-fast all-in-one FASTQ preprocessor. *Bioinformatics* **2018**, *34*, i884–i890. [CrossRef] [PubMed]

12. Overbeek, R.; Olson, R.; Pusch, G.D.; Olsen, G.J.; Davis, J.J.; Disz, T.; Edwards, R.A.; Gerdes, S.; Parrello, B.; Shukla, M.; et al. The SEED and the Rapid Annotation of microbial genomes using Subsystems Technology (RAST). *Nucleic Acids Res.* **2014**, *42*, D206–D214. [CrossRef] [PubMed]

13. Li, W.; O'Neill, K.R.; Haft, D.H.; DiCuccio, M.; Chetvernin, V.; Badretdin, A.; Coulouris, G.; Chitsaz, F.; Derbyshire, M.K.; Durkin, A.S.; et al. RefSeq: Expanding the Prokaryotic Genome Annotation Pipeline reach with protein family model curation. *Nucleic Acids Res.* **2021**, *49*, D1020–D1028. [CrossRef] [PubMed]

14. Brettin, T.; Davis, J.J.; Disz, T.; Edwards, R.A.; Gerdes, S.; Olsen, G.J.; Olson, R.; Overbeek, R.; Parrello, B.; Pusch, G.D.; et al. RASTtk: A modular and extensible implementation of the RAST algorithm for building custom annotation pipelines and annotating batches of genomes. *Sci. Rep.* **2015**, *10*, 8365. [CrossRef] [PubMed]

15. Wattam, A.R.; Davis, J.J.; Assaf, R.; Boisvert, S.; Brettin, T.; Bun, C.; Conrad, N.; Dietrich, E.M.; Disz, T.; Gabbard, J.L.; et al. Improvements to PATRIC, the all-bacterial Bioinformatics Database and Analysis Resource Center. *Nucleic Acids Res.* **2017**, *45*, D535–D542. [CrossRef]

16. Dieckmann, M.A.; Beyvers, S.; Nkouamedjo-Fankep, R.C.; Hanel, P.H.G.; Jelonek, L.; Blom, J.; Goesmann, A. EDGAR3.0: Comparative genomics and phylogenomics on a scalable infrastructure. *Nucleic Acids Res.* **2021**, *49*, W185–W192. [CrossRef]

17. Bertelli, C.; Laird, M.R.; Williams, K.P.; Simon Fraser University Research Computing Group; Lau, B.Y.; Hoad, G.; Winsor, G.L.; Brinkman, F.S.L. IslandViewer 4: Expanded prediction of genomic islands for larger-scale datasets. *Nucleic Acids Res.* **2017**, *45*, W30–W35. [CrossRef]

18. Cui, Y.; Chen, X.; Luo, H.; Fan, Z.; Luo, J.; He, S.; Yue, H.; Zhang, P.; Chen, R. BioCircos.js: An interactive Circos JavaScript library for biological data visualization on web applications. *Bioinformatics* **2016**, *32*, 1740–1742. [CrossRef]

19. Meier-Kolthoff, J.P.; Göker, M. TYGS is an automated high-throughput platform for state-of-the-art genome-based taxonomy. *Nat. Commun.* **2019**, *10*, 2182. [CrossRef]

20. Meier-Kolthoff, J.P.; Sardà Carbasse, J.; Peinado-Olarte, R.L.; Göker, M. TYGS and LPSN: A database tandem for fast and reliable genome-based classification and nomenclature of prokaryotes. *Nucleic Acid Res.* **2022**, *50*, D801–D807. [CrossRef]

21. Ondov, B.D.; Treangen, T.J.; Melsted, P.; Mallonee, A.B.; Bergman, N.H.; Koren, S.; Phillippy, A.M. Mash: Fast genome and metagenome distance estimation using MinHash. *Genome Biol.* **2016**, *17*, 132. [CrossRef] [PubMed]

22. Chaudhari, N.M.; Gupta, V.K.; Dutta, C. BPGA- an ultra-fast pan-genome analysis pipeline. *Sci. Rep.* **2016**, *6*, 24373. [CrossRef] [PubMed]

23. Jain, C.; Rodriguez-R, L.M.; Phillippy, A.M.; Konstantinidis, K.T.; Aluru, S. High throughput ANI analysis of 90K prokaryotic genomes reveals clear species boundaries. *Nat. Commun.* **2018**, *9*, 5114. [CrossRef] [PubMed]

24. Lewis, J.A.; Fleming, J.T. Basic culture methods. In *Caenorhabditis Elegans: Modern Biological Analysis of an Organism*; Epstein, H.F., Shakes, D.C., Eds.; Academic: San Diego, CA, USA, 1995; pp. 3–29.

25. Liu, Z.; Budiharjo, A.; Wang, P.; Shi, H.; Fang, J.; Borriss, R.; Zhang, K.; Huang, X. The highly modified microcin peptide plantazolicin is associated with nematicidal activity of Bacillus amyloliquefaciens FZB42. *Appl. Microbiol. Biotechnol.* **2013**, *97*, 10081–10090. [CrossRef]

26. Hooper, D.J.; Hallmann, J.; Subbotin, S.A. Methods for extraction, processing and detection of plant and soil nematodes. In *Plant Parasitic Nematodes in Subtropical and Tropical Agriculture*; Luc, M., Sikora, R., Bridge, J., Eds.; CAB International: Wallingford, UK, 2005; pp. 53–86.

27. Bridge, J.; Page, S.L.J. Estimation of Root Knot Nematode Infestation Levels in Roots Using a Rating Chart. *Trop. Pest Manag.* **1980**, *26*, 296–298. [CrossRef]

28. Budiharjo, A.; Chowdhury, S.P.; Dietel, K.; Beator, B.; Dolgova, O.; Fan, B.; Bleiss, W.; Ziegler, J.; Schmid, M.; Hartmann, A.; et al. Transposon mutagenesis of the plant-associated *Bacillus amyloliquefaciens* ssp. *plantarum* FZB42 revealed that the nfrA and RBAM17410 genes are involved in plant-microbe-interactions. *PLoS ONE* **2014**, *9*, e98267. [CrossRef]

29. Blin, K.; Shaw, S.; Kloosterman, A.M.; Charlop-Powers, Z.; van Wezel, G.P.; Medema, M.H.; Weber, T. antiSMASH 6.0: Improving cluster detection and comparison capabilities. *Nucleic Acids Res.* **2021**, *49*, W29–W35. [CrossRef]

30. van Heel, A.J.; de Jong, A.; Song, C.; Viel, J.H.; Kok, J.; Kuipers, O.P. BAGEL4: A user-friendly web server to thoroughly mine RiPPs and bacteriocins. *Nucleic Acids Res.* **2018**, *46*, W278–W281. [CrossRef]

31. Makarova, K.S.; Grishin, N.V.; Koonin, E.V. The HicAB cassette, a putative novel, RNA-targeting toxin-antitoxin system in archaea and bacteria. *Bioinformatics* **2006**, *22*, 2581–2584. [CrossRef]

32. Min, Y.N.; Tabuchi, A.; Womble, D.D.; Rownd, R.H. Transcription of the stability operon of IncFII plasmid NR1. *J. Bacteriol.* **1991**, *173*, 2378–2384. [CrossRef]

33. Marsin, S.; Forterre, P. A rolling circle replication initiator protein with a nucleotidyl-transferase activity encoded by the plasmid pGT5 from the hyperthermophilic archaeon Pyrococcus abyssi. *Mol. Microbiol.* **1998**, *6*, 1183–1192. [CrossRef] [PubMed]

34. Monzingo, A.F.; Ozburn, A.; Xia, S.; Meyer, R.J.; Robertus, J.D. The structure of the minimal relaxase domain of MobA at 2.1 A resolution. *J. Mol. Biol.* **2007**, *366*, 165–178. [CrossRef] [PubMed]

35. Lu, J.; Holmgren, A. The thioredoxin antioxidant system. *Free. Radic. Biol. Med.* **2014**, *66*, 75–87. [CrossRef] [PubMed]
36. Baxter, J.C.; Waples, W.G.; Funnell, B.E. Nonspecific DNA binding by P1 ParA determines the distribution of plasmid partition and repressor activities. *J. Biol. Chem.* **2020**, *295*, 17298–17309. [CrossRef] [PubMed]
37. Dong, M.J.; Luo, H.; Gao, F. Ori-Finder 2022: A Comprehensive Web Server for Prediction and Analysis of Bacterial Replication Origins. *Genom. Proteom. Bioinform. Biorxiv* **2022**. [CrossRef]
38. Meier-Kolthoff, J.P.; Auch, A.F.; Klenk, H.-P.; Göker, M. Genome sequence-based species delimitation with confidence intervals and improved distance functions. *BMC Bioinform.* **2013**, *14*, 60. [CrossRef]
39. Dunlap, C.A. *Lysinibacillus mangiferihumi*, *Lysinibacillus tabacifolii* and *Lysinibacillus varians* are later heterotypic synonyms of *Lysinibacillus sphaericus*. *Int. J. Syst. Evol. Microbiol.* **2019**, *69*, 2958–2962. [CrossRef]
40. Chaumeil, P.A.; Mussig, A.J.; Hugenholtz, P.; Parks, D.H. GTDB-Tk v2: Memory friendly classification with the Genome Taxonomy Database. *Bioinformatics* **2022**, *38*, 5315–5316. [CrossRef]
41. Parks, D.H.; Chuvochina, M.; Rinke, C.; Mussig, A.J.; Chaumeil, P.A.; Hugenholtz, P. GTDB: An ongoing census of bacterial and archaeal diversity through a phylogenetically consistent, rank normalized and complete genome-based taxonomy. *Nucleic Acids Res.* **2022**, *50*, D785–D794. [CrossRef]
42. Richter, M.; Rosselló-Móra, R.; Glöckner, F.O.; Peplies, J. JSpeciesWS: A web server for prokaryotic species circumscription based on pairwise genome comparison. *Bioinformatics* **2016**, *32*, 929–931. [CrossRef]
43. Goris, J.; Konstantinidis, K.T.; Klappenbach, J.A.; Coenye, T.; Vandamme, P.; Tiedje, J.M. DNA-DNA hybridization values and their relationship to whole-genome sequence similarities. *Int. J. Syst. Evol. Microbiol.* **2007**, *57*, 81–91. [CrossRef] [PubMed]
44. Lefort, V.; Desper, R.; Gascuel, O. FastME 2.0: A comprehensive, accurate, and fast distance-based phylogeny inference program. *Mol. Biol. Evol.* **2015**, *32*, 2798–2800. [CrossRef] [PubMed]
45. Richter, M.; Rosselló-Móra, R. Shifting the genomic gold standard for the prokaryotic species definition. *Proc. Natl. Acad. Sci. USA* **2009**, *106*, 19126–19131. [CrossRef] [PubMed]
46. Arie, T. *Fusarium* diseases of cultivated plants, control, diagnosis, and molecular and genetic studies. *J. Pestic. Sci.* **2019**, *44*, 275–281. [CrossRef]
47. Chi, N.M.; Thu, P.Q.; Nam, H.B.; Quang, D.Q.; Phong, L.V.; Van, N.D.; Trang, T.T.; Kien, T.T.; Tam, T.T.T.; Dell, B. Management of *Phytophthora palmivora* disease in *Citrus reticulata* with chemical fungicides. *J. Gen. Plant Pathol.* **2020**, *86*, 494–502. [CrossRef]
48. Eisenback, J.D.; Triantaphyllou, H.H. Root-knot nematodes: Meloidogyne species and races. In *Manual of Agricultural Nematology*; Nickle, W.R., Ed.; Marcell Dekker: New York, NY, USA, 1991; pp. 191–274.
49. Hussey, P.S.; Barker, K.R. A comparison of methods of collecting inocula of *Meloidogyne* spp., including a new technique. *Plant Dis. Rep.* **1973**, *57*, 1025–1028.
50. Anastasiadis, I.A.; Giannakou, I.O.; Prophetou-Athanasiadou, D.A.; Gowen, S.R. The combined effect of the application of a biocontrol agent *Paecilomyces lilacinus*, with various practices for the control of root-knot nematodes. *Crop Prot.* **2008**, *27*, 352–361. [CrossRef]
51. Engelbrecht, G.; Claassens, S.; Mienie, M.S.; Fourie, H. Filtrates of mixed *Bacillus* spp. inhibit second-stage juvenile motility of root-knot nematodes. *Rhizosphere* **2022**, *22*, 100528. [CrossRef]
52. Fan, B.; Wang, C.; Song, X.; Ding, X.; Wu, L.; Wu, H.; Gao, X.; Borriss, R. *Bacillus velezensis* FZB42 in 2018: The Gram-Positive Model Strain for Plant Growth Promotion and Biocontrol. *Front. Microbiol.* **2018**, *9*, 2491. [CrossRef]
53. Tracanna, V.; de Jong, A.; Medema, M.H.; Kuipers, O.P. Mining prokaryotes for antimicrobial compounds: From diversity to function. *FEMS Microbiol. Rev.* **2017**, *41*, 417–429. [CrossRef]
54. Medema, M.H.; Kottmann, R.; Yilmaz, P.; Cummings, M.; Biggins, J.B.; Blin, K.; de Bruijn, I.; Chooi, Y.H.; Claesen, J.; Coates, R.C.; et al. Minimum Information about a Biosynthetic Gene cluster. *Nat. Chem. Biol.* **2015**, *11*, 625–631. [CrossRef] [PubMed]
55. Liu, Q.; Zhang, L.; Wang, Y.; Zhang, C.; Liu, T.; Duan, C.; Bian, X.; Guo, Z.; Long, Q.; Tang, Y.; et al. Enhancement of edeine production in Brevibacillus brevis X23 via in situ promoter engineering. *Microb. Biotechnol.* **2022**, *15*, 577–589. [CrossRef] [PubMed]
56. Vater, J.; Herfort, S.; Doellinger, J.; Weydmann, M.; Borriss, R.; Lasch, P. Genome Mining of the Lipopeptide Biosynthesis of *Paenibacillus polymyxa* E681 in Combination with Mass Spectrometry: Discovery of the Lipoheptapeptide Paenilipoheptin. *ChemBioChem* **2018**, *19*, 744–753. [CrossRef] [PubMed]
57. Khosla, C.; Gokhale, R.S.; Jacobsen, J.R.; Cane, D.E. Tolerance and specificity of polyketide synthases. *Annu. Rev. Biochem.* **1999**, *68*, 219–253. [CrossRef]
58. Austin, M.B.; Noel, J.P. The chalcone synthase superfamily of type III polyketide synthases. *Nat. Prod. Rep.* **2003**, *20*, 79–110. [CrossRef]
59. Funa, N.; Ohnishi, Y.; Fujii, I.; Shibuya, M.; Ebizuka, Y.; Horinouchi, S. A new pathway for polyketide synthesis in microorganisms. *Nature* **1999**, *400*, 897–899. [CrossRef]
60. Nakano, C.; Ozawa, H.; Akanuma, G.; Funa, N.; Horinouchi, S. Biosynthesis of aliphatic polyketides by type III polyketide synthase and methyltransferase in *Bacillus subtilis*. *J. Bacteriol.* **2009**, *191*, 4916–4923. [CrossRef]
61. Finking, R.; Marahiel, M.A. Biosynthesis of nonribosomal peptides. *Annu. Rev. Microbiol.* **2004**, *58*, 453–488. [CrossRef]
62. Mootz, H.D.; Marahiel, M.A. The tyrocidine biosynthesis operon of *Bacillus brevis*: Complete nucleotide sequence and biochemical characterization of functional internal adenylation domains. *J. Bacteriol.* **1997**, *179*, 6843–6850. [CrossRef]

63. Wenzel, M.; Rautenbach, M.; Vosloo, J.A.; Siersma, T.; Aisenbrey, C.H.M.; Zaitseva, E.; Laubscher, W.E.; van Rensburg, W.; Behrends, J.C.; Bechinger, B.; et al. The Multifaceted Antibacterial Mechanisms of the Pioneering Peptide Antibiotics Tyrocidine and Gramicidin S. *mBio* **2018**, *9*, e00802-18. [CrossRef]
64. Kessler, N.; Schuhmann, H.; Morneweg, S.; Linne, U.; Marahiel, M.A. The linear pentadecapeptide gramicidin is assembled by four multimodular nonribosomal peptide synthetases that comprise 16 modules with 56 catalytic domains. *J. Biol. Chem.* **2004**, *279*, 7413–7419. [CrossRef] [PubMed]
65. Zhou, X.; Huang, H.; Chen, Y.; Tan, J.; Song, Y.; Zou, J.; Tian, X.; Hua, Y.; Ju, J. Marthiapeptide A, an anti-infective and cytotoxic polythiazole cyclopeptide from a 60 L scale fermentation of the deep sea-derived *Marinactinospora thermotolerans* SCSIO 00652. *J. Nat. Prod.* **2012**, *75*, 2251–2255. [CrossRef]
66. Kurylo-Borowska, Z.; Szer, W. Inhibition of bacterial DNA synthesis by edeine. Effect on *Escherichia coli* mutants lacking DNA polymerase I. *Biochim. Biophys. Acta* **1972**, *287*, 236–245. [CrossRef] [PubMed]
67. Westman, E.L.; Yan, M.; Waglechner, N.; Koteva, K.; Wright, G.D. Self resistance to the atypical cationic antimicrobial peptide edeine of *Brevibacillus brevis* Vm4 by the N-acetyltransferase EdeQ. *Chem Biol.* **2013**, *20*, 983–990. [CrossRef] [PubMed]
68. Arnison, P.G.; Bibb, M.J.; Bierbaum, G.; Bowers, A.A.; Bugni, T.S.; Bulaj, G.; Camarero, J.A.; Campopiano, D.J.; Challis, G.L.; Clardy, J.; et al. Ribosomally synthesized and post-translationally modified peptide natural products: Overview and recommendations for a universal nomenclature. *Nat. Prod. Rep.* **2013**, *30*, 108–160. [CrossRef] [PubMed]
69. Burkhart, B.J.; Hudson, G.A.; Dunbar, K.L.; Mitchell, D.A. A prevalent peptide-binding domain guides ribosomal natural product biosynthesis. *Nat. Chem. Biol.* **2015**, *11*, 564–570. [CrossRef]
70. Kloosterman, A.M.; Shelton, K.E.; van Wezel, G.P.; Medema, M.H.; Mitchell, D.A. RRE-Finder: A Genome-Mining Tool for Class-Independent RiPP Discovery. *mSystems* **2020**, *5*, e00267-20. [CrossRef]
71. Valdés-Stauber, N.; Scherer, S. Isolation and characterization of Linocin M18, a bacteriocin produced by *Brevibacterium linens*. *Appl. Environ. Microbiol.* **1994**, *60*, 3809–3814. [CrossRef]
72. Hegemann, J.D.; Süssmuth, R.D. Matters of class: Coming of age of class III and IV lanthipeptides. *RSC Chem. Biol.* **2020**, *1*, 110–127. [CrossRef]
73. Anthony, T.; Chellappa, G.S.; Rajesh, T.; Gunasekaran, P. Functional analysis of a putative holin-like peptide-coding gene in the genome of *Bacillus licheniformis* AnBa9. *Arch. Microbiol.* **2010**, *192*, 51–56. [CrossRef]
74. Aunpad, R.; Panbangred, W. Evidence for two putative holin-like peptides encoding genes of *Bacillus pumilus* strain WAPB4. *Curr. Microbiol.* **2012**, *64*, 343–348. [CrossRef] [PubMed]
75. Lee, J.Y.; Janes, B.K.; Passalacqua, K.D.; Pfleger, B.F.; Bergman, N.H.; Liu, H.; Håkansson, K.; Somu, R.V.; Aldrich, C.C.; Cendrowski, S.; et al. Biosynthetic analysis of the petrobactin siderophore pathway from *Bacillus anthracis*. *J. Bacteriol.* **2007**, *189*, 1698–1710. [CrossRef] [PubMed]
76. Hagan, A.K.; Carlson, P.E., Jr.; Hanna, P.C. Flying under the radar: The non-canonical biochemistry and molecular biology of petrobactin from Bacillus anthracis. *Mol. Microbiol.* **2016**, *102*, 196–206. [CrossRef] [PubMed]
77. Koppisch, A.T.; Dhungana, S.; Hill, K.K.; Boukhalfa, H.; Heine, H.S.; Colip, L.A.; Romero, R.B.; Shou, Y.; Ticknor, L.O.; Marrone, B.L.; et al. Petrobactin is produced by both pathogenic and non-pathogenic isolates of the *Bacillus cereus* group of bacteria. *Biometals* **2008**, *21*, 581–589. [CrossRef] [PubMed]
78. Shen, Q.; Zhou, H.; Dai, G.; Zhong, G.; Huo, L.; Li, A.; Liu, Y.; Yang, M.; Ravichandran, V.; Zheng, Z.; et al. Characterization of a Cryptic NRPS Gene Cluster in *Bacillus velezensis* FZB42 Reveals a Discrete Oxidase Involved in Multithiazole Biosynthesis. *ACS Catal.* **2022**, *12*, 3371–3381. [CrossRef]

**Disclaimer/Publisher's Note:** The statements, opinions and data contained in all publications are solely those of the individual author(s) and contributor(s) and not of MDPI and/or the editor(s). MDPI and/or the editor(s) disclaim responsibility for any injury to people or property resulting from any ideas, methods, instructions or products referred to in the content.

 *microorganisms*

*Article*

# New Potential Biological Limiters of the Main Esca-Associated Fungi in Grapevine

**Francesco Mannerucci, Giovanni D'Ambrosio, Nicola Regina, Domenico Schiavone and Giovanni Luigi Bruno ***

Department of Soil, Plant and Food Sciences, University of Bari Aldo Moro, Via Amendola 165/A, 70126 Bari, Italy; francesco.mannerucci@uniba.it (F.M.); domenico.schiavone1@uniba.it (D.S.)
* Correspondence: giovanniluigi.bruno@uniba.it; Tel.: +39-080-544-3085

**Abstract:** The strains *Trichoderma harzianum* TH07.1-NC (TH), *Aphanocladium album* MX95 (AA), *Pleurotus eryngii* AL142PE (PE) and *Pleurotus ostreatus* ALPO (PO) were tested as biological limiters against *Fomitiporia mediterranea* Fme22.12 (FM), *Phaeoacremonium minimum* Pm22.53 (PM) and *Phaeomoniella chlamydospora* Pc22.65 (PC). Pathogens were obtained from naturally Esca-affected 'Nero di Troia' vines cropped in Grumo Appula (Puglia region, Southern Italy). The antagonistic activity of each challenge organism was verified in a dual culture. TH and PO completely overgrew the three pathogens. Partial replacement characterized PE-FM, PE-PM, PE-PC and AA-PC interactions. Deadlock at mycelial contact was observed in AA-FM and AA-PM cultures. The calculated antagonism index (AI) indicated TH and PE as moderately active antagonists (10 < AI < 15), while AA and PO were weakly active (AI < 10). The maximum value of the re-isolation index (s) was associated with deadlock among AA-PM, AA-PC and PE-FM dual cultures. The tested biological limiters were always re-isolated when PO and TH completely replaced the three tested pathogens. TH and AA confirmed their efficiencies as biological limiters when inoculated on detached canes of 'Nero di Troia' in dual combination with FM, PC and PM. Nevertheless, additional experiments should be performed for a solid conclusion, along with validation experiments in the field.

**Keywords:** *Aphanocladium album*; *Pleurotus ostreatus*; *Pleurotus eryngii*; *Trichoderma harzianum*; antagonist index; hyphal interaction; deadlock; inhibition; re-isolation; biocontrol

**Citation:** Mannerucci, F.; D'Ambrosio, G.; Regina, N.; Schiavone, D.; Bruno, G.L. New Potential Biological Limiters of the Main Esca-Associated Fungi in Grapevine. *Microorganisms* **2023**, *11*, 2099. https://doi.org/10.3390/microorganisms11082099

Academic Editors: Chetan Keswani and Rainer Borriss

Received: 19 July 2023
Revised: 10 August 2023
Accepted: 13 August 2023
Published: 17 August 2023

**Copyright:** © 2023 by the authors. Licensee MDPI, Basel, Switzerland. This article is an open access article distributed under the terms and conditions of the Creative Commons Attribution (CC BY) license (https://creativecommons.org/licenses/by/4.0/).

## 1. Introduction

The disease known as 'Esca' is one of the longest-recognized and most devastating threats to grapevines (*Vitis vinifera*) and viticulture around the world [1]. The most recent studies consider this mycopathy as different diseases that overlap in the same vine or develop at different stages of plant life [1–5]. Cross-sections of rooted cuttings, trunk, branches, and shoots show brown to black spots often accompanied by a dark, viscous exudate ('black goo'). Longitudinally, xylem necrosis extends in columnar strips, called 'brown wood streaking' [1–5]. Brown wood streaking affects rooted cuttings (also named brown wood streaking of grapevine cuttings) and young (2–7 years) vines (defined as Petri disease, formerly also known as black goo, slow dieback, and Phaeoacremonium grapevine decline) [1–3]. Inside the trunk or the branches, the wood shows 'white rot' (which is the origin of the name Esca), as well as cracking or fissuring of the bark and wood (known as 'mal dello spacco'—cracking disease—in Italy) [1,3,6]. On leaves, small pale green or chlorotic zones, roundish or irregular, occur dispersed to the veins or adjacent the margin. Progressively, these areas increase, merge, and, in part, necrotize, and at last leave only a narrow strip of unaffected green tissue along the main veins. Depending on the cultivar, dead tissues appear dark brown to red-brown and the diseased leaves develop a 'tiger stripes' appearance. Occasionally, the necrotic zones of the lamina desiccate and detach and leaf margins become irregular [7]. These symptoms characterize grapevine leaf stripe disease, previously named 'young Esca' [1–4]. Brown wood streaking of rooted cuttings,

Petri disease, and grapevine leaf stripe disease manifest in newly grafted plant material, young vines, and adult plants, respectively [1,3]. The concomitant presence of grapevine leaf stripe disease and white rot is described as 'Esca proper'. Berries display tiny brown spotting ('black measles'), shriveling, and wilting [1–4]. Within the Esca complex, white rot and Esca proper can also show apoplectic symptoms: the sudden wilting of whole vines or individual vine arms within a few days, with leaves and grape clusters that usually remain attached to the plant [1,5].

Members of the basidiomycetous genus *Fomitiporia* (*F. mediterranea* in Europe and the Mediterranean area) are mainly associated with white rot [8]. *Phaeomoniella chlamydospora* and *Phaeoacremonium minimum* (syn. *P. aleophilum*) are the most important etiological agents of brown wood streaking in cuttings, Petri disease, and grapevine leaf stripe disease [1,3,5,7–10]. No pathogens have been associated with symptomatic leaves or berries of diseased plants. Symptoms on leaves and berries are thought to be related to vine susceptibility and oldness, the implicated pathogens, and pedoclimatic and physiological features [2,5,9,11]. From pycnidia, perithecia or basidiomata embedded in the bark of infected grapevines, and/or from other infected woody hosts surrounding vineyards, produce conidia, ascospores, or basidiospores produced by Esca-associated pathogens spread by rain droplets and wind, land on receptive pruning wounds, and initiate infection [1,5,12]. Infected propagating material also causes disease spread in the field [7,12].

Despite the amount of research devoted to the Esca complex, the frequency of grapes showing Esca-related symptoms has increased worldwide [11]. Thus far, no curative methods are available. In the past, arsenite (revoked because of carcinogens in humans and toxicity towards the environment) spray application alleviated leaf symptoms' expression [13]. Different strategies are applied, both in nurseries and fields, to limit the occurrence of 'Esca' symptoms [1]. Preventive practices, e.g., pruning wound protection and infected stock elimination, are recommended to reduce Esca complex spread [1,4,14] but are insufficient to guarantee effective control [13–16]. On the other hand, invasive methods (e.g., trunk renewal or surgery, regrafting, and dry rot removal) mitigate the loss of productivity over the years [1,12–16]. No efficient chemical control is yet available. Trunk applications of fosetyl-Al provide reliable results in the reduction of browning and leaf symptom manifestation [13,17]. In vitro treatments with carbendazim, flusilazole, or tebuconazole induced 50% inhibition of *P. chlamydospora* and *P. minimum* mycelia growth, while azoxystrobin, carbendazim, and thiram reduced conidia germination [18]. Hot water (50 °C) treatment for 30 min reduced the *P. chlamydospora* load in propagation material [12]. Different concentrations of cysteine, $FeSO_4$, salicylic acid, and fosetyl-Al, alone or in combination, showed antifungal activity against *P. chlamydospora* and *P. minimum*, as well as systemic action and bioactive stimulation on vines [14]. Preventive pruning wound protection seems the most effective strategy [1,12].

Several studies place in biological control the possibility of counteracting the deleterious effects of Esca complex pathogens [1,15–22]. Promising is the application of biocontrol agents, including bacteria, oomycetes, and fungi. Favorable are bacteria of the genera *Acinetobacter*, *Bacillus*, *Brevibacillus*, *Curtobacterium*, *Enterobacter*, and *Paenibacillus*, the oomycote *Pythium oligandrum*, and mycoparasite fungi of the genus *Trichoderma*, still with innovative, eco-friendly hybrid nanomaterials [1,15,19,20].

Strains of *Bacillus subtilis* have revealed in vitro and in planta antagonistic traits against grapevine trunk pathogens, including Esca-associated fungi, both in pruning wound protection and in nurseries [21]. *Bacillus subtilis* and *Bacillus amyloliquefaciens* interact with *P. chlamydospora* and *P. minimum* hyphae via antibiosis [15,22,23]. *Bacillus pumilus* (S32) and *Paenibacillus* isolates S18 and S19 exhibit action by producing volatile compounds and diffusible antibiotic substances and inducing grapevine defense-related gene expression [15,24]. *Enterobacter* isolate S24, *Bacillus reuszeri* strains S28, S31, and S27, *Bacillus* isolate S34, *Pantoea illinoisensis* strain S13, *Pantoea agglomerans* strains S1 and S3, and *Bacillus firmus* strain S41 are efficient biological control agents for Esca-associated fungi on rooted cuttings under greenhouse conditions [15,24]. *Pseudomonas protegens* strain MP12, obtained from a forest soil sample, and *Pseudomonas protegens* strain DSM 19095T significantly inhibited *P. chlamydospora* and *P. minimum* in vitro growth [15,25]. The volatile compound produced by *Pseudomonas* isolate S45, *Stenotrophomonas* isolate S180, and *Novosphingobium* isolate S112 induces deleterious effects on *F. mediterranea* mycelia, while *Enterobacter* isolate S11, *Paenibacillus* isolates S150 and S270, *Weeksellaceae* S259, and *Bacillus* strain S5 even promoted *F. mediterranea* growth in a dual culture [26]. Strains of the cellulolytic and xylanolytic *Paenibacillus* sp. display a synergistic interaction with *F. mediterranea* and enhance wood degradation [26].

Strains of *Actinobacteria* and *Streptomyces* isolates E1 and R4 induced a significant reduction in infection rates at the lower end of the rootstock in the context of Petri disease [15,27,28].

*Streptomyces plymuthica, Bacillus velezensis, Pseudomonas chlororaphis,* and isolates of *Micromonospora* sp. strongly inhibited *P. minimum* in vitro growth [15,27,29].

The application of *Pythium oligandrum* at the root level reduced *P. chlamydospora* and *P. minimum* necroses in the stem and triggered the plant defense pathways, including PR proteins, phenylpropanoid, oxylipin, and oxidoreduction systems [15,30,31].

*Epicoccum layuense,* an ascomycete often associated with the mycobiomes of grapevines, through in vitro dual culture, inhibited the growth of *P. chlamydospora, F. mediterranea,* and some *P. minimum* strains without colony contact, suggesting the production of inhibitory compounds [32]. *E. layuense* strain E24 colonized the rooted grapevine cuttings of Cabernet Sauvignon and Touriga Nacional cultivars under greenhouse conditions and decreased (31–82%) the wood symptomatology via chemical interaction and competition for space, depending on the pathogen and grapevine cultivar [32]. *Epicoccum mezzettii* E17 overgrew *F. mediterranea* in vitro and competed for space and nutrients [32].

Encouraging results regarding the management of Esca-associated fungi have been obtained with the application of Remedier® (ISAGRO S.p.A., Milan, Italy) [1,15,33] and Ecofox Life® (ISAGRO S.p.A., Milan, Italy), bio-fungicides containing *Trichoderma asperellum* and *T. gamsii.* Strains of different species of *Trichoderma,* including *T. atroviride, T. harzianum, T. hamatum, T. longibrachiatum,* and *T. gamsii,* effectively controlled *P. chlamydospora* and *P. minimum* in vitro and under greenhouse, field, and nursery conditions, [1,15,22,34]. Hyphae of *T. atroviride* ATCC74058 and *T. harzianum* ATCC 26799 overgrew *P. chlamydospora* and *P. minimum,* competed for nutrients, utilizing carbon and nitrogen sources, induced direct antagonism, and allowed a 90% growth reduction [15,35]. Commercial strains of *T. harzianum* and *T. atroviride* overgrew *P. chlamydospora* [15,22,35]. *T. atroviride* strains USPP-T1 and USPP-T2 stopped *P. minimum* growth by coiling or disintegrating hyphae [35]. Strains of different *Trichoderma* spp. protected grapevine pruning wounds in nurseries and reduced the incidence of *P. chlamydospora* and *P. minimum* after inoculation under field conditions [22,28]. The application of *T. harzianum* (Trichodex®, Fertiberia, Sevilla, España) at rooting under organic nursery conditions reduced *P. chlamydospora* infection [36–38]. *T. harzianum* T39 (Trichodex®) and *T. longibrachiatum* treatments on cuttings reduced the necrosis length, caused by *P. chlamydospora* inoculation at the rootstock via the enhancement of the grapevine defenses [15,36–38]. A *Trichoderma koningii* strain TK7 suspension's dip application on roots reduced the incidence of *P. chlamydospora* infection in the field on a young grafted Spanish Tempranillo cultivar [15,28].

Strains of *Clonostachys rosea,* in vitro, overgrew and inhibited *P. chlamydospora* and *P. minimum* through antibiosis and mycoparasitism [15,39]. Under greenhouse conditions, amended soil with the endophytic *C. rosea* strain 19B/1 significantly decreased the length of the necrotic lesions caused by *P. chlamydospora* [39]. *Lecanicillium lecanii* (ATCC 46578) reduced in vitro the growth of *P. chlamydospora* and *P. minimum* by carbon and nitrogen competition [34] or by the production of antibiotic compounds [39]. *Fusarium oxysporum* strain F2 reduced *P. chlamydospora* in vitro growth by 43% and sporulation by 90%; nonetheless, the strain showed no reduction in the discoloration length inside the trunk, despite an 82% reduction in the *P. chlamydospora* DNA amount [40].

This work aimed to assess the antagonistic activity of four new biological limiters against the three most important Esca-associated fungi of grapevine. The challenge organisms applied included *Trichoderma harzianum*, the chitinolytic fungus *Aphanocladium album*, and two xylotrophic fungi belonging to the genus *Pleurotus*. Radial growth inhibition, the morphology of the interaction, and the antagonism index were used to evaluate the in vitro dual interactions. The number of positive re-isolations after pathogen and biological limiter interaction was calculated. *T. harzianum* and *A. album* were additionally evaluated on *V. vinifera* cv Nero di Troia detached canes against the three main fungal species associated with the Esca complex.

## 2. Materials and Methods

### 2.1. Strains, Media, and Growth Conditions

Target organisms (Table 1) were isolated from wood samples of 12-year-old cv Nero di Troia vines, grafted on 157-11 located in the countryside of Grumo Appula (Puglia region, southern Italy) and trained to a 'tendone' not irrigated.

**Table 1.** Strains used in this study.

| Organism | Code | Acronym |
|---|---|---|
| Potential biological limiters | | |
| *Aphanocladium album* | DiSSPA [1] MX95 | AA |
| *Pleurotus ostreatus* | DiSSPA ALPO | PO |
| *Pleurotus eryngii* | DiSSPA AL142 | PE |
| *Trichoderma harzianum* | DiSSPA TH07.1-NC | TH |
| Target pathogens | | |
| *Fomitiporia mediterranea* | DiSSPA Fme22.12 | FM |
| *Phaeoacremonium minimum* | DiSSPA Pm22.53 | PM |
| *Phaeomoniella chlamydospora* | DiSSPA Pc22.65 | PC |

[1] DiSSPA: Department of Soil, Plant and Food Sciences, University of Bari Aldo Moro, Bari, Italy.

In the 2020 and 2021 growing seasons, the vineyard was surveyed, and vines with typical Esca symptoms were selected (Figure 1). During the winter pruning operations (January 2022), samples of branches were taken from five symptomatic vines. Internal symptoms through the wood, revealed after cutting, included vascular streaking, light brown to black discoloration, and white rot (Figure 1). Wood samples were wetted with denatured ethanol and surface-sterilized by a flame. From each sample, 15 wood chips (2 mm × 2 mm × 2 mm) were aseptically cut from the margins of diseased wood and plated (5 chips per dish) either on potato dextrose agar (PDA; from Oxoid Part of Thermo Fisher Scientific—Microbiology, Hampshire, UK) amended with 100 mgL$^{-1}$ of each streptomycin and ampicillin (PDA-SA) or PDA modified by adding 1 mL L$^{-1}$ of thiabendazole lactate (PDA-T: 2.3 g of thiabendazole in 10 mL of lactic acid). The incubation of Petri dishes occurred at 24 ± 1 °C in the dark for up to 28 days. Emerging colonies were hyphal-tip-purified. Isolates were identified based on their microscopic morphological characteristics. Strains CBS 229.95, CBS 631.94, and DBPV-1 of *P. chlamydospora*, *P. minimum*, and *F. mediterranea*, respectively [41], were used as a reference for morphological features. The obtained strains were stored on PDA slants in Wheaton bottles at 4 ± 1 °C in the fungal collection of the Department of Soil, Plant and Food Sciences (Di.S.S.P.A.), Plant Pathology Section, University of Bari.

*Aphanocladium album* MX95 (Patent N° 00041374382), *Pleurotus ostreatus* ALPO, *Pleurotus eryngii* AL142, and *Trichoderma harzianum* TH07.1-NC strains (Table 1), available at the fungal collection of the Di.S.S.P.A. Plant Pathology Section, were revitalized on PDA at 25 ± 1 °C in the dark. All fungi were routinely grown on PDA at 25 ± 1 °C in the dark.

**Figure 1.** Symptoms on *Vitis vinifera* cv Nero di Troia plants used for isolation experiments: 'tiger-stripes' (**A**) and apoplexy (**B**), brown wood streaking (**C–E**), and white rot (**C**).

## 2.2. Growth Rate

Potential biological limiters and target organisms were singly grown in a 90 mm Petri dish containing 18 mL of PDA medium. A plug (3 mm in diameter) of each strain, collected from the margins of actively grown cultures, was placed 1 cm from the border of the plate on the line of the dish diameter. Inoculated plates were sealed with Parafilm M and incubated in darkness at 25 ± 1 °C. Radius measurements were performed every eight hours following the line of the dish diameter. The average of the daily radius increment was calculated. All strains were tested in triplicate and the experiment was repeated at least two times.

## 2.3. In Vitro Dual Culture Interactions

Dual interactions between the tested strains (Table 1) were performed in plastic Petri dishes (diameter 90 mm, height 15 mm) containing PDA (18 mL per plate). Mycelium plugs (3 mm in diameter) were cut from the edge side of the actively growing pure culture and used as an inoculant. The target organism and potential biological limiter plugs were placed together on the same plate on opposite sides, 1 cm from the border of the dish (Figure 2a). As a control, the target organism and potential biological limiter plugs were placed alone. Inoculate plates were sealed with Parafilm M and incubated at 25 ± 1 °C, in the dark.

Based on the differences in the growth rates of each fungal strain, the slower PC was placed 8 days in advance of the faster-growing TH.

Three replicates were maintained for each treatment. The experiment was conducted twice for reproducibility. Plates were observed every eight hours to record the time of the first contact between the two mycelia.

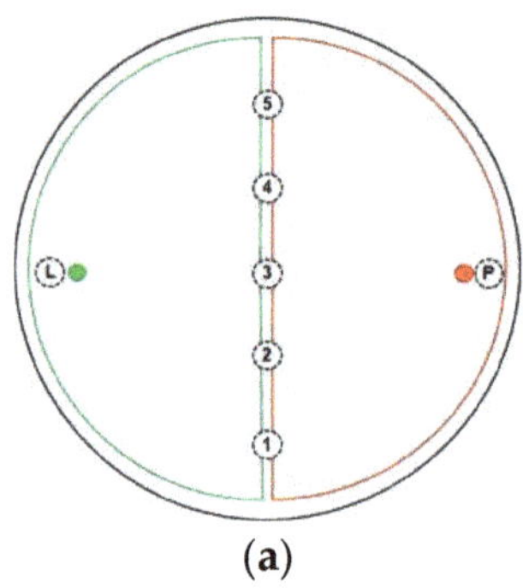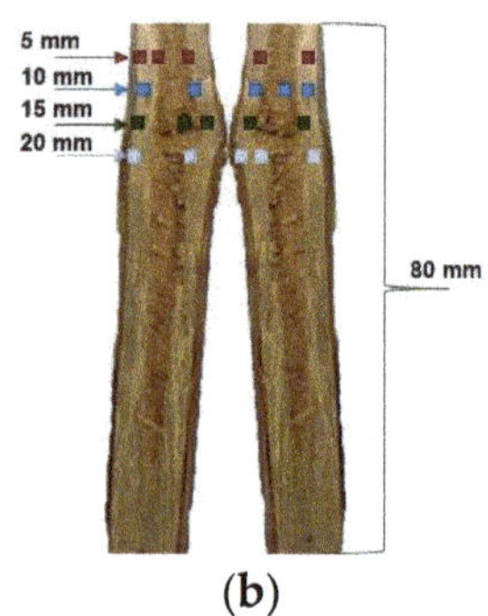



**(a)**  **(b)**

**Figure 2.** Schemes showing the positions of (**a**) inoculation plugs (• and •) and discs taken from the interaction zone (1–5), the potential biological limiter (L, •), and the pathogen (P, •); (**b**) approximate sites of chip recovery (■, ■, ■, ■) for grapevine detached cane re-isolation experiments.

Every day, the colony radii of the target organism were assessed in dual culture and on the control plate. In plates with dual culture, radii were measured in the direction of the potential biological limiter. The measurements evaluated 10 days after inoculation were used to determine the percentage of radial growth inhibition as $100 \times [(C - T)/C]$, where C and T are the mean radii (mm) of the pathogen in the absence of a biological limiter (control) and dual culture, respectively. Based on the radial growth inhibition rates, antagonist activity was considered [42] very high (>75%), high (61–75%), moderate (51–60%), or low (<50%).

Mycelial interactions in dual culture were examined daily and scored with the rating proposed by Badalyan et al. [43,44]. The antagonism index, which defines the ability of a fungus to compete with another species, was calculated as $\sum (n \times i)$, where n is the number of each type or subtype of interaction and i is the corresponding score [43,44]. Biological limiters were considered active (antagonism index > 15), moderately active (antagonism index = 10–15), or weak (antagonism index < 10) antagonists [43,44].

Twenty days after mycelia interaction, the observed results of the competition were quantified after fungal re-isolation. From each plate, seven agar discs (5 mm in diameter) were cut (Figure 2a): five from the interaction zone and two (one per strain) from areas with the presumed growth of only one individual fungus (positions labeled "L" and "P" in Figure 2a). Discs were placed on PDA plates and incubated at $25 \pm 1\ °C$ in the dark. The score of the re-isolation success of potential biocontrol agents against each pathogen was quantified as $\ln[(A + 1)/(P + 1)]$, where A and P are the mean numbers of the successfully re-isolated biological limiter and target organism, respectively (n = max 5). The estimated value of s ranges between +1.79 and −1.79. Higher values of s indicate greater re-isolation success [45,46].

### 2.4. Interaction on Detached Grapevine Canes

From 20 healthy 'Nero di Troia' dormant vines, canes (diameter 10–12 mm) were collected in January 2023. In the laboratory, the canes were cut into one node segment (80 mm length) each 2 cm above the bud to simulate a fresh pruning wound. Cane segments were autoclaved (121 °C, 40 min) and distributed, respecting the polarity, in sterile 50 mL Falcon-type plastic tubes (one per tube) containing 20 mL of sterile distilled water.

For AA, TH, PM, and PC, conidia suspensions were prepared in 0.2% agar water as a bio-adhesive. For FM, a mycelia homogenate was obtained in 0.2% agar water using a Sorvall model 17105 Omni-Mixer homogenizer (DuPont de Nemours, Inc., Wilmington, DE, USA). The viability of conidia and mycelia fragments was assessed on PDA by the decimal dilutions method, adjusted to $10^6$ colony-forming units per milliliter, and used (10 µ) to inoculate the apical end of each cane. AA, TH, PC, PM, and FM were assayed singly and in dual pathogen–limiter arrangements. In dual combinations, target organisms were inoculated 5 days before or after the biological limiters. Canes treated with 0.2% agar water and untreated canes were used as controls. A total of 20 canes per treatment were used. All tubes were incubated at $25 \pm 1$ °C, in the dark, for 12 weeks. Re-isolations were performed to verify the vitality of the fungal strains inoculated and their progression inside the cane tissue. Each apical end was gently pressed on the PDASA medium surface to create 5 different imprints. Then, every cane was divided longitudinally. Chips (2 mm × 2 mm × 2 mm) were aseptically taken at approximately 5, 10, 15, and 20 mm (Figure 2b) from the upper end and plated (5 per dish) on PDASA. Inoculated dishes were incubated at $25 \pm 1$ °C, in the dark, for 28–45 d. The developed fungal colonies were identified based on their macro- and microscopic morphological characters. The percentage of colonization frequency of re-isolated fungi was estimated as $100 \times (N \times n)$, where N is the number of colonies of each species developed from each chip, and n is the total number of plated fragments (n = 15).

## 2.5. Statistical Analysis

Plates and tubes were allotted in a randomized design. Normality and homogeneity of variances were verified with Shapiro–Wilk's test and Levene's test, respectively. The standard deviations (sd) were calculated for all mean values. The experimental data obtained were compared using an analysis of variance (ANOVA), followed by the Fisher least significant difference or Kruskal–Wallis test ($p = 0.05$). The data of inhibition radial growth were analyzed as radius values and expressed as a percentage. The frequency of re-isolated fungi percentage was transformed to arcsine before analysis.

## 3. Results

### 3.1. Growth Rate

All the tested strains developed a different growth rate (Figure 3).

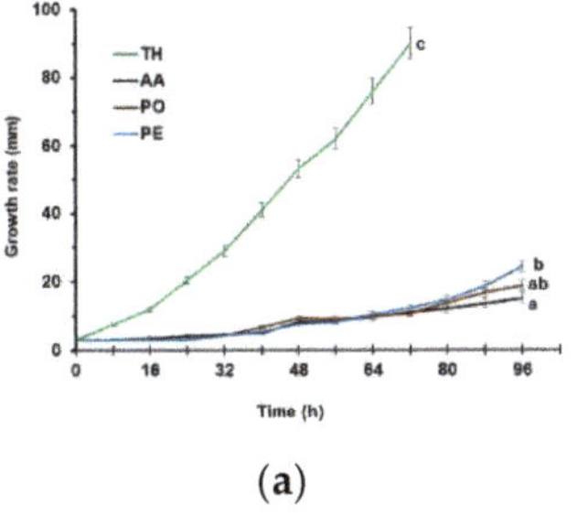


(a)

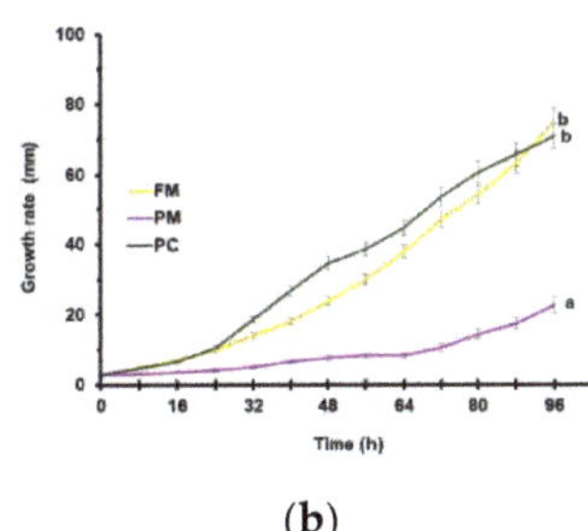


(b)

**Figure 3.** Growth rate at $25 \pm 1$ °C in the dark on 90 mm Petri dishes containing potato dextrose agar of potential biological limiters (**a**) and target pathogens (**b**). Data are the means of six replicates $\pm$ sd. For acronym definitions, see Table 1. Growth curves with different letters are significantly different according to Fisher's least significant difference test at $p \leq 0.05$.

TH colonized the entire Petri dish over 72 h (30.0 mm d$^{-1}$ radius increment). PO (6.1 mm d$^{-1}$), PE (4.7 mm d$^{-1}$), and AA (3.7 mm d$^{-1}$) reached, after 96 h, 15, 19, and 24 mm, respectively (Figure 3a). Among the tested pathogens (Figure 3b), PC (5.7 mm d$^{-1}$) was the slowest, while the fastest growth rate was recorded for PM (17.7 mm d$^{-1}$) and FM (18.7 mm d$^{-1}$).

### 3.2. In Vitro Dual Culture Interactions

In the dual cultures, pathogens and potential biological limiters required 56 to 96 h for the first colony contact (Table 2). Short times were associated with organisms showing fast mycelial growth (e.g., TH against FM, PM, or PC). On the contrary, long times affected the interactions between slow-growth organisms (e.g., AA against PM and PC).

**Table 2.** Time (hours) required for the first contact between the tested potential biological limiters and phytopathogenic organisms [1].

| Phytopathogenic Organisms | Biological Limiters [2] | | | |
|---|---|---|---|---|
| | **AA** | **PO** | **PE** | **TH** |
| FM | 72 ±5.06 a | 88 ± 5.06 a | 88 ± 5.06 a | 56 ± 5.06 a |
| PM | 96 ± 5.06 b | 88 ± 5.06 a | 88 ± 5.06 a | 64 ± 5.06 b |
| PC [3] | 96 ± 5.06 b | 88 ± 7.16 a | 88 ± 7.16 a | 64 ± 7.16 b |
| PC8 [3] | NT [4] | NT | NT | 64 ± 5.06 b |

[1] For acronym definitions, see Table 1. [2] Data are the means of six replicates ± sd. Within each column, data with different letters are significantly different according to Fisher's least significant difference test at $p \leq 0.05$. [3] PC was inoculated concomitantly (PC) or 8 days in advance (PC8) of TH. [4] NT = not tested.

The percentage of radial growth inhibition (Figure 4) ranged from 48 to 87% in the PO-FM and TH-PC8 interactions, respectively.

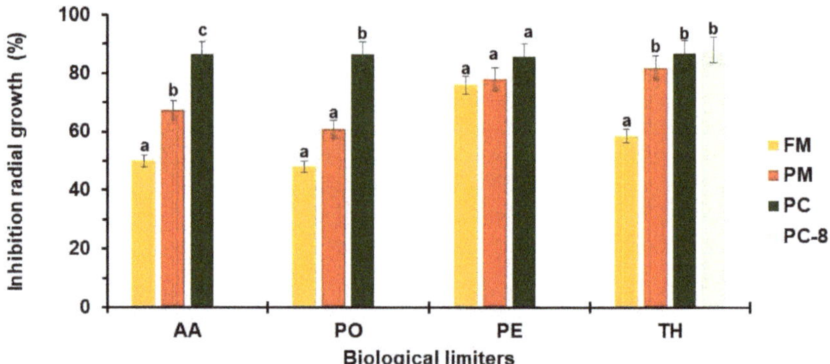


**Figure 4.** Percentage of inhibition radial growth in dual cultures on potato dextrose agar of AA, PO, PE, and TH against FM (■), PM (■), and PC, inoculated concomitantly (■) or 8 days in advance of TH (■). Data are the means of six replicates ± sd. For each potential biological limiter, values accompanied by the same letters are not significantly different ($p \leq 0.05$) according to Fisher's LSD test. For acronym definitions, see Table 1.

Very high radial growth inhibition was calculated in the AA-PC, PO-PC, PE-FM, PE-PC, PE-PM, TH-PM, TH-PC, and PE-PC8 interactions. High antagonist activity was recovered in the AA-PM and PO-PM relations. Moderate antagonist activity (51–60%) was recovered in the AA-PM and PO-PC groups, while low inhibition rates (about 49%) were retrieved in the PO-FM and AA-FM dual cultures.

The dual culture assays displayed a miscellaneous pattern of mycelial interaction (Figure 5). Following the definition of mycelial interaction proposed by Badalyan et al. [43,44], deadlock (mutual inhibition in which neither strain was able to overgrow the other) at mycelial contact, replacement (overgrowth without initial deadlock), and partial replacement after an initial deadlock with mycelial contact occurred among the pairings tested.

AA showed deadlock at mycelial contact (interaction type A by Badalyan et al. [43,44]) during the interactions with FM and PM. Partial replacement after initial deadlock with mycelial contact (interaction type $C_{A1}$ as suggested by Badalyan et al. [43,44]) was exhibited during AA-PC dual cultures and PE towards FM, PC, and PM. PO and TH completely replaced (interaction type C as indicated by Badalyan et al. [43,44]) the three tested phytopathogenic organisms, including PC inoculated 8 days in advance of TH. Among the 78 pairings tested, replacement of the pathogen by the potential biological limiter was more frequent (53.8%) than partial replacement after initial deadlock with mycelial contact (30.8%) and deadlock (15.4%).

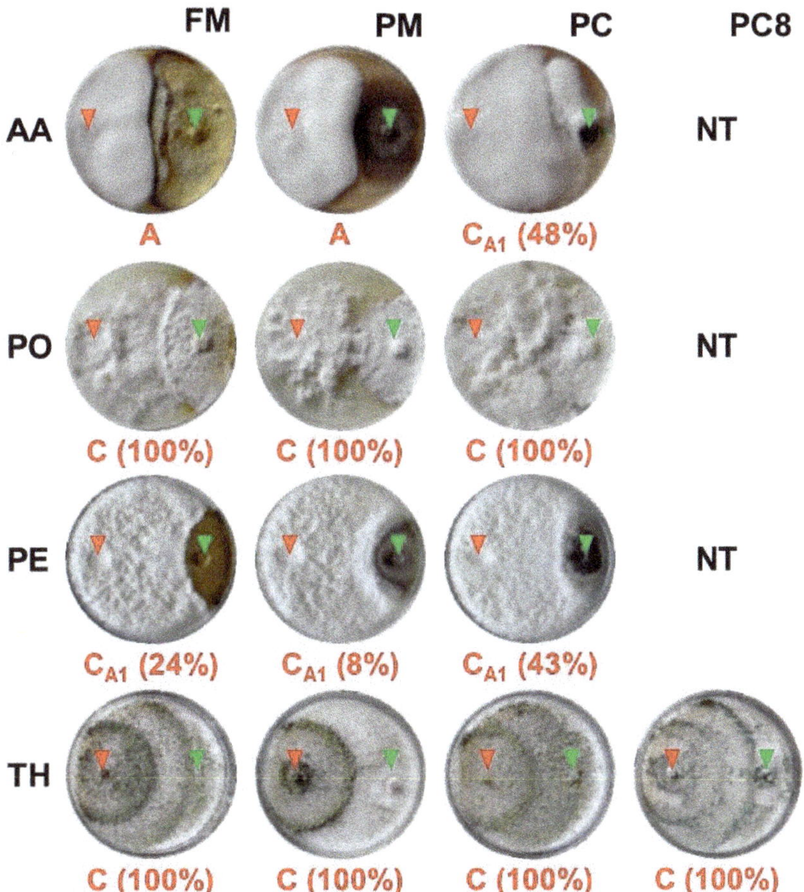



**Figure 5.** Mycelial interactions after 28 days of dual culture on potato dextrose agar between the potential biological limiters (AA, PO, PE, or TH) and the target pathogens (FM, PM, or PC). PC was inoculated concomitantly (PC) or 8 days in advance of TH (PC8). For acronym definitions, see Table 1. Red letters show the type of interaction as proposed by Badalyan et al. [43,44]: A = deadlock (mutual inhibition in which neither strain was able to overgrow the other) at mycelial contact, C = replacement (overgrowth without initial deadlock), $C_{A1}$ = partial replacement after an initial deadlock with mycelial contact. In brackets is the percentage of overgrowth 20 days after inoculation. NT = not tested. Red and green arrows indicate the sites where biological limiters and pathogens were inoculated, respectively.

Based on the calculated antagonism index values, AA and PO were weak antagonists, while PE and TH were moderately active antagonists, reaching antagonism indexes of 5.5, 9.0, 10.5, and 12.0, respectively.

The outcomes of re-isolation success in the dual cultures are shown in Table 3. No target pathogens were re-isolated after the interactions PO-FM, PO-PM, PO-PC, TH-FM, TH-PM, TH-PC, and TH-PC8. In all the other combinations, the pathogen was always re-isolated from the agar disc cut at the P position (Figure 2a). FM was re-isolated from the connective line with AA, while PM and PC were obtained from the deadlock lines with PO.

**Table 3.** Mean scores of re-isolation success of potential biological limiters against each pathogen from in vitro dual cultures [1,2].

| Biological | Phytopathogens [3] | | | |
|---|---|---|---|---|
| Limiters | FM | PM | PC [4] | PC8 [4] |
| AA | $1.09 \pm 0$ a | $1.79 \pm 0$ b | $1.79 \pm 0$ b | NT [5] |
| PO | * | * | * | NT |
| PE | $1.79 \pm 0$ b | $1.09 \pm 0$ a | $1.09 \pm 0$ a | NT |
| TH | * | * | * | * |

[1] The score of re-isolation success was quantified as $\ln[(A + 1)/(P + 1)]$, where A and P are the mean numbers of the successfully re-isolated biological limiter and target organism, respectively (n = max 5). [2] For species acronym definitions, see Table 1. [3] The data are the means of six replicates $\pm$ sd. Within each row, values with different letters are significantly different according to the Kruskal–Wallis test at $p \leq 0.05$. [4] PC was inoculated concomitantly (PC) or 8 days in advance of TH (PC8). [5] NT = not tested. * = re-isolated with only the biological limiter.

### 3.3. Interaction on Detached Grapevine Canes

Abundant mycelial efflorescence appeared on the cutting surfaces of the detached 'Nero di Troia' canes inoculated with AA, FM, FM5-AA, FM5-TH, AA5-FM, AA5-PM, AA5-PC, and TH5-PC (Figure 6).

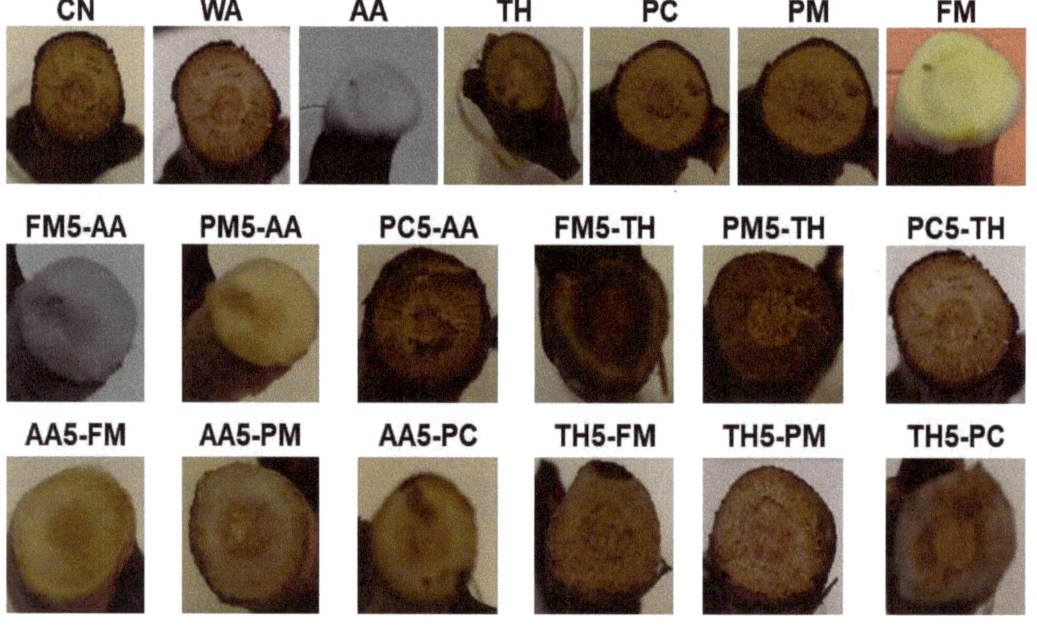


**Figure 6.** Aspect of detached 'Nero di Troia' canes inoculated with 100 µL of $10^6$ mL$^{-1}$ viable conidia (AA, TH, PM, and PC) or mycelium fragment (FM) suspensions in 0.2% agar in water (WA) as the adhesive medium. Strains were inoculated individually (AA, TH, PC, PM, FM) or in dual combination, in which the pathogens were distributed 5 days before the biological limiters (FM5-AA, PM5-AA, PC5-AA, FM5-TH, PM5-TH, PC5-TH) or biological limiters were deposited 5 days before the pathogens (AA5-FM, AA5-PM, AA5-PC, TH5-FM, TH5-PM, TH5-PC). CN = non-inoculated control. Pictures were taken 35 days after the first deposition.

From the single inoculations with AA, TH, PM, PC, or FM, the inoculated fungi were recovered from all the imprints (Table 4). In dual inoculations AA5-FM, AA5-PM, AA5-PC, FM5-AA, PM5-AA, and PC5-AA, only AA was re-isolated, whereas TH was obtained from the TH5-FM, TH5-PM, TH5-PC, FM5-TH, PM5-TH, and PC5-TH combinations.

**Table 4.** Colonization frequency of re-isolated fungi (%) [1] from 'Nero di Troia' detached canes inoculated with AA, TH, FM, PM, PC, and their dual combination with pathogens deposited 5 days before (FM5-AA, PM5-AA, PC5-AA, FM5-TH, PM5-TH, PC5-TH) or after (AA5-FM, AA5-PM, AA5-PC, TH5-FM, TH5-PM, TH5-PC) the biological limiters [2].

| Treatments | Strains | Position [4] | | | | |
|---|---|---|---|---|---|---|
| | | Cutting Surface [3] | Longitudinal Section (mm below the surface) | | | |
| | | | 5 | 10 | 15 | 20 |
| CN [5] | | NFI [7] | NFI | NFI | NFI | NFI |
| WA [6] | | NFI | NFI | NFI | NFI | NFI |
| AA | AA | 100 | 100 | NFI | NFI | NFI |
| TH | TH | 100 | 100 | NFI | NFI | NFI |
| FM | FM | 100 | 100 | NFI | NFI | NFI |
| PM | PM | 100 | 100 | NFI | NFI | NFI |
| PC | PC | 100 | 100 | NFI | NFI | NFI |
| FM5-AA | AA | 100 | 100 | NFI | NFI | NFI |
| | FM | 0 | 0 | | | |
| PM5-AA | AA | 100 | 100 | NFI | NFI | NFI |
| | PM | 0 | 0 | | | |
| PC5-AA | AA | 100 | 100 | NFI | NFI | NFI |
| | PC | 0 | 0 | | | |
| FM5-TH | TH | 100 | 100 | NFI | NFI | NFI |
| | FM | 0 | 0 | | | |
| PM5-TH | TH | 100 | 100 | NFI | NFI | NFI |
| | PM | 0 | 0 | | | |
| PC5-TH | TH | 100 | 100 | NFI | NFI | NFI |
| | PC | 0 | 0 | | | |
| AA5-FM | AA | 100 | 100 | NFI | NFI | NFI |
| | FM | 0 | 0 | | | |
| AA5-PM | AA | 100 | 100 | NFI | NFI | NFI |
| | PM | 0 | 0 | | | |
| AA5-PC | AA | 100 | 100 | NFI | NFI | NFI |
| | PC | 0 | 0 | | | |
| TH5-FM | TH | 100 | 100 | NFI | NFI | NFI |
| | FM | 0 | 0 | | | |
| TH5-PM | TH | 100 | 100 | NFI | NFI | NFI |
| | PM | 0 | 0 | | | |
| TH5-PC | TH | 100 | 100 | NFI | NFI | NFI |
| | PC | 0 | 0 | | | |

[1] Colonization frequency of re-isolated fungi was estimated as $100 \times (N \times n)$, where N is the number of colonies of each species developed from each chip, and n is the total number of plated fragments (n = 15). [2] For acronym definitions, see Table 1. [3] The data are the means of fifteen imprints per treatment. [4] Data are the means of fifteen chips for treatment. [5] CN = non-inoculated control. [6] WA = 0.2% agar in water. [7] NFI = no fungal isolation.

A similar pattern was observed from the re-isolations performed with the fragments cut at 5 mm below the upper end of the cutting surface (Table 4).

Negative was the isolation from the longitudinal sections of the canes performed at approximately 10, 15, and 20 mm below the upper end of the cutting surface for all the tested detached canes (Table 4). No organisms were obtained from the re-isolations performed with the imprints and the chips cut at 5, 10, 15, and 20 mm below the apical end of the detached cane used as a non-inoculated control (CN) and the control inoculated with the adhesive suspension WA (Table 4).

## 4. Discussion

The use of microbial antagonists to manage disease agents is one of the current frontiers in reducing the worrying dependence on pesticides, ensuring food safety and security, and protecting the environment and consumers [47–49].

*T. harzianum* TH1, *A. album* MX95, *P. ostreatus* ALPO, and *P. eryngii* AL142PE were tested as potential biological limiters against *F. mediterranea*, *P. minimum*, and *P. chlamydospora*, the main pathogens worldwide associated with grapevine Esca complex.

The isolations here carried out from 'Nero di Troia' vines confirmed the involvement of *P. minimum* and *P. chlamydospora* in brown wood streaking and *F. mediterranea* in white rot, while their mixture in the wood led to foliar symptoms [1–10,50].

A dual culture evaluates the antagonistic properties of microorganisms. This technique shows rapidly and clearly the mutual effects of the paired organisms and their interactions. However, the method excludes the host plant and cannot be used for biocontrol agents that allow disease control through tissue colonization, systemic resistance induction, and/or niche competition [43–45,51,52].

The growth rate on the PDA medium differentiates the behavior of tested organisms and could support potential biological limiters and pathogens during dual culture interaction.

Following the suggestions of Bell et al. [53], TH and PO are valid antagonists, because they covered completely the surfaces of the three tested pathogens (type C interaction following the Badalyan et al. [43,44] score rating).

This was supported by the antagonism index and the quantification of the re-isolation success. Low values of the antagonism index indicate a weak response of the strain in terms of inhibition, while high antagonism indices are associated with highly competitive and inhibitory properties in the antagonist [43,44]. The quantification of the re-isolation success of a potential biological control agent against a pathogen is a reliable tool to confirm the success of an antagonist in the interaction with the mycelium of a competing pathogen and represents an important parameter in dual culture assays [45]. The re-isolation of both the antagonist and pathogen from the interaction zone allowed us to assume that the challenge strains were not highly active competitors. In contrast, the re-isolation of only the antagonist from the interaction zone suggested the elimination of the challenger organism and obviously higher success in competition [45]. During the interactions labeled as "type c" (replacement), shown by TH and PO against the three tested pathogens, the biological limiter was always re-isolated from the overlapping zone and the area of pathogen presence (e.g., letter P in Figure 2a). This confirms the antagonistic aptitude of TH and PO and their ability to stop or inhibit the growth of challenged organisms [43,44,53]. Re-isolations of pathogens and biological limiters, even from the same discs, taken from the interface zone after type A and type $C_{A1}$ interactions, were obtained in AA-FM, PE-PM, and PE-PC. In the other combinations, the pathogenic organism was replaced by the biological limiter. The outcomes of these experiments allowed us to assume the weak competitive capacity of FM, PM, and PC, as hypothesized for *Scleroconidioma sphagnicola* [45] and *Eutypella parasitica* [46]. A similar re-isolation result after deadlock at a distance (interaction type B) was reported by Koukol et al. [45], which was not recorded in our study.

The detached cane assay, a modification of the single node cutting technique [54], assesses cane fruitfulness, reduces the time needed to evaluate pruning wound treatments and screenings of vine cultivars for grapevine trunk disease development [55], and allows potential biocontrol activity characterization [56]. Here, we adapted the technique as a model for the study of Esca-associated fungi in wood colonization and the interactions with AA and TH, two probable biological limiters. Due to the results of the study of the interaction among *P. chlamydospora*, *P. minumum*, and *F. mediterranea* performed in vitro and on grapevines cv. Italia and Matilde [57], single interaction pathogens vs. biological limiters were here evaluated. During previous in vitro and in planta interactions, *P. chlamydospora* and *P. minumum* competed for the substratum, albeit not directly challenging: *F. mediterranea* overgrew *P. chlamydospora*, and the interaction of *P. minumum* vs. *F. mediterranea* showed a deadlock at mycelia contact. In triple cultures, *P. minumum* in some way "prevented" *F. mediterranea* from overgrowing *P. chlamydospora* [57]. A similar pattern was seen on the woody tissue of both tested grapevine cultivars [57]. Furthermore, in naturally infected vine material, *F. mediterranea* is commonly associated with decayed wood, while *P. chlamydospora* and *P. minimum* are linked to brown wood streaking [1,3,5,7–10]. Inside the trunks of Esca-affected vines, rotted tissues are often bordered by a brown line that separates rotted from nondecayed wood [2]. It is probable that when colonizing a vacant resource, such as a fresh wound, each invading fungus became competitive and could exclude other microorganisms. Another theory of wood colonization during Esca development considers *P. chlamydospora* able to reduce plant resistance due to its toxic activity, *P. minimum* affects cell wall integrity through its enzymatic activity, and *F. mediterranea* enables the complete degradation of wood tissues, resulting in white rot formation [1–3,6,15,57].

To avoid a competitive association among *P. chlamydospora*, *P. minumum*, and *F. mediterranea*, single pathogen inoculation was here considered.

The detached cane assay confirmed the ability of PM, PC, and FM to colonize pruning wounds and to use this as a penetration method to colonize the wood tissues of the vine [1–3,49,50,57]. The experiments showed that the pathogens and potential biological limiters were effective in colonizing the apical end of the cane and progressing inside the woody tissue when singly inoculated. TH and AA also effectively protected pruning wounds against FM, PM, and PC for at least 3 months after treatment and confirmed the antagonistic effect exhibited in the dual cultures.

The four species used in the present study are well-known antagonists of plant-pathogenic fungi and nematodes.

*A. album* (*Fungi, Ascomycota, Pezizomycotina, Sordariomycetes, Hypocreales, Nectriaceae*) is a necrotrophic mycoparasite of *Puccinia coronata*, *Puccinia hordei*, *Puccinia graminis* f. sp. *avenue*, *Puccinia recondita* f. sp. *triticina*, *Golovinomyces* (*Oidium*) *lycopersici*, *Podosphaera* (*Sphaerotheca*) *fusca*, *Pseudopyrenochaeta* (*Pyrenochaeta*) *lycopersici*, *Meloidogyne incognita*, *Meloidogyne javanica*, and soil-borne plant pathogenic fungi [58–62]. The necrotrophic effects of *A. album* are associated with the production of hydrolytic enzymes (e.g., protease, glucanase) and chitinases [58].

*P. eryngii* and *P. ostreatus* (*Fungi, Basidiomycota, Agaricomycotina, Agaricomycetes, Agaricales, Pleurotaceae*) are xylotrophic mushrooms extensively cultivated around the world and evaluated as biological control agents for sugar beet nematode *Heterodera schachtii* [63]. In a dual culture, *P. ostreatus* strongly inhibited the mycelia growth of cereal-pathogenic fungi such as *Ceratobasidium cereale* (anamorph *Rhizoctonia cerealis*), *Gaeumannomyces tritici* (formerly *Gaeumannomyces graminis* var. *tritici*), *F. culmorum*, and *Bipolaris sorokiniana* [43]. *P. ostreatus* also combats the mycoparasites *Clonostachys rosea*, *T. harzianum*, *Trichoderma pseudokoningii*, and *Tichoderma viride* [44]. Furthermore, in dual culture assays, strains of *P. ostreatus* and *P. eryngii* exhibited strong inhibitory activity against *F. solani*, *F. oxysporum* f. sp. *lycopersici*, *V. dahliae*, *P. nicotianae*, and *S. sclerotiorum* [62].

*Trichoderma* is a genus of fungi from the family *Hypocreaceae*, commonly associated with the rhizosphere [64]. Several species of this genus and their *Hypocrea* teleomorphs are used in agriculture as bio-regulators of plant growth and biocontrol agents for the management of nematodes and plant diseases [64–66]. Among the *Trichoderma* species useful as biocontrol agents, *T. harzianum* is widely used in plant disease control [67]. The species is proposed as a complex of species based on secondary metabolite production, the target pathogen, the host range, and the distribution area [66–69]. Morphological and molecular characterizations of commercial *T. harzianum* biological control products show the presence of nine new species [67]. Different *Trichoderma* species, including *T. atroviride*, *T. harzianum*, *T. asperellum*, *T. gamsii*, and *T. longibrachiatum*, have been widely used against fungi in the Esca complex during grapevine nursery propagation processes and for pruning wound protection [1,12,15,19,20,33,36–38]. Different species of *Trichoderma* provide efficacy against fungi in the *Botryosphaeriaceae* and the *Diatrypaceae* associated with grapevine trunk disease [56]. Antibiosis, mycoparasitism, and nutrient and/or space competition are mechanisms of action associated with *Trichoderma*'s antifungal activity [15,19,20,69]. Therefore, effectiveness in pruning wound protection requires the establishment of a biological agent. For this, *Trichoderma*-based commercial products give the greatest protective action several days after application.

In the present study, TH and AA effectively prevented pruning wound colonization by *F. mediterranea*, *P. minimum*, and *P. chlamydospora* mycelia and conidia spread. They effectively also stopped the colonization developed by each pathogen.

*P. chlamydospora*, *P. minimum*, and *F. mediterranea*, the main fungal species involved in the Esca complex of grapevine, were successfully added to the target organisms of these interesting biological limiters.

Among the four species here tested, *T. harzianum* and *A. album* demonstrate a long history as fungal antagonists. Strains of *T. harzianum*, including the commercial one (Trichodex®), effectively controlled *P. chlamydospora* and *P. minimum* in vitro and under greenhouse, field, and nursery conditions [1,15,22,34]. During antagonistic interaction, *T. harzianum* explores different modes of action: it competes for nutrients, utilizing carbon and nitrogen sources, induces direct antagonism, overgrows its "hosts", induces growth reductions [15,22,35], and reduces infection [36–38] and the necrosis length on rootstock and nursery material [15,36–38]. The strain DiSSPA TH07.1-NC, tested in this study, was a moderately active antagonist (antagonism index of 12.0), showed overgrowth, inhibited radial growth, and quickly interacted with the tested Esca-associated pathogens. A direct effect of this strain on PC, PM, and FM can be assumed from the score of the re-isolation success and the colonization frequency. No target pathogens were re-isolated after the interactions with TH on PDA plates and from dual-inoculated detached 'Nero di Troia' canes. Furthermore, the strain DiSSPA TH07.1-NC was able to colonize the apical end of each treated cane, to produce abundant mycelial efflorescence on the cutting surfaces and penetrate the first 5 mm of canes during 30 days.

Strains of the necrotrophic mycoparasite *A. album* inhibit the sporulation and the growth of several powdery mildew and rust agents, soil-borne plant pathogens, and nematodes [59–62]. This antagonistic efficiency is supported by the production of hydrolytic enzymes such as protease, glucanase, and several chitinases involved in the cell wall degradation of many phytopathogenic fungi [58].

The strain DiSSPA MX95, tested in this study, was a weak antagonist (antagonism index of 5.5), showing deadlock (mutual inhibition in which neither strain was able to overgrow the other) at mycelial contact against FM and PM and partial replacement after an initial deadlock with mycelial contact against PC. Despite its slow growth rate, DiSSPA MX95 inhibits radial growth and greater re-isolation success on PDA plates. During the experiments on detached 'Nero di Troia' canes, the strain DiSSPA MX95 was able to colonize the apical end of each treated cane, to produce abundant mycelial efflorescence on the cutting surfaces and penetrate the first 5 mm of canes.

This was the first application of *P. ostreatus* and *P. eryngii* strains against fungal species involved in the Esca complex of grapevine. The strains DiSSPA ALPO and DiSSPA AL142, tested in this study, were moderately active antagonists (antagonism index of 9.0 and 10.5 for PO and PE, respectively), showed overgrowth, inhibited radial growth, and quickly interacted with the tested Esca-associated pathogens. A direct effect of these strains on PC, PM, and FM can be assumed from the score of the re-isolation success: no target pathogens were re-isolated after the interactions with PO and PE on PDA plates.

Further screening of the strains used in this study should be conducted to assess their efficacy against other grapevine trunk disease fungi, including *Ilyonectria* spp., *Cadophora luteo-olivacea*, *Diplodia seriata*, and *Neofusicoccum parvum*, commonly found in nursery-propagated material and responsible for young grapevines' decline and death.

A shortcoming of our study is related to the slight repetitions performed. Additional experiments, testing a broader range of *P. chlamydospora*, *P. minimum*, and *F. mediterranea* strains, are needed for the confirmation and clarification of our results and a solid conclusion. However, further testing should be performed to optimize the application methods (e.g., dose and time), together with validation experiments in the field. Moreover, combinations of AA, PO, PM, and TH will be tested.

## 5. Conclusions

The interactions between the three most important species isolated from the wood of Esca-complex-affected vines and the four strains tested as biological limiters in the dual cultures, here analyzed in terms of the first contact between the two mycelia, the percentage of radial growth inhibition, the type of mycelial interaction, the antagonism index, and the quantification of fungal re-isolation, suggested that the tested strains *P. ostreatus* ALPO, *P. eryngii* AL142, *A. album* MX-95, and *T. harzianum* TH07.1-NC were competitive and caused the greatest inhibition of the three pathogens. Interactions on detached grapevine canes among TH and AA against FM, PC, and PM confirmed the in vitro effects. In particular, *A. album* MX-95 and *T. harzianum* TH07.1-NC are the most promising strains to be used as biological control agents against the three main pathogens associated with the Esca complex of grapevine.

**Author Contributions:** Conceptualization, G.L.B., F.M., G.D. and N.R.; methodology, N.R., F.M., D.S., G.D. and G.L.B.; validation, G.L.B. and F.M.; formal analysis, G.L.B. and F.M.; investigation, N.R., F.M., D.S., G.D. and G.L.B.; data curation, N.R., F.M., G.D., D.S. and G.L.B.; writing—original draft preparation, N.R., F.M., G.D., D.S. and G.L.B.; writing—review and editing, N.R., F.M., D.S., G.D. and G.L.B.; visualization, N.R., F.M., D.S., G.D. and G.L.B.; supervision, F.M. and G.L.B. All authors have read and agreed to the published version of the manuscript.

**Funding:** This research was in part funded by the University of Bari Aldo Moro, under the Fondo ordinario per la Ricerca Scientifica (ex 60%) years 2017 and 2018, grant number Disspa.RicercaLocale.Bruno01.

**Data Availability Statement:** Data are contained within the article.

**Conflicts of Interest:** The authors declare no conflict of interest.

## References

1. Mondello, V.; Songy, A.; Battiston, E.; Pinto, C.; Coppin, C.; Trotel-Aziz, P.; Clément, C.; Mugnai, L.; Fontaine, F. Grapevine Trunk Diseases: A review of fifteen years of trials for their control with chemicals and biocontrol agents. *Plant Dis.* **2018**, *102*, 1189–1217. [CrossRef]
2. Graniti, A.; Surico, G.; Mugnai, L. Esca of grapevine: A disease complex or a complex of diseases? *Phytopathol. Mediterr.* **2000**, *39*, 16–20. [CrossRef]
3. Surico, G. Towards a redefinition of the diseases within the esca complex of grapevine. *Phytopathol. Mediterr.* **2009**, *48*, 5–10. [CrossRef]
4. Bruno, G.L.; Ippolito, M.P.; Mannerucci, F.; Bragazzi, L.; Tommasi, F. Physiological responses of 'Italia' grapevines infected with Esca pathogens. *Phytopathol. Mediterr.* **2021**, *60*, 321–336. [CrossRef]
5. Bertsch, C.; Ramírez-Suero, M.; Magnin-Robert, M.; Larignon, P.; Chong, J.; Abou-Mansour, E.; Spagnolo, A.; Clément, C.; Fontaine, F. Grapevine trunk diseases: Complex and still poorly understood. *Plant Pathol.* **2012**, *62*, 243–265. [CrossRef]
6. Surico, G.; Mugnai, L.; Marchi, G. The Esca Disease Complex. In *Integrated Management of Diseases Caused by Fungi, Phytoplasma and Bacteria*; Ciancio, A., Mukerji, K.G., Eds.; Springer: Dordrecht, The Netherlands, 2008; pp. 119–136.
7. Mugnai, L.; Graniti, A.; Surico, G. Esca (black measles) and brown wood-streaking: Two old and elusive diseases of grapevines. *Plant Dis.* **1999**, *83*, 404–418. [CrossRef]
8. Fischer, M. Biodiversity and geographic distribution of basidiomycetes causing esca-associated white rot in grapevine: A worldwide perspective. *Phytopathol. Mediterr.* **2006**, *45*, 30–42. [CrossRef]
9. Mostert, L.; Groenewald, J.Z.; Summerbell, R.C.; Gams, W.; Crous, P.W. Taxonomy and pathology of *Togninia* (*Diaporthales*) and its *Phaeoacremonium* anamorphs. *Stud. Mycol.* **2006**, *54*, 1–113. [CrossRef]
10. Mostert, L.; Hallen, F.; Fourie, P.; Crous, P.W. A review of *Phaeoacremonium* species involved in Petri disease and esca of grapevines. *Phytopathol. Mediterr.* **2006**, *45*, S12–S29. [CrossRef]
11. Claverie, M.; Notaro, M.; Fontaine, F.; Wery, J. Current knowledge on Grapevine Trunk Diseases with complex etiology: A systemic approach. *Phytopathol. Mediterr.* **2020**, *59*, 29–53. [CrossRef]
12. Gramaje, D.; Armengol, J.; Salazar, D.; López-Cortés, I.; García-Jiménez, J. Effect of hot-water treatments above 50 °C on grapevine viability and survival of Petri disease pathogens. *Crop Prot.* **2009**, *28*, 280–285. [CrossRef]
13. Di Marco, S.; Mazzullo, A.; Calzarano, F.; Cesari, A. The control of esca: Status and perspectives. *Phytopathol. Mediterr.* **2000**, *39*, 232–240. [CrossRef]
14. Roblin, G.; Luinia, E.; Fleurat-Lessarda, P.; Larignon, P.; Berjeauda, J.M. Towards a preventive and/or curative treatment of esca in grapevine trunk disease: General basis in the elaboration of treatments to control plant pathogen attacks. *Crop Prot.* **2019**, *116*, 156–169. [CrossRef]

15. Mesguida, O.; Haidar, R.; Yacoub, A.; Dreux-Zigha, A.; Berthon, J.-Y.; Guyoneaud, R.; Attard, E.; Rey, P. Microbial Biological Control of Fungi Associated with Grapevine Trunk Diseases: A Review of Strain Diversity, Modes of Action, and Advantages and Limits of Current Strategies. *J. Fungi* **2023**, *9*, 638. [CrossRef]

16. Rolshausen, P.E.; Urbez-Torres, S.; Rooney-Latham, S.; Eskalen, A.; Smith, R.J.; Gubler, W.D. Evaluation of pruning wound susceptibility and protection against fungi associated with grapevine trunk diseases. *Am. J. Enol. Vitic.* **2010**, *61*, 113–119. [CrossRef]

17. Di Marco, S.; Osti, F.; Calzarano, F.; Roberti, R.; Veronesi, A.; Amalfitano, C. Effects of grapevine applications of fosetyl-aluminium formulations for downy mildew control on "esca" and associated fungi. *Phytopathol. Mediterr.* **2011**, *50*, S285–S299. [CrossRef]

18. Gramaje, D.; Aroca, A.; Raposo, R.; García-Jiménez, J.; Armengol, J. Evaluation of fungicides to control Petri disease pathogens in the grapevine propagation process. *Crop Prot.* **2009**, *28*, 1091–1097. [CrossRef]

19. Chervin, J.; Romeo-Oliván, A.; Fournier, S.; Puech-Pages, V.; Dumas, B.; Jacques, A.; Marti, G. Modification of Early Response of *Vitis vinifera* to Pathogens Relating to Esca Disease and Biocontrol Agent Vintec® Revealed By Untargeted Metabolomics on Woody Tissues. *Front. Microbiol.* **2022**, *13*, 835463. [CrossRef]

20. Spasova, M.; Manolova, N.; Rashkov, I.; Naydenov, M. Eco-Friendly Hybrid PLLA/Chitosan/*Trichoderma asperellum* Nanomaterials as Biocontrol Dressings against Esca Disease in Grapevines. *Polymers* **2022**, *14*, 2356. [CrossRef]

21. Daraignes, L.; Gerbore, J.; Yacoub, A.; Dubois, L.; Romand, C.; Zekri, O.; Roudet, J.; Chambon, P.; Fermaud, M. Efficacy of *P. oligandrum* affected by its association with bacterial BCAs and rootstock effect in controlling grapevine trunk diseases. *Biol. Control* **2018**, *119*, 59–67. [CrossRef]

22. Kotze, C.; Van Niekerk, J.; Halleen, F.; Mostert, L.; Fourie, P. Evaluation of biocontrol agents for grapevine pruning wound protection against trunk pathogen infection. *Phytopathol. Mediterr.* **2011**, *50*, 247–263. [CrossRef]

23. Alfonzo, A.; Conigliaro, G.; Torta, L.; Burruano, S.; Moschetti, G. Antagonism of *Bacillus subtilis* Strain AG1 against Vine Wood Fungal Pathogens. *Phytopathol. Mediterr.* **2009**, *48*, 155–158. [CrossRef]

24. Haidar, R.; Roudet, J.; Bonnard, O.; Dufour, M.C.; Corio-Costet, M.F.; Fert, M.; Gautier, T.; Deschamps, A.; Fermaud, M. Screening and modes of action of antagonistic bacteria to control the fungal pathogen *Phaeomoniella chlamydospora* involved in grapevine trunk diseases. *Microbiol. Res.* **2016**, *192*, 172–184. [CrossRef] [PubMed]

25. Andreolli, M.; Zapparoli, G.; Angelini, E.; Lucchetta, G.; Lampis, S.; Vallini, G. *Pseudomonas protegens* MP12: A Plant Growth-Promoting endophytic bacterium with broad-spectrum antifungal activity against grapevine phytopathogens. *Microbiol. Res.* **2018**, *219*, 123–131. [CrossRef] [PubMed]

26. Haidar, R.; Yacoub, A.; Vallance, J.; Compant, S.; Antonielli, L.; Saad, A.; Habenstein, B.; Kauffmann, B.; Grélard, A.; Loquet, A.; et al. Bacteria associated with wood tissues of Esca-diseased grapevines: Functional diversity and synergy with *Fomitiporia mediterranea* to degrade wood components. *Environ. Microbiol.* **2021**, *23*, 6104–6121. [CrossRef] [PubMed]

27. Álvarez-Pérez, J.M.; González-García, S.; Cobos, R.; Olego, M.Á.; Ibañez, A.; Díez-Galán, A.; Garzón-Jimeno, E.; Coque, J.J.R. Use of endophytic and rhizosphere actinobacteria from grapevine plants to reduce nursery fungal graft infections that lead to young grapevine decline. *Appl. Environ. Microbiol.* **2017**, *83*, e01564-17. [CrossRef]

28. del Pilar Martínez-Diz, M.; Díaz-Losada, E.; Andrés-Sodupe, M.; Bujanda, R.; Maldonado-González, M.M.; Ojeda, S.; Yacoub, A.; Rey, P.; Gramaje, D. Field evaluation of biocontrol agents against black-foot and Petri diseases of grapevine. *Pest Manag. Sci.* **2021**, *77*, 697–708. [CrossRef]

29. Bustamante, M.I.; Elfar, K.; Eskalen, A. Evaluation of the Antifungal Activity of Endophytic and Rhizospheric Bacteria against Grapevine Trunk Pathogens. *Microorganisms* **2022**, *10*, 2035. [CrossRef]

30. Yacoub, A.; Gerbore, J.; Magnin, N.; Chambon, P.; Dufour, M.C.; Corio-Costet, M.F.; Guyoneaud, R.; Rey, P. Ability of *Pythium oligandrum* strains to protect *Vitis vinifera* L., by inducing plant resistance against *Phaeomoniella chlamydospora*, a pathogen involved in Esca, a grapevine trunk disease. *Biol. Control* **2016**, *92*, 7–16. [CrossRef]

31. Yacoub, A.; Magnin, N.; Gerbore, J.; Haidar, R.; Bruez, E.; Compant, S.; Guyoneaud, R.; Rey, P. The biocontrol Root-Oomycete, *Pythium oligandrum*, triggers grapevine resistance and shifts in the transcriptome of the trunk pathogenic fungus, *Phaeomoniella chlamydospora*. *Int. J. Mol. Sci.* **2020**, *21*, 6876. [CrossRef]

32. Del Frari, G.; Cabral, A.; Nascimento, T.; Boavida Ferreira, R.; Oliveira, H. *Epicoccum layuense* a potential biological control agent of esca-associated fungi in grapevine. *PLoS ONE* **2019**, *14*, e0213273. [CrossRef]

33. Aloi, F.; Reggiori, G.; Bigot, A.; Montermini, P.; Bortolotti, R.; Nannini, F.; Osti, L.; Mugnai, L.; Di Marco, S. Remedier® (*Trichoderma asperellum* and *Trichoderma gamsii*): A new opportunity to control the esca disease complex. Five years of results of field trials in Italy. *Phytopathol. Mediterr.* **2015**, *54*, 420–436.

34. Silva-Valderrama, I.; Toapanta, D.; Miccono, M.d.l.A.; Lolas, M.; Díaz, G.A.; Cantu, D.; Castro, A. Biocontrol Potential of Grapevine Endophytic and Rhizospheric Fungi Against Trunk Pathogens. *Front. Microbiol.* **2021**, *11*, 614620. [CrossRef]

35. Wallis, C.M. Nutritional Niche Overlap Analysis as a method to identify potential biocontrol fungi against trunk pathogens. *BioControl* **2021**, *66*, 559–571. [CrossRef]

36. Di Marco, S.; Osti, F. Applications of *Trichoderma* to prevent *Phaeomoniella chlamydospora* infections in organic nurseries. *Phytopathol. Mediterr.* **2007**, *46*, 11. [CrossRef]

37. Di Marco, S.; Osti, F.; Cesari, A. Experiments on the control of Esca by *Trichoderma*. *Phytopathol. Mediterr.* **2004**, *43*, 108–115. [CrossRef]

38. Martínez-Diz, M.d.P.; Díaz-Losada, E.; Díaz-Fernández, Á.; Bouzas-Cid, Y.; Gramaje, D. Protection of grapevine pruning wounds against *Phaeomoniella chlamydospora* and *Diplodia seriata* by commercial biological and chemical methods. *Crop Prot.* **2021**, *143*, 105465. [CrossRef]

39. Geiger, A.; Karácsony, Z.; Geml, J.; Váczy, K.Z. Mycoparasitism capability and growth inhibition activity of *Clonostachys rosea* isolates against fungal pathogens of grapevine trunk diseases suggest potential for biocontrol. *PLoS ONE* **2022**, *17*, e0273985. [CrossRef]

40. Gkikas, F.-I.; Tako, A.; Gkizi, D.; Lagogianni, C.; Markakis, E.A.; Tjamos, S.E. *Paenibacillus alvei* K165 and *Fusarium oxysporum* F2: Potential Biocontrol Agents against *Phaeomoniella chlamydospora* in Grapevines. *Plants* **2021**, *10*, 207. [CrossRef]

41. Sparapano, L.; Bruno, G.L.; Campanella, A. Interactions between three fungi associated with esca of grapevine, and their secondary metabolites. *Phytopathol. Mediterr.* **2001**, *40*, S417–S422. [CrossRef]

42. Sharfuddin, C.; Mohanka, R. In vitro antagonism of indigenous *Trichoderma* isolates against phytopathogen causing wilt of lentil. *Int. J. Life Sci. Pharma Res.* **2012**, *2*, 195–202.

43. Badalyan, S.M.; Innocenti, G.; Garibyan, N.G. Antagonistic activity of xylotrophic mushrooms against pathogenic fungi of cereals in dual culture. *Phytopathol. Mediterr.* **2002**, *41*, 200–225. [CrossRef]

44. Badalyan, S.M.; Innocenti, G.; Garibyan, N.G. Interactions between xylotrophic mushrooms and mycoparasitic fungi in dual culture. *Phytopathol. Mediterr.* **2004**, *43*, 44–48. [CrossRef]

45. Koukol, O.; Mrnka, L.; Kulhankova, A.; Vosatka, M. Competition of *Scleroconidioma sphagnicola* with fungi decomposing spruce litter needles. *Can. J. Bot.* **2006**, *84*, 469–476. [CrossRef]

46. Brglez, A.; Piškur, B.; Ogris, N. In vitro Interactions between *Eutypella parasitica* and some frequently isolated fungi from the wood of the dead branches of young sycamore maple (*Acer pseudoplatanus*). *Forests* **2020**, *11*, 1072. [CrossRef]

47. Thambugala, K.M.; Daranagama, D.A.; Phillips, A.J.L.; Kannangara, S.D.; Promputtha, I. Fungi vs. fungi in biocontrol: An overview of fungal antagonists applied against fungal plant pathogens. *Front. Cell. Infect. Microbiol.* **2020**, *10*, 604923. [CrossRef]

48. Lahlali, R.; Ezrari, S.; Radouane, N.; Kenfaoui, J.; Esmaeel, Q.; El Hamss, H.; Belabess, Z.; Barka, E.A. Biological Control of Plant Pathogens: A Global Perspective. *Microorganisms* **2022**, *10*, 596. [CrossRef]

49. Kothari, I.L.; Patel, M. Plant immunization. *Indian J. Exp. Biol.* **2004**, *42*, 244–252.

50. Graniti, A.; Bruno, G.; Sparapano, L. Three-Year Observation of Grapevines Cross-Inoculated with Esca-Associated Fungi. *Phytopathol. Mediterr.* **2001**, *40*, 376–386. [CrossRef]

51. Kloepper, J.W.; Ryu, C.M.; Zhang, S. Induce systemic resistance and promotion of plant growth by *Bacillus* spp. *Phytopathology* **2004**, *94*, 1259–1266. [CrossRef]

52. Morón-Ríos, A.; Gómez-Cornelio, S.; Ortega-Morales, B.O.; De la Rosa-García, S.; Partida-Martínez, L.P.; Quintana, P.; Alayón-Gamboa, A.; Cappello-García, S.; Gonzáles-Gómez, S. Interactions between abundant fungal species influence the fungal community assemblage on limestone. *PLoS ONE* **2017**, *12*, e0188443. [CrossRef]

53. Bell, D.K.; Wells, H.D.; Markham, C.R. In vitro antagonism of *Trichoderma* spp. against six fungal plant pathogens. *Phytopathology* **1982**, *72*, 379–382. [CrossRef]

54. Sosnowski, M.; Emmett, B.; Clarke, K.; Wicks, T. Susceptibility of table grapes to black spot (anthracnose) disease. *Aust. N. Z. Grapegrow. Winemak.* **2007**, *521a*, 8–11.

55. Mundy, D.C.; Robertson, S.M. Evaluation of single-node plantlets as a model system for grapevine trunk diseases. *N. Z. Plant Protect.* **2010**, *63*, 167–173. [CrossRef]

56. Úrbez-Torres, J.R.; Tomaselli, E.; Pollard-Flamand, J.; Boulé, J.; Gerin, D.; Pollastro, S. Characterization of *Trichoderma* isolates from southern Italy, and their potential biocontrol activity against grapevine trunk disease fungi. *Phytopathol. Mediterr.* **2020**, *59*, 425–439. [CrossRef]

57. Sparapano, L.; Bruno, G.; Ciccarone, C.; Graniti, A. Infection of grapevines by some fungi associated with esca. II. Interaction among *Phaeoacremonium chlamydosporum*, *P. aleophilum* and *Fomitiporia punctata*. *Phytopathol. Mediterr.* **2000**, *39*, 53–58. [CrossRef]

58. Kunz, C.; Sellam, O.; Bertheau, Y. Purification and characterization of a chitinase from the hyperparasitic fungus *Aphanocladium album*. *Physiol. Mol. Plant Pathol.* **1992**, *40*, 117–131. [CrossRef]

59. Biali, M.; Dinoor, A.; Eshed, N.; Kenneth, R. *Aphanocladium album*, a fungus inducing teliospore production in rusts. *Ann. Appl. Biol.* **1972**, *72*, 37–42. [CrossRef]

60. Sasanelli, N.; Ciccarese, F.; Papajová, I. *Aphanocladium album* by via sub-irrigation in the control of *Pyrenochaeta lycopersici* and *Meloidogyne incognita* on tomato in a plastic-house. *Helminthologia* **2008**, *45*, 137–142. [CrossRef]

61. Leoni, C.; Piancone, E.; Sasanelli, N.; Bruno, G.L.; Manzari, C.; Pesole, G.; Ceci, L.R.; Volpicella, M. Plant Health and Rhizosphere Microbiome: Effects of the Bionematicide *Aphanocladium album* in Tomato Plants Infested by *Meloidogyne javanica*. *Microorganisms* **2020**, *8*, 1922. [CrossRef]

62. D'Ambrosio, G.; Cariddi, C.; Mannerucci, F.; Bruno, G.L. In vitro screening of new biological limiters against some of the main soil-borne phytopathogens. *Sustainability* **2022**, *14*, 2693. [CrossRef]

63. Palizi, P.; Goltapeh, E.M.; Pourjam, E.; Safaie, N. Potential of *Oyster mushrooms* for the biocontrol of sugar beet nematode (*Heterodera schachtii*). *J. Plant Prot. Res.* **2009**, *49*, 27–33. [CrossRef]

64. TariqJaveed, M.; Farooq, T.; Al-Hazmi, A.S.; Hussain, M.D.; Rehman, A.U. Role of *Trichoderma* as a biocontrol agent (BCA) of phytoparasitic nematodes and plant growth inducer. *J. Invertebr. Pathol.* **2021**, *183*, 107626. [CrossRef] [PubMed]

65. Vinale, F.; Sivasithamparam, K.; Ghisalberti, L.E.; Marra, R.; Woo, L.S.; Lorito, M. *Trichoderma*-plant-pathogen interactions. *Soil Biol. Biochem.* **2008**, *40*, 1–10. [CrossRef]
66. Zin, N.A.; Badaluddin, N.A. Biological functions of *Trichoderma* spp. For agriculture applications. *Ann. Agric. Sci.* **2020**, *65*, 168–178. [CrossRef]
67. Chaverri, P.; Branco-Rocha, F.; Jaklitsch, W.; Gazis, R.; Degenkolb, T.; Samuels, G.J. Systematics of the *Trichoderma harzianum* species complex and the re-identification of commercial biocontrol strains. *Mycologia* **2015**, *107*, 558–590. [CrossRef]
68. Ahluwalia, V.; Kumar, J.; Rana, V.S.; Sati, O.P.; Walia, S. Comparative evaluation of two *Trichoderma harzianum* strains for major secondary metabolite production and antifungal activity. *Nat. Prod. Res.* **2015**, *29*, 914–920. [CrossRef]
69. Schuster, A.; Schmoll, M. Biology and biotechnology of *Trichoderma*. *Appl. Microbiol. Biot.* **2010**, *87*, 787–799. [CrossRef]

**Disclaimer/Publisher's Note:** The statements, opinions and data contained in all publications are solely those of the individual author(s) and contributor(s) and not of MDPI and/or the editor(s). MDPI and/or the editor(s) disclaim responsibility for any injury to people or property resulting from any ideas, methods, instructions or products referred to in the content.

*microorganisms*

*Communication*

# Potential Biocontrol Agents of Corn Tar Spot Disease Isolated from Overwintered *Phyllachora maydis* Stromata

Eric T. Johnson *, Patrick F. Dowd, José Luis Ramirez and Robert W. Behle

Crop Bioprotection Research Unit, National Center for Agricultural Utilization Research, Agricultural Research Service, United States Department of Agriculture, 1815 N University Street, Peoria, IL 61604, USA; patrick.dowd@usda.gov (P.F.D.); jose.ramirez2@usda.gov (J.L.R.); robert.behle@usda.gov (R.W.B.)
* Correspondence: eric.johnson2@usda.gov

**Abstract:** Tar spot disease in corn, caused by *Phyllachora maydis*, can reduce grain yield by limiting the total photosynthetic area in leaves. Stromata of *P. maydis* are long-term survival structures that can germinate and release spores in a gelatinous matrix in the spring, which are thought to serve as inoculum in newly planted fields. In this study, overwintered stromata in corn leaves were collected in Central Illinois, surface sterilized, and caged on water agar medium. Fungi and bacteria were collected from the surface of stromata that did not germinate and showed microbial growth. Twenty-two *Alternaria* isolates and three *Cladosporium* isolates were collected. Eighteen bacteria, most frequently *Pseudomonas* and *Pantoea* species, were also isolated. Spores of *Alternaria*, *Cladosporium*, and *Gliocladium catenulatum* (formulated as a commercial biofungicide) reduced the number of stromata that germinated compared to control untreated stromata. These data suggest that fungi collected from overwintered tar spot stromata can serve as biological control organisms against tar spot disease.

**Keywords:** *Phyllachora maydis*; tar spot disease; corn pathogen; biological control; biofungicide

**Citation:** Johnson, E.T.; Dowd, P.F.; Ramirez, J.L.; Behle, R.W. Potential Biocontrol Agents of Corn Tar Spot Disease Isolated from Overwintered *Phyllachora maydis* Stromata. *Microorganisms* **2023**, *11*, 1550. https://doi.org/10.3390/microorganisms11061550

Academic Editors: Michael J. Bidochka, Chetan Keswani and Rainer Borriss

Received: 20 April 2023
Revised: 6 May 2023
Accepted: 6 June 2023
Published: 10 June 2023

**Copyright:** © 2023 by the authors. Licensee MDPI, Basel, Switzerland. This article is an open access article distributed under the terms and conditions of the Creative Commons Attribution (CC BY) license (https://creativecommons.org/licenses/by/4.0/).

## 1. Introduction

Corn is one of the most important grain crops in the world, but production is limited by a variety of pests and diseases [1]. One devastating disease is caused by *Phyllachora maydis* and is commonly referred to as tar spot because of its characteristic black, shiny, raised leaf spots which generally range from 2–4 mm in diameter. Infection by this pathogenic fungus can result in significant yield losses, and even death of plants if infection occurs early in a susceptible variety [2]. Tar spot disease has been endemic in much of Central and South America for several decades [2]. It was first reported in two U.S. "corn belt" states, Illinois and Indiana, in 2015 [3], and since then there have been major outbreaks in several regions in both 2018 and 2021 [4].

Management strategies for tar spot disease include use of resistant varieties and application of fungicides [2]. However, under high inoculum pressure, significant yield losses can still occur, even with the use of resistant varieties [5]. Fungicide application has not been very effective because once symptoms are noticed, further infection becomes difficult to control [6].

The black structures characteristic of tar spot disease are long term survival structures called stromata, which are similar to the sclerotia produced by the vegetable pathogen *Sclerotinia sclerotiorum* and other fungi [7]. These stromata are the overwintering structures for tar spot, and the source of initial inoculum. However, reports of spore production and percent germination by overwintered stromata vary considerably [8], suggesting the presence of natural biocontrol organisms. Mycoparasites may contribute to reduced survival of stromata, as reported for other stromata or sclerotia-like structures in several fungal species [7,9] For example, several species of biocontrol organisms have been isolated from the sclerotia of *S. sclerotiorum*, some of which have been commercialized as biological control products [10,11]. Mycoparasites of stromata of *Coccodiella miconiae*, a relative of

*P. maydis*, included fungi in the genera *Cladosporium, Cornespora.* and *Sagenomella* [12]. To investigate the potential for biocontrol of tar spot, stromata from overwintered corn leaves collected in Illinois were examined for the presence of mycoparasites or other biocontrol agents. This study identifies several bacterial and fungal species infecting tar spot stromata and describes bacterial and fungal species with potential to serve as biocontrol agents against tar spot disease of corn.

## 2. Materials and Methods

### 2.1. Isolation of Organisms

Several overwintered corn leaves of inbred GE440 (planted from seed increased from plants originally obtained from the USDA-ARS Plant Introduction Center) were collected in late April 2022 from a 2021-planted research plot. This site in Peoria, IL, has had continuous corn production for several years and is thus more ecologically stable and likely to yield biocontrol agents for tar spot than commercial fields where crop rotation typically occurs. Stromata were cut from leaves along with a small leaf piece "handle". Approximately 50 stromata were surface sterilized with 70% ethanol and then blotted dry as described previously [8]. Stromata were placed in Petri dishes with tight fitting lids (Falcon$^R$ 351006, Corning Inc., Corning, NY, USA) containing 5 mL of 3% water agar to induce rehydration and stimulate growth of mycoparasites; in some cases, a few μL of sterile water was added to help with rehydration. Plates were held at 25–27 °C. Stromata were examined daily for outgrowth of organisms that were not visually the same as any seen from the attached leaf material, and organisms noted were photographed with cameras equipped with macro lenses. Microorganisms were isolated using two different methods. In the first method, fungal mycelium growing out of a single stroma was transferred to a potato dextrose agar (PDA, Difco potato dextrose broth, Becton Dickinson Company, Sparks, MD, USA, with bacto agar at 20 g/L) plate using a sterile metal probe. A few days later, bacteria were observed growing together with the fungus. The bacteria were transferred to a new PDA plate, whereas the bacterium-contaminated fungus was transferred to a PDA plus 0.01% chloramphenicol plate. In the second method, petroleum jelly was used to stick leaves with stromata to the inside of the top lid over a 3% water agar plate. After 2–3 days, microorganisms growing on the surface of the agar were transferred to nutrient media plates (Luria broth (LB) or tryptone glucose yeast extract (TGY) for bacteria or PDA for fungi). LB agar plates consisted of tryptone (10 g/L), sodium chloride (10 g/L), yeast extract (5 g/L), and bacto agar at 15 g/L. TGY agar plates consisted of tryptone (5 g/L), yeast extract (5 g/L), $K_2HPO_4$ (1 g/L), and glucose (1 g/L); the pH of the final mixture was adjusted to 7.0 and bacto agar was added at 15 g/L before autoclaving.

### 2.2. Identification of Organisms

Genomic DNA was isolated from fungi as previously described [13] with a modification: a small fragment of fungal mycelia from the culture plate was pulverized with a small amount of 800 μm silica beads using a Minibead Beater (Biospec Products, Bartlesville, OK, USA). Genomic DNA was isolated from bacteria as described in [14] with several modifications. A 1 mL aliquot of bacterial overnight culture was centrifuged at $16,000 \times g$; the bacterial pellet was first suspended in 480 μL of 50 mM EDTA and 120 μL of 5 mg/mL lysozyme was then added. The suspended pellet was incubated at 37 °C for 30–60 min and the suspension was centrifuged at $16,000 \times g$ for 2 min. The supernatant was moved to a new tube and 600 μL of nuclei lysis solution was added. The lysate was incubated at 80 °C for 5 min. After cooling to room temperature, 12 μg of ribonuclease A was added to the lysate and the mixture was incubated at 37 °C for 15–60 min. Then, 250 μL of 5 M NaCl was added to the lysate and the mixture was vortexed for 20 s. The mixture was kept on ice for 5 min and then centrifuged at $16,000 \times g$ for 3 min. The method was continued as previously described [14]. Genomic DNA was stored at $-20$ °C until processing. PCR amplification of various gene products from genomic DNA was performed using Platinum™ SuperFi™ II DNA polymerase (Thermo Fisher Scientific, Waltham, MA, USA) according to the manufacturer's instructions; primers used for PCR product amplification are listed in Supplementary File S1. PCR products were sequenced using the BigDye Terminator Cycle Sequencing Kit (Version 3.1, Applied Biosystems, Foster City, CA, USA) and BLAST analysis (National Center for Biotechnology Information (NCBI) was used to determine potential identities [15]. For bacteria, conserved gene sequences targeting the *16S rDNA* gene were amplified by PCR, sequenced, and utilized for identification of each organism. In some cases of bacterial identification, a portion of the *gyrB* gene was also amplified from genomic DNA because the *16S rDNA* gene sequence was not suitable to identify the bacterium to the species level. The threshold for assigning a bacterial sequence to the species level was 98% or greater similarity of the PCR product sequence with a GenBank accession from the nucleotide collection or the whole genome shotgun contig database. For fungi, conserved gene sequences targeting the *rDNA* gene were amplified from genomic DNA using PCR and sequenced; these PCR products could not identify the fungi to the species level. Therefore, a portion of the actin gene was amplified by PCR in the *Cladosporium* fungi [16,17], whereas portions of three genes (RNA polymerase second largest subunit, *rpb2*; *Alternaria* major allergen, *Alt-a1*; glyceraldehyde 3-phosphate dehydrogenase, *gapdh*) were amplified via PCR in the *Alternaria* fungi. These *Alternaria* genes were sufficient to distinguish new species in a recent paper [18]. Phylogenetic analysis (see below) was used to identify *Cladosporium* and *Alternaria* fungi to the species level. All the sequencing results were supplied in the Supplementary File S2.

### 2.3. Repression of Stromata Germination by Representative Fungal Potential Biocontrol Agents

Corn leaves with tar spot stromata were collected from a field in Peoria County, IL, in late July 2022. Leaves were rubbed gently while held under flowing deionized water to dislodge debris and blotted dry with Wipeall$^R$ L40 wipes (Kimberly ClarkRoswell, GA, USA). Leaves containing approximately 50 to 100 *P. maydis* stromata in a 100 cm$^2$ square were trimmed with the midvein central so they would fit snugly and flat in 100 × 15 mm Petri dishes containing 25 mL of 3% water agar. Plates were sealed with 3 M micropore™ surgical tape (3 M Company, ST. Paul, MN, USA) and allowed to incubate at room temperature overnight. The experiment was set up with three treatments, including two representatives of the most common genera of potential biocontrol fungi isolated from overwintered stromata (*Alternaria alternata/arborescens* and *Cladosporium rectoides*), and the commercial biofungicide LALstop G46wg, which contains *Gliocladium catenulatum* strain J1446, obtained from Lallemand Specialties (Milwaukee, WI, USA). Spores were obtained from fresh cultures of potential biocontrol fungi that were grown on S-medium [19]. Solutions of spores from the candidate biocontrol fungi were diluted to 10 colony forming units per µL in 0.01% Triton X 100 (Sigma Chemical, St. Louis, MO, USA). The commercial biofungicide LALstop was also diluted to 10 CFU per µL to match that for the candidate biocontrol fungi from the stromata, which was comparable to the recommended application rate. The next day, stromata that had not germinated were treated with 1 µL per mm diameter of control or spore solution, with 10 stromata treated on each side of the leaf midrib with either control (Triton X 100 solution alone) or spore solution. Plates were allowed to incubate at room temperature for 4 days, and then the number of treated germinated stromata was determined. Each treatment was replicated 3 times.

### 2.4. Phylogenetic Analysis

Partial gene sequences from *Alternaria* (*Alt-a1*, *gapdh*, and *rpb2* genes) and *Cladosporium* (actin gene) ex-type species were retrieved from NCBI. Sequences of *Cladosporium* ex-type strains, ex-epitype strains, and the strains from this study were aligned (using the Muscle alignment tool) and phylogenetic tree constructed using MEGA 11 [20]. Sequences of *Alternaria* ex-type strains and the strains from this study were aligned (using the "very accurate" algorithm in the classical sequence analysis menu) in CLC Genomics Workbench Version 22.0.2 (Qiagen, Valencia, CA, USA). Each *Alternaria* gene was aligned individually and trimmed at the edges; nucleotides within the gene that had low consensus among all the sequences were removed from the alignment. A multi-locus alignment of the three genes was made using the "join alignments" function. The alignment was imported into MEGA 11 and the phylogenetic tree constructed.

*2.5. Statistical Analysis*

Significant differences in frequency of tar spot stromata germination between control treated stromata and biological control organism treated stromata were determined using weighted Chi Square analysis with the SAS Version 7.1 Proc Freq (SAS Institute, Cary, NC, USA).

## 3. Results

No germinating stromata were observed in any collected overwintered tar spot infected material. However, several different types of organisms were observed growing out from stromata and not the leaf "handle" (Figure 1). We recovered 43 isolates of bacteria and fungi from approximately 50 of the *P. maydis* stromata (Table 1). Bacteria isolates included four species of *Pantoea*, including *P. agglomerans* (3 isolates), as well as two isolates of *Priestia* (formerly known as *Bacillus*) *megaterium*. Other bacteria isolated included *Curtobacterium faccumfaciens*, and multiple species of *Pseudomonas*, including *P. graminis*, *P. prosekii*, and *P. quercus*. One *Pseudomonas* isolate is either *P. fluorescens* or *P. shahriarae*.

**Figure 1.** Several examples of a possible biological control organism emerging from *P. maydis* overwintered stromata.

**Table 1.** Identities of organisms recovered from overwintered tar spot stromata.

| Organisms Directly Removed from Stromata | Number of Individual Organisms | Biological Control Properties Reported |
|---|:---:|:---:|
| Fungi | | |
| *Alternaria* | 1 | |
| *A. alternata/A. arborescens* | 11 | Yes * |
| *Alternaria ovoidea* | 6 | Yes |
| *Cladosporium rectoides* | 1 | Yes * |
| *Cladosporium subuliforme* | 1 | Yes |
| Bacteria | | |
| *Curtobacterium flaccumfaciens* | 2 | |
| *Pantoea agglomerans* | 3 | Yes |
| *Pantoea ananatis* | 1 | Yes |
| *Pantoea anthophila* | 1 | Yes |
| *Pantoea eucalypti* | 2 | Yes |
| *Priestia megaterium* | 2 | Yes |
| *Pseudomonas graminis* | 3 | Yes |
| *Pseudomonas prosekii* | 1 | Yes |
| *Pseudomonas quercus* | 1 | |
| **Organisms that fell onto water agar plates from stromata** | | |
| Fungi | | |
| *A. alternata/A. arborescens* | 4 | Yes |
| *Cladosporium crousii* | 1 | |
| Bacteria | | |
| *Priestia flexa* | 1 | |
| *Pseudomonas fluorescens* or *shahriarae* | 1 | |

* Reported in this study for the first time; isolates *A. alternata/A. arborescens* 11C + F and *Cladosporium rectoides* 10RDF exhibited biological control properties.

*Alternaria* was the most common fungus isolated from the stromata (Table 1). The *Alternaria* fungi could not be identified to the species level using the amplified region of the *rDNA* gene. PCR amplification of three gene fragments and subsequent phylogenetic analysis of the multi-locus alignment indicated that 6 isolates were *A. ovoidea* and 15 were possibly *A. alternata* or *A. arborescens* (Supplementary File S3). Three putative *Cladosporium* strains were isolated. The phylogenetic analysis (Supplementary File S4) indicated the three *Cladosporium* isolates were *C. crousii*, *C. rectoides*, and *C. subuliforme*.

The number of germinating stromata as indicated by exudate presence was significantly reduced by spore suspensions of two representative fungal species isolated from tar spot stromata compared to control applications (Table 2). However, none of these fungi were as effective as the commercially available *G. catenulatum*.

**Table 2.** Inhibition of tar spot stromata germination by putative biocontrol fungi.

| Organism | Source | % Germination Control | % Germination with Organism | $X^2$ Value | $p$ Value |
|---|---|---|---|---|---|
| *Gliocladium catenulatum* | Commercial | 83.3 | 10.0 | 32.4107 | <0.0001 |
| *Alternaria alternata/ arborescens 11C + F* | Present study | 66.7 | 36.7 | 5.4060 | 0.0201 |
| *Cladosporium rectoides 10RDF* | Present study | 50.0 | 26.7 | 3.4548 | 0.0653 |

% germination values are based on three replicates of 10 stromata for each treatment per replicate, with each replicate having the control and fungal treatments on opposites sides of the same leaf.

## 4. Discussion

### 4.1. Biocontrol Potential of Bacteria Isolated from Tar Spot Stromata

Of the 18 bacteria isolated from tar spot stromata, only *P. megaterium* has been reported from the corn microbiome [21]. Some bacterial species that we isolated from tar spot stromata have previously been reported as biocontrol agents or have properties of biocontrol agents. Due to limited available leaf material with the needed density of stromata, bioassay evaluation of representative bacteria was not performed. Although the *Curtobacterium* species we isolated from tar spot stromata has not previously been reported as a biocontrol agent, *Curtobacterium* spp. can be present in plants as pathogens or endophytes. An undescribed species of *Curtobacterium* was chitinolytic and had potential as a biocontrol agent [22]. *Pantoea* (formerly *Enterobacter*) *agglomerans* is reported as a biocontrol for many fungal pathogens such as *Botrytis cinera*, *Pythium* sp., and *Sclerotinia sclerotiorum* in several fruit, vegetable, and row crops [23]. Although many strains of *Pantoea ananatis* are plant pathogens, one has been reported as mycoparasite of wheat leaf rust, *Puccinia graminis* [24]. A strain of *P. anthophola* antagonized the growth of *Ralstonia solanacearum* [25]. A non-nitrogen-fixing strain of *P. eucalypti* promotes the growth of the pine *Pinus massoniana* [26], which could be due to inhibition of pathogens. Other species of *Pantoea* have also been reported as biocontrol agents [27].

*Priestia megaterium* has been reported to control septoria tritici blotch [28] of wheat caused by ascomycete fungus, *Mycophaerella graminicola*, multiple species of mycotoxigenic fungi [29], and other fungal pathogens in diverse crops [23]. Other *Priestia* species have also been reported as biocontrol agents [30]. *Pseudomonas graminis* has been reported as an antagonist of the fire blight causal organism *Erwinia amylovora* [31]. *Pseudomonas prosekii* inhibits plant pathogenic strains of *P. fluorescens* and *P. viridiflava* [32]. *Pseudomonas fluorescens* has been widely reported as a biological control agent effective against both bacterial and fungal plant pathogens, and different strains have been commercialized in several instances [33]. For example, pathogen inhibition by *P. fluorescens* has been reported for *Venturia inaequalis* in apple, *Macrophomina phaseolina* in mung bean, *Fusarium moniliforme* in cauliflower, *Xanthomonas campestris* pv. *malvacearum* in cotton, *Puccinia arachidis* in peanut, *Magnaporthe grisea* in rice, *Rhizoctonia solani* in tomato, and *Helminthosporium sativum* in wheat [33]. No biological control properties have been described for *Pseudomonas shahriarae*, which was isolated from the rhizosphere of wheat growing in Iran [34].

### 4.2. Biocontrol Potential of Fungi Isolated from Tar Spot Stromata

The commercial fungus formulation of *G. catenulatum* was very effective at inhibiting germination of *P maydis* stromata in corn leaves. *Alternaria alternata/arborescens* strain 11C + F, which was isolated from an overwintered *P. maydis* stroma, reduced germination of stromata in growing season leaves at statistically significant levels compared to the control leaves but was not as effective at reducing germination rates as *G. catenulatum*. *Cladosporium rectoides* strain 10RDF reduced germination rates of stromata in growing season leaves but the germination rate was not lower than the germination rate in the control leaves at a statistically significant level ($p = 0.0653$). Of the fungi isolated from tar spot stromata, only *A. alternata* has been reported from the corn microbiome [21]. Because different taxonomic approaches can identify the same fungus as two different species [35], it is challenging to compare past reports of fungal biocontrol agents that were predominately based on morphological identifications with isolated species identified using molecular methods in the present study. However, several species in the two genera of fungi we isolated from tar spot stromata have been reported as biocontrol agents. Eight species of *Alternaria* are reported as biocontrol agents [36], one of which, *A. alternata*, we identified in the present study. *A. alternata* is reported as a biocontrol agent for tan spot of wheat caused by *Pyrenophora tritici-repentis* [37], verticillium wilt of cotton caused by *Verticillium dahlia* [38], and white mold of bean caused by *Sclerotinia sclerotiorum* [39]. *Alternaria ovoidea*, which we isolated in the present study, has been reported as a saprophyte of the grass *Dactylis glomerata* [18]. Based on the saprophytic capabilities of some biocontrol agents of sclerotia described previously, it would not be surprising that *A. ovoidea* could also function as a mycoparasite. The other species of *Alternaria* reported previously as biocontrol agents have been used on banana, cyclamen, cotton, grape onion, roses, and strawberry, primarily against *Botrytis* and *Verticillium* spp. pathogens [36].

One species of *Cladosporium*, *C. subuliforme*, which we isolated from tar spot stromata in the present study, has previously been reported as a biocontrol agent. *C. subuliforme* from the rice phylloplane inhibited the growth of four rice fungal pathogens in dual culture plate assays [40]. Five other species of *Cladosporium* have been reported as biocontrol agents [36], although none of the species were the same as those identified in the present study. These additional species of *Cladosporium* have been used as biocontrol agents on chrysanthemum, cotton, guava, and grape for control of *Colletotrichum*, *Eutypa*, *Puccinia*, and *Verticillium* spp. pathogens [36].

### 4.3. Potential Role of Biocontrol in Tar Spot Management

The disease cycle of the tar spot pathogen is not well understood, although it is thought that the overwintered stromata are the source of initial infective material each growing season, and that the disease can spread through a corn field when spores are released by germinating stromata [2]. Plant resistance has been suggested and utilized to help reduce tar spot disease [2]. However, under certain conditions, even resistant varieties can be severely infected by the tar spot pathogen [5]. Although some fungicides are labeled for tar spot disease control (e.g., Veltyma), only spores and mycelia are listed as targets, not stromata. Stromata from fields treated with fungicide labeled for tar spot will still sporulate a few weeks after treatment (authors, personal observation). However, there are commercial products labeled to control species of sclerotia forming fungi, including *Coniothyrium minitans* (Contans WG), *G. castenulatum* (LALstopG46), and *Trichoderma harzianum* (Trianum-G); it should be noted that sclerotia are analogous to *P. maydis* stromata. Previous reports indicated these species can control several species of sclerotia-forming fungi, although the tar spot organism is not mentioned as one of them; *G. castenulatum* and *T. harzianum* are described as "ecologically facultative" [7] and thus can also survive as saprophytes and potentially persist to exert long-term control. The commercial strain of *G. castanulatum*, which is labeled for use in many crops, including corn, has saprophytic capabilities [41], and this strain was able to prevent germination of tar spot stromata as indicated in the present study. *Trichoderma harzianum* has been used to control foliar disease in corn [42].

The present study indicates that overwintered tar spot stromata can also be a source of useful biological control organisms, as two representative isolates significantly reduced the percentage of stromata germination compared to the control treatment. Stromatal isolates may have advantages over existing commercial materials such as the ability to grow and infect tar spot stromata under cooler conditions, making them more suitable for applications in the fall season to reduce the viability of overwintering stromata. However, the present study also suggests that overwintering fungi can be applied during the growing season and provide some control of stromata to augment fungicides that target spores and mycelia of the tar spot pathogen. Infected nongerminating stromata have also been collected during the growing season at different locations in 2022 and were common in late season stromata collected from the site investigated in the present study indicating growing season applications could also be used to reduce overwintering survival of tar spot stromata. However, further study is needed to identify the most effective organisms/strains under proposed field application conditions and determine whether undesirable traits, such as plant pathogenicity or other genes that produce harmful proteins, are present.

## 5. Conclusions

The present study suggests that survival of overwintering tar spot stromata can be greatly reduced by both bacterial and fungal mycoparasites. Some of the species of organisms that we recovered from stromata have previously been reported to be biocontrol agents. Three example microorganisms that we tested significantly reduced the percent sporulation of growing season stromata, suggesting that mycoparasites can be used as part of an integrated management plan for both early season preventative and growing season control measures for tar spot disease in corn.

**Supplementary Materials:** The following supporting information can be downloaded at: https://www.mdpi.com/article/10.3390/microorganisms11061550/s1. File S1: A list of the PCR primers used in this study. File S2: Sequences of all of the PCR products used in this study for identification of bacteria and fungi. File S3: Phylogenetic analysis of the *Alternaria* fungi. File S4: Phylogenetic analysis of the *Cladosporium* fungi. Refs. [43–50] are cited in the Supplementary Materials.

**Author Contributions:** Conceptualization, P.F.D. and E.T.J.; methodology, all authors; data analysis, P.F.D. and E.T.J.; writing—original draft preparation, P.F.D. and E.T.J.; writing—review and editing, all authors; project administration, J.L.R. All authors have read and agreed to the published version of the manuscript.

**Funding:** Funding for this work was provided to E.T.J. using USDA ARS in-house project 5010-22410-024-00-D, and to P.D., R.W.B. and J.L.R. using USDA ARS in-house project 5010-22410-023-00D.

**Data Availability Statement:** All data generated in this study are available in the Tables and Supplementary Materials.

**Acknowledgments:** We thank Mark Doehring and David Lee for excellent technical assistance. We appreciate the comments from Kirk Broders and Ephantus Muturi on a draft of this manuscript. The mention of firm names or trade products does not imply that they are endorsed or recommended by the USDA over other firms or similar products not mentioned. USDA is an equal opportunity provider and employer.

**Conflicts of Interest:** The authors declare no conflict of interest.

## References

1. Savary, S.; Willocquet, L.; Pethybridge, S.J.; Esker, P.; McRoberts, N.; Nelson, A. The global burden of pathogens and pests on major food crops. *Nat. Ecol. Evol.* **2009**, *3*, 430–439. [CrossRef]
2. Valle-Torres, J.; Ross, T.J.; Plewa, D.; Avellaneda, M.C.; Check, J.; Chilvers, M.I.; Cruz, A.P.; Dalla Lana, F.; Groves, C.; Gongora-Canul, C.; et al. Tar spot: An understudied disease threatening corn production in the Americas. *Plant Dis.* **2020**, *104*, 2541–2550. [CrossRef]
3. Rocco da Silva, C.; Check, J.; MacCready, J.S.; Alakonya, A.E.; Beiriger, R.; Bissonnette, K.M.; Collins, A.; Cruz, C.D.; Esker, P.D.; Goodwin, S.B.; et al. Recovery plan for tar spot of corn, caused by *Phyllachora maydis*. *Plant Health Prog.* **2021**, *22*, 596–616. [CrossRef]
4. AgWeb. Tar Spot Found in New States, Severe Infestation Slashes Yield. 2021. Available online: https://www.agweb.com/news/crops/corn/tar-spot-found-new-states-severe-infestation-slashes-yield (accessed on 28 February 2023).
5. Chilvers, M.I.; Byrne, P.M.; Widdicombe, W.D.; Williams, L.; Singh, M.P. Corn tar spot: Hybrid resistance and susceptibility. 2018 Michigan Corn Hybrids Compared MSU Extension Bulletin E-Coutinho T, Ventr SN (2009) *Pantoea ananatis*: An unconventional plant pathogen. *Mol. Plant Pathol.* **2018**, *10*, 325–335.
6. Brandt Company. Tar Spot. 2022, p. 16. Available online: https://brandt.co/tar-spot (accessed on 28 February 2023).
7. Whipps, J.M. Effects of mycoparasites on sclerotia-forming fungi. *Dev. Agric. Manag. For. Ecol.* **1991**, *23*, 129–140.

8.  Groves, C.L.; Kleczewski, N.M.; Telenko, D.E.P.; Chilvers, M.L.; Smith, D.L. *Phyllachora maydis* ascospore release and germination from overwintered corn residue. *Plant Health Prog.* **2020**, *21*, 26–30. [CrossRef]
9.  Karlsson, M.; Atanasova, L.; Jensen, D.F.; Seilinger, S. Necrotrophic mycoparasites and their genomes. *Microbiol. Spectr.* **2017**, *5*. [CrossRef]
10. Jones, E.E.; Steward, A. Selection of mycoparasites of sclerotia of *Sclerotinia sclerotiorum* isolated from New Zealand soils. *N. Z. J. Crop Hortic. Sci.* **2000**, *28*, 105–114. [CrossRef]
11. de Vrije, T.; Antoine, N.; Buitelaar, R.M.; Bruckner, S.; Dissevelt, M.; Durand, A.; Gerlagh, M.; Jones, E.E.; Lüth, P.; Oostra, J.; et al. The fungal biocontrol agent *Coniothyrium minitans*: Production by solid-state fermentation, application and marketing. *Appl. Microbiol. Biotechnol.* **2001**, *56*, 58–68. [CrossRef]
12. Seixas, C.D.S.; Barreto, R.W.; Beserra, J.L.; David, J. Mycoparasites of *Coccodiella miconiae* (Ascomycota: Phyllachoraceae) a potential biocontrol agent for *Miconia calvescens* (Melastomataceae). In *Pacific Cooperative Studies Unit University of Hawaii at Manoa Technical Report 133*; Pacific Cooperative Studies Unit, University of Hawaii at Manoa, Department of Botany: Honolulu, HI, USA, 2005.
13. Johnson, E.T.; Dowd, P.F. A quantitative method for determining relative colonization rates of maize callus by *Fusarium graminearum* for resistance gene evaluations. *J. Microbiol. Methods* **2016**, *130*, 73–75. [CrossRef]
14. Johnson, E.T.; Dowd, P.F.; Hughes, S.R. Expression of a wolf spider toxin in tobacco inhibits the growth of microbes and insects. *Biotechnol. Lett.* **2014**, *36*, 1735–1742. [CrossRef] [PubMed]
15. Johnson, M.; Zaretskaya, I.; Raytselis, Y.; Merezhuk, Y.; McGinnis, S.; Madden, T.L. NCBI BLAST: A better web interface. *Nucleic Acids Res.* **2008**, *36*, W5–W9. [CrossRef] [PubMed]
16. El-Dawy, E.G.A.E.M.; Gherbawy, Y.A.; Hussein, M.A. Morphological, molecular characterization, plant pathogenicity and biocontrol of *Cladosporium* complex groups associated with faba beans. *Sci. Rep.* **2021**, *11*, 14183. [CrossRef] [PubMed]
17. Iturrieta-González, I.; García, D.; Gené, J. Novel species of *Cladosporium* from environmental sources in Spain. *MycoKeys* **2021**, *77*, 1–25. [CrossRef]
18. Li, J.; Phookamsak, R.; Jiang, H.; Bhat, D.J.; Camporesi, E.; Lumyong, S.; Kumla, J.; Hongsanan, S.; Mortimer, P.E.; Xu, J.; et al. Additions to the inventory of the genus *Alternaria* section *Alternaria* (*Pleosporaceae, Pleosporales*) in Italy. *J. Fungi* **2022**, *24*, 898. [CrossRef]
19. Shahin, E.A.; Shepard, J.F. An efficient technique for inducing profuse sporulation of *Alternaria* species. *Phytopathology* **1979**, *69*, 618–620. [CrossRef]
20. Tamura, K.; Stecher, G.; Kumar, S. MEGA11, molecular evolutionary genetics analysis version 11. *Mol. Biol. Evol.* **2021**, *38*, 3022–3027. [CrossRef]
21. Singh, R.; Goodwin, S.B. Exploring the corn microbiome: A detailed review on current knowledge, techniques, and future directions. *PhytoFrontiers* **2022**, *2*, 158–175. [CrossRef]
22. Dimkic', I.; Bhardwaj, V.; Carpentieri, P.; Polo, V.; Kuzmanovic', N.; Degrassi, G. The chitinolytic activity of the *Curtobacterium* S1 isolated from field-grown soybean and analysis of its genomic sequence. *PLoS ONE* **2021**, *16*, e0259465.
23. Walterson, A.M.; Stravrinides, J. *Pantoea*: Insights into a highly versatile and diverse genus within the Enterobacteriaceae FEMS *Microbiol. Rev.* **2015**, *39*, 968–974.
24. Coutinho, T.; Venter, S.N. *Pantoea ananatis*: An unconventional plant pathogen. *Mol. Plant Pathol.* **2009**, *10*, 325–335. [CrossRef] [PubMed]
25. He, Q.; Li, J.; Liu, Q.; Li, Y.; Liao, X.; Shou, J.; Zhou, Z. Complete genome sequence resource of *Pantoea anthophila* CL1 causing soft rot disease in *Claosena lansium* (Wampee) in China. *Plant Dis.* **2023**, *107*, 1613–1616. [CrossRef] [PubMed]
26. Song, Z.; Lu, Y.; Liu, X.; Wei, C.; Oladipo, A.; Fan, B. Evaluation of *Pantoea eucalypti* FBS135 for pine (*Pinus massoniana*) growth promotion and its genome analysis. *J. Appl. Microbiol.* **2020**, *129*, 958–970. [CrossRef] [PubMed]
27. Al-Nadabi, H.H.; Al-Boraiki, N.S.; Al-Nabhani, A.A.; Maharachchikumbura, S.N.; Velazhaben, R.; Al-Sadi, A.M. In vitro antifungal activity of endophytic bacteria isolated from date palm (*Phoenix dactylifera* L) against fungal pathogens causing leaf spot of date. *Egypt. J. Biol. Pest Control* **2021**, *31*, 65. [CrossRef]
28. Kildea, S.; Ransbotyn, V.; Hkan, M.R.; Fagan, B.; Leonard, G.; Mullins, E.; Doohan, F.M. *Bacillus megaterium* shows potential for the biocontrol of *Septoria tritici* blotch of wheat. *Biol. Control* **2008**, *47*, 37–45. [CrossRef]
29. Saleh, A.E.; Ul-Hassan, Z.; Seidan, R.; Al-Shamary, N.; Al-Yafei, T.; Alnaimi, H.; Higazy, N.S.; Migheli, Q.; Jaoua, S. Biocontrol activity of *Bacillus megaterium* BM344-1 against toxigenic fungi. *ACS Omega* **2021**, *6*, 10984–10990. [CrossRef]
30. Bashir, S.; Iqbal, A.; Hasnain, S.; White, J.F. Screening of sunflower associated bacteria as biocontrol agents for plant growth promotion. *Arch. Microbiol.* **2021**, *203*, 4901–4912. [CrossRef]
31. Mikiciński, A.; Sobiczewski, P.; Pulawska, J.; Malusa, E. Antagonistic potential of *Pseudomonas graminis* 49M against *Erwinia amylovora*, the causal agent of fire blight. *Arch. Microbiol.* **2016**, *198*, 531–539. [CrossRef]
32. Snopková, K.; Dufková, K.; Šmajs, D. *Pseudomonas prosekii* isolated in Antarctica inhibits plant-pathogenic strains of *Pseudomonas fluorescens* and *Pseudomonas viridiflava*. *Czech Polar. Rep.* **2021**, *11*, 270–278. [CrossRef]
33. David, B.V.; Chandrasehar, G.; Selvam, P.N. Pseudomonas fluorescens: A plant-growth-promoting Rhizobacterium (PGPR) with potential role in biocontrol of pests of crops. In *Crop Improvement through Microbial Biotechnology*; Prasad, R., Gill, S.S., Tuteja, N., Eds.; Elsevier: New York, NY, USA, 2018; pp. 221–243.

34. Girard, L.; Lood, C.; Höfte, M.; Vandamme, P.; Rokni-Zadeh, H.; van Noort, V.; Lavigne, R.; De Mot, R. The ever-expanding *Pseudomonas* genus: Description of 43 new species and partition of the *Pseudomonas putida* group. *Microorganisms* **2021**, *9*, 1766. [CrossRef]

35. Lücking, R.; Aime, M.C.; Robbertse, B.; Miller, A.N.; Aoki, T.; Ariyawansa, H.A.; Cardinali, G.; Crous, P.W.; Druzhinina, I.S.; Geiser, D.M.; et al. Fungal taxonomy and sequence-based nomenclature. *Nat. Microbiol.* **2021**, *6*, 540–548. [CrossRef]

36. Thambugala, K.M.; Daranagama, D.A.; Phillips, A.J.L.; Kannangara, S.D.; Promputtha, I. Fungi vs. fungi in biocontrol: An overview of fungal antagonists applied against fungal plant pathogens. *Front. Cell Infect. Microbiol.* **2020**, *10*, 604923. [CrossRef] [PubMed]

37. Li, B.C.; Sutton, J.C. Evaluation of leaf-associated microorganisms for biocontrol of tar spot in wheat foliage. *Fitopatol. Bras.* **1995**, *20*, 545–552.

38. Li, Z.F.; Wang, L.F.; Feng, Z.L.; Zhao, L.H.; Shi, Y.Q.; Shu, H.Q. Diversity of endophytic fungi from different *Verticillium*-wilt-resistant *Gossypium hirsutum* and evaluation of antifungal activity against *Verticillium dahlia* in vitro. *J. Microbiol. Biotechnol.* **2014**, *24*, 1149–1161. [CrossRef] [PubMed]

39. Boland, G.J.; Inglis, G.D. Antagonism of white mold (*Sclerotinia sclerotiorum*) of bean by fungi from bean and rapeseed flowers. *Can. J. Bot.* **1989**, *67*, 1775–1781. [CrossRef]

40. Chaibub, A.A.; de Sousa, T.P.; de Araújo, L.G.; de Filippi, M.C. Molecular and morphological characterization of rice phylloplane fungi and determination of the antagonistic activity against rice pathogens. *Microbiol. Res.* **2020**, *231*, 126353. [CrossRef]

41. United States Environmental Protection Agency. Gliocladium catenulatum strain J. In *Biopesticides Registration Action Document*; United States Environmental Protection Agency: Washington, DC, USA, 2002; p. 1446.

42. Savaravanakumar, K.; Fan, L.; Fu, K.Y.C.; Wang, M.; Xia, H.; Sun, J.; Li, Y.; Chen, J. Cellulase from *Trichoderma harzianum* interacts with roots and triggers induced systemic resistance to foliar disease in maize. *Sci. Rep.* **2016**, *6*, 35543. [CrossRef] [PubMed]

43. Gerrits van den Ende, A.H.G.; de Hoog, G.S. Variability and molecular diagnostics of the neurotropic species *Cladophialophora bantiana*. *Stud. Mycol.* **1999**, *43*, 151–162.

44. Liu, Y.J.; Whelen, S.; Hall, B.D. Phylogenetic relationships among ascomycetes: Evidence from an RNA polymerse II subunit. *Mol. Biol. Evol.* **1999**, *16*, 1799–1808. [CrossRef]

45. Hong, S.G.; Cramer, R.A.; Lawrence, C.B.; Pryor, B.M. *Alt a 1* allergen homologs from *Alternaria* and related taxa: Analysis of phylogenetic content and secondary structure. *Fungal Genet. Biol.* **2005**, *42*, 119–129. [CrossRef]

46. Berbee, M.L.; Pirseyedi, M.; Hubbard, S. *Cochliobolus* phylogenetics and the origin of known, highly virulent pathogens, inferred from ITS and glyceraldehyde-3-phosphate dehydrogenase gene sequences. *Mycologia* **1999**, *91*, 964–977. [CrossRef]

47. Carbone, I.; Kohn, L.M. A method for designing primer sets for speciation studies in filamentous ascomycetes. *Mycologia* **1999**, *91*, 553–556. [CrossRef]

48. Lane, D.J. 16S/23S rRNA sequencing, p. 115-147. In *Nucleic Acid Techniques in Bacterial Systematics*; Stackebrandt, E., Goodfellow, M., Eds.; John Wiley & Sons: New York, NY, USA, 1991.

49. Stackebrandt, E.; Liesack, W. Nucleic acids and classification. In *Handbook of New Bacterial Systematics*; Goodfellow, M., O'Donnell, A.G., Eds.; Academic Press: London, UK, 1993; pp. 152–189.

50. Yamamoto, S.; Harayama, S. PCR amplification and direct sequencing of *gyrB* genes with universal primers and their application to the detection and taxonomic analysis of *Pseudomonas putida* strains. *Appl. Environ. Microbiol.* **1995**, *61*, 1104–1109. [CrossRef] [PubMed]

**Disclaimer/Publisher's Note:** The statements, opinions and data contained in all publications are solely those of the individual author(s) and contributor(s) and not of MDPI and/or the editor(s). MDPI and/or the editor(s) disclaim responsibility for any injury to people or property resulting from any ideas, methods, instructions or products referred to in the content.

 *microorganisms*

*Article*

# A Genetically Engineered *Escherichia coli* for Potential Utilization in Fungal Smut Disease Control

Guobing Cui [1,2,3], Xinping Bi [3], Shan Lu [4], Zide Jiang [3] and Yizhen Deng [1,3,*]

1   Guangdong Laboratory for Lingnan Modern Agriculture, South China Agricultural University, Guangzhou 510642, China
2   Henry Fork School of Biology and Agriculture, Shaoguan University, Shaoguan 512000, China
3   State Key Laboratory for Conservation and Utilization of Subtropical Agro-Bioresources, Guangdong Province Key Laboratory of Microbial Signals and Disease Control, Integrative Microbiology Research Centre, South China Agricultural University, Guangzhou 510642, China
4   State Key Laboratory for Conservation and Utilization of Subtropical Ago-Bioresouces Ministry and Province Co-Sponsored Collaborative Innovation Center for Sugarcane and Sugar Industry, Nanning 530004, China
*   Correspondence: dengyz@scau.edu.cn

**Abstract:** *Sporisorium scitamineum*, the basidiomycetous fungus that causes sugarcane smut and leads to severe losses in sugarcane quantity and quality, undergoes sexual mating to form dikaryotic hyphae capable of invading the host cane. Therefore, suppressing dikaryotic hyphae formation would potentially be an effective way to prevent host infection by the smut fungus, and the following disease symptom developments. The phytohormone methyl jasmonate (MeJA) has been shown to induce plant defenses against insects and microbial pathogens. In this study, we will verify that the exogenous addition of MeJA-suppressed dikaryotic hyphae formation in *S. scitamineum* and *Ustilago maydis* under in vitro culture conditions, and the maize smut symptom caused by *U. maydis*, could be effectively suppressed by MeJA in a pot experiment. We constructed an *Escherichia coli*-expressing plant *JMT* gene, encoding a jasmonic acid carboxyl methyl transferase that catalyzes conversion from jasmonic acid (JA) to MeJA. By GC-MS, we will confirm that the transformed *E. coli*, designated as the pJMT strain, was able to produce MeJA in the presence of JA and S-adenosyl-L-methionine (SAM as methyl donor). Furthermore, the pJMT strain was able to suppress *S. scitamineum* filamentous growth under in vitro culture conditions. It waits to further optimize *JMT* expression under field conditions in order to utilize the pJMT strain as a biocontrol agent (BCA) of sugarcane smut disease. Overall, our study provides a potentially novel method for controlling crop fungal diseases by boosting phytohormone biosynthesis.

**Keywords:** filamentous growth; jasmonic acid carboxyl methyl transferase (JMT); methyl jasmonate (MeJA); pathogenicity; *Sporisorium scitamineum*; sugarcane smut

**Citation:** Cui, G.; Bi, X.; Lu, S.; Jiang, Z.; Deng, Y. A Genetically Engineered *Escherichia coli* for Potential Utilization in Fungal Smut Disease Control. *Microorganisms* 2023, 11, 1564. https://doi.org/10.3390/microorganisms11061564

Academic Editors: Rainer Borriss and Chetan Keswani

Received: 15 May 2023
Revised: 3 June 2023
Accepted: 9 June 2023
Published: 13 June 2023

**Copyright:** © 2023 by the authors. Licensee MDPI, Basel, Switzerland. This article is an open access article distributed under the terms and conditions of the Creative Commons Attribution (CC BY) license (https://creativecommons.org/licenses/by/4.0/).

## 1. Introduction

*Sporisorium scitamineum* is the fungal pathogen that causes sugarcane smut [1]. Diploid teliospores of *S. scitamineum* are airborne and can survive in harsh environmental conditions. The smut teliospores germinate on sugarcane buds, forming promycelium from which two pairs of haploid sporidia (basidiospores) with opposite mating types, *MAT-1* and *MAT-2*, are generated via a round of meiosis. The sporidia of opposite mating types can recognize each other and fuse, undergoing sexual mating, and afterwards switch from a yeast-like growth style to dikaryotic hyphae, undergoing a so-called dimorphic switch. Dikaryotic hyphae are capable of infecting sugarcane, likely through buds. The smut disease symptoms are characterized by the emergence of a typical structure produced on the apex or side shoots of sugarcane stalks, called "smut whip", which contains teliospores formed from a fusion of two nuclei of the dikaryotic hyphae [2,3].

Given that a post-mating dimorphic switch is critical for host infection, molecules controlling this step are potential disease management targets. The mechanisms of sexual mating and pathogenicity of smut fungi were well-studied on the model fungus *Ustilago maydis* [4]. The processes of infection and sexual mating are similar in smut fungus, and the research on *U. maydis* provides an important reference for *S. scitamineum* [5]. Different mating types are determined by two alleles, named *MAT* loci which include a locus and b locus, in *S. scitamineum* and *U. maydis* [5]. In *S. scitamineum*, the *MFA* genes of the mating locus encode pheromone precursors (including a factor and α factor), which undergo a series of post-translational modifications and are secreted out of sporidial cells, serving as signaling molecules [6–8]. The a or α pheromone is recognized by the pheromone receptor Pra of the opposite-mating type, which transmits the signal into the cell and through the MAPK or cAMP-PKA, and thus signaling cascades to the global transcriptional factor Prf1, which in turn induces dikaryotic hyphae growth via activating a and b locus [8–11]. The conserved farnesyltransferase (FTase) β subunit SsRam regulates the yeast-to-hyphae dimorphic switch, stress resistance, and pathogenicity of *S. scitamineum*, likely via catalyzing farnesylation of the pheromone precursor Mfa1 [8]. Besides conveying the sexual pheromone signal, the MAPK SsKpp2 and SsHog1 are involved in the response to extracellular osmotic stress, and thus regulate the dimorphic switch and stress resistance [9,11]. An investigation of the molecular mechanism underlying the *S. scitamineum* dimorphic switch/filamentous growth would provide a theoretical basis for developing novel and effective management strategies of sugarcane smut, which is largely lacking and urgently needed in agricultural practices.

Jasmonic acid (JA) and its methylated product, methyl jasmonate (MeJA), are important phytohormones and play an important role in plant resistance to biotic and abiotic stresses [12]. MeJA treatment with the tomato before sowing can reduce the attack of the soil-born fungal pathogen *Fusarium oxysporum* f.sp. *lycopersici* [13]. Endogenous MeJA content of plants increases significantly in response to mechanical damage, pathogen infection, ozone and metal stress in *Brassica napus* L. [14]. In *Arabidopsis thaliana*, gaseous MeJA can effectively reduce the harm caused by fungal pathogens *Alternaria brassicicola*, *Botrytis cinerea*, or *Plectosphaerella cucumerina* [15]. Jasmonic acid carboxyl methyl transferase (JMT) catalyzes MeJA formation by using JA as a precursor [16]. Over-expression of *AtJMT* in *A. thaliana* induces the constitutive expression of JA biosynthesis-related genes and confers enhanced resistance towards the fungal pathogen *B. cinerea* [17]. Rice *OsJMT* gene is induced by infestation with brown planthopper (*Nilaparvata lugens*), catalyzing JA to MeJA and thus regulating herbivore-induced defense responses [18]. Taken together, enhanced levels of MeJA by JMT function can confer plant immunity against pathogenic fungi and pests.

In this study, MeJA displayed a significant inhibitory effect on *S. scitamineum* mating/filamentation, while JA did not. In addition, exogenous application of MeJA to maize seedlings grown in pots could significantly improve the ability of maize to resist the occurrence of smut disease. Through heterologous expression of *A. thaliana JMT* gene in *Escherichia coli*, we generated a strain, designated as pJMT strain, capable of converting JA to MeJA under in vitro culture conditions. To test the capacity of pJMT strain in controlling smut disease, we applied it to maize or sugarcane plants exposed to smut fungi (*U. maydis* and *S. scitamineum*), respectively. In summary, enhancing plant resistance to pathogens by microbes-dependent phytohormone biosynthesis may serve as a biological control method pending necessary optimization for utilization in field conditions.

## 2. Materials and Methods

### 2.1. Fungal Strains and Culture Conditions

The *S. scitamineum* strains *MAT-1* (eGFP) and *MAT-2* (dsRED) used in this study were generated by Yan et al. [19], and stored in Deng Y's lab. The haploid *MAT-1* (U9) and *MAT-2* (U10) strains of *U. maydis* were provided by Jiang Z's lab. The *S. scitamineum* and *U. maydis* sporidia were cultured in YePS medium [19], and the sexual mating medium of *U. maydis* was YePS medium supplemented with 1% activated carbon. Fungal cells were cultured in 2 mL YePS medium at 28 °C, shaken at 200 rpm for 2 days, then collected by centrifugation and washed with 2 mL sterilized water before being re-suspended in sterilized water to reach $OD_{600} \approx 1.0$. Equal volumes of *S. scitamineum* or *U. maydis* sporidia, of opposite mating types, were mixed to induce sexual mating and the following filamentous growth. For the suppression assay, sporidia of wild-type *MAT-1* and *MAT-2* sporidia of *S. scitamineum* ($OD_{600} \approx 1.0$ for both) were mixed (1 μL for each) and allowed to grow on YePS solid medium containing JA or MeJA of different concentrations. The filamentation of the fungal colonies was assessed visually, and the experiments were repeated three times.

### 2.2. Construction of JMT Expression Plasmid

The *A. thaliana* JMT (GeneBank: AY008435.1) nucleotide sequence was synthesized by Suzhou Genewiz Biological Technology Co., Ltd. (Suzhou, China), and inserted into the *EcoR* I-*Sal* I site of the expression vector pRSFDUET-1 [20]. The resultant plasmid carries the *JMT* gene driven by T7 promoter following the lacI repressor, and therefore is inducible by Isopropyl β-D-Thiogalactoside (IPTG) [21]. The expression plasmid was transformed into *E. coli* BL21(DE3) strain, and verified by PCR amplification using the primers as listed in Table 1. The verified strain was named pJMT strain. To test whether it could produce MeJA, JA (as the precursor) and SAM (S-adenosyl-L-methionine as methyl donor) were applied to the culture medium.

**Table 1.** Primer sequences.

| Primer | Sequence (5′-3′) | Notes |
| --- | --- | --- |
| Actin-F | CAGCTCGATGAAGGTCAAGAT | |
| Actin-R | CAGCTCGATGAAGGTCAAGAT | |
| Q-bE1-F | CCAACGACGAAAGCGCGACG | |
| Q-bE1-R | GACTCTCTGCGAGCGGGCAT | |
| Q-bE2-F | CCAACGACGAAAGCGCGACG | |
| Q-bE2-R | GACTCTCTGCGAGCGGGCAT | |
| Q-bW1-F | CGAGAAAGGCACACAACGTC | |
| Q-bW1-R | CACCTTTTGGGGAGTTCCGA | |
| Q-bW2-F | TGTTGATGAGCCAGTGCCTT | |
| Q-bW2-R | AGTTCCGACTGGCTGAAGTG | qRT-PCR |
| Pra2-P1F | GAAGAGCCTCAGCCGTTATAC | |
| Pra2-P1R | GGGTTCCCTTACTGAACCTTAG | |
| Q-PCR-Mfa1-F | ATGCTTTCCATCTTTACCCAGA | |
| Q-PCR-Mfa1-R | GTGCAGCTAGAGTAGCCAAG | |
| Mfa2-P1F | CGTCCAGGCCATTGTTTCT | |
| Mfa2-P1R | TAGGCCACGGTGCAGTA | |
| Q-PCR-Pra1-F | GGACGCTATCACCCAATCTTAC | |
| Q-PCR-Pra1-R | TCTCCAACATGGCAACACTC | |
| pJMT-F | GAATTCATGGAGGTAATGCGA | Red front denotes restriction enzyme sites used |
| pJMT-R | GTCGACTCAACCGGTTCTAAC | for plasmid construction |

### 2.3. Total RNA Extraction and Quantitative Real-Time PCR (qRT-PCR)

A RNeasy mini kit (Qiagen (Hilden, Germany), 74104) was used for total RNA extraction. A TURBO DNA-free kit (Invitrogen (Waltham, MA, USA), AM1907) and TransScript first-strand cDNA synthesis supermix (TransGen Biotech (Beijing, China), AT301) were, respectively, used for gDNA removal and first strand cDNA synthesis, following manufacturer's instructions. PowerUp SYBR green master mix (Applied Biosystems (Waltham, MA, USA), A25742) was used for qRT-PCR, with the primers as listed in Table 1, and the reaction was run on a Quant Studio 6 Flex Real-Time PCR system (Applied Biosystems). Relative expression folds were calculated by the $2^{-\Delta\Delta CT}$ method [22] with *ACTIN* as an internal control. Statistical analysis was performed by one-way ANOVA and Duncan's multiple analyses.

### 2.4. JMT Gene Induction

Induced expression of *JMT* gene: The pJMT-carrying strain (hereafter named as pJMT strain) was cultured in 5 mL LB broth containing 50 μg/mL kanamycin (Dingguo, AK177), at 37 °C and shaken at 200 rpm, until $OD_{600} \approx 1.0$. IPTG (Genview, CI175) was added to the pJMT strain culture to reach the final concentration of 1 mM, and then it was cultured at 28 °C and 200 rpm for another 12 h. The bacterial cells were collected by centrifugation at $1000 \times g$ and transferred to 50 mL of LB medium containing 200 μM jasmonic acid (JA, Sigma-aldrich (St. Louis, MO, USA), J2500) and 200 μM SAM (Sigma-aldrich, A4377), and incubated at 28 °C, 200 rpm for 12 h. The *E. coli* expressing the vector pRSFDUET-1 was used as a negative control.

### 2.5. MeJA or JA Detection

For the extraction of MeJA from the induced pJMT strain cultured in the presence of JA and SAM, the supernatant was collected after centrifuge and mixed with ethyl acetate of equal volume. After phase separation, the ethyl acetate phase was collected and dried by rotary evaporation (EYELA, OSB-2100). Such extracts were dissolved in 1 mL of methanol and filtered through a 0.22-μm microporous filter (Nylon6). For detection of JA from the extracts, high performance liquid chromatography (HPLC) was performed using the following settings: 0.1% formic acid; water $= 65/35$ $(v/v)$; flow rate $= 0.3$ mL/min; detection wavelength $= 205$ nm; detection time $= 20$ min; and column temperature $= 30$ °C. The column is Agilent HC-C18(2) $250 \times 4.6$ mm (5 μm).

MeJA was identified by gas chromatography/mass spectrometry (GC/MS) on an Agilent 7890B/5977B (Agilent Technologies Inc., Santa Clara, CA, USA) series GC system equipped with an Agilent HP-5MS capillary column (30 m $\times$ 250 μm $\times$ 0.25 μm) in scan mode and split mode. The following temperature program was used: initial temperature of 60 °C (5 min hold), increase to 270 °C at 30 °C/min. Ion source temperature was 220 °C, and transfer line temperature was 270 °C. Electron ionization energy was 70 eV, and m/z $= 30 \sim 450$. MeJA was identified through standard compounds from the National Institute of Standards and Technology (NIST, Gaithersburg, MD, USA) library database.

### 2.6. Pathogenicity Assays

Pot experiment: Maize (Hua Meitian 9; DYQS-16 $\times$ TG-9) was grown in pots until the four-leaf stage, under the growth condition set as follows: humidity: 82%; light duration: 12 h. The mixed *U. maydis MAT-1* and *MAT-2* cells ($OD_{600} \approx 1.0$, 1:1 in volume), with or without 400 μM MeJA, were injected to the base of maize leaves. Injection with sterile water served as blank control. The injected maize seedlings were allowed to grow under natural conditions for 3 weeks, before examination of disease symptoms and photographing. The pJMT strain's effect on maize smut disease was tested following the same procedure. The pJMT strain was pre-treated with IPTG to induce expression of the *JMT* gene.

Field experiment: Approximately 290 seedlings of ratoon cane (ROC 22) were used for pJMT treatment, or untreated as control, respectively. The sugarcane cultivar ROC 22 was established as a susceptible cultivar towards smut fungus, as reported [23]. The pathogenicity and biocontrol assay were carried out on an established diseased field (Experimental fields of South China Agricultural University, 23.23 E, 113.63 W), which steadily causes smut disease to the susceptible cane (ROC 22) to a rate above 20%. The tested canes are 3rd year ratoon canes. Liquid-cultured pJMT (pre-treated with IPTG; the concentration of pJMT fermentation product to the soil is approximately $4 \times 10^5$ cell/cm$^2$) was applied within three days from germination of ratoons after cutting off the above-ground part of the canes, following the established procedure [24].

## 3. Results

### 3.1. MeJA Suppresses S. scitamineum Mating/Filamentation

We first tested the effect of MeJA, a phytohormone used by plants against biotic or abiotic stresses [12], on *S. scitamineum* mating/filamentation, which is a critical step determining fungal pathogenicity. The result showed that, with 400–800 µM of MeJA, the filamentation (dikaryotic hyphae growth) of *S. scitamineum* was completely suppressed, and 200 µM MeJA caused reduced filamentation (Figure 1A). In contrast, treatment of JA, the precursor of MeJA, did not affect *S. scitamineum* filamentation even under high concentration (800 µM; Figure 1A). We conclude that MeJA specifically suppressed *S. scitamineum* dikaryotic hyphae growth.

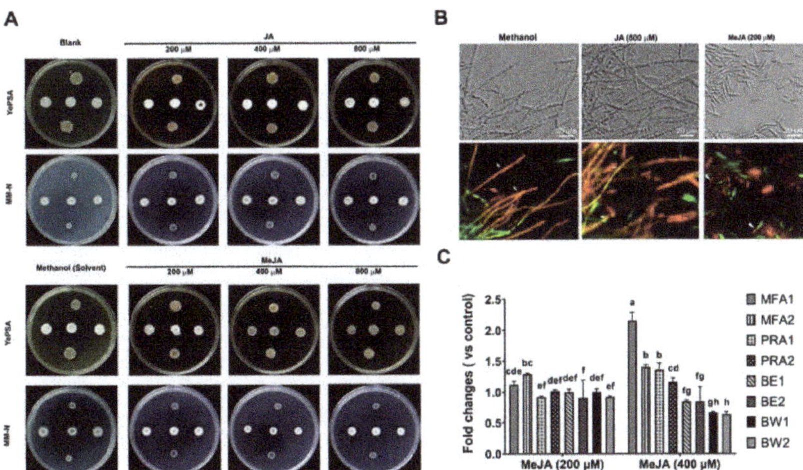

**Figure 1.** MeJA suppressed *S. scitamineum* dikaryotic hyphae growth after sexual mating. (**A**) The *MAT-1* (*eGFP*) and *MAT-2* (*dsRED*) sporidia were cultured in liquid YePS medium for 2 d and adjusted to $OD_{600} \approx 1.0$, before mixed as 1:1 ratio and inoculated on YePSA or MM-N medium containing the indicated chemicals of various concentrations. The mating cultures were kept in the dark, at 28 °C, for 5 d before photographing. *MAT-1* sporidia were spotted in the upper panel, while *MAT-2* in the lower panel, of each plate. Three colonies in the middle of each plate are the mixed *MAT-1* and *MAT-2* sporidia to induce mating and filamentation. (**B**) Microscopic observation of fluorescent protein-tagged *S. scitamineum MAT-1* and *MAT-2* sporidia [19] during sexual mating/filamentous growth, using ZEISS Observer Z1. Arrows denote un-mating sporidia or hyphae. Scale bars = 20 µm. (**C**) qPCR analysis to assess transcription of a locus gene *MFA 1/2* and *PRA 1/2*, and b locus *BE 1/2* and *BW 1/2* during *S. scitamineum* filamentation with or without MeJA treatment. The 1:1 ratio mixed *MAT-1* and *MAT-2* sporidia was allowed to grow on YePS medium under 28 °C, for 2 d, before total RNA extraction and qRT-PCR analysis. Relative gene expression level was calculated by the $2^{-\Delta\Delta CT}$ method [22] with *ACTIN* as an internal control. Primers used for transcriptional profiling were listed in Table 1. Different letters indicate significant differences, *n* = 3. Methanol (20 µL per 40 mL medium) served as solvent control in (**A–C**). The experiments were repeated three times.

We further assessed whether MeJA affected sexual mating by observing sporidia conjugation between two fluorescent proteins tagged sporidia of opposite mating types, and the following dikaryotic hyphae formation [19]. Under MeJA treatment, we observed abundant sporidia staying as monocyte status, failing to conjugate with the sporidia of the opposite mating type even though they were close to each other (Figure 1B). In contrast, the untreated or the JA-treated sporidia can form the healthy hyphal, displaying mixed fluorescent signals, suggesting cellular fusion between two sporidia (Figure 1B). This confirms that MeJA is able to suppress *S. scitamineum* sexual mating. We further measured the transcriptional levels of a and b locus genes, which control sexual mating and filamentous growth, respectively [25], under different treatment conditions. The results showed that the pheromone genes *mfa1/2* and pheromones receptor gene *pra1/2* increased their transcription significantly in the presence of MeJA. In contrast, the transcription levels of b locus genes controlling filamentous growth and pathogenicity were suppressed by MeJA (Figure 1C), likely accounting for reduced filamentous growth. Overall, MeJA is a phytohormone with potential in suppressing the pathogenic development of the sugarcane smut fungus.

### 3.2. MeJA Suppresses Maize Smut Disease Symptom in Pot Experiment

As it takes a long time for the systematic infection of *S. scitamineum* on sugarcane before whip symptoms develop, we instead used maize smut fungus *U. maydis* to test the effect of MeJA on fungal pathogenicity. *U. maydis* is a model fungus belonging to basidiomycetes and closely related to *S. scitamineum*. As it takes long time for typical "whip" symptoms to develop when caused by *S. scitamineum* infection to the sugarcane, processing a timely application of MeJA would be difficult. Here, we intend to test the potential effect of MeJA on smut disease suppression by using the *U. maydis*-maize system. First, we tested the effect of MeJA on *U. maydis* sexual mating and filamentous growth. As is similarly observed in *S. scitamineum*, MeJA was effective in suppressing *U. maydis* filamentous growth after sexual mating (Figure 2A). Overall, MeJA caused reduced dimorphic switch in *U. maydis*.

We next tested the effect of MeJA on controlling maize smut disease. Approximately 60.87% of *U. maydis* infected seedlings displayed tumor symptoms on the stems or leaves, and some seedlings were even dead (Figure 2B,C). MeJA treatment could effectively reduce tumor formation rate, to 41.38% (Figure 2D). The death rate of *U. maydis*-infected seedling under MeJA treatment was significantly reduced as compared to un-treated seedlings (Figure 2D). Therefore, we conclude that MeJA has potential in preventing smut disease caused by fungi.

### 3.3. Construction of Escherichia coli Expressing Plant JMT Gene

We next generated a plasmid for expressing the plant *JMT* gene (GenBank: AY008434.1), encoding the jasmonic acid carboxyl methyl transferase that catalyzes conversion from JA to MeJA [17]. The *JMT* gene was driven by a T7 promoter following the lacI repressor; therefore, it is inducible by isopropyl-β-D-thiogalactoside (IPTG) [21]. We transformed this constructed vector into *E. coli* (hereafter named as pJMT strain) and tested whether it could produce MeJA, by using JA as the precursor and SAM (S-adenosyl-L-methionine) as a methyl donor. Under in vitro culture conditions, IPTG induction led to a significant decrease in JA (Figure 3A, retention time ≈ 18.5 min) content detected in the supernatant of pJMT strain when compared to what was detected in the supernatant from un-treated pJMT strain (Figure 3A). This indicates that induced expression of *JMT* gene catalyzed conversion from JA to MeJA (Figure 3B). Furthermore, we used GC/MS to detect the presence of MeJA in the supernatant of pJMT strain, with or without IPTG inducion. The results showed that MeJA was detected in IPTG-induced pJMT strain, while not in un-induced strain (Figure 3C). Overall, the result showed that the induced JMT enzyme was capable of catalyzing conversion from JA to MeJA, likely using SAM as a methyl donor, under in vitro culture conditions.

We further tested whether the pJMT strain was able to suppress *S. scitamineum* filamentous growth by producing MeJA. By either confrontation, or supplying the supernatant of pJMT culture, the dikaryotic hyphae growth of *S. scitamineum* was significantly suppressed, when the *JMT* gene was induced by IPTG (Figure 4A,B). This indicates that the pJMT strain was able to produce and secrete MeJA, and has potential as a biocontrol agent against the sugarcane smut fungus.

### 3.4. Utilization of E. coli pJMT Strain in Controlling Smut Diseases

We next tested the effect of pJMT strain in controlling smut diseases, caused by *U. maydis* on maize or by *S. scitamineum* on sugarcane. As sugarcane smut caused by *S. scitamineum* takes a longer time (usually 3–6 months) to develop the typical "whip" symptom, we firstly tested the potential disease control capacity of pJMT on maize seedlings grown in pots. The results showed that *U. maydis* infection caused tumors formed on the stem of maize seedlings, either treated with pJMT strain or the control strain carrying vector only (Figure 5). The disease incidence rate seems to show no obvious difference between pJMT or the control strain-treated plants (Figure 5). Therefore, the pJMT strain did not seem to be effective in suppressing tumor formation on the maize stems or leaves.

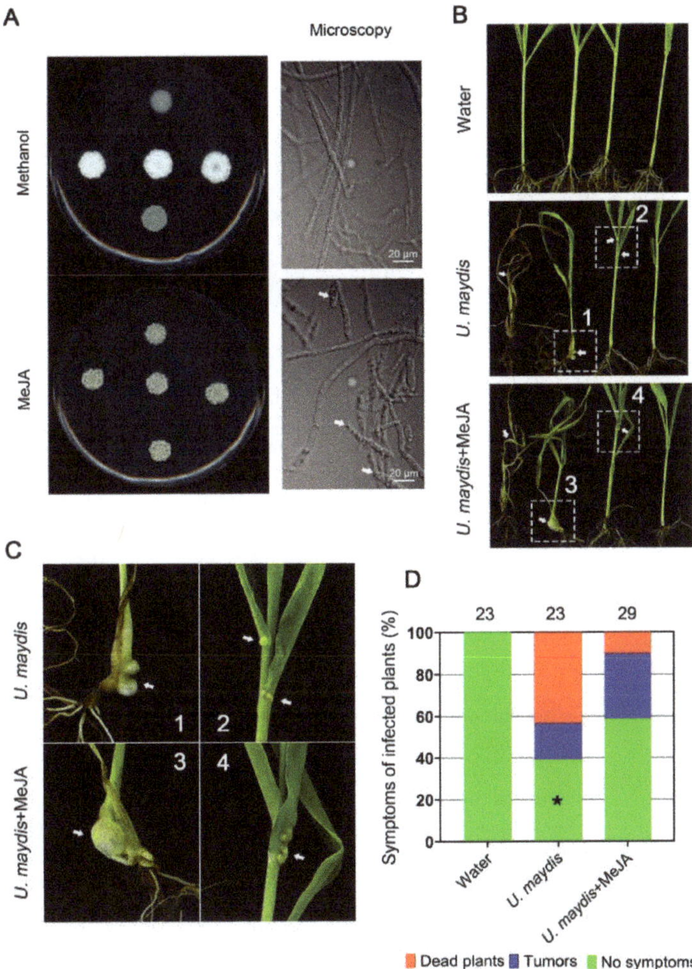


**Figure 2.** MeJA suppressed *U. maydis* sexual mating/filamentation. (**A**) The *MAT-1* and *MAT-2* sporidia of $OD_{600} \approx 1.0$ were mixed as 1:1 (*v/v*) and inoculated on the YePSA medium containing MeJA (400 µM) or methanol (20 µL per 40 mL medium; solvent control). The mating cultures were kept in the dark, at 28 °C, for 5 d before photographing and microscopic observation. Scale bar = 20 µm. (**B–D**): Mixed *U. maydis* sporidia (1 mL; $OD_{600} \approx 1.0$; *MAT-1: MAT-2*= 1:1, *v/v*), with or without MeJA (400 µM) treatment, were injected to maize seedlings at the four-leaf stage (Hua Meitian 9; DYQS-16 × TG-9). Injection with water served as blank control. The injected seedlings were allowed to grow in pots under natural conditions for 21 d, before photographing the tumor symptoms, as denoted by the arrows. The histogram was drawn by GraphPad Prism (Version: 8.0.2). The experiments were repeated three times, containing at least 23 seedlings for each treatment. Significant differences in no-symptom plants between *U. maydis*-infected seedlings with or without MeJA treatment were determined by a two-tailed Student's *t*-test (* $p$ = 0.0175).

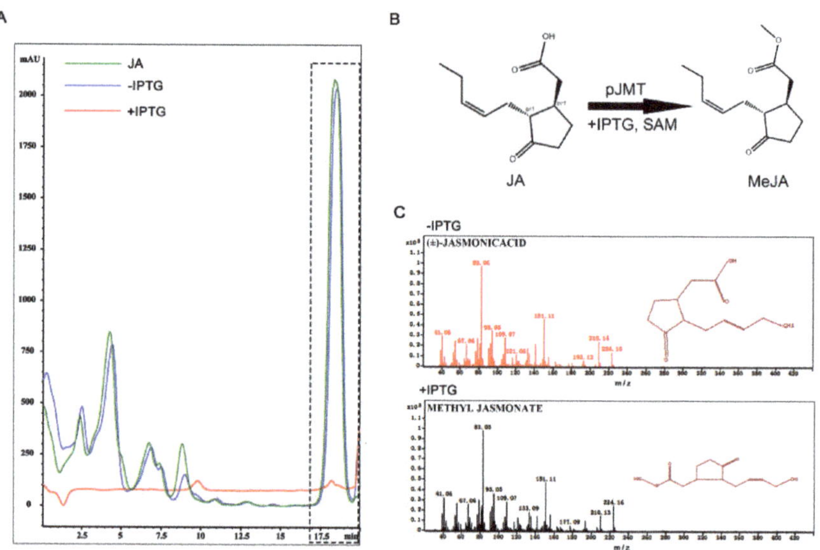

**Figure 3.** *E. coli* pJMT strain was able to convert JA to MeJA under in vitro culture conditions. (**A**) Detection of JA by HPLC. The region in the dashed box denotes JA, detected as retention time = 18.5 min, wavelength = 205 nm. The green line denotes JA standard, while the red line and green line, respectively, denote the supernatant extracted from the pJMT culture with or without IPTG induction. Peak area for the green line is 107,510.0, and 1815.1 for red line. (**B**) Schematic representation of pJMT catalyzing conversion of JA to MeJA under in vitro culture conditions. (**C**) GC-MS analysis to detect JA and MeJA from supernatant of the pJMT culture without IPTG induction (upper panel) or under IPTG induction (lower panel).

Under laboratory conditions, pJMT strain showed a significant inhibitory effect on the dimorphic switch of *S. scitamineum* (Figure 4B). We assumed that it could be a potential biocontrol agent for protecting sugarcane from smut disease. A field experiment was conducted to evaluate the control ability of pJMT strain against sugarcane smut. Approximately 290 sugarcanes from 90 clumps were grown from ratoons in an established diseased field and used for untreated control and pJMT treatment plots, respectively. Liquid-cultured pJMT (pre-treated with IPTG; the concentration of pJMT fermentation product to the soil is approximately $4 \times 10^5$ cell/cm$^2$) were applied within three days from the germination of ratoons. The disease canes grew short and slim, and the whips appeared from the trunks of a very early stage, in either untreated or pJMT-treated plots (Figure 6). There was no significant difference in the disease-occurring rate between the untreated control plot (20.6% diseased seedlings out of a total 294 seedlings) and the pJMT strain treatment (18.6% diseased out of a total of 294 seedlings).

In summary, the engineered *E. coli* pJMT strain—which can catalyze a JA-to-MeJA conversion—displayed a mild suppression effect on maize smut in the pot experiment and had no obvious effect on sugarcane smut under field conditions. This strain needs further optimization before it can be applied as a biocontrol agent against smut diseases.

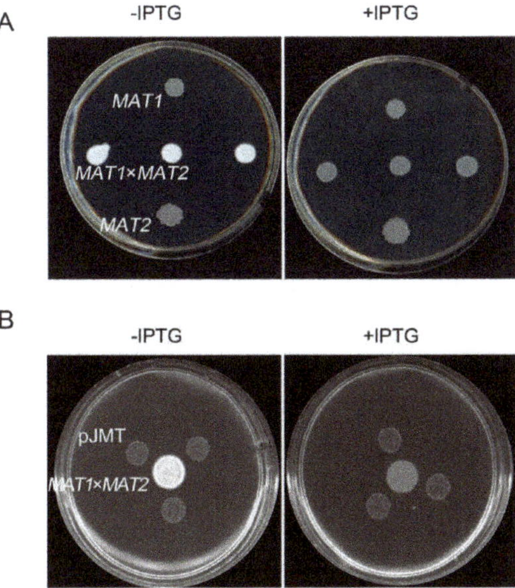



**Figure 4.** *E. coli* pJMT strain suppressed *S. scitamineum* filamentation under in vitro culture conditions. The pJMT strain was able to suppress filamentous growth of *S. scitamineum* after sexual mating by supplying the supernatent to the fungal colonies (**A**), or in confrontation assay (**B**). (**A**) JA was added in the liquid LB to reach a final concentration of 200 µM, with or without IPTG (1 mM) to induce *JMT* induction. A total of 4 mL supernatant of pJMT culture was mixed with 6 mL YePS medium, on which *MAT-1*, *MAT-2* sporidia, or mixed sporidia (to induce mating and filamentation) were allowed to grow for 3 d before photographing. (**B**) A total of 200 µM JA was added into YePS medium, on which mating *MAT-1* × *MAT-2* culture was confronted with three colonies of pJMT, with or without IPTG induction. The results were observed after 3 d of incubation.

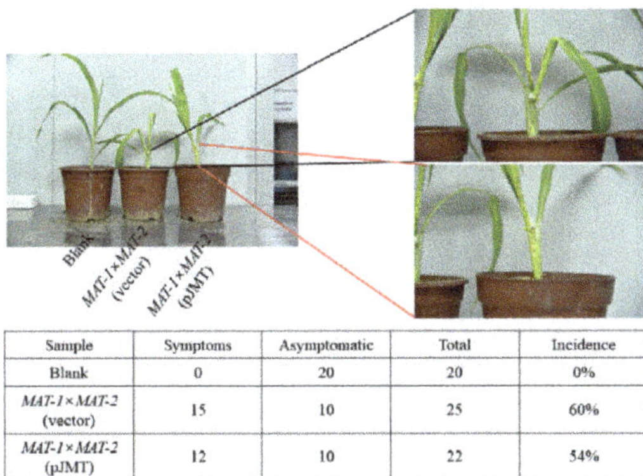



| Sample | Symptoms | Asymptomatic | Total | Incidence |
| --- | --- | --- | --- | --- |
| Blank | 0 | 20 | 20 | 0% |
| *MAT-1×MAT-2* (vector) | 15 | 10 | 25 | 60% |
| *MAT-1×MAT-2* (pJMT) | 12 | 10 | 22 | 54% |

**Figure 5.** *E. coli* pJMT applied in controlling maize smut disease in pot experiment. The *U. maydis MAT-1* and *MAT-2* sporidia (1:1, *v/v*, $OD_{600} \approx 1.0$) were mixed, and 1 mL of such mixture was injected into the the stem of four-leaf stage maize seedlings. In the treatment group, the *U. maydis* sporidia mixture was suspended with liquid-cultured pJMT ($OD_{600} \approx 1.0$). The *E. coli* harboring the vector pRSFDUET-1 served as an untreated control, and the seedlings injected with water as the blank control. The tumor symptoms were examined and documented at 21 d post infection.

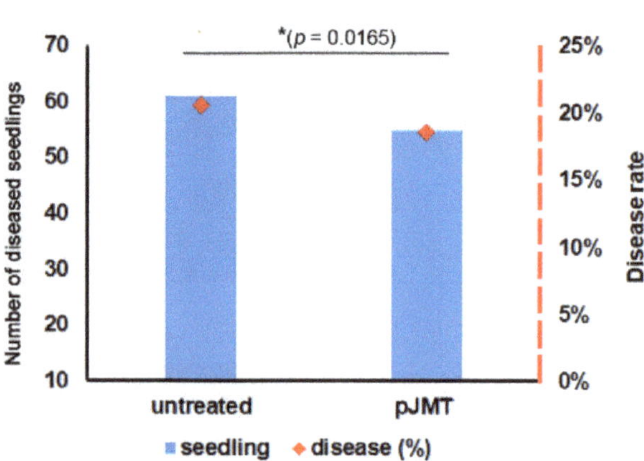


**Figure 6.** Black whip symptom of sugarcane smut in the sugarcane with or without pJMT treatment under field conditions. Ratoons of the canes grown in the diseased field in the experimental fields of South China Agricultural University (23.23 E, 113.63 W) were used for testing smut control capacity of pJMT strain. Three plots containing approximately 290 seedlings, out of 90 clumps, were, respectively, set as control or pJMT treatment plots. The pJMT culture was pre-treated with IPTG before being applied to the ratoons within 3 days after cutting the above-ground parts. Disease symptoms, including typical symptom and obvious growth defects of the infected seedlings, were evaluated at three months post growth of the ratoons. A seedling with typical black whip and slim stem symptoms is displayed in left panel. The disease occurring rate based on the number of black whips is quantified and presented in the right panel. A significant difference (* $p$ = 0.0165) in the number of diseased seedlings in comparison between the pJMT treatment and untreated (control) plot was determined by a two-tailed Student's *t*-test.

## 4. Discussion

Sugarcane smut caused by *Sporisorium scitamineum* is considered as the most serious and widespread disease of sugarcane [2], severely affecting sugarcane yields and quality, and thus causes significant economic loss [26–28]. The disease control methods for sugarcane smut include physical control, fungicide control, and disease-resistant breeding [29–31]. The application of fungicides is not only expensive, but also leads to serious environmental pollution problems and drug-resistance to the pathogens with increasing fungicide doses in pursuit of the desired control effects. Although disease-resistant breeding is the most effective method to control sugarcane smut in agricultural practices, the breeding process of disease-resistant varieties is difficult and time-consuming [32,33], and ever-evolving pathogens cause the loss of resistance of disease-resistant varieties [34]. At present, a highly efficient, sustainable, and environmentally friendly control strategy for sugarcane smut is still lacking, and is in urgent need.

Phytohormones not only regulate plant growth, but also precipitate/regulate plant immune responses to pathogens. Auxin regulates plant cell division, enlargement, and differentiation [35]. Exogenous addition of auxin increases the tolerance of *Medicago truncatula* to *Macrophomina phaseolina* infection [36], indicating a positive regulation of plant immunity. In *Arabidopsis thaliana*, the exogenous addition of auxin affects the proliferation of *Pseudomonous syringae* [37], but does not affect *Fusarium oxysporum* [38]. Salicylic acid (SA) and jasmonic acid (JA) are important defense hormones. Necrotrophic pathogens can activate JA-related defense responses [39]. SA plays an important role in defending against infection by biotrophic and hemibiotroph pathogens [40]. Moreover, as a plant hormone, SA can interact with a variety of plant hormone-related signaling pathways to activate the immune response and disease resistance of plants [41]. The application of MeJA to roots or foliar tissues can enhance transcriptional expression levels of resistance genes in sugarcane towards various classes of soil-borne pests and pathogens [42]. MeJA increases plant resistance by regulating antioxidant defense systems in sugarcane seedlings [43], *Panax ginseng* C.A. Mey. roots [44], and *Glycine max* L. leaves [45]. Furthermore, foliar application of MeJA could improve the grape and wine quality [46].

In this study, we found that the phytohormone MeJA could inhibit the dimorphic transition, a critical step in pathogenic development, of smut fungi *S. scitamineum* and *U. maydis* (Figures 1 and 2A). The a and b locus play an important role in sexual mating and hypha formation [3,5,24]. The qRT-PCR results of the coding genes showed that MeJA inhibits *S. scitamineum* sexual mating while causing a significant increase in the transcriptional levels of the mating type genes, pheromone genes mfa1/2, and pheromones receptor gene pra1/2. We infer that suppressed transcription of b locus genes is more important for regulating filamentation, which is inhibited by MeJA. Up-regulation of the a locus genes does not guarantee promoted sexual mating, as the encoded pheromone peptides need post-translational modification [8] for activation, and the pheromone receptors Pra1/2 need to recognize the activated pheromone peptide from the opposite mating-type [47]. In addition, we noticed a limited difference in the transcription of the a and b locus genes in presence of MeJA, which would almost completely suppress mating/filamentous growth. We hypothesize that in addition to suppression of a and b locus gene transcription, MeJA may target other pathways critical to the filamentation of *S. scitamineum*, which were not identified in this study.

Furthermore, we found that the exogenous addition of MeJA suppressed disease symptoms (tumors formation) caused by a *U. maydis* infection in maize seedlings, under the pot experiment (Figure 2B). The pJMT strain, heterogeneously expressing the plant *JMT* gene, was able to catalyze MeJA production using JA as a precursor (Figure 3). Although, under in vitro culture conditions, the formation of *S. scitamineum* dikaryotic hyphae could be effectively inhibited by confrontation or by supplying the supernatant of the pJMT culture (Figure 4), suggesting that such a MeJA-producing pJMT strain could be potentially used as a biocontrol agent against smut diseases caused by the fungal pathogens. However, direct application of this pJMT strain to the infected maize or sugarcane seedlings, in pot experiments or in field experiments, did not show an obvious disease control ability (Figures 5 and 6). When performing the pot and field experiments, we pre-treated the pJMT culture with IPTG to induce JMT expression. A possible reason for the failure of the pJMT strain in controlling smut diseases may be the fact that the symptom development process in the field is long and complicated. In the future, we would like to further modify this strain by using a strong, constitutive, and native promoter of *E. coli* to drive the JMT gene. Alternatively, a pathogen-inducible promoter would be screened and tested.

In conclusion, our results demonstrate that the phytohormone MeJA could suppress a dimorphic switch, which was established as a key step of pathogenicity for smut fungi (*S. scitamineum* and *U. maydis*). We further constructed a pJMT strain by expressing the plant's *JMT* gene, and verified that the engineered strain could catalyze a conversion from JA to MeJA. At present, this pJMT construct showed no capacity in suppressing smut disease either in maize (pot condition) or sugarcane (field condition). However, the idea that a bacteria strain expressing a JMT gene could utilize a plant-source JA to facilitate MeJA production was shown to suppress the dikaryotic hyphae growth of smut fungi in this study. Future attempts will be conducted to monitor and enhance the colonization ability and/or strain stability of the pJMT strain, with the aim to develop it as a useful tool in the prevention and management of smut diseases.

**Author Contributions:** Conceptualization, S.L., Z.J. and Y.D.; Investigation, G.C. and X.B.; Writing—original draft, G.C.; Writing—review & editing, Y.D.; Supervision, Y.D. All authors have read and agreed to the published version of the manuscript.

**Funding:** This work was supported by the Key Projects of Guangzhou Science and Technology Plan (201904020010) and Key Realm R&D Program of Guangdong Province (2020B0202090001).

**Data Availability Statement:** No data available.

**Acknowledgments:** This work was supported by the Key Projects of Guangzhou Science and Technology Plan (201904020010) and Key Realm R&D Program of Guangdong Province (2020B0202090001). The funders had no role in study design, data collection and interpretation, or the decision to submit the work for publication.

**Conflicts of Interest:** The authors declare no conflict of interest.

## References

1. Bhuiyan, S.A.; Magarey, R.C.; McNeil, M.D.; Aitken, K.S. Sugarcane Smut, Caused by *Sporisorium scitamineum*, a Major Disease of Sugarcane: A Contemporary Review. *Phytopathology* **2021**, *111*, 1905–1917. [CrossRef]
2. Rajput, M.A.; Rajput, N.A.; Syed, R.N.; Lodhi, A.M.; Que, Y. Sugarcane Smut: Current Knowledge and the Way Forward for Management. *J. Fungi* **2021**, *7*, 1095. [CrossRef]
3. Agisha, V.N.; Nalayeni, K.; Ashwin, N.M.R.; Vinodhini, R.T.; Jeyalekshmi, K.; Suraj Kumar, M.; Ramesh Sundar, A.; Malathi, P.; Viswanathan, R. Molecular Discrimination of Opposite Mating Type Haploids of *Sporisorium scitamineum* and Establishing Their Dimorphic Transitions During Interaction with Sugarcane. *Sugar Tech.* **2022**, *24*, 1430–1440. [CrossRef]
4. Zuo, W.; Okmen, B.; Depotter, J.R.L.; Ebert, M.K.; Redkar, A.; Misas Villamil, J.; Doehlemann, G. Molecular Interactions Between Smut Fungi and Their Host Plants. *Annu. Rev. Phytopathol.* **2019**, *57*, 411–430. [CrossRef]
5. Kijpornyongpan, T.; Aime, M.C. Investigating the Smuts: Common Cues, Signaling Pathways, and the Role of MAT in Dimorphic Switching and Pathogenesis. *J. Fungi* **2020**, *6*, 368. [CrossRef]
6. Spellig, T.; Bolker, M.; Lottspeich, F.; Frank, R.W.; Kahmann, R. Pheromones trigger filamentous growth in *Ustilago maydis*. *Embo J.* **1994**, *13*, 1620–1627. [CrossRef]
7. Yan, M.; Dai, W.; Cai, E.; Deng, Y.Z.; Chang, C.; Jiang, Z.; Zhang, L.H. Transcriptome analysis of *Sporisorium scitamineum* reveals critical environmental signals for fungal sexual mating and filamentous growth. *BMC Genom.* **2016**, *17*, 354. [CrossRef]
8. Sun, S.; Deng, Y.; Cai, E.; Yan, M.; Li, L.; Chen, B.; Chang, C.; Jiang, Z. The Farnesyltransferase Beta-Subunit Ram1 Regulates *Sporisorium scitamineum* Mating, Pathogenicity and Cell Wall Integrity. *Front. Microbiol.* **2019**, *10*, 976. [CrossRef]
9. Deng, Y.Z.; Zhang, B.; Chang, C.Q.; Wang, Y.X.; Lu, S.; Sung, S.Q.; Zhang, X.M.; Chen, B.S.; Jiang, Z.D. The MAP Kinase SsKpp2 Is Required for Mating/Filamentation in *Sporisorium scitamineum*. *Front. Microbiol.* **2018**, *9*, 2555. [CrossRef]
10. Chang, C.; Cai, E.; Deng, Y.Z.; Mei, D.; Qiu, S.; Chen, B.; Zhang, L.H.; Jiang, Z. cAMP/PKA signalling pathway regulates redox homeostasis essential for *Sporisorium scitamineum* mating/filamentation and virulence. *Environ. Microbiol.* **2019**, *21*, 959–971. [CrossRef]
11. Cai, E.; Li, L.; Deng, Y.; Sun, S.; Jia, H.; Wu, R.; Zhang, L.; Jiang, Z.; Chang, C. MAP kinase Hog1 mediates a cytochrome P450 oxidoreductase to promote the *Sporisorium scitamineum* cell survival under oxidative stress. *Environ. Microbiol.* **2021**, *23*, 3306–3317. [CrossRef]
12. Mousavi, S.R.; Niknejad, Y.; Fallah, H.; Tari, D.B. Methyl jasmonate alleviates arsenic toxicity in rice. *Plant Cell Rep.* **2020**, *39*, 1041–1060. [CrossRef]
13. Krol, P.; Igielski, R.; Pollmann, S.; Kepczynska, E. Priming of seeds with methyl jasmonate induced resistance to hemi-biotroph *Fusarium oxysporum* f.sp. *lycopersici* in tomato via 12-oxo-phytodienoic acid, salicylic acid, and flavonol accumulation. *J. Plant Physiol.* **2015**, *179*, 122–132. [CrossRef]
14. Farooq, M.A.; Gill, R.A.; Islam, F.; Ali, B.; Liu, H.; Xu, J.; He, S.; Zhou, W. Methyl Jasmonate Regulates Antioxidant Defense and Suppresses Arsenic Uptake in *Brassica napus* L. *Front. Plant Sci.* **2016**, *7*, 468. [CrossRef]
15. Thomma, B.P.H.J.; Eggermonta, K.; Broekaerta, W.F.; Cammueab, B.P.A. Disease development of several fungi on *Arabidopsis* can be reduced by treatment with methyl jasmonate. *Plant Physiol. Biochem.* **2000**, *38*, 421–427. [CrossRef]
16. Santino, A.; Taurino, M.; De Domenico, S.; Bonsegna, S.; Poltronieri, P.; Pastor, V.; Flors, V. Jasmonate signaling in plant development and defense response to multiple (a)biotic stresses. *Plant Cell Rep.* **2013**, *32*, 1085–1098. [CrossRef]
17. Seo, H.S.; Song, J.T.; Cheong, J.J.; Lee, Y.H.; Lee, Y.W.; Hwang, I.; Lee, J.S.; Choi, Y.D. Jasmonic acid carboxyl methyltransferase: A key enzyme for jasmonate-regulated plant responses. *Proc. Natl. Acad. Sci. USA* **2001**, *98*, 4788–4793. [CrossRef]
18. Qi, J.; Li, J.; Han, X.; Li, R.; Wu, J.; Yu, H.; Hu, L.; Xiao, Y.; Lu, J.; Lou, Y. Jasmonic acid carboxyl methyltransferase regulates development and herbivory-induced defense response in rice. *J. Integr. Plant Biol.* **2016**, *58*, 564–576. [CrossRef]
19. Yan, M.; Cai, E.; Zhou, J.; Chang, C.; Xi, P.; Shen, W.; Li, L.; Jiang, Z.; Deng, Y.Z.; Zhang, L.H. A Dual-Color Imaging System for Sugarcane Smut Fungus *Sporisorium scitamineum*. *Plant Dis.* **2016**, *100*, 2357–2362. [CrossRef]
20. Yang, B.; Zheng, P.; Wu, D.; Chen, P. Efficient Biosynthesis of Raspberry Ketone by Engineered *Escherichia coli* Coexpressing Zingerone Synthase and Glucose Dehydrogenase. *J. Agric. Food Chem.* **2021**, *69*, 2549–2556. [CrossRef]
21. Zhang, C.; Liu, L.; Teng, L.; Chen, J.; Liu, J.; Li, J.; Du, G.; Chen, J. Metabolic engineering of *Escherichia coli* BL21 for biosynthesis of heparosan, a bioengineered heparin precursor. *Metab. Eng.* **2012**, *14*, 521–527. [CrossRef]
22. Livak, K.J.; Schmittgen, T.D. Analysis of relative gene expression data using real-time quantitative PCR and the 2(-Delta Delta C(T)) Method. *Methods.* **2001**, *25*, 402–408. [CrossRef]
23. Huang, N.; Ling, H.; Su, Y.; Liu, F.; Xu, L.; Su, W.; Wu, Q.; Guo, J.; Gao, S.; Que, Y. Transcriptional analysis identifies major pathways as response components to *Sporisorium scitamineum* stress in sugarcane. *Gene.* **2018**, *678*, 207–218. [CrossRef]

24. Cui, G.; Yin, K.; Lin, N.; Liang, M.; Huang, C.; Chang, C.; Xi, P.; Deng, Y.Z. *Burkholderia gladioli* CGB10: A Novel Strain Biocontrolling the Sugarcane Smut Disease. *Microorganisms.* **2020**, *8*, 1943. [CrossRef]
25. Yan, M.; Zhu, G.; Lin, S.; Xian, X.; Chang, C.; Xi, P.; Shen, W.; Huang, W.; Cai, E.; Jiang, Z.; et al. The mating-type locus b of the sugarcane smut *Sporisorium scitamineum* is essential for mating, filamentous growth and pathogenicity. *Fungal. Genet. Biol.* **2016**, *86*, 1–8. [CrossRef]
26. Comstock, J.C. Sugarcane Smut: Comparison of Natural Infection Testing and Artificial Inoculation. *Hawaii. Plant. Rec.* **1987**, *60*, 17.
27. Khan, H.M.W.A.; Chattha, A.A.; Munir, M.; Zia, A. Evaluation of resistance in sugarcane promising lines against whip smut. *Pak. J. Phytopathol.* **2009**, *21*, 92–93.
28. Tadesse, A. *Increasing Crop Prodcution through Improved Plant Protection*; Ethiopian Institute of Agricultural Research: Addis Ababa, Ethiopia, 2009.
29. Ming, R.; Moore, P.H.; Wu, K.K.; D'Hont, A.; Glaszmann, J.C.; Tew, T.L.; Mirkov, T.E.; da Silva, J.; Jifon, J.; Rai, M.; et al. Sugarcane Improvement through Breeding and Biotechnology. *Plant Breed. Rev.* **2010**, *27*, 15–118.
30. Sundar, A.R.; Barnabas, E.L.; Malathi, P.; Viswanathan, R. A Mini-Review on Smut Disease of Sugarcane Caused by *Sporisorium scitamineum*. *InTech Open.* **2012**, *5*, 108–128.
31. Shailbala; Sharma, S.K. Effect of fungicides and hot water treatment on control of sugarcane smut. *Pestology* **2013**, *37*, 29–32.
32. Silva, J.A.; Sorrells, M.E.; Burnquist, W.L.; Tanksley, S.D. RFLP linkage map and genome analysis of *Saccharum spontaneum*. *Genome* **1993**, *36*, 782–791. [CrossRef] [PubMed]
33. Burnquist, W.L.; Sorrelles, M.E. *Characterization of Genetic Variability in Saccharum germplasm by Means of Restriction Fragment Length Polymorphism (RFLP) Analysis*; FAO: Rome, Italy, 1995; pp. 355–366.
34. Nelson, R.; Wiesner-Hanks, T.; Wisser, R.; Balint-Kurti, P. Navigating complexity to breed disease-resistant crops. *Nat. Rev. Genet.* **2018**, *19*, 21–33. [CrossRef] [PubMed]
35. Sundberg, E.; Ostergaard, L. Distinct and dynamic auxin activities during reproductive development. *Cold Spring Harb. Perspect. Biol.* **2009**, *1*, a001628. [CrossRef] [PubMed]
36. Mah, K.M.; Uppalapati, S.R.; Tang, Y.H.; Allen, S.; Shuai, B. Gene expression profiling of *Macrophomina phaseolina* infected *Medicago truncatula* roots reveals a role for auxin in plant tolerance against the charcoal rot pathogen. *Physiol. Mol. Plant Pathol.* **2012**, *79*, 21–30. [CrossRef]
37. Chen, Z.; Agnew, J.L.; Cohen, J.D.; He, P.; Shan, L.; Sheen, J.; Kunkel, B.N. *Pseudomonas syringae* type III effector AvrRpt2 alters *Arabidopsis thaliana* auxin physiology. *Proc. Natl. Acad. Sci. USA.* **2007**, *104*, 20131–20136. [CrossRef]
38. Kidd, B.N.; Kadoo, N.Y.; Dombrecht, B.; Tekeoglu, M.; Gardiner, D.M.; Thatcher, L.F.; Aitken, E.A.; Schenk, P.M.; Manners, J.M.; Kazan, K. Auxin signaling and transport promote susceptibility to the root-infecting fungal pathogen *Fusarium oxysporum* in *Arabidopsis*. *Mol. Plant Microbe Interact.* **2011**, *24*, 733–748. [CrossRef]
39. Katagiri, F.; Tsuda, K. Understanding the plant immune system. *Mol. Plant Microbe Interact.* **2010**, *23*, 1531–1536. [CrossRef]
40. Al-Daoude, A.; Al-Shehadah, E.; Shoaib, A.; Jawhar, M.; Arabi, M.I.E. Salicylic Acid Pathway Changes in Barley Plants Challenged with either a Biotrophic or a Necrotrophic Pathogen. *Cereal. Res. Commun.* **2019**, *47*, 324–333. [CrossRef]
41. An, C.; Mou, Z. Salicylic acid and its function in plant immunity. *J. Integr. Plant Biol.* **2011**, *53*, 412–428. [CrossRef]
42. Bower, N.I.; Casu, R.E.; Maclean, D.J.; Reverter, A.; Chapman, S.C.; Manners, J.M. Transcriptional response of sugarcane roots to methyl jasmonate. *Plant Sci.* **2005**, *168*, 761–772. [CrossRef]
43. Seema, G.; Srivastava, M.K.; Srivastava, S.; Shrivastava, A.K. Effect of methyl jasmonate on sugarcane seedlings. *Sugar Tech.* **2003**, *5*, 189–191. [CrossRef]
44. Ali, M.B.; Yu, K.W.; Hahn, E.J.; Paek, K.Y. Methyl jasmonate and salicylic acid elicitation induces ginsenosides accumulation, enzymatic and non-enzymatic antioxidant in suspension culture *Panax ginseng* roots in bioreactors. *Plant Cell Rep.* **2006**, *25*, 613–620. [CrossRef] [PubMed]
45. Keramat, B.; Kalantari, K.; Arvin, M. Effects of methyl jasmonate in regulating cadmium induced oxidative stress in soybean plant (*Glycine max* L.). *Afr. J. Microbiol. Res.* **2009**, *3*, 240–244.
46. Garde-Cerdan, T.; Portu, J.; Lopez, R.; Santamaria, P. Effect of methyl jasmonate application to grapevine leaves on grape amino acid content. *Food Chem.* **2016**, *203*, 536–539. [CrossRef]
47. Lu, S.; Shen, X.; Chen, B. Development of an efficient vector system for gene knock-out and near in-*cis* gene complementation in the sugarcane smut fungus. *Sci. Rep.* **2017**, *7*, 3113. [CrossRef] [PubMed]



**Disclaimer/Publisher's Note:** The statements, opinions and data contained in all publications are solely those of the individual author(s) and contributor(s) and not of MDPI and/or the editor(s). MDPI and/or the editor(s) disclaim responsibility for any injury to people or property resulting from any ideas, methods, instructions or products referred to in the content.

 *microorganisms*

*Review*

# Towards Sustainable Green Adjuvants for Microbial Pesticides: Recent Progress, Upcoming Challenges, and Future Perspectives

Fuyong Lin, Yufei Mao, Fan Zhao, Aisha Lawan Idris, Qingqing Liu, Shuangli Zou, Xiong Guan and Tianpei Huang *

State Key Laboratory of Ecological Pest Control for Fujian and Taiwan Crops & Key Laboratory of Biopesticide and Chemical Biology of Ministry of Education & Biopesticide Research Center, College of Life Sciences & College of Plant Protection, Fujian Agriculture and Forestry University, Fuzhou 350002, China
* Correspondence: tianpeihuang@fafu.edu.cn; Tel.: +86-591-83789258

**Abstract:** Microbial pesticides can be significantly improved by adjuvants. At present, microbial pesticide formulations are mainly wettable powders and suspension concentrates, which are usually produced with adjuvants such as surfactants, carriers, protective agents, and nutritional adjuvants. Surfactants can improve the tension between liquid pesticides and crop surfaces, resulting in stronger permeability and wettability of the formulations. Carriers are inert components of loaded or diluted pesticides, which can control the release of active components at appropriate times. Protective agents are able to help microorganisms to resist in adverse environments. Nutritional adjuvants are used to provide nutrients for microorganisms in microbial pesticides. Most of the adjuvants used in microbial pesticides still refer to those of chemical pesticides. However, some adjuvants may have harmful effects on non-target organisms and ecological environments. Herein, in order to promote research and improvement of microbial pesticides, the types of microbial pesticide formulations were briefly reviewed, and research progress of adjuvants and their applications in microbial pesticides were highlighted, the challenges and the future perspectives towards sustainable green adjuvants of microbial pesticides were also discussed in this review.

**Keywords:** surfactant; carrier; protective agent; nutritional adjuvants

**Citation:** Lin, F.; Mao, Y.; Zhao, F.; Idris, A.L.; Liu, Q.; Zou, S.; Guan, X.; Huang, T. Towards Sustainable Green Adjuvants for Microbial Pesticides: Recent Progress, Upcoming Challenges, and Future Perspectives. *Microorganisms* **2023**, *11*, 364. https://doi.org/10.3390/microorganisms11020364

Academic Editors: Chetan Keswani and Rainer Borriss

Received: 28 December 2022
Revised: 28 January 2023
Accepted: 29 January 2023
Published: 31 January 2023

**Copyright:** © 2023 by the authors. Licensee MDPI, Basel, Switzerland. This article is an open access article distributed under the terms and conditions of the Creative Commons Attribution (CC BY) license (https://creativecommons.org/licenses/by/4.0/).

## 1. Introduction

In modern agriculture and forestry, pesticides have always played important roles in ensuring agricultural production as a necessity to increase grain production, regulate crop growth, and control plant pests and diseases [1,2]. Following economic globalization, the problems of chemical pesticide residue have attracted the attention of many countries. Many of them have increasingly strict trade standards for pesticide residue in agricultural products. Ideally, pesticides must be lethal to target pests, but not to non-target species, including humans [1,3,4]. Microbial pesticides, mainly produced with naturally occurring bacteria, fungi (including some protozoa and yeasts), and viruses, have been attracting widespread attention due to their advantages in target specificity, environmental safety, efficacy, biodegradability, and applicability in integrated pest management programs [5–7].

The United States is the first country in the world to manage pesticide additives. In 1954, the United States FDA implemented the ADI limit. In 1987, according to the policy of "Reducing the potential adverse effects of the use of pesticide products containing toxic inert ingredients", the United States EPA conducted classified list management on the toxicity and exposure hazards of additives (divided into four categories) [8,9]. The European Union, through the implementation of REACH system, divides pesticide additives into three categories of management, requiring that the limits of 227 substances with potential risks among more than 3200 agricultural chemicals should be marked on the labels, and has clearly stipulated that benzene organic solvents, nonylphenol/octylphenol polyoxyethylene ether, and bovine ester amine surfactants should be restricted or prohibited in pesticide preparations [10]. At present, China has not implemented systematic management of pesticide adjuvants, But a series of regulations have been put in place to regulate some adjuvants [11].

Commercial pesticide preparations are always a mixture of "active ingredients" and "other ingredients" (adjuvants) [12,13]. Biocontrol agents are formulations produced from living organisms or substances produced by them for the control of pests or diseases. Biopesticides are formulations that use living organisms (fungi, bacteria, insect viruses, genetically modified organisms, natural enemies, etc.) or their metabolites (pheromones and auxin, etc.) to kill or inhibit agricultural pests [14,15]. The proportion of bacteria as the main active ingredient of microbial pesticides is the largest, followed by fungi, viruses, and genetically modified microorganisms. *Bacillus thuringiensis* (Bt) bacterial insecticide is the most widely used and most productive microbial insecticide in the world, accounting for 95% of microbial insecticides [16–18].

Since the active ingredients of microbial pesticides are mainly microorganisms and their bioactive compounds, their growth and reproduction in the field are affected by environmental factors such as temperature, humidity, and ultra-violet (UV) ray radiation [19]. The microorganisms have strong hydrophobicity and are likely to form large particles in solutions, which make the pesticides difficult to use to solve field application problems such as wettability and suspension [20]. The drawbacks may be overcome by the application of adjuvants to deliver the active ingredients to targets because adjuvants can effectively improve physical and chemical properties of pesticide preparations. There are more than three thousand kinds of pesticide adjuvants in common use. The value of pesticide adjuvants market in 2015 was USD 2.51 billion [19]. To date, adjuvants are developing towards multi-function adjuvants with characteristics that make them labor-saving, low consumption, easy degradation, and low toxicity. They are often applied as wetting agents and emulsifiers in the production of pesticide formulations [21,22].

Green pesticide is harmless to human health, friendly to the environment, ultra-low dosage, high selectivity, and through the green process to produce pesticides [23]. At present, most of the adjuvants used in microbial pesticides still refer to those of chemical pesticides [24]. A green adjuvant is a kind of non-toxic, harmless, and good biodegradable adjuvant added in the process of pesticide production [25]. Biogenic adjuvants are traditionally green, but adjuvants synthesized from fine chemicals are also partially green. Some traditional adjuvants are mostly chemical adjuvants, which have different effects on the active ingredients and the environment. Compared with green adjuvants, it can improve the performance of the preparation and has no impact on the environment. With the development of green agriculture, the safety of pesticides and green pesticide adjuvants is particularly important. [26–28]. Herein, in order to promote the research and development of better microbial pesticides, types of microbial pesticide formulations were briefly reviewed, and research progress of adjuvants and their applications in microbial pesticides were highlighted, while the challenges and the future perspectives towards sustainable green adjuvants of microbial pesticides were also discussed.

## 2. Types of Microbial Pesticide Formulations

There are many formulations of microbial pesticides such as wettable powders, suspensions, oil suspensions, water dispersible granules, and suspension seed coatings (Table 1) [29–31]. However, the production of microbial pesticides is relatively more complicated than those of chemical pesticides [32,33]. During the preparation process, the microorganisms in the pesticides are more sensitive to external environmental factors such as UV rays, temperature, and humidity [34]. Compared with chemical pesticides, relatively slow action affects the application efficiency of microbial pesticides. The microorganisms, which are mainly bacteria, fungi, or viruses, are insoluble biological particulate matters from 0.5 μm to 1000 μm. They will directly affect the physical properties of the pesticide preparations such as suspension, dispersibility, and wettability [35].

**Table 1.** Main formulations of microbial pesticides.

| Main Formulations of Microbial Pesticides | Types of Microbial Pesticides | Main Registered Formulations |
| --- | --- | --- |
| *Bacillus thuringiensis* | Bacterium | WP, SC, and WG |
| *Bacillus licheniformis* | Bacterium | SL |
| *Bacillus cereus* | Bacterium | WP and SC |
| *Bacillus sphaericus* | Bacterium | SC |
| *Bacillus subtilis* | Bacterium | WP and FS |
| *Bacillus amyloliquefaciens* | Bacterium | WP, SC, and WG |
| *Bacillus velezensis* | Bacterium | WG and TK |
| *Bacillus firmus* | Bacterium | WP |
| *Bacillus marine* | Bacterium | WP |
| *Brevibacterium brevis* | Bacterium | SC |
| *Paenibacillus polymyxa* | Bacterium | WP and SC |
| *Pseudomonas fluorescens* | Bacterium | WP and WG |
| *Beauveria bassiana* | Fungus | WP |
| *Locust Microsporidia* | Fungus | SC |
| *Metarhizium anisopliae* | Fungus | WP |
| *Paecilomyces lilacinus* | Fungus | DP and GR |
| *Trichoderma* | Fungus | WP and GR |
| *Verticillium pachyspora* | Fungus | WP |
| *Autographa californica* nuclear polyhedrosis virus | Virus | SE |
| *Ectropis obliqua* nuclear polyhedrosis virus | Virus | TK |
| *Helicoverpa armigera* nuclear polyhedrosis virus | Virus | WP and SC |
| *Mamestra brassicae* nuclear polyhedrosis virus | Virus | WP, SC, GR, and TK |
| *Plutella xylostella* granulosa virus | Virus | WP |
| *Pierisrapae granulosis* | Virus | TK |
| *Spodoptera exigua* nuclear polyhedrosis virus | Virus | SC, WG, and TK |
| *Spodoptera litura* nuclear polyhedrosis virus | Virus | WP, SC, and WG |

Wettable powder(WP), suspension concentration(SC), dispersible granules(WG), soluble liquid(SL), seed treatment(FS), granules(GR), powder(DP), technical concentration(TK).

## 3. Classifications and Functions of Adjuvants Used for Microbial Pesticides

The adjuvants used in the production of microbial pesticides are mainly classified into four categories: surfactants, carriers, protective agents, and nutritional adjuvants.

### 3.1. Surfactants

Surfactants can improve cost-effectiveness, increase processing efficiency, and save energy and raw materials for the production of microbial pesticides (Figure 1). They play important roles in maintaining long-term physical stability and improving biological functions of the pesticides [36]. About 3.4 million tons of surfactants were produced in 2019, and microbial pesticides usually contain one or more surfactants [37,38].

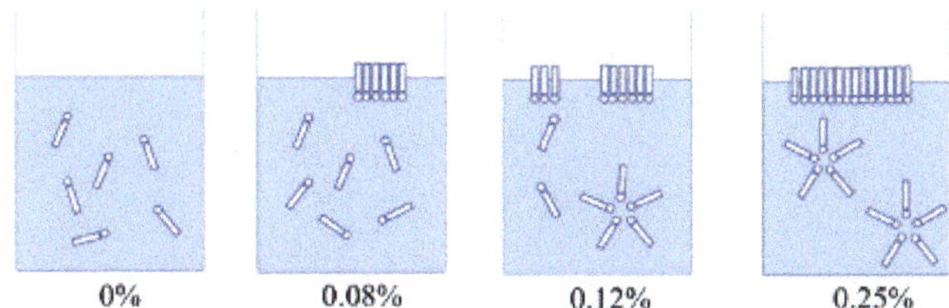


**Figure 1.** Micelle formation and surface adsorption of surfactants at different concentrations.

Surfactants have special molecular structures (amphiphilic structures) consisting of hydrophobic and hydrophilic groups [39,40]. They can be divided into cationic, anionic, nonionic, and amphoteric types by their hydrophilic groups (Table 2 and Figure 2) [41]:

(1)  Cationic surfactants: The hydrophilic part is mainly quaternary ammonium ions with strong antibacterial properties [36];
(2)  Anionic surfactants: They mainly include ionic sulfonates, sulfates, and carboxylate, most of which contain sodium or calcium. Straight chain alkyl sulfonates occupy the largest market share of all anionic surfactants [42];
(3)  Nonionic surfactants: They have polymerized glycol ether and glucose units, which are mainly used as emulsifiers and wetting agents [43,44];
(4)  Amphoteric surfactants: Contain both cationic and anionic groups in their structures. They are not only soluble in water, but also highly compatible with other surfactants to form mixed micelles. Their electric charges vary with pH, thereby affecting pesticide properties such as wetting, sedation, and foaming [36].

**Table 2.** Main surfactants used for the production of microbial pesticides.

| Types | Main Products | Features | References |
|---|---|---|---|
| Cationic surfactants | Stearyl trimethyl ammonium chloride (STAC) and Hexadecyl trimethyl ammonium bromide (TTAB) | Good water solubility, strong bactericidal power, and adsorption power; cannot be used with anionic surfactants | [45] |
| Anionic surfactants | Calcium dodecylben-zenesulfonate (ABSCa) and Sodium 2-butyl-1-naphthalenesulfonate (BX) | Good solubility, safety and low toxicity, and strong stability | [46] |
| Nonionic surfactants | Dibenzyl biphenyl polyoxyethylene ether, Defoamer GP330 and tween-80 | Good environmental adaptability, strong stability; cannot be affected by strong electrolytes, acids, and bases | [47] |
| Amphoteric surfactants | Dodecyl dimethyl betaine (BS-12) | High cost with strong permeability, flocculation, adhesion, resistance reduction, and thickening | [48] |

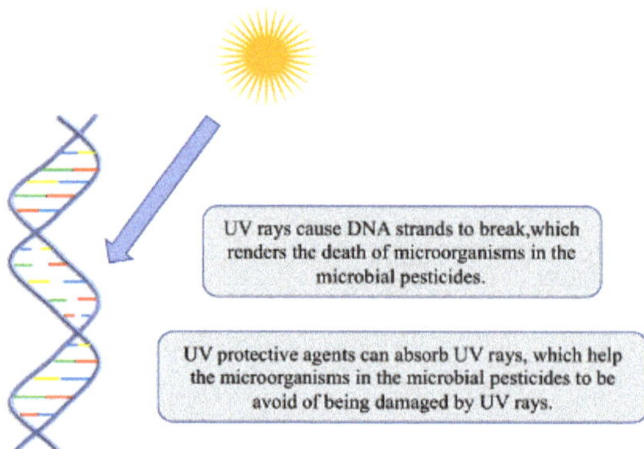


**Figure 2.** Added protective agents in the microbial pesticides to avoid damage by UV rays.

### 3.2. Carriers

The main function of a carrier is to act as a tiny container or diluent for active ingredients of microbial pesticides [20]. The mass percentage of a carrier may exceed that of the active ingredient in the pesticide [49,50]. Carriers such as diatomite, attapulgite, silica, and bentonite with strong adsorption capacities can be used to manufacture high-concentration powders, wettable powders, or granules (Table 3) [51]. Carriers such as talc, pyrophyllite, sepiolite, and clay materials with low or moderate adsorption capacities are generally used as diluents and fillers to produce low-concentration powders (Table 3) [52]. Their pore structures and specific surface areas enable the pesticides to be released into the environments at relatively slow rates [53,54]. Carriers made with biodegradable materials also enable targeted and controlled release of active ingredients from microbial pesticides [55]. For example, chitosan/carbon nanotube nanocomposites were used as the carrier of controlled-release pesticides and prepared, uniformly dispersed carbon nanotube-enhanced Cs films were applied to reduce the harm of pesticides to the environment during the release process [56,57]. Biodiesel may be used as a dilution carrier in liquid pesticide formulations [58]. Some natural plants and synthetic substances are also commonly used as carriers (Table 3) [59].

**Table 3.** Carriers commonly used for the production of microbial pesticides.

| Types of Carriers | Features | References |
|---|---|---|
| Bentonite | Its main component is montmorillonite, which has strong adsorption and large specific surface area. After absorbing a large number of water molecules, it expands and splits into extremely fine particles. | [60] |
| Diatomite | Its main component is SiO2, which harbors many micropores and has a capacity of low relative density, high porosity, and strong adsorption. It is widely used in the manufacture of high-concentration powder carriers. | [61] |
| Attapulgite clay | Its main component is attapulgite, which has a capacity of strong adsorption and a property of large specific surface area and unique thickening. It is widely used in the manufacture of high-concentration powder carriers and granule substrates, as well as thickeners for suspensions. | [57] |
| Kaolin | Its main component is kaolinite, which has a relatively compact structure, a small specific surface area, and an adsorption capacity. It is often used as a carrier for low-concentration powders, and the price is relatively cheap. | [57] |
| Zeolite | Active ingredients in forms of porous hydrous aluminosilicate crystals are highly adsorbable to certain polar molecules such as zeolite and then slowly released. It is often used as a carrier for sustained-release granules. | [62] |
| Sepiolite | Its main component is magnesium-rich fibrous clay minerals with large porosity and specific surface area. It can absorb liquid and low-melting pesticides. Because of its light weight, it can float on the water surface. | [63] |
| Synthetic vectors | Most of them are made with light calcium carbonate and white carbon black, which have strong specific surface area and adsorption capacity, and can be used as a carrier for high concentration powders. | [64] |
| Plant vectors | Most of them are made with bagasse, corn bagasse, chaff powder, tobacco powder, and walnut shell powder. Plant-based carriers are rarely used at present. Some of them have special properties such as absorbing ultraviolet (UV) rays. | [65] |
| Nanostructured Lipid Carriers | They have excellent permeability, retention, targeting, stability, and can reduce or eliminate the side effects of active ingredients, with good slow release and controlled release properties. | [66] |

In the screening process of microbial pesticide carriers, the biocompatibility of carriers and microorganisms in microbial pesticides is the primary concern [67]. Herein, a typical microbial pesticide formulation will select a carrier with good biocompatibility. And the formulation may also rely on wetting and dispersing agents to further improve its wetting and dispersing properties [68]. However, a suitable carrier can confer enough wettability for microbial pesticide formulations without the help of other adjuvants [69]. In fact, carriers greatly affect the wetting, suspending, and dispersing properties of microbial pesticide formulations [35].

*3.3. Protective Agents*

They are mainly divided into the following two categories:

(1)   UV protective agents

UV radiation causes passivation, degradation, or damage to microbial pesticides (Figure 2) [70]. UV rays can inhibit the growth of microorganisms, and even cause them to die in severe cases. UV protection agents used for the production of microbial pesticides are mainly divided into UV ray absorbers and anti-oxidative UV protection agents. UV ray absorbers such as locust toxin, optical brighteners, lignosulfonates, and milk can absorb the UV part of sunlight and fluorescent light sources without changing itself [71]. Anti-oxidative UV protection agents such as Rubus oil, hemp seed oil, kojic acid, hydroxykynurenic acid, flavonoids, and lecithin have strong antioxidant effects to help microbial pesticides to avoid being easily oxidized and degraded into other substances that are ineffective against harmful organisms under the irradiation of UV rays. Recently, the development of nanomaterials also provides new ideas for novel UV protective agents [72]. Although UV protective agents such as melanin, berberine, fluorescent whitening agents, and Congo red have significant photoprotective effects on Bt, they are cytotoxic to Bt, and may have a certain impact on the environment [73]. Herein, environmental friendliness should also be considered when UV protective agents are developed.

(2)   Other protective agents

Microorganisms in microbial pesticide preparations are easily affected by adverse factors such as temperature, humidity, and oxidation during storage and transportation, which reduce the viable microorganism rate in the formulations and affect their field control effects. Protective agents may also improve the control effect of microbial pesticides in the field and prolong their shelf lives [74]. For example, a protective agent comprised of 8.00% NaCl and 1.00% sodium acetate can increase the spore survival rate of a *Bacillus subtilis* preparation by 22.53% [75]. The addition of protective adjuvants depends on the quality of the biomaterial. In addition to benzoic acid or other chemicals, Bacillus spores can be kept for many years without other protective substances to prevent contamination [76].

*3.4. Nutritional Adjuvants*

During the growth of microorganisms, they need water, inorganic salts, carbon sources, nitrogen sources, and growth factors to grow normally. Nutritional adjuvants can provide nutrients to microorganisms of microbial pesticides to improve their reproductive capacity and promote their proliferation and growth in the field. At present, the research on nutritional adjuvants mainly focuses on the supplementation of carbon and nitrogen sources [77].

## 4. Application of Adjuvants in Microbial Pesticide Formulations

The majority of microbial pesticide formulations are wettable powders and suspension concentrations. Therefore, this section first introduces the applications of adjuvants in them and takes into account other applications.

*4.1. Application of Adjuvants in Wettable Powders of Microbial Pesticides*

A wettable powder refers to a pesticide formulation produced with active ingredients and adjuvants such as carriers, wetting agents, stabilizers, and UV protective agents [78]. The average particle size of a wettable powder is about 44 µm with a $\geq$ 75% suspension rate and a < 2 min wetting time [79,80]. It can form a stable suspension with good dispersibility after being dissolved and stirred in water and is generally used for spraying in the field [81]. Microbial pesticides, comprised of both microorganisms insoluble in water and organic solvents as the active ingredients, are suitable for processing into wettable powders [82].

Transportation and packaging costs of wettable powders are relatively low. The percentage of active ingredients in the pesticides is higher than that of powders and renders good adhesion to the crops after spraying. However, the powders may not be uniformly dispersed and suspended in the solution during the application, causing problems such as blockage of the nozzle and uneven spraying [83]. Half of registered Bt formulations are wettable powders, which play important roles in the formulations of microbial pesticides [78,84].

Carriers are crucial to wettable powders because the first step of the production is to culture microorganisms to obtain fermentation broths. Substances with strong adsorption properties such as silica, bentonite, attapulgite, and diatomite may be selected as carriers, whose effects on the survival states of microorganisms should also be considered [85]. Biocompatibility is the primary consideration to select a microbial pesticide carrier. At present, the primary evaluation method of biocompatibility is to mix carriers with microorganisms to investigate their survival rates [86]. There are some limitations to this approach. When encountering an adverse external environment, microorganisms will enter dormancy or spore state and survive well, and the carriers cannot show incompatibility with them [87].

Wetting agents such as sodium lauryl sulfate, stretch powder, polyethylene glycol, calcium lignosulfonate, sodium carboxymethyl cellulose, sodium dodecylbenzene sulfonate, sodium tripolyphosphate, lignin, and sodium sulfonate can also be added in the formulations to reduce the interfacial tension between liquid and solid to increase the adhesion of liquid to solid surface [88,89]. In the storage and transportation of microbiological pesticides, the influence of a wetting agent on the water activity of microbiological pesticides should be considered. Under certain water activity conditions, the active microorganism can maintain a relatively good storage state [90].

Stabilizers such as sodium carboxymethyl cellulose, potassium dihydrogen phosphate, calcium carbonate, and methyl cellulose are also key adjuvants for wettable powders [91,92]. Taking *Bacillus* as an example, spores are an important factor to help the bacteria resist adverse environments by protecting the spores from decomposition caused by high temperature [70].

The microorganisms in the pesticides are very sensitive to UV rays in a natural environment. Herein, it is necessary to add UV protective agents to the production of wettable powders [93].

*4.2. Application of Adjuvants in Suspension Concentrations of Microbial Pesticides*

A suspension concentrate, which is formed with active ingredients, carriers, and other adjuvants, is super-pulverized by a sander wet method. Generally, its melting point is higher than 60 °C, which is similar with the characteristics of both emulsifiable concentrates and wettable powders [94]. Suspension concentrates can be mainly classified into water suspension concentrates, oil suspension concentrates, and dry suspension concentrates. They do not use organic substances as solvents and have no risk of inflammability and explosion [82]. They have low toxicities to humans and animals. They are convenient for storage and transportation and can be better sprayed by aircrafts than wettable powders. When applied in the field, they can be mixed with any proportion of water, and easily to adhere to crop surfaces. The production process of a suspension concentrate is relatively simple and environmentally friendly, making it the mainstream formulation for microbial pesticides [95].

The fermentation broth of microorganisms is directly used for the production of suspension concentrates, which are prone to agglomeration, freezing, and precipitation due to the fluidity of the fermentation broth. Herein, adjuvants such as preservatives, dispersants, and stabilizers may be included in the formulations. Preservatives such as potassium sorbate, ethyl p-hydroxybenzoate, and sodium benzoate need to have a good inhibitory effect on mold, yeast, and aerobic bacteria in order to store the suspension concentrations for a long time [96]. The particles of microbial pesticides can automatically aggregate in the solvent, making the surface free energy of the formulations decrease, thereby forming thermodynamically stable systems. A dispersant can make the pesticide particles become smaller, preventing the particles from sedimentation and agglomeration. Dispersants such as calcium lignosulfonate and carboxymethyl cellulose play important roles in maintaining the formulation stability [97]. Stabilizers such as xanthogen glue and magnesium aluminum silicate can control the sizes of the formulation droplets and prevent them from drifting during spraying [98].

### 4.3. Application of Adjuvants in Other Formulations of Microbial Pesticides

Besides the above two formulations, granules, water-dispersible granules, and suspended seed coatings are also often applied in the production of microbial pesticides [99].

Granules are solid and granulated with active ingredients and adjuvants. Their diameter is 300–1700 microns. Their application is convenient, and their absorption and dissolution efficiency are fast. Compared with wettable powders and suspension concentrates, the granules need adjuvants such as binders, colorants, lubricants, and disintegrating agents [75]. Binders, connecting same or different solid materials, are hydrophilic or hydrophobic and indispensable to the formulation. Colorants have functions of warning and classification for microbial pesticides. Lubricants are used to reduce the excessive pressure in the manufacturing process of granules. Disintegrating agents such as $NaCl$, $AlCl_3$, and $CaCl_2$ are applied to accelerate the disintegration rates of the granules [100].

Water-dispersible granules are granular formulations that can be rapidly disintegrated and dispersed into a suspension after adding water. Dispersants, binders, and disintegrating agents are indispensable for accelerating their disintegration rate in water [101]. They also need wetting agents, anti-caking agents, and defoamers. The wetting agents are intended to increase the wetting rates of pesticide particles into water and to improve the penetration of water into the granules [102]. Anti-caking agents such as silica gel prevent the formulation from caking by acting as a layer of sliding balls [99].

Suspension seed coatings are flowable and stable uniform suspensions made with active ingredients, adjuvants, and water by wet grinding. The production process is simple and the coating efficiency is high. However, they also have some disadvantages, including aggregation of pesticide particles rendered by long-term storage, and poor coating caused by sedimentation [103]. The formulation needs wetting agents such as alkyl sulfate, lignin sulfur salt, and fatty amine polyoxyethylene ether to reduce the surface tension of solid-liquid interfaces. Xanthan gum can be used to adjust the viscosity of the formulation [104]. Film-forming agents are indispensable to the formulation in that they can bond together and evenly wrap around the outside of the seeds. They are polymer composite materials such as PVA series, PEG macromolecule series, and gum arabic with good air permeability and water permeability. Colorants such as rhododendron in red are also applied to prevent from misuse during storage [105].

## 5. Challenges towards Sustainable Green Adjuvants for Microbial Pesticides

At present, the adjuvants of microbial pesticides are facing many challenges. The author summarizes the following points:

First of all, adjuvants of microbial pesticides are commonly considered to be inert adjuvants without activities. However, some adjuvants may be toxic to non-target organisms and environment [19]. For example, the U.S. Environmental Agency has conducted toxicity analyses on more than 2000 kinds of adjuvants, of which 26% are mutagenic, teratogenic, and carcinogenic (triple effects). They may also have neurotoxicity, endocrine disrupting effects, and cause harmful reproductive damage [59,106].

Secondly, wettable powder formulations of microbial pesticides on the market are very popular with the help of adjuvants [91,107]. However, their particles are relatively coarse, which is not conducive to coverage of crop surfaces, and absorption and utilization of crops.

Finally, the production process of suspension concentrations is relatively simple and cost-effective, but the liquid formulations are not conducive to the storage of microorganisms compared with wettable powders. Drawbacks such as layering, creaming, crystallization, heat storage solidification, room temperature thickening, and short shelf life also need to be solved during the development of suspension concentrations [71,108].

## 6. Future Perspectives

The concept of green pesticides has been gradually accepted by the pesticide industry; consequently, they are actively promoting the use of more environmentally friendly solvents and substances [12,109]. Compared to chemical pesticides, desirable properties of microbial pesticides include target specificity, low environmental persistence, and low non-target biological toxicities [110].

At present, most microbial pesticide adjuvants are very similar to those of chemical pesticides [111]. It is necessary to establish standard administration systems for various adjuvants used for microbial pesticides. That is, all adjuvants of microbial pesticides should be subject to the same risk assessments as that of the active ingredients.

Since the active ingredients of microbial pesticides are microorganisms, the effects of adjuvants on the survival and proliferation of the microorganisms should be considered [59]. That is, biocompatibility between the adjuvants and microorganisms used for microbial pesticides should be determined in storage periods. The effects of adjuvants on the proliferation of microorganisms in crop fields should also be investigated after the application of microbial pesticides.

Focus should also be placed on ensuring adjuvants have stronger adsorption capacity, higher dispersion performance, and better safety in order to enhance biocontrol efficiency of microbial pesticides by improving physical and chemical properties of adjuvants.

Microbial pesticides are key products for the development of sustainable and efficient green agriculture [112]. They will replace highly toxic and highly residual chemical pesticides with the help of sustainable green adjuvants.

**Author Contributions:** Conceptualization, F.L., Y.M., A.L.I. and T.H.; writing—original draft preparation, F.L.; literature searching, F.L., Q.L. and Y.M.; figure preparation, F.L., S.Z. and F.Z.; writing—review and editing, all authors; supervision, T.H.; funding acquisition, X.G. and T.H. All authors have read and agreed to the published version of the manuscript.

**Funding:** This project is supported by the Fujian Science and Technology Planning Project (No. 2020N5014), the Fujian Agriculture and Forestry University Projects (No. KFb22056XA, No. K1520005A03, the First-class Ecology Discipline Project, the "Research Advances in Life Science" Project and the "Special Topics on Bioscience" Project), and the National Natural Science Foundation of China (No. 31672084). The authors sincerely thank the laboratory members for their help.

**Institutional Review Board Statement:** Not applicable.

**Informed Consent Statement:** Not applicable.

**Data Availability Statement:** Not applicable.

**Conflicts of Interest:** The authors declare no conflict of interest.

## References

1. Aktar, W.; Sengupta, A. Impact of pesticides use in agriculture: Their benefits and hazards. *Interdiscip. Toxicol.* **2009**, *2*, 1–12. [CrossRef]
2. Blair, A.; Ritz, B.; Wesseling, C.; Freeman, L.B. Pesticides and human health. *Occup. Environ. Med.* **2015**, *72*, 81–82. [CrossRef] [PubMed]
3. Kim, K.H.; Kabir, E.; Jahan, S.A. Exposure to pesticides and the associated human health effects. *Sci. Total Environ.* **2017**, *575*, 525–535. [CrossRef] [PubMed]
4. Pan, X.; XU, J.; Liu, X. 2Progress of the discovery, application, and control technologies of chemical pesticides in China. *J. Integr. Agric.* **2019**, *18*, 840–853. [CrossRef]
5. Aldas-Vargas, A.; van der Vooren, T.; Rijnaarts, H.H.M.; Sutton, N.B. Biostimulation is a valuable tool to assess pesticide biodegradation capacity of groundwater microorganisms. *Chemosphere* **2021**, *280*, 130793. [CrossRef] [PubMed]
6. Arthurs, S.; Dara, S.K. Microbial biopesticides for invertebrate pests and their markets in the United States. *J. Invertebr. Pathol.* **2019**, *165*, 13–21. [CrossRef]
7. Kumar, S.; Singh, A. Biopesticides: Present Status and the Future Prospects. *J. Fertil. Pestic.* **2015**, *6*, e129. [CrossRef]

8.  Cox, C.; Surgan, M. Unidentified inert ingredients in pesticides: Implications for human and environmental health. *Environ. Health Perspect.* **2006**, *114*, 1803–1806. [CrossRef]
9.  Mohammad, N.; Abidin, E.Z.; How, V.; Praveena, S.M.; Hashim, Z. Pesticide management approach towards protecting the safety and health of farmers in Southeast Asia. *Rev. Environ. Health* **2018**, *33*, 123–134. [CrossRef]
10. Parinya, P.; Siriwong, W.; Prapamontol, T.; Ryan, P.B.; Fiedler, N.; Robson, M.G.; Barr, D.B. Agricultural pesticide management in Thailand: Status and population health risk. *Environ. Sci. Policy* **2012**, *17*, 72–81.
11. Wang, T.; Zhong, M.; Lu, M.; Xu, D.; Xue, Y.; Huang, J.; Blaney, L.; Yu, G. Occurrence, spatiotemporal distribution, and risk assessment of current-use pesticides in surface water: A case study near Taihu Lake, China. *Sci. Total Environ.* **2021**, *782*, 146826. [CrossRef] [PubMed]
12. Fan, T.; Chen, C.; Fan, T.; Liu, F.; Peng, Q. Novel surface-active ionic liquids used as solubilizers for water-insoluble pesticides. *J. Hazard. Mater.* **2015**, *297*, 340–346. [CrossRef] [PubMed]
13. Góngora-Echeverría, V.R.; García-Escalante, R.; Rojas-Herrera, R.; Giácoman-Vallejos, G.; Ponce-Caballero, C. 2020Pesticide bioremediation in liquid media using a microbial consortium and bacteria-pure strains isolated from a biomixture used in agricultural areas. *Ecotoxicol. Environ. Saf.* **2020**, *11*, 07–34. [CrossRef]
14. Do Nascimento, J.; Cristina Goncalves, K.; Pinto Dias, N.; Luiz de Oliveira, J.; Bravo, A.; Antonio Polanczyk, R. Adoption of *Bacillus thuringiensis*-based biopesticides in agricultural systems and new approaches to improve their use in Brazil. *Biol. Control* **2022**, *165*, 104792. [CrossRef]
15. Dwivedi, S.A.; Singh, R.S. Synthetic insecticides and bio-pesticide affect natural enemies of aphid (*Lipaphis erysimi kalt*) in mustard. *Indian J. Agric. Res.* **2022**, *56*, 717–725. [CrossRef]
16. Bravo, A.; Likitvivatanavong, S.; Gill, S.S.; Soberón, M. *Bacillus thuringiensis*: A story of a successful bioinsecticide. *Insect Biochem. Mol. Biol.* **2011**, *41*, 423–431. [CrossRef] [PubMed]
17. Duke, S.O.; Powles, S.B. Glyphosate: A once-in-a-century herbicide. *Pest Manag. Sci.* **2008**, *63*, 1100–1106. [CrossRef]
18. Ramakrishnan, B.; Venkateswarlu, K.; Sethunathan, N.; Megharaj, M. Local applications but global implications: Can pesticides drive microorganisms to develop antimicrobial resistance? *Sci. Total Environ.* **2019**, *654*, 177–189. [CrossRef]
19. Mesnage, R.; Antoniou, M.N. Ignoring Adjuvant Toxicity Falsifies the Safety Profile of Commercial Pesticides. *Front. Public Health* **2018**, *45*, 361. [CrossRef]
20. Borger, C.P.D.; Riethmuller, G.P.; Ashworth, M.; Minkey, D.; Hashem, A.; Powles, S.B. Increased Carrier Volume Improves Preemergence Control of Rigid Ryegrass (*Lolium rigidum*) in Zero-Tillage Seeding Systems. *Weed Technol.* **2013**, *27*, 649–655. [CrossRef]
21. Gouli, V.V.; Provost, C.; Parker, B.L.; Skinner, M.; Gouli, S.Y. Comparison of two methods for estimation of spray deposits after application of microbial pesticides. *Biocontrol Sci. Technol.* **2010**, *20*, 331–337. [CrossRef]
22. Nobels, I.; Spanoghe, P.; Haesaert, G.; Robbens, J.; Blust, R. Toxicity ranking and toxic mode of action evaluation of commonly used agricultural adjuvants on the basis of bacterial gene expression profiles. *PLoS ONE* **2011**, *6*, e24139. [CrossRef]
23. Khursheed, A.; Rather, M.A.; Jain, V.; Wani, A.R.; Rasool, S.; Nazir, R.; Malik, N.A.; Majid, S.A. Plant based natural products as potential ecofriendly and safer biopesticides: A comprehensive overview of their advantages over conventional pesticides, limitations and regulatory aspects. *Microb. Pathog.* **2022**, *173 Pt A*. [CrossRef] [PubMed]
24. Borges, S.; Alkassab, A.T.; Collison, E.; Hinarejos, S.; Jones, B.; McVey, E.; Roessink, I.; Steeger, T.; Sultan, M.; Wassenberg, J. Overview of the testing and assessment of effects of microbial pesticides on bees: Strengths, challenges and perspectives. *Apidologie* **2021**, *52*, 1256–1277. [CrossRef] [PubMed]
25. Beck, B.; Steurbaut, W.; Spanoghe, P. How to define green adjuvants. *Pest Manag. Sci.* **2012**, *68*, 1107–1110. [CrossRef] [PubMed]
26. Dechesne, A.; Badawi, N.; Aamand, J.; Smets, B.F. Fine scale spatial variability of microbial pesticide degradation in soil: Scales, controlling factors, and implications. *Front. Microbiol.* **2014**, *5*, 667. [CrossRef] [PubMed]
27. Idris, A.L.; Fan, X.; Muhammad, M.H.; Guo, Y.; Guan, X.; Huang, T. Ecologically controlling insect and mite pests of tea plants with microbial pesticides: A review. *Arch. Microbiol.* **2020**, *202*, 1275–1284. [CrossRef] [PubMed]
28. Pickard, F.C.; Simmonett, A.C.; Rigoberto Brooks, B. Efficient High Accuracy Non-Bonded Interactions in the CHARMM Simulation Package. *Biophys. J.* **2015**, *108*, 159a. [CrossRef]
29. Kumar, K.K.; Sridhar, J.; Murali-Baskaran, R.K.; Senthil-Nathan, S.; Kaushal, P.; Dara, S.K.; Arthurs, S. Microbial biopesticides for insect pest management in India: Current status and future prospects. *J. Invertebr. Pathol.* **2019**, *165*, 74–81. [CrossRef] [PubMed]
30. Mnif, I.; Ghribi, D. Potential of bacterial derived biopesticides in pest management. *Crop Prot.* **2015**, *77*, 52–64. [CrossRef]
31. Zhu, L.; Wang, Z.; Zhang, S.; Long, X. Fast microencapsulation of chlorpyrifos and bioassay. *J. Pestic. Sci.* **2010**, *35*, 339–343. [CrossRef]
32. Jamshidnia, A.; Abdoli, S.; Farrokhi, S.; Sadeghi, R. Efficiency of spinosad, *Bacillus thuringiensis* and Trichogramma brassicae against the tomato leafminer in greenhouse. *BioControl* **2018**, *63*, 619–627. [CrossRef]
33. Vemmer, M.; Patel, A. v. Review of encapsulation methods suitable for microbial biological control agents. *Biol. Control* **2013**, *67*, 380–389. [CrossRef]
34. Jalali, E.; Maghsoudi, S.; Noroozian, E. Ultraviolet protection of *Bacillus thuringiensis* through microencapsulation with Pickering emulsion method. *Sci. Rep.* **2020**, *10*, 20633. [CrossRef] [PubMed]
35. Shukla, P.G.; Kalidhass, B.; Shah, A.; Palaskar, D. v. Preparation and characterization of microcapsules of water-soluble pesticide monocrotophos using polyurethane as carrier material. *J. Microencapsul.* **2002**, *19*, 293–304. [CrossRef] [PubMed]

36. Castro, M.J.L.; Ojeda, C.; Cirelli, A.F. Advances in surfactants for agrochemicals. *Environ. Chem. Lett.* **2014**, *12*, 85–95. [CrossRef]
37. Edser, C. Multifaceted role for surfactants in agrochemicals. *Focus Surfactants* **2007**, *3*, 1–2. [CrossRef]
38. Hernández-Soriano, M.C.; Mingorance, M.D.; Peña, A. Interaction of pesticides with a surfactant-modified soil interface: Effect of soil properties. *Colloids Surf. A Physicochem. Eng. Asp.* **2007**, *306*, 49–55. [CrossRef]
39. Janků, J.; Bartovská, L.; Soukup, J.; Jursík, M.; Hamouzová, K. Densitysity and surface tension of aqueous solutions of adjuvants used for tank-mixes with pesticides. *Plant Soil Environ.* **2012**, *58*, 568–572. [CrossRef]
40. Krogh, K.A.; Halling-Sørensen, B.; Mogensen, B.B.; Vejrup, K. v. Environmental properties and effects of nonionic surfactant adjuvants in pesticides: A review. *Chemosphere* **2003**, *50*, 871–901. [CrossRef]
41. Farming, O. CO and Ni: A Review. *Media* **2009**, *136*, 2–13. [CrossRef]
42. Zhao, X.; Chen, Z.; Yu, L.; Hu, D.; Song, B. Investigating the antifungal activity and mechanism of a microbial pesticide *Shenqinmycin* against *Phoma* Sp. *Pestic. Biochem. Physiol.* **2018**, *147*, 46–50. [CrossRef] [PubMed]
43. Mascarin, G.M.; Jaronski, S.T. The production and uses of *Beauveria bassiana* as a microbial insecticide. *World J. Microbiol. Biotechnol.* **2016**, *32*, 177. [CrossRef] [PubMed]
44. Tang, H.; Zhao, L.; Sun, W.; Hu, Y.; Han, H. Surface characteristics and wettability enhancement of respirable sintering dust by nonionic surfactant. *Colloids Surf. A Physicochem. Eng. Asp.* **2016**, *509*, 323–333. [CrossRef]
45. Beekman, P.; Enciso-Martinez, A.; Pujari, S.P.; Terstappen, L.; Zuilhof, H.; le Gac, S.; Otto, C. Author Correction: Organosilicon uptake by biological membranes. *Commun. Biol.* **2021**, *4*, 704, Erratum in *Commun. Biol.* **2021**, *4*, 812. [CrossRef]
46. Yuan, M.; Nie, W.; Zhou, W.; Yan, J.; Bao, Q.; Guo, C.; Tong, P.; Zhang, H.; Guo, L. Determining the effect of the non-ionic surfactant AEO9 on lignite adsorption and wetting via molecular dynamics (MD) simulation and experiment comparisons. *Fuel* **2020**, *278*, 118–339. [CrossRef]
47. Banat, I.M.; Franzetti, A.; Gandolfi, I.; Bestetti, G.; Martinotti, M.G.; Fracchia, L.; Smyth, T.J.; Marchant, R. Microbial biosurfactants production, applications and future potential. *Appl. Microbiol. Biotechnol.* **2010**, *87*, 427–444. [CrossRef] [PubMed]
48. Chen, L.; Shi, H.; Wu, H.; Xiang, J. Synthesis and combined properties of novel fluorinated anionic surfactant. *Colloids Surf. A Physicochem. Eng. Asp.* **2011**, *384*, 331–336. [CrossRef]
49. Huang, G.; Deng, Y.; Zhang, Y.; Feng, P.; Xu, C.; Fu, L.; Lin, B. Study on long-term pest control and stability of double-layer pesticide carrier in indoor and outdoor environment. *Chem. Eng. J.* **2021**, *403*, 126–342. [CrossRef]
50. Singh, A.; Dhiman, N.; Kar, A.K.; Singh, D.; Purohit, M.P.; Ghosh, D.; Patnaik, S. Advances in controlled release pesticide formulations: Prospects to safer integrated pest management and sustainable agriculture. *J. Hazard. Mater.* **2020**, *385*, 121525. [CrossRef] [PubMed]
51. Kashyap, P.L.; Xiang, X.; Heiden, P. Chitosan nanoparticle based delivery systems for sustainable agriculture. *Int. J. Biol. Macromol.* **2015**, *77*, 36–51. [CrossRef]
52. Patra, J.K.; Das, G.; Fraceto, L.F.; Campos, E.V.R.; Rodriguez-Torres, M.D.P.; Acosta-Torres, L.S.; Diaz-Torres, L.A.; Grillo, R.; Swamy, M.K.; Sharma, S.; et al. Nano based drug delivery systems: Recent developments and future prospects. *J. Nanobiotechnology* **2018**, *16*, 71. [CrossRef]
53. Mohammadzadeh Kakhki, R.; Karimian, A.; Saadati Rad, M. Highly efficient removal of paraquat pesticide from aqueous solutions using a novel nano Kaolin modified with sulfuric acid via host–guest interactions. *J. Incl. Phenom. Macrocycl. Chem.* **2020**, *96*, 307–313. [CrossRef]
54. Shan, Y.; Xu, C.; Zhang, H.; Chen, H.; Bilal, M.; Niu, S.; Cao, L.; Huang, Q. Polydopamine-modified metal–organic frameworks, NH2-Fe-MIL-101, as pH-sensitive nanocarriers for controlled pesticide release. *Nanomaterials* **2020**, *10*, 2000. [CrossRef] [PubMed]
55. Tong, Y.; Wu, Y.; Zhao, C.; Xu, Y.; Lu, J.; Xiang, S.; Zong, F.; Wu, X. Polymeric Nanoparticles as a Metolachlor Carrier: Water-Based Formulation for Hydrophobic Pesticides and Absorption by Plants. *J. Agric. Food Chem.* **2017**, *65*, 7371–7378. [CrossRef] [PubMed]
56. Xu, C.; Qi, J.; Yang, W.; Chen, Y.; Yang, C.; He, Y.; Wang, J.; Lin, A. Immobilization of heavy metals in vegetable-growing soils using nano zero-valent iron modified attapulgite clay. *Sci. Total Environ.* **2019**, *686*, 476–483. [CrossRef] [PubMed]
57. Xu, L.; Cao, L.D.; Li, F.M.; Wang, X.J.; Huang, Q.L. Utilization of Chitosan-Lactide Copolymer Nanoparticles as Controlled Release Pesticide Carrier for Pyraclostrobin Against Colletotrichum gossypii Southw. *J. Dispers. Sci. Technol.* **2014**, *35*, 544–550. [CrossRef]
58. Purkait, A.; Hazra, D.K. Biodiesel as a carrier for pesticide formulations: A green chemistry approach. *Int. J. Pest Manag.* **2020**, *66*, 341–350. [CrossRef]
59. Rota, E.; Healy, B. A taxonomic study of some swedish enchytraeidae (Oligochaeta), with descriptions of four new species and notes on the genus fridericia. *J. Nat. Hist.* **1999**, *33*, 29–64. [CrossRef]
60. Zhang, H.; Chen, W.; Zhao, B.; Phillips, L.A.; Zhou, Y.; Lapen, D.R.; Liu, J. Sandy soils amended with bentonite induced changes in soil microbiota and fungistasis in maize fields. *Appl. Soil Ecol.* **2020**, *146*, 103–378. [CrossRef]
61. Lei, Y.; Liu, X.; Li, S.; Jiang, L.; Zhang, C.; Li, S.; He, D.; Wang, S. High stability of palladium/kieselguhr composites during absorption/desorption cycling for hydrogen isotope separation. *Fusion Eng. Des.* **2016**, *113*, 260–264. [CrossRef]
62. Maghsoodi, M.R.; Najafi, N.; Reyhanitabar, A.; Oustan, S. Hydroxyapatite nanorods, hydrochar, biochar, and zeolite for controlled-release urea fertilizers. *Geoderma* **2020**, *379*, 114644. [CrossRef]
63. Dutta, J.; Devi, N. Preparation, optimization, and characterization of chitosan-sepiolite nanocomposite films for wound healing. *Int. J. Biol. Macromol.* **2021**, *186*, 244–254. [CrossRef]

64. Xiang, Y.; Lu, X.; Yue, J.; Zhang, Y.; Sun, X.; Zhang, G.; Cai, D.; Wu, Z. Stimuli-responsive hydrogel as carrier for controlling the release and leaching behavior of hydrophilic pesticide. *Sci. Total Environ.* **2020**, *722*, 137811. [CrossRef]
65. Heris Anita, S.; Mangunwardoyo, W.; Yopi, Y. Sugarcane Bagasse as a Carrier for the Immobilization of *Saccharomyces cerevisiaein* Bioethanol Production. *Makara J. Technol.* **2016**, *20*, 73. [CrossRef]
66. Haider, M.; Abdin, S.M.; Kamal, L.; Orive, G. Nanostructured Lipid Carriers for Delivery of Chemotherapeutics: A Review. *Pharmaceutics* **2020**, *12*, 288. [CrossRef] [PubMed]
67. Dos Santos, C.A.M.; do Nascimento, J.; Gonçalves, K.C.; Smaniotto, G.; de Freitas Zechin, L.; da Costa Ferreira, M.; Polanczyk, R.A. Compatibility of Bt biopesticides and adjuvants for Spodoptera frugiperda control. *Sci. Rep.* **2021**, *11*, 5271. [CrossRef]
68. Li, J.; Yao, J.; Li, Y.; Shao, Y. Controlled release and retarded leaching of pesticides by encapsulating in carboxymethyl chitosan /bentonite composite gel. *J. Environ. Sci. Health - Part B Pestic. Food Contam. Agric. Wastes* **2012**, *47*, 795–803. [CrossRef]
69. Voinova, O.N.; Kalacheva, G.S.; Grodnitskaya, I.D.; Volova, T.G. Microbial polymers as a degradable carrier for pesticide delivery. *Appl. Biochem. Microbiol.* **2009**, *45*, 384–388. [CrossRef]
70. Jarzębski, M.; Smułek, W.; Siejak, P.; Kobus-Cisowska, J.; Pieczyrak, D.; Baranowska, H.M.; Jakubowicz, J.; Sopata, M.; Białopiotrowicz, T.; Kaczorek, E. *Aesculus hippocastanum* L. extract as a potential emulsion stabilizer. *Food hydrocolloids.* **2019**, *54*, 68. [CrossRef]
71. Jadhav, N.; Pantwalawalkar, J.; Sawant, R.; Attar, A.; Lohar, D.; Kadane, P.; Ghadage, K. Development of Progesterone Oily Suspension Using Moringa Oil and Neusilin US2. *J. Pharm. Innov.* **2021**, *21*, 79. [CrossRef]
72. Kaur, R.; Thakur, N.S.; Chandna, S.; Bhaumik, J. Development of agri-biomass based lignin derived zinc oxide nanocomposites as promising UV protectant-cum-antimicrobial agents. *J. Mater. Chem.* **2020**, *8*, 260–269. [CrossRef] [PubMed]
73. Sukirno, S.; Lukmawati, D.; Hanum, S.S.L.; Ameliya, V.F.; Sumarmi, S.; Purwanto, H.; Suparmin, S.; Sudaryadi, I.; Soesilohadi, R.C.H.; Aldawood, A.S. The effectiveness of Samia ricini Drury (*Lepidoptera: Saturniidae*) and Attacus atlas L. (*Lepidoptera: Saturniidae*) cocoon extracts as ultraviolet protectants of Bacillus thuringiensis for controlling Spodoptera litura Fab. (*Lepidoptera: Noctuidae*). *Int. J. Trop. Insect Sci.* **2022**, *42*, 255–260. [CrossRef]
74. Pershakova, T.V.; Gorlov, S.M.; Lisovoy, V.V.; Mikhaylyuta, L.V.; Babakina, M.V.; Aleshin, V.N. Influence of electromagnetic fields and microbial pesticide Vitaplan on stability of apples during storage. *IOP Conf. Ser. Earth Environ. Sci.* **2021**, *640*, 022053. [CrossRef]
75. Zhang, J.; Wang, W.; Pei, Z.; Wu, J.; Yu, R.; Zhang, Y.; Sun, L.; Gao, Y. Mutagenicity assessment to pesticide adjuvants of toluene, chloroform and trichloroethylene by ames test. *Int. J. Environ. Res. Public Health* **2021**, *18*, 8095. [CrossRef]
76. Wang, G.; Chen, H.; Wang, X.; Peng, L.; Peng, Y.; Li, Y.-Q. Probing the germination kinetics of ethanol-treated Bacillus thuringiensis spores. *Appl. Opt* **2017**, *56*, 3263–3269. [CrossRef]
77. Ruiu, L.; Satta, A.; Floris, I.; Floris, I. Administration of Brevibacillus laterosporus spores as a poultry feed additive to inhibit house fly development in feces: A new eco-sustainable concept. *Poult. Sci.* **2014**, *93*, 519–526. [CrossRef] [PubMed]
78. Whalon, M.E.; Wingerd, B.A. Bt: Mode of action and use. *Arch. Insect Biochem. Physiol.* **2003**, *54*, 200–211. [CrossRef] [PubMed]
79. Gómez, J.; Guevara, J.; Cuartas, P.; Espinel, C.; Villamizar, L. *Microencapsulated Spodoptera frugiperda* nucleopolyhedrovirus: Insecticidal activity and effect on arthropod populations in maize. *Biocontrol Sci. Technol.* **2013**, *23*, 829–846. [CrossRef]
80. Taylor, P.; Soe, K.T.; De Costa, D.M. Development of a spore-based formulation of microbial pesticides for control of rice sheath blight. *Biocontrol Sci. Technol.* **2012**, *22*, 633–657.
81. Pratap, A.P.; Bhowmick, D.N. Pesticides as microemulsion formulations. *J. Dispers. Sci. Technol.* **2008**, *29*, 1325–1330. [CrossRef]
82. Liu, J.; He, Y.; Chen, S.; Xiao, Y.; Hu, M.; Zhong, G. Development of a freeze-dried fungal wettable powder preparation able to biodegrade chlorpyrifos on vegetables. *PLoS ONE* **2014**, *9*, e103558. [CrossRef] [PubMed]
83. Cheng, H.; Li, L.; Hua, J.; Yuan, H.; Cheng, S. A preliminary preparation of endophytic bacteria CE3 wettable powder for biological control of postharvest diseases. *Not. Bot. Horti Agrobot.* **2015**, *43*, 159–164. [CrossRef]
84. Leng, P.; Zhang, Z.; Pan, G.; Zhao, M. Applications and development trends in biopesticides. *Afr. J. Biotechnol.* **2011**, *10*, 19864–19873. [CrossRef]
85. Samada, L.H.; Tambunan, U.S.F. Biopesticides as promising alternatives to chemical pesticides: A review of their current and future status. *OnLine J. Biol. Sci.* **2020**, *20*, 66–76. [CrossRef]
86. Zulfitri, A.; Krishanti, N.P.R.A.; Lestari, A.S.; Meisyara, D.; Zulfiana, D. Efficacy of several entomopathogenic microorganism as microbial insecticide against insect pest on chili (*Capsicum annum* L.). *IOP Conf. Ser. Earth Environ. Sci.* **2020**, *572*, 012020. [CrossRef]
87. Zhao, M.; Li, S.; Zhou, Q.; Zhou, D.; He, N.; Qian, Z. Safety evaluation of microbial pesticide (HaNPV) based on PCR method. *Front. Chem. Sci. Eng.* **2019**, *13*, 377–384. [CrossRef]
88. Shao, H.; Zhang, Y. Non-target effects on soil microbial parameters of the synthetic pesticide carbendazim with the biopesticides cantharidin and norcantharidin. *Sci. Rep.* **2017**, *7*, 5521. [CrossRef] [PubMed]
89. Vehapi, M.; Özçimen, D. Antimicrobial and bacteriostatic activity of surfactants against B. subtilis for microbial cleaner formulation. *Arch. Microbiol.* **2021**, *203*, 3389–3397. [CrossRef]
90. Oliveira-Hofman, C.; Cottrell, T.E.; Bock, C.; Mizell, R.F.; Wells, L.; Shapiro-Ilan, D.I. Impact of a biorational pesticide on the pecan aphid complex and its natural enemies. *Biol. Control* **2021**, *161*, 104–709. [CrossRef]
91. Du, W.; Zhou, J.; Jiang, P.; Yang, T.; Bu, Y.Q.; Liu, C.H.; Dai, C.C. Effects of Beauveria bassiana and acephate on enzyme activities and microbial diversity in paddy soil. *Plant Soil Environ.* **2013**, *59*, 562–567. [CrossRef]

92. Kumar, J.; Parmar, B.S. Stabilization of azadirachtin A in neem formulations: Effect of some solid carriers, neem oil, and stabilizers. *J. Agric. Food Chem.* **1999**, *47*, 1735–1739. [CrossRef]
93. Brownbridge, M.; Buitenhuis, R. Integration of microbial biopesticides in greenhouse floriculture: The Canadian experience. *J. Invertebr. Pathol.* **2019**, *165*, 4–12. [CrossRef]
94. Aslam, S.; Jahan, N.; Khalil-Ur-Rehman; Ali, S. Formulation, optimisation and in-vitro, in-vivo evaluation of surfactant stabilised nanosuspension of *Ginkgo biloba*. *J. Microencapsul.* **2019**, *36*, 576–590. [CrossRef]
95. Hong, N.; Li, Y.; Qiu, X. A highly efficient dispersant from black liquor for carbendazim suspension concentrate: Preparation, self-assembly behavior and investigation of dispersion mechanism. *J. Appl. Polym. Sci.* **2016**, *133*, 1–9. [CrossRef]
96. Vineela, V.; Nataraj, T.; Reddy, G.; Vimala Devi, P.S. Enhanced bioefficacy of Bacillus thuringiensis var. kurstaki against Spodoptera litura (*Lepidoptera: Noctuidae*) through particle size reduction and formulation as a suspension concentrate. *Biocontrol Sci. Technol.* **2017**, *27*, 58–69. [CrossRef]
97. Feng, N.; Zhang, B.; Xin, X.; Li, H.; Zhao, Y. Role of aliphatic alcohol polyoxyethylene ether phosphate in 25 wt% tebuconazole suspension concentrate: Dispersion and wetting. *Colloids Surf. A Physicochem. Eng. Asp.* **2021**, *628*, 127350. [CrossRef]
98. Wang, T.; Chen, L. Dynamic complexity of microbial pesticide model. *Nonlinear Dyn.* **2009**, *58*, 539–552. [CrossRef]
99. Vimala Devi, P.S.; Duraimurugan, P.; Poorna Chandrika, K.S.V.; Vineela, V.; Hari, P.P. Novel formulations of *Bacillus thuringiensis* var. kurstaki: An eco-friendly approach for management of lepidopteran pests. *World J. Microbiol. Biotechnol.* **2020**, *36*, 78. [CrossRef] [PubMed]
100. Kaczmarek, D.K.; Rzemieniecki, T.; Marcinkowska, K.; Pernak, J. Synthesis, properties and adjuvant activity of docusate-based ionic liquids in pesticide formulations. *J. Ind. Eng. Chem.* **2019**, *78*, 440–447. [CrossRef]
101. Egamberdieva, D.; Jabbarov, Z.; Arora, N.K.; Wirth, S.; Bellingrath-Kimura, S.D. Biochar mitigates effects of pesticides on soil biological activities. *Environ. Sustain.* **2021**, *4*, 335–342. [CrossRef]
102. Knowles, A. Adjuvants for agrochemicals. *Pestic. Outlook* **2001**, *12*, 183–184. [CrossRef]
103. Shenoy, A.; Reddy, C.M.; Shree Padma, M.; Niranjan, V.; Rao, N.N. Evaluation of alternative bio-receptors for pesticide detection. *Mater. Today Proc.* **2018**, *5*, 20977–20980. [CrossRef]
104. Fine, J.D.; Cox-Foster, D.L.; Mullin, C.A. An inert pesticide adjuvant synergizes viral pathogenicity and mortality in honey bee larvae. *Sci. Rep.* **2017**, *7*, 40499. [CrossRef] [PubMed]
105. Partners, C.R. Kline analyses US pesticide adjuvants market. *Focus Surfactants* **2018**, *2018*, 5–6. [CrossRef]
106. Abbott, W.S. The Value of the Dry Substitutes for Liquid Lime. *J. Econ. Entomol.* **1925**, *18*, 265–267. [CrossRef]
107. Ali, R.A.; Hasan, M.; Sagheer, M.; Sahi, S.T.; Rasul, A. Factors influencing the combined efficacy of microbial insecticides and inert dusts for the control of Trogoderma granarium. *Int. J. Trop. Insect Sci.* **2022**, *42*, 425–433. [CrossRef]
108. Yanagisawa, K.; Muroi, T.; Ohtsubo, T.; Watano, S. Effect of binder composition on physicochemical properties of water dispersible granules obtained through direct granulation of agrochemical suspension using fluidized bed. *J. Pestic. Sci.* **2017**, *42*, 112–115. [CrossRef]
109. Nile, A.S.; Kwon, Y.D.; Nile, S.H. Horticultural oils: Possible alternatives to chemical pesticides and insecticides. *Environ. Sci. Pollut. Res.* **2019**, *26*, 21127–21139. [CrossRef]
110. Grillo, R.; Abhilash, P.C.; Fraceto, L.F. Nanotechnology applied to bio-encapsulation of pesticides. *J. Nanosci. Nanotechnol.* **2016**, *16*, 1231–1234. [CrossRef]
111. Ryckaert, B.; Spanoghe, P.; Haesaert, G.; Heremans, B.; Isebaert, S.; Steurbaut, W. Quantitative determination of the influence of adjuvants on foliar fungicide residues. *Crop Prot.* **2007**, *26*, 1589–1594. [CrossRef]
112. Mansour, R.; Biondi, A. Releasing natural enemies and applying microbial and botanical pesticides for managing Tuta absoluta in the MENA region. *Phytoparasitica* **2021**, *49*, 179–194. [CrossRef]

**Disclaimer/Publisher's Note:** The statements, opinions and data contained in all publications are solely those of the individual author(s) and contributor(s) and not of MDPI and/or the editor(s). MDPI and/or the editor(s) disclaim responsibility for any injury to people or property resulting from any ideas, methods, instructions or products referred to in the content.

MDPI
St. Alban-Anlage 66
4052 Basel
Switzerland
www.mdpi.com

*Microorganisms* Editorial Office
E-mail: microorganisms@mdpi.com
www.mdpi.com/journal/microorganisms

Disclaimer/Publisher's Note: The statements, opinions and data contained in all publications are solely those of the individual author(s) and contributor(s) and not of MDPI and/or the editor(s). MDPI and/or the editor(s) disclaim responsibility for any injury to people or property resulting from any ideas, methods, instructions or products referred to in the content.

www.ingramcontent.com/pod-product-compliance
Lightning Source LLC
LaVergne TN
LVHW070115170726

843515LV00007B/1696